HISTORICAL SURVEY

OF THE

ASTRONOMY OF THE ANCIENTS.

BY THE RIGHT HON.

SIR GEORGE CORNEWALL LEWIS.

LONDON:
PARKER, SON, AND BOURN, WEST STRAND.
1862.

The right of Translation is reserved.

**Kessinger Publishing's Rare Reprints
Thousands of Scarce and Hard-to-Find Books!**

We kindly invite you to view our extensive catalog list at:
http://www.kessinger.net

CONTENTS.

CHAPTER I.

Primitive Astronomy of the Greeks and Romans.

§ 1 Introduction. Object and nature of the work . . . p. 1
2 Original conception of the world by the Greeks . . 2
3 The primitive year of the Greeks was a solar year . . 9
4 The month of the Greeks as determined by the moon. The lunar year 16
5 Discordance of the Greek national calendars . . . 22
6 Absence of a common chronological era in Greece . . 25
7 Divergent years—Carian, Acarnanian, and Arcadian years 30
8 Short Egyptian year 32
9 Roman year of ten months, instituted by Romulus . . 34
10 Early observation and nomenclature of the fixed stars in Greece. Ignorance of the planets 58
11 The religion and mythology of the Greeks had scarcely any reference to the heavenly bodies 62
12 The divination of the early Greeks was unconnected with Astrology 70

CHAPTER II.

Philosophical Astronomy of the Greeks from the time of Thales to that of Democritus.

§ 1 Mythological stories respecting the origin of astronomy in Greece 72
2 Thales, the earliest cultivator of astronomy in Greece . 78
3 The eclipse of Thales 85
4 Reform of the Athenian calendar by Solon . . . 90
5 Astronomical doctrines of Anaximander 91
6 „   „   of Anaximenes 94
7 „   „   of Heraclitus 96
8 „   „   of Xenophanes 99

CONTENTS.

§ 9   Astronomical doctrines of Parmenides . . . . p. 99
10   „   „   of Empedocles 100
11   „   „   of Anaxagoras 102
12   „   „   of Diogenes of Apollonia . . 109
13   Opinions of Socrates on astronomy . . . . ib.
14   Reform of the Athenian calendar by Meton . . . 113
15   The triëteric and octaëteric intercalary cycles . . 114
16   Reform of the Metonic cycle by Callippus . . . 122
17   Astronomical speculations of Pythagoras and of his school. Philolaic system of the universe . . . . ib.
18   Pythagorean cycles 135
19   Astronomical doctrines of Leucippus 136
20   And of Democritus 137

CHAPTER III.

Scientific Astronomy of the Greeks from Plato to Eratosthenes.

§ 1   Astronomical doctrines of Plato 141
2   Eudoxus of Cnidos. His astronomical observations, and writings 145
3   His hypothesis of the planetary motions . . . 151
4   Astronomical doctrines and writings of Aristotle . . 158
5   Theories of Callippus and Aristotle respecting the movements of the planets 163
6   Other astronomical doctrines of Aristotle . . . 165
7   Hypothesis of the rotation of the earth on its axis, propounded by Hicetas, Heraclides of Pontus, and Ecphantus 170
8   Histories of astronomy by Theophrastus and Eudemus . 174
9   Measures of daily time in Greece and Italy. The sun-dial and the clepsydra. Division of the day into hours . 174
10   Astronomical treatises of Autolycus 183
11   The Phænomena of Euclid 180
12   Aristarchus of Samos. His heliocentric system of the universe 189
13   Archimedes. His orrery 194
14   The Alexandrine school of astronomers: Aristyllus and Timocharis, Conon of Samos, Eratosthenes, Dositheus, Apollonius of Perga 195
Note A, on the passage in the Timæus respecting the supposed motion of the earth 202

CONTENTS. v

Note B, on the passage of Archimedes respecting the heliocentric theory of Aristarchus . . . . . p. 203
Note C, on the dials of the ancients 205

CHAPTER IV.

Scientific Astronomy of the Greeks and Romans from Hipparchus to Ptolemy.

(160 B.C. to 160 A.D.)

§ 1 Hipparchus: his astronomical observations and discoveries 207
2 Astronomical writings of Posidonius. Extant treatises of Cleomedes and Geminus. Pseud-Aristotelic treatise de Mundo. Astronomy of the Epicurean school. . 214
3 The practical astronomy of the Greeks did not keep pace with their scientific astronomy 220
4 Prediction of eclipses by the Greeks and Romans . . 222
5 Absence of a uniform calendar in Greece and Italy. Connexion of the calendar with the religious festivals of each state 234
6 Confusion of the Roman calendar. Its reform by Julius Cæsar 236
7 Roman measures of daily time 241
8 Measures of daily time in the middle ages . . . 242
9 Later hypotheses as to the order of the planets, and as to the revolution of the inferior planets round the sun . 245
10 Ptolemy: his astronomical treatise. 248
11 The ancient astronomers held to the geocentric system. Comparison with the modern heliocentric astronomy . 252

CHAPTER V.

Astronomy of the Babylonians and Egyptians.

§ 1 The Babylonians and the Egyptians are stated to have originated astronomical observation 256
2 Statements respecting the high antiquity of astronomical observation in Babylon and Egypt 262

CONTENTS.

§ 3 The Chaldæan and Egyptian priests are reported to have been astronomical observers . . . . p. 265
4 The Egyptians are stated to have first determined the solar year. Year ascribed to the Babylonians . . 266
5 The Egyptian priests are related to have taught astronomy to the Greeks. Alleged visits of the Greek philosophers to Egypt 268
6 True character and extent of the astronomical knowledge of the Babylonian and Egyptian priests . . . 277
7 Length of the Egyptian year 278
8 The Canicular or Sothiac period 281
9 Further examination of the statements as to the scientific knowledge of astronomy possessed by the Babylonian and Egyptian priests 285
10 Astrology was invented by the Chaldæans . . 291
11 Its introduction into Greece by Berosus the Chaldæan . 296
12 Its diffusion among the Greeks and Romans . . 298
13 Its introduction into Egypt 301
14 The astrological divination of the Chaldæans was founded on the connexion of a star with the time of the child's birth 306
15 Causes which promoted the diffusion of astrology in Greece and Italy 309
16 Intricacy of the astrological system . . . 314

CHAPTER VI.

Early History and Chronology of the Egyptians.

§ 1 Chronology of Egypt and reigns of Egyptian kings, from 670 to 525 B.C. 315
2 Egyptian chronology for the period anterior to 670 B.C.; Egyptian chronology of Herodotus for this period . 320
3 Plato's Egyptian chronology 326
4 Manetho's ancient history and dynasties of Egypt . . ib.
5 Egyptian dynasties of Eratosthenes . . . 331
6 Ancient Egyptian history and chronology of Diodorus . ib.
7 Chronological scheme of the 'Ancient Chronicle' . . 338
8 Credibility of the different schemes of Egyptian chronology for the early period, examined and questioned . . 339
9 Attempts to diminish the improbability of the traditional schemes of Egyptian chronology. (1) By reducing the length of the period designated as a year . . . 362

§ 10 (2) By placing the dynasties in parallel lines; and by other contrivances . . . . p. 366
11 Arbitrary methods of transmuting the ancient chronology of Egypt employed by the modern Egyptological school 367
12 Statements as to an early Egyptian literature examined and rejected 375
13 Arguments in favour of the Egyptian chronology derived from the hieroglyphical inscriptions 377

CHAPTER VII.

Early History and Chronology of the Assyrians.

§ 1 Brief account of early Assyrian history by Herodotus . 397
2 Ctesias first gave the Greeks a copious account of early Assyrian history 398
3 Babylonian history of Berosus 400
4 Babylonian kings of the Astronomical Canon . . . 404
5 Assyrian chronology of Syncellus . . . . . *ib.*
6 And of Eusebius 405
7 Examination of the different schemes of Assyrian chronology; their discrepancies 406
8 Separate kingdoms of Nineveh and Babylon . . . 423
9 Notices of Assyrian kings in the books of the Old Testament 425
10 Examination of the series of Assyrian kings in the Astronomical Canon 428
11 Evidence of the age of the world furnished by the great buildings of Assyria and Egypt 433

CHAPTER VIII.

Navigation of the Phœnicians.

§ 1 The Phœnicians were sometimes considered as the originators of astronomy. Connexion of astronomy with their skill in navigation 446
2 Foundation of Gades by the Phœnicians. Their voyages from Gades to the north of Europe 448

CONTENTS.

§ 3   Tin trade of the Phœnicians . . . . . . p. 450
   4   Amber trade of the Phœnicians 457
   5   Supposed voyage of Pytheas to the Island of Thule . . 466
   6   Imperfect knowledge of Britain before the time of Cæsar.
        Earliest mention of Ireland 481
   7   Sacred Islands of the North 489
   8   Circumnavigation of Africa by the Phœnicians under Neco   497

INDEX 516

A HISTORICAL SURVEY OF THE ASTRONOMY OF THE ANCIENTS.

CHAPTER I.

PRIMITIVE ASTRONOMY OF THE GREEKS AND ROMANS.

§ 1 THE history of sciences has in general been written with a scientific purpose, and with a view of throwing light upon the special science of which the origin and progress are described. In such a work as Dr. Whewell's History of the Inductive Sciences, the historical form is subsidiary to the scientific end: it is a book intended for the use and instruction of the professors of the several sciences which it successively passes in review. In like manner, Delambre's elaborate histories of Ancient, Mediæval, and Modern Astronomy,(¹) are works, composed by an astronomer, principally for the use of astronomers. No one can master them who is not versed in the modern mathematical astronomy.

But astronomy has this peculiarity, that it is conversant with subjects which from the earliest ages have attracted the daily attention of mankind, and which gave birth to observation and speculation before they were treated by strictly scientific methods. Chronology, moreover, without which political history cannot exist, is dependent upon astronomical determinations. The year and the month are measured by the

---

(1) A just character of Delambre's History of Ancient Astronomy is given by Martin, Études sur le Timée de Platon, tom. ii. p. 424. M. Martin remarks that the work of Delambre must be considered rather as materials for the history of ancient astronomy, than as a history itself.

motions of the sun and moon; and in order to secure the accuracy of the necessary measurements, the assistance of the astronomer must be obtained. The history of astronomy has numerous points of contact with the general history of mankind; and it concerns questions which interest a wider class than professed astronomers, for whose benefit the existing histories have been mainly composed. This remark applies with especial force to the early periods of astronomical observation and science, and to the ages anterior to the formation of a calendar recognised by all civilized nations. It has, in consequence, appeared to the author of the following work that an attempt might advantageously be made to treat the history of ancient astronomy, without exclusive reference to physical science, and without any pretension on his part to that profound and comprehensive knowledge of modern mathematical astronomy which some of his predecessors in the treatment of this subject have possessed. The works by English authors on the history of ancient astronomy are little known, and are not sufficient to meet the existing demands of historical criticism.(²)

§ 2 As the accounts of the Greek astronomy, though often meagre and fragmentary, afford in general a firm footing to the historian, and enable him to fix with confidence all the principal stages in its progress, it will be desirable to begin with this branch of the subject, and afterwards to attempt to determine how far the Greeks derived their astronomical knowledge from foreign nations, and were learners, instead, as was their custom, of being teachers.

---

(2) The two principal works are—1. The History of Astronomy, with its application to Geography, History, and Chronology, by the Rev. George Costard, 1 vol. 4to. London, 1767.—2. An Historical Account of the Origin and Progress of Astronomy, by John Narrien, 1 vol. 8vo. London, 1833. There is a life of Mr. Costard in Chalmers' Biographical Dictionary. He contributed some papers on questions of ancient astronomy to the Philosophical Transactions. Mr. Narrien was mathematical teacher at the Military College of Sandhurst. His work betrays an imperfect knowledge of the classical languages, of which some examples are given in Notes and Queries, 2nd ser. vol xi. p. 426.

The original idea of the earth, as we find it in the Homeric poems, and as it still continued to be entertained, after a lapse of five centuries, in the time of Herodotus, was that it was a circular plane, surmounted(³) and bounded by the heaven, which was a solid vault, or hemisphere, with its concavity turned downwards.(⁴)

That the earth was a plane, appeared to result from the evidence of the senses. The belief that this plane was circular, seems to have had its origin in two causes.

First, the fact, that whether the spectator is on a high eminence, or in a large plain, or at sea, the horizon appears everywhere equidistant from the eye, naturally led, in the infancy of geographical knowledge, to a rude induction that the entire earth was circular. When geographers began to construct maps, by adding the form and size of one country to another, they arrived at a different conception of the figure of the earth. Hence Herodotus, who attempted to solve this problem by personal observation and inquiry during his travels, and by subsequent reflection upon the information which he had thus obtained, ridicules the idea of the circularity of the earth, and treats it as childish. 'Many even now (he says) commit the ludicrous and ignorant error of drawing a map of the earth, in which it is represented of a circular form, as if its outline were traced with a compass, and the ocean is made to flow round it.'(⁵) The ancients (says Agathemerus, in his

---

(3) Γαῖα δέ τοι πρῶτον μὲν ἐγείνατο ἶσον ἑαυτῇ
Οὐρανὸν ἀστερόενθ', ἵνα μιν περὶ πάντα καλύπτοι.
                    Hesiod, Theog. 126-7.

(4) See Völcker's Homerische Geographie, pp. 97-101, and the authorities cited by him. Homer calls the heaven χάλκεος, πολύχαλκος, and σιδήρεος, Il. v. 504; xvii. 425; Od. iii. 2, xv. 328, xvii. 565, which epithets must express its solidity, though Völcker, p. 5, refers them to its imperishableness. Even Empedocles considered the heaven as solid: Ἐμπεδοκλῆς στερέμνιον εἶναι τὸν οὐρανόν. Stob. Ecl. Phys. i. 23.

(5) γελῶ δὲ ὁρέων γῆς περιόδους γράψαντας πολλοὺς ἤδη, καὶ οὐδένα νόον ἔχοντας ἐξηγησάμενον· οἳ ὠκεανόν τε ῥέοντα γράφουσι πέριξ τὴν γῆν ἐοῦσαν κυκλοτερέα ὡς ἀπὸ τόρνου, iv. 36. A τόρνος was probably a string fastened to a pin or peg, with which either a circle or a straight line could be drawn on a flat surface. It is used in the latter sense by Theognis, v. 803. Eurip. Bacch. 1066, describes a tree bent down to the ground by force as follows:

treatise on geography,) represented the earth with a circular figure: they placed Greece in its centre, and made Delphi the central point of Greece.(⁶) The omphalos, or navel-stone, which marked Delphi as the centre of Greece and of the earth, existed in the Delphian temple during the historical period.(⁷) It was illustrated by an etiological legend, that Jupiter sent out two eagles, one from the east, the other from the west, and that they met at this spot: thus determining it as the centre of the earth.(⁸)

Geminus, a scientific Greek writer upon astronomy, thus describes the ancient ideas on the subject. 'Homer (he says) and nearly all the ancient poets conceive the earth to be a plane; they likewise suppose the ocean to encircle it, as a horizon; and the stars to rise from and set in the ocean. Hence they

---

κυκλοῦτο δ' ὥστε τόξον, ἢ κυρτὸς τροχὸς
τόρνῳ γραφόμενος περιφορὰν ἕλκει δρομόν.

Τόρνος, in Plat. Phil. § 116, is used for an instrument by which geometrical figures are described, both plane and solid. Ib. § 131. it is placed in company with the διαβήτης and other instruments used in the work of a carpenter. Hesych. Τόρνος, ἐργαλεῖον τεκτονικὸν ᾧ τὰ στρογγύλα σχήματα περιγράφεται. A sphere is described as fashioned by a τόρνος in Pseud-Aristot. de Mundo, c. 2. Theodorus of Samos was the mythical inventor of the tornus. 'Normam et libellam et tornum et clavem Theodorus Samius;' Plin. vii. 57. *Libella* is an instrument for levelling; *norma* is a square. A compass used by a geometer is called by Aristophanes, Nub. 178, διαβήτης. Compare Schol. ad loc. In Latin it is called *circinus*. See Dr. Smith's *Dict. of Ant.* in v., where there is an engraving of a compass from an ancient tomb.

Herodotus points out the error as to the size and limits of Europe, Asia, and Africa in the maps of his day, iv. 42. He states that the boundaries of Europe to the west and north were unknown, iv. 45, iii. 115, v. 9. The extent of his geographical knowledge was considerable; see Rawlinson's *Herodotus*, vol. i. p. 116.

Anaximander is said to have made the first map, and to have been followed by Hecatæus. Agathemer. i. 1; Strab. i. 1, § 11; Ukert, *Geogr. der Gr. und Röm.*, 1, 2, p. 169. Aristagoras of Miletus exhibited at Sparta in 500 B.C., a brazen plate, on which was designed a map of the entire earth.—Herod. v. 49. This event was only sixteen years before the birth of Herodotus.

(6) i. 1. His exact date is unknown, but he is subsequent to Ptolemy, and he is placed at the beginning of the 3rd century after Christ.

(7) Æsch. Choëph. 1036; Eum. 40; Soph. Œd. T. 480-898; Eurip. Ion. 461; Orest. 331-591; Phœn. 237; Plat. Rep. iv. § 5, p. 427; Livy, xxxviii, 48; Cic. de Div. ii. 56; Ovid, Met. x. 167, xv. 630.

(8) Pindar, Pyth. iv. 131, vi. 3; Strab. ix. 3, § 6; Paus. x. 16, § 2.
Homer makes the island of Ogygia the seat of the ὀμφαλὸς θαλάσσης, Od. i. 50.

believed the Æthiopians, who dwelt in the remote east and west, to be scorched by the vicinity of the sun. This belief is consistent with the ancient conception of the world, but is irreconcilable with its real spherical form.'(9)

Even in the time of Aristotle, a century later than Herodotus,(10) the error in question was not eradicated; for this philosopher, in his Meteorologics, speaks of geographers still constructing maps in which the inhabited world was represented as circular in form. He declares this notion to be inconsistent both with reason and with observation. The measures obtained by navigation and by land journeys, show (he says) that the distance from the Pillars of Hercules to India, compared with the distance from Æthiopia to the Lake Mæotis and Scythia, is more than 5 to 3.(11)

The other cause of the belief in the circularity of the earth was the apparent shape of the heaven, which seemed like a solid cupola, the hemispherical outline of which defined the extremity of the earth in every direction. The sun, the moon, and the stars appeared to move upon, or with, the inner surface of this hemisphere; and hence, as the ocean was supposed to flow in a stream round the outer margin of the earth, the heavenly bodies were believed to emerge from ocean at their rising, and to sink into it at their setting.(12)

---

(9) Elem. Astr. c. 13.

(10) The birth of Aristotle is exactly a hundred years after the birth of Herodotus: the latter was born in 484, the former in 384 B.C.

(11) διὸ καὶ γελοίως γράφουσι νῦν τὰς περιόδους τῆς γῆς· γράφουσι γὰρ κυκλοτερῆ τὴν οἰκουμένην. Meteor. ii. 5, § 13. Compare i. 9. The doctrine of Eratosthenes, that the entire earth is σφαιροειδὴς, οὐχ ὡς ἐκ τόρνου δὲ, ap. Strab. i. 3, § 3, alludes to the words of Herodotus, but has a different meaning. Eratosthenes considered the earth as a sphere, not as a circular plane.

(12) See Völcker, ib. pp. 93-97, where the Homeric conception of the ocean-stream is copiously illustrated. Homer places the sea within the shield of Achilles, but makes the circumfluous ocean run along the outward rim, Il. xviii. 483, 607. Hesiod assigns to it the same position, Scut. Herc. 314. The ocean surrounds the earth, Æsch. Prom. 141. Herodotus rejects the idea of an ocean-stream flowing round the earth, and he attributes the invention of it to Homer or to some other early poet, ii. 21, 23, iv. 36.

Thus Homer describes the sun as rising out of ocean, and ascending the heaven;(13) again, as plunging into ocean, passing under the earth, and producing darkness.(14) He likewise speaks of the stars as bathed in the waters of the ocean;(15) and he distinguishes the Great Bear as the only constellation which never sinks into the ocean-stream.(16) A fabulous reason for this immunity was devised by the mythologists: it was founded upon the metamorphosis of Callisto, and her transfer to the starry heaven, which was as ancient as the poems of Hesiod.

Other nations, as was natural, entertained a similar belief with respect to the sun setting in the waters of the far west. The ancients inform us that the Iberians supposed themselves to hear the hissing of the sea when the burning sun plunged into

---

(13) Il. vii. 422; Od. iii. 1, xix. 433. See Völcker, ib. pp. 20-23.

(14) Il. viii. 485, xviii. 239; Od. x. 191; Hymn. Merc. 68.
The language of subsequent poets generally followed that of Homer; thus Virgil—
   'Nec quum invectus equis magnum petit æthera; nec quum
   Præcipitem oceani rubro lavit æquore currum.'—Georg. iii. 358.
   'Quid tantum oceano properent se tingere soles Hyberni.'
   Hyberni.   Georg. ii. 481.
Compare Ovid, Met. ii. 68, 157; Lucan, vii. 1, ix. 625. Æn. i. 749.
Virgil, Æn. viii. 589, speaks of the morning star as rising from ocean.
   'Taygete simul os terris ostendit honestum
   Pleias, et oceani spretos pede reppulit amnes;
   Aut eadem sidus fugiens ubi piscis aquosi
   Tristior hibernas cœlo descendit in undas.'—Georg. iv. 232-5.
   'Tingitur oceano custos Erymanthidos Ursæ,
   Æquoreasque suo sidere turbat aquas.'—Ovid, Trist. i. 4, 1.

(15) Iliad v. 6. Ovid, Met. ii. 171, describing the conflagration of Phaethon, says:
   'Tum primum radiis gelidi caluere Triones,
   Et vetito frustra tentarunt æquore tingi.'
In Trist. iv. 3. 2, he characterizes the two bears as 'utraque sicca.' Lucan speaks of Carmania as a country situated so far to the south that the Great Bear partly sets:
   'Carmanosque duces, quorum devexus in austrum
   Æther, non totam mergi tamen aspicit arcton.—iii. 250.

(16)   Οἴη δ' ἄμμορός ἐστι λοετρῶν ὠκεανοῖο.
Il. xviii. 489; Od. v. 275. Compare Kruse's Hellas, vol. i. p. 245-273. Thus, Aratus, v. 48, ἄρκτοι κυανέου πεφυλαγμέναι ὠκεανοῖο, and Virgil, Georg. i. 246:
   'Arctos, oceani metuentes æquore tingi.'

the western ocean.(17) A similar report is mentioned by Tacitus with respect to the Northern Germans.(18)

The Greeks even conceived the sun in the personified form of a divine charioteer, who drove his fiery steeds over the steep of heaven, until he bathed them at evening in the western wave.(19) The story of Phaethon, who aspired to drive the chariot of the sun, who was hurled by the lightning of Jupiter into the river Eridanus, and whose sisters, metamorphosed into poplar-trees, shed tears at his death, which were hardened into amber, is a poetical figment, belonging to the early Greek mythology, though apparently later than Homer.(20) The personification of the sun led to his being regarded as a universal witness, not only of what could be seen, but also of what could be heard.(21) It is this notion of the sun as an all-seeing witness which induces Homer to represent him as giving information to Vulcan of the infidelity of Venus;(22) and which causes the author of the Homeric hymn to represent Ceres as applying to the sun for information respecting her lost daughter Proserpine.(23) In like manner the comic poet Philemon describes the air as knowing everything, because it is everywhere present.(24)

In the later classical mythology, Apollo or Phœbus had become the god of the sun, and Artemis, or Diana, the goddess

---

(17) Posidonius ap. Strab. iii. 1, 5; Cleomed. ii. 1, p. 109, et not. p. 423, ed. Bake; Flor. ii. 17; Juv. xiv. 279; cum. Schol. Stat. Sylv. ii. 7, v. 24; Auson. Epist. xix. 2; Claudian in Laude Serenæ, v. 52.

(18) Germ. 45. Compare Grimm, Deutsche Mythologie, p. 429.

(19) See Grote, Hist. of Gr. vol. i. p. 465. Ovid represents the god of the sun as saying that none of the gods but himself, not even Jupiter, is able to drive his chariot, Met. ii. 59.

(20) Ovid, Met. ii.—See below, ch. viii. § 4.

(21)   'Ηέλιός θ' ὃς πάντ' ἐφορᾷς καὶ πάντ' ἐπακούεις.

Iliad, iii. 277, repeated in Od. xi. 109, xii. 323. Respecting the personification of the sun, see Völcker, ib. p. 26.

(22) Od. viii. 302. Compare the remark of Jupiter to Juno, Il. xiv. 344.

(23) Hom. Hymn. Cer. 69. Æschylus likewise speaks of the sun as an all-seeing witness, Agam. 632.

(24) Ap. Stob. Ecl. Phys. i. 10, § 10. Compare Meineke, Fragm. Com. Gr. vol. iv. p. 31.

of the moon; but this identification had not been made by Homer and the early poets.(25)

The infantine astronomy of the Greeks did not explain how the sun found his way from the west, after his daily course through the heaven, back to the east and the region of the dawn. The mythological account of this journey was that the sun was carried along the external ocean in a golden goblet fabricated for him by Vulcan; and that he reclined during the night upon this vessel, while he performed the navigation from his western goal back to his starting-point in the east.(26)

According to Aristotle, many of the ancient meteorologers believed that the course of the sun was not under the earth, but that after its setting in the west it travelled round the north to the east, and that night was caused by the elevation of the northern part of the earth, which intercepted the sun's rays during this transit.(27)

But however this transport might be effected, the light and warmth of the sun were supposed to be confined to the upper surface of the earth. Underneath, all was gloomy and cold. In this part of the world Hades, the receptacle of the departed souls, was placed. It was conceived as a subterranean abode, excluded from the aspect of the heavenly bodies, and therefore dark, frigid, and cheerless. Its only communications with the upper earth were through the mouths of caverns.(28)

---

(25) See below, § 11.

(26) See the passages of Mimnermus, Stesichorus, Æschylus, Pherecydes, and others in Athen. xi. pp. 469-70. The words of Pherecydes are quite distinct : "Ἥλιος δὲ δίδωσιν αὐτῷ [Hercules] τὸ δέπας τὸ χρυσίον, ὅ αὐτὸν ἐφόρει σὺν ταῖς ἵπποις, ἐπὴν δύνῃ, διὰ τοῦ ὠκεανοῦ τὴν νύκτα πρὸς ἕω, ἵν' ἀνίσχει ὁ ἥλιος.

(27) Meteorol. ii. 1. Compare Avienus, Ora Maritima, v. 647.

(28) The descent of Orpheus through the Tænarian cave illustrates this belief:

'Tænarias etiam fauces, alta ostia Ditis,
Et caligantem nigrâ formidine lucum
Ingressus.'—Georg. iv. 467.

In the Axiochus, p. 371, the earth is described as a plane, dividing the spherical heaven into two hemispheres. The upper hemisphere belongs to the celestial, the lower to the infernal gods.

See the declaration of the Sun in Od. xii. 383, and the answer of Jupiter, 386.

Although the ancient Greeks and Romans admitted the doctrine of posthumous punishment, the predominant idea of their Hades was simply that of a general receptacle of the souls of the dead. The hell of the Christian world is a place in which the wicked are punished with fire; and hence the modern Sicilians consider the crater of Etna as a spiracle of hell.[29] The ancients saw nothing in a burning mountain to remind them of Hades: they found, however, a mythological explanation for this natural phenomenon, and they supposed that both Etna and the Lipari Islands were the chimneys of Vulcan's smithy.[30]

§ 3 The sun exercises so decisive and conspicuous an influence upon the actions of men during every hour of their life, that all nations, from their earliest existence, must have found it necessary to watch his movement; and from their observations to form certain measures of time. The diurnal course of the sun, and the alternation of day and night, would be easily observed, and would lead to simple rules of conduct. The number of the seasons, and their regular succession, are facts almost equally obvious, though a longer time is requisite for their notation and their reduction to a regular series. Hence the annual course of the sun—corresponding to the earth's motion round its orbit—must have been soon determined within certain narrow limits of error. It must soon have been perceived that the lengthening and shortening of the days and nights—such as occurred in the latitudes of Greece and Western Asia—was subject to a fixed law; and means must have been found for determining the recurrence of the equinoxes and

---

(29) See Brydone's Travels in Sicily, Letter ix. He heard from the inhabitants of the mountain, that an English queen, Anna, who had made her husband a heretic, was burning in the interior of the volcano. This Dantesque idea is a singular counterpart to Gray's couplet:

'Twas love that taught a monarch to be wise,
And gospel light first beamed from Boleyn's eyes.'

(30) See Callim. Hymn. Dian. 46—61. This fable is rejected in the Ætna of Lucilius, v. 29 (Wernsdorf, Poet. Lat. Min. vol. iv), and a physical cause is substituted.

solstices. The points of the compass would thus be given. The conception of the ecliptic, of a great circle in the heaven formed by the sun's annual course, and of its obliquity when compared with the equator—the great circle deduced from the sun's daily course—would be of later introduction, and is not necessary to the formation of an annual period, and of a rude calendar sufficient for the guidance of agriculture and navigation, and for the other ordinary purposes of life.[31]

Annus was synonymous with *annulus*, and originally meant a ring or circle, like *circus* and *circulus*, the former of which words is identical with the Greek κρίκος, or κίρκος, a ring. It denoted a cycle of time; that is, a fixed and recurrent period.[32]

The succession of the seasons forms a natural cycle,[33] which must, from the earliest formation of civil communities, have led to the establishment of a customary annual period, defined with greater or less exactness. The cultivation of the soil, the breeding of sheep and cattle, and the hunting of wild animals, were dependent on the season. The same was the case with navigation, which ceased during the winter. Amusements in

---

[31] Annus vertens est naturâ, dum sol percurrens duodecim signa eodem unde profectus est redit. Censorin. de D. N. c. 19.

ἐνιαυτὸς, ἔτος, δωδεκάμηνος χρόνος, ἡλίου περίοδος, περιελθόντος ἐξ ὡρῶν εἰς ὥρας τοῦ θεοῦ, τὸν κύκλον τοῦ ἄστρου περιδραμόντος.—Pollux, i. 54.

[32] 'Ut parvi circuli annuli, sic magni dicebantur circites anni, unde annus.' Varro, L. L. vi. 8.

Compare Virgil, Georg. ii. 401:—

Redit agricolis labor actus in orbem,
Atque in se sua per vestigia volvitur annus.

The following verses of Hermippus, a poet of the old comedy, indicate the recurrent character of the year:—

ἐκεῖνός ἐστι στρογγύλος τὴν ὄψιν, ὦ πονηρὲ,
ἐντὸς δ' ἔχων περιέρχεται κύκλῳ τὰ πάντ' ἐν αὐτῷ,
ἡμᾶς δὲ τίκτει περιτρέχων τὴν γῆν ἀπαξάπασαν·
ὀνομάζεται δ' ἐνιαυτὸς, ὢν δὲ περιφερὴς τελευτὴν
οὐδεμίαν οὐδ' ἀρχὴν ἔχει, κυκλῶν δ' ἀεὶ τὸ σῶμα
οὐ παύσεται δι' ἡμέρας ὁσημέραι τροχάζων.

Ap. Stob. Ecl. i. 8. Meineke, Fragm. Com. Gr. vol. ii. p. 380. The word ἡμᾶς in v. 3, seems to show that those verses were spoken by a chorus of Ὧραι.

[33] See Gen. i. 14; viii. 22.

the open air, such as national games, must likewise have been fixed with reference to the weather. War was waged during the fine part of the year. At a period of semi-barbarism, men were more dependent on the seasons than at a more civilized stage of society. Houses were ill constructed, clothing was scanty, and even scarcity of food was a constantly recurring evil. Roads and bridges did not exist. Navigation was timid and unskilful.

The division of the four seasons, though of considerable antiquity,[34] is not decisively indicated in nature. The antithesis between summer and winter is obvious; the revival of nature in spring, after the torpor of winter, is also a marked epoch.[35] But autumn is a less definite season. Hence a division of the year into the three seasons of spring, summer, and winter, is attributed to the ancient Egyptians.[36] The year of the ancient Germans consisted only of these three seasons, according to the testimony of Tacitus;[37] and in English, the same three seasons are denoted by Anglo-Saxon words; whereas the word 'autumn' is borrowed from the Latin.[38] The recurrence of the seasons at fixed intervals is one of the most remarkable laws of physical

---

(34) See Galen. ad Hippocrat. Epidem. i. vol. xvii. 1. p. 18, ed. Kühn. The treatise of Hippocrates de Diætâ, iii. 68 (vol. vi. p. 594, ed. Littré), says that the year is generally divided into four seasons: that the winter lasts from the setting of the Pleiads to the spring equinox; the spring from the equinox to the rising of the Pleiads; the summer from the rising of the Pleiads until the rising of Arcturus; and the autumn from the rising of Arcturus to the setting of the Pleiads. The treatise De Diætâ is either genuine, or as ancient as Hippocrates, Littré, vol. i. p. 356. Manilius, ii. 656, considers the four seasons as marked by nature.

(35) See Macrob. Sat. i. 12, § 14. Virgil supposes the spring to have immediately followed the creation of the world; as being the season of renewal of vegetable and animal life, Georg. ii. 336.

(36) Diod. i. 26. Æschylus, in describing the division of the year by Prometheus, enumerates only the three seasons of winter, summer, and spring, Prom. 454—6.

(37) Germ. 26. Compare Grimm, Deutsche Mythologie, p. 435.

(38) See Dean Trench, Select Glossary of English Words (Lond. 1859), v. Harvest.

The Hippocratean treatise περὶ ἑβδομάδων distributed the year into seven seasons. It left the spring and autumn undivided; but made the winter consist of three seasons, and the summer of two. See Galen, ib.; Hippocrates, vol. viii. p. 635, ed. Littré.

nature, and it is often employed by the ancients as a proof of the superintendence of the world by a divine Governor.(39)

Besides the recurrence of the seasons, and the general phenomena of animal and vegetable life which severally accompanied them, there were certain special and local phenomena which returned at annual periods. Such were certain winds, called by the Greeks the Etesian, from their annual recurrence, migrations of certain birds,(40) and the inundation of the Nile; which last phenomenon was of a peculiarly striking and important character, and occupied the attention of Greek observers and speculators from an early date.(41)

Homer frequently mentions a definite number of years; which he conceives as composed of months, and as recurring in a constant cycle. This fact proves that the year was in his time a fixed and recognised period of time.(42) Now, the poems of Homer were at an early period recited and read over every part of the Greek world. Each Hellenic city and community, from Sinope to Massilia, must have understood the same period to be

---

(39) See the celebrated passage of Claudian, at the beginning of the poem against Rufinus:—

> Nam cum dispositi quæsissem foedera mundi,
> Præscriptosque mari fines, *annisque meatus*,
> Et lucis noctisque vices, tunc omnia rebar
> Consilio firmata Dei, qui lege moveri
> Sidera, *qui fruges diverso tempore nasci*,
> Qui variam Phœben alieno jusserit igni
> Compleri, solemque suo; porrexerit undis
> Litora; tellurem medio libraverit axe.

The mutability of human affairs is contrasted with the stability of the universe by Manilius, i. 522—530.

(40) Hesiod speaks of the annual cry of the crane, as marking the time for ploughing and the winter season. Op. 446. See likewise Aristoph. Av. 710.

(41) See below, ch. 2, § 2.

(42) The third and fourth years are mentioned in Od. ii. 89, xxiv. 141-2; four years in Od. ii. 107, xix. 152; five years in Il. xxiii. 833, Od. xxiv. 309; seven years, Od. xiv. 285, vii. 259; eight years, Od. vii. 259, 261; xix. 287; nine years, Il. ii. 134, 295, Od. v. 107; ten years, Il. viii. 404, 418, xii. 15, Od. v. 107, xv. 18; eleven years, Od. iii. 391; twenty years, Il. xxiv. 765, Od. ii. 175, xvi. 205, xvii. 327, xix. 222, 484, xxi. 208, xxiii. 102, 170, xxiv. 322: on the year as a recurrent period, Il. ii. 295, Od. xiv. 294. The word λυκάβας is used for *year* in Od. xiv. 161, xix. 306, *i.e.* 'the course of the sun.' That Homer's was the tropical year is held by Ideler, Chron. vol. i. p. 260.

designated in his poetry by a year. Each, for example, must have conceived the siege of Troy to have occupied ten tropical years; each must have conceived Ulysses to have passed eight tropical years in the island of Calypso; each must have conceived the dog Argus to have died in the twentieth tropical year from his master's departure for Troy.

It is clear that from an early period there must have been a measure for the age of man. Husbands and wives must have known each other's age. Parents must have known the age of their children.

Hesiod advises a man to marry about the age of thirty years.[43] His wife is to be nineteen years old at her marriage.[44] The same early poet mentions a boy of twelve months, and also of twelve years.[45]

Homer speaks of Nestor having outlived two generations, and ruling over the third.[46] Hesiod says that the raven lives nine generations of man; the stag four generations of the raven; the crow three generations of the stag; the phœnix nine generations of the crow; and the nymphs ten generations of the phœnix.[47] These passages imply a conventional number of years for the duration of a generation of mankind. Mimnermus and Solon mention the ages of sixty and eighty years.[48] Xenophanes reckons his age at $67 + 25 = 92$ years.[49] The life of man is computed by Herodotus, as by the Psalmist, at seventy years.[50]

The successive ages of man are enumerated by Solon, who

---

[43] Op. 694.

[44] Ib. She is to be five years after ἥβη, which is the age of fourteen. Pollux, i. 58, explains ἥβη to be fourteen years. See K. F. Hermann, Privatalt. der Gr. § 4, n. 19.

[45] Ib. 749, 50.   [46] Il. i. 250.

[47] Fragm. 50, ed. Gaisford.

[48] Mimn. fr. 6, Solon, fr. 20, ed. Schneidewin. They avoid the number ἑβδομήκοντα, because it will not go into elegiac verse. Mr. Gladstone, Studies on Homer, vol. iii. p. 441, remarks that all the multiples of ten up to one hundred occur in the Homeric catalogue, except the intractable ἑβδομήκοντα.

[49] Diog. Laert. ix. 19.

[50] i. 32; compare Psalm xc.

measures them by periods of seven years. He places the perfection of a man's bodily strength in the fourth hebdomad, from twenty-eight to thirty-five years of age, and the perfection of his mental powers in the seventh and eighth hebdomads, from forty-nine to sixty-three years of age. He considers life to have reached its natural term at the age of seventy.[51]

Even in the rudest systems of jurisprudence and government, there must have been legal definitions dependent upon age. The age at which a man became master of his property, could enjoy his civic rights, was liable to or exempt from military service, must have been defined.[52] Rules as to prescription, more or less precise, must likewise have found their way, at an early period, into the judicial practice of all intelligent nations, and would naturally be founded upon the year.[53]

The accounts of the early history of the Greek States represent annual magistracies as created immediately after the abolition of the heroic royalty. Thus the annual prytanes are stated to have ruled at Corinth from 747 to 657 B.C.; the decennial archons are stated to have governed Athens from 752 to 684, and the annual archons from 683 B.C. The Spartan ephors, the date of whose institution is uncertain, were likewise annual officers. The traditions of annual Corinthian prytanes and decennial Athenian archons do not indeed rest on a firm basis; but it cannot be doubted that the determination of high magistracies by yearly measurements was of great antiquity in the Greek commonwealths. The Roman consuls were not established

---

[51] Sol. fr. 23; Censorin. c. 14; Macrob. Com. Somn. Scip. i. 5. The genuineness of these verses has been suspected by some critics.

[52] Concerning the period of ἥβη, see K. F. Hermann, Staatsalt. der Gr. § 123; Privatalt. § 35, n. 13, § 56, n. 11. As to the Roman law respecting *pubertas*, see Rein, Röm. Privatrecht, p. 113.

[53] The German word for prescription is *Verjährung*.
The idea of prescription is expressed in the following passage of Æschylus, referring to the combat of Eteocles and Polynices:—

ἀλλ' ἄνδρας Ἀργείοισι Καδμείους ἅλις
ἐς χεῖρας ἐλθεῖν· αἷμα γὰρ καθάρσιον.
ἀνδροῖν δ' ὑμαίμοιν θάνατος ὧδ' αὐτοκτόνος,
οὐκ ἔστι γῆρας τοῦδε τοῦ μιάσματος.

Sept. Theb. 679—82.

till 509 B.C.; but the custom for officers of State to mark the year by annually driving a nail into the wall of a temple, seems to have been ancient in Italy.(54)

Tyrtæus, whose lifetime is placed, on sufficient evidence, in the 7th century B.C., speaks of the First Messenian War having lasted nineteen years, and being terminated by the flight of the Messenians from Ithome in the twentieth year of the war.(55) Both ancients and moderns have concurred in understanding that in this passage ordinary solar years are meant.(56)

Homer does not mention the equinoxes; and he alludes to the solstices only in a fabulous manner. In the 15th book of the Odyssey (v. 403), the swineherd speaks of the island of Syria as situated beyond Ortygia, 'the seat of the turns of the sun.' By Ortygia it has been supposed that Delos is meant; by Syria the island of Syros. The Scholiast thinks that the allusion is to a cave of the sun, by means of which the solstices were noted.(57)

The solstices are mentioned three times by Hesiod, in his Works and Days.(58) In the first passage he warns the husbandman to plough before the winter solstice. In the second, he speaks of Arcturus rising fifty days after the winter solstice. In the third, he says that the period of fifty days after the summer solstice is favourable to navigation.

Thucydides is studious to mark that his chronological notation is defined by natural periods, and not by any civil calendar. He determines his year by the seasons—that is to say, his historical year is the solar or tropical year.(59) Notwithstanding the absence of any calendar of authority, recognised by all the

---

(54) See Müller's Etrusker, vol. ii. p. 329; Inquiry into the Cred. of the Early Rom. Hist., vol. i. c. 5. § 13.

(55) Fragm. 4, ed. Gaisford.

(56) See Grote, Hist. of Gr. vol. ii. p. 563.

(57) Homer describes the island of Ææa as the seat of the rising of the sun, Od. xii. 4.

(58) See vv. 477, 562, 661.

(59) γέγραπται δὲ ἑξῆς ὡς ἕκαστα ἐγίγνετο κατὰ θέρος καὶ χειμῶνα, ii. 1. Compare v. 20, on the inaccuracy of reckoning by magistrates. The annual computation of Thucydides is illustrated by Dodwell, Annales Thucydidei, § 6, 14.

Greek States, in the Peloponnesian War, yet the period of a year was fixed by usage, and is mentioned as a known period in treaties between different States made for a definite number of years.(60)

Biot supposes the points of the compass to have been originally determined by observing the places of the rising and setting of the sun on any day. The bisection of the angle made by lines drawn from these points to the place of the spectator would give the south and north, with a close approach to accuracy; and when the east and west points were known, the equinoxes might be determined by watching the days when a wall running due east and west cast no shade at sunset and sunrise.(61)

It may be considered as certain that the solar year, with its solstitial and equinoctial points, was determined without astronomical precision, but sufficiently for practical purposes, at a remote period, among the Greek communities.(62)

§ 4. A lunation, or synodical month, being the interval between two conjunctions of the sun and moon, is equal to 29d. 12h. 44m. It was founded on the most obvious determination of the moon's course, and furnished the original month of the Greeks, which was taken, in round numbers, at 30 days.(63) By

---

(60) A five years' truce was made between the Peloponnesians and Athenians in 450 B.C., Thuc. i. 112; a thirty years' truce in 445 B.C., ib. 115; a fifty years' truce, v. 18, 23 in 421 B.C.; a hundred years' truce, v. 47; a treaty between the Lacedæmonians and Argives for fifty years, v. 79. The ancient treaty between the Eleans and a neighbouring state, preserved on a brazen plate in the British Museum (Bœckh, Corp. Inscript. Gr. n. 11) is for a hundred years.
Solon is said to have enacted his laws for the term of a hundred years, Plut. Sol. 25.

(61) See Biot, Restes de l'ancienne Uranographie égyptienne, Jour. des Sav. août, 1854; and, Détermination de l'équinoxe vernal effectuée en Egypte, Jour. des Sav. mai, juin, juillet, 1855.

(62) The celebration of the summer solstice by bonfires is considered by Grimm as of great antiquity among the northern nations, Deutsche Mythol. p. 349, 413.

(63) μήν ἐστι χρόνος ἀπὸ συνόδου ἐπὶ σύνοδον, ἢ ἀπὸ πανσελήνου ἐπὶ πανσέληνον. ἐστι δὲ σύνοδος μὲν ὅταν ἐν τῇ αὐτῇ μοίρᾳ γένηται ὁ ἥλιος καὶ ἡ σελήνη· τοῦτ' ἐστὶ περὶ τὴν τριακάδα σελήνης, Gemin. c. 6. καλεῖται μὴν τὸ

SECT. 4.] THE GREEKS AND ROMANS. 17

combining the course of the sun with that of the moon, the tropical year was assumed, at a rough and approximate computation, to consist of 12 lunations, or 360 days.(64)

The month is often mentioned by Homer, and is treated as a constituent part of the year.(65) Hesiod gives detailed precepts respecting the character of its days, and speaks of it as containing thirty days.(66) The riddle ascribed to Cleobulus, one of the Seven Sages, or to his daughter Cleobulina, assigns twelve months to the year, and thirty days to each month,

---

ἀπὸ συνόδου ἐπὶ σύνοδον χρονικὸν διάστημα, καὶ λοιπὸν, ὁ τριακονθήμερος χρόνος, Cleomed. ii. 5, p. 135. Luna singulos suos menses conficit diebus undetriginta circiter et dimidiato. Censorin. D. N. c. 22. Romulus, cum ingenio acri quidem sed agresti statum proprii ordinaret imperii, initium cujusque mensis ex illo sumebat die quo novam lunam contigisset videri. Macrob. Sat. i. 15, § 5.

νὺξ μὲν οὖν ἡμέρα τε γέγονεν οὕτω καὶ διὰ ταῦτα, ἡ τῆς μιᾶς καὶ φρονιμωτάτης κυκλήσεως περίοδος· μεὶς δὲ ἐπειδὰν σελήνη περιελθοῦσα τὸν ἑαυτῆς κύκλον ἥλιον ἐπικαταλάβῃ, ἐνιαυτὸς δὲ ὁπόταν ἥλιος τὸν ἑαυτοῦ περιέλθῃ κύκλον. Plat. Tim. § 14, p. 39.

(64) Dein morata in coitu solis biduo, cum tardissime, a tricesimâ luce rursus ad easdem vices exit, haud scio an omnium quæ in cœlo pernosci potuerunt magistra. In duodecim mensum spatia oportere dividi annum, quando ipsum toties solem redeuntem ad principia consequitur. Plin. ii. 6. Compare Petav. de Doctr. Temp. i. 5.

(65) One month is mentioned in Il. ii. 292, Od. x. 14, xii. 325, xiv. 244, xxiv. 118; three months, Od. xvii. 408; thirteen months, Il. v. 387; τοῦ μὲν φθίνοντος μηνὸς, τοῦ δ' ἱσταμένοιο, Od. xiv. 162, xix. 306; a year consists of months and days, Od. xi. 294, xiv. 293. The moon is called μήνη in Il. xix. 374, xxiii. 455. The Greek word for month was originally μένς, which form is preserved in the Latin *mensis*. As the Greek language did not tolerate this termination, the word μένς was in some dialects softened into μὴν, in others into μείς. The word σελήνη, derived from σέλας, denotes the moon's brightness; because, though less luminous than the sun, it is devoid of heat. ἠέλιος or ἥλιος is allied with ἕλη or εἴλη, which expresses heat. The Latin *luna* is contracted from *lucina*, which is a derivative of *lux* or *luceo*.

(66) Op. et Di. 764; Pollux, i. 63, states that the month consists of three decads, or δεχήμερα, and that the last day is called τριακάς. The Athenians said thirty days rather than a month, Elmsley ad Aristoph. Acharn. 858.

According to Suidas, in γεννῆται, the ancient Attic tribes were four in number; each of them contained three φρατρίαι or τριττύες, making twelve; and each φρατρία or τριττὺς contained thirty γένη, making three hundred and sixty γένη. Suidas adds, that these numbers were derived from the four seasons, the twelve months, and the three hundred and sixty days of the year. This numerical symmetry, however, is probably unhistorical, and the reference to the calendar fanciful. See Grote, vol. iii. p. 73.

Strabo, ix. 1, § 20, states that more than three hundred statues were erected to Demetrius Phalereus at Athens, whose administration lasted ten years from 317 B.C. Nepos, Miltiad. 6, and Plutarch, Rep. Ger. Præc.

making a year of 360 days.([67]) Hippocrates assumes the month to consist of thirty days.([68]) Herodotus, in like manner, reckons the month at thirty, and the year at 360 days.([69]) Even Aristotle, in more than one place, gives this duration both to the month and the year.([70]) According to Plutarch the primitive Roman year consisted of 360 days.([71])

The following account of the ancient Greek calendar is given by Geminus:—

'The system pursued by the ancient Greeks was to determine their months by the moon, and their years by the sun. The laws of the State and the oracles of the gods concurred in prescribing that sacrifices should be regulated according to three established standards—namely, months, days, and years. This precept all the Greeks reduced into practice by distinguishing between their measures of time; and by regulating the years according to the sun, and the days and months according to the moon. By the regulation of the year

---

27, fix the number at three hundred. Varro, ap. Non. c. 12 (in luces), p. 361, ed. Gerlach et Roth.

'Hic Demetrius æneas tot aptu'st,
Quot luces habet annus absolutas.'

Lastly, Diog. Laert. v. 75, says that three hundred and sixty statues were dedicated to him. The number of statues fluctuates, and the reference to the days of the year is doubtless fanciful and fabulous.

([67]) Diog. Laert. i. § 91; Suidas in Κλεοβουλίνη, Stob. Ecl. Phys. i. 8, 37; Anth. Pal. xiv. 101. Stobæus has ἑξήκοντα in the second line; the other three authorities have τριάκοντα. The sense is the same; but τριάκοντα destroys the metre. By ἑξήκοντα, thirty days and thirty nights,= thirty νυχθήμερα, are meant. Cleobulus was contemporary with Solon. The genuineness of this riddle cannot be safely assumed, but it cannot be disproved. There is nothing in the language or contents inconsistent with the time to which it is referred. On Cleobulina, see Plutarch, Sept. Sap. Conviv. 3.

([68]) De Carn. vol. xxi. p. 442. ed. Kühn; he makes 9 months 10 days =280 days, and De Epidem. ii. 3. vol. xxiii. p. 454, 9 months =270 days.

([69]) He reckons 35 months as equal to 1050 days, and 70 years as equal to 25,200 days, i. 32. In iii. 90, he states that the annual tribute yielded by the Cilicians to Darius was 360 white horses, being one horse for each day.

([70]) Aristot. H. A. ii. 17, states that the ribs of serpents are equal in number to the days of the month, being 30. He states, in vi. 20, that 60 days is the sixth part of the year; 72 days the fifth part; three entire months the fourth part.

([71]) Num. 18.

according to the sun was meant such an arrangement as secured that the same periodical sacrifices should always be performed to the gods at the same seasons of the year; that the vernal sacrifices, for example, should always fall in the spring, and the summer sacrifices in the summer. This accordance of the sacrifices with the season of the year was considered as acceptable and gratifying to the gods; but it could not be brought about, unless the solstices and equinoxes fell at the same times of the civil year. The regulation of days by the moon consisted in counting them by the moon's phases—the last day of the month being called the *triacas*, or thirtieth.'[72]

Plato states that the months and years are regulated in order that the sacrifices and festivals may correspond with the natural seasons;[73] and Cicero remarks that the system of intercalation was introduced with this object.[74]

'All nations (says Newton) before the just length of the solar year was known, reckoned months by the course of the moon, and years by the returns of winter and summer, spring and autumn; and in making calendars for their festivals, they reckoned thirty days to a lunar month, and twelve lunar months to a year, taking the nearest round numbers; whence came the division of the ecliptic into 360 degrees.'[75]

---

[72] c. 6. Compare Ovid, Fast. iii., ad fin.
'Luna regit menses. Hujus quoque tempora mensis
Finit Aventino Luna colenda jugo.'

[73] Leg. vii. § 14, p. 809.

[74] De Leg. ii. 12.

Speaking of the Jewish festivals, Michaelis, Comment. on the Laws of Moses, art. 199, vol. iii. p. 206, remarks:—'The festivals appointed for a certain day of the moon, had all a reference to the beginning and end of harvest, and to the vintage, and could not by any means have been celebrated in an appropriate manner, nor even with the shadow of an allusion to these seasons, if the lunar year [of 354 days] had been allowed to fall behind (as it really does) about 33 days in the course of 33 moons.' He proceeds to show that the ancient Jews rectified this defect in their calendar by a process of intercalation.

Michaelis cites a statement from a traveller, that the negroes in Western Africa had a year composed of lunations, which they corrected by a reference to the harvests.

[75] Chronology of Ancient Kingdoms, Works, vol. v. p. 55.

It is stated that Solon made a regulation of the Attic calendar, intended to remedy the inaccuracy caused by treating the synodical lunar month as exactly equal to thirty days: whereas it wants 11h. 16m., or nearly half a day, of that period. This contrivance consisted in dividing the thirtieth day (which he called ἕνη καὶ νέα) between the old and new month; and in making some of the months *full* and others *hollow*: the former consisting of thirty, the latter of twenty-nine days.[76] If this regulation was synchronous with the legislation of Solon, it took place about 594 B.C., near the middle of the lifetime of Thales.

The periodical, as distinguished from the synodical month, is the time during which the moon makes a revolution from any point in the zodiac back to the same point; and it consists of 27d. 7h. 43m.[77] Hence a lunar month is sometimes taken in round numbers at twenty-eight days: and this is the length of a lunar month according to the law of England.[78]

The lunar year, measured by twelve periodical lunations, is 354d. 8h. 48m. 36s., being nearly eleven days shorter than the solar tropical year.

Allusions to the year and month, as fixed by these lunar standards, sometimes occur in antiquity. Thus the passage of the

---

[76] Plut. Solon, 25; Aristoph. Nub. 1178—97; Diog. Laert. i. § 57, 59. The word ἕνος appears to mean 'that which belongs to the former of two consecutive periods.' Thus it is synonymous with περύσινος, 'belonging to the previous year.' In the phrase ἕνη καὶ νέα, it means the last day of the first of two consecutive months, as opposed to the first day of the second of those months.

Minime videntur errasse, qui ad lunæ cursum menses civiles accommodarunt, ut in Græciâ plerique, apud quos alterni menses ad tricenos dies sunt facti. Censorin. c. 22. Compare Ideler, Chron. vol. i. p. 266.

In the apocryphal book of Enoch, written about 130 B.C., alternate months of 30 and 29 days are mentioned, c. 78. v. 15, ed. Dillmann, Leipzig, 1853. Concerning the date of its composition, see pref. p. xliv.

[77] Macrob. in Somn. Scip. i. 6, § 49, gives about 28 days to the periodical lunar month, and 30 days to the synodical lunar month. Cleomedes, i. 3, ii. 3, 5, gives $28\frac{1}{2}$ days to the former, and 30 to the latter. Geminus, c. 6, states the duration of the synodical lunar month at $29\frac{1}{2}\frac{1}{33}$ days.

[78] Blackstone, Com. vol. ii. p. 141.

Odyssey which describes seven herds of oxen and seven flocks of sheep, each containing fifty head, in the island of Thrinacia, as belonging to the Sun, and as tended by Phaëthusa and Lampetia, daughters of Hyperion, was explained by Aristotle to refer to the lunar year of 350 days.([79]) This round number designated the lunar year, as the round number of 360 days designated the solar year. According to Macrobius, the common year of the Greeks consisted of 354 days.([80])

The period of child-bearing in women, which is 280 days, is usually spoken of by the ancient writers as consisting of ten months. This seems to assume a month of twenty-eight days. The moderns, who reckon by calendar months of thirty and thirty-one days, commonly designate this period by nine months.([81])

([79]) Od. xii. 129, cum Schol. See Nitzsch, ad loc. vol. iii. p. 387. Homer, Od. xiv. 13, says that in Ithaca, belonging to the palace of Ulysses, were twelve enclosures, each containing fifty breeding sows (= 600). The males slept outside, and were much fewer, for their number was diminished by the suitors, to whom the swineherd supplied the fattest: their number was 360. Mr. Gladstone, Studies on Homer, vol. iii. p. 435, thinks that this number alludes to the number of days in the year; but the coincidence seems merely accidental. If the poet had described the consumption of hogs for an exact year, he might have specified 360, meaning that one was killed each day, as in the case of the Cilician tribute of 360 horses, which, as Herodotus remarks, iii. 90, was a horse for every day: but the number in question represents the live hogs which were left after the consumption of the suitors had been provided for. The word ἑξήκοντα was probably inserted for the same reason that ἑβδομήκοντα was omitted in the Catalogue, viz. metrical convenience.

(80) Saturn. i. 12, § 2.

(81) Aristot. Hist. An. vii. 4, states that children are born at the 7th, 8th, and 9th month, but that most are born at the 10th, some even at the eleventh.

In Hippocrat. Endem. ii. 3, 17, it seems to be assumed that the period of gestation is 9 months, or 270 days. But in the Hippocratean Treatises, περὶ ἑπταμήνου, 7, and περὶ ὀκταμήνου, 13, it is laid down that the period is seven quarantines, or 280 days, and that the children born at this time are called ten months' children. Nevertheless, the month is here taken at 30 days. See vol. vii. p. 443, 460, ed. Littré. Compare Galen, vol. xvii. part i. p. 450, Kühn.

Menander ap. Meineke, Fragm. Com. Gr. vol. iv. p. 192, γυνὴ κυεῖ δεκάμηνος, followed by Terent. Adelph. iii. 4, 29, iv. 5, 57; Hec. v. 3, 24. The fabulous story respecting the generation and birth of Orion in Schol. Il. xviii. 486, alludes to ten months as the period of pregnancy. Ten months is given as the ordinary period of pregnancy in Iamblich. vit. Pythag. 192.

The Decemvirs stated this period at ten months, Gell. iii. 16; Dirksen,

§ 5 All the authentic evidence which is now extant, and all the indications contained in the earliest writers, concur in proving that there was a general agreement as to the length of the year and the month from a remote period of antiquity. The duration of the year, as marked by the apparent course of the sun, and by the real revolution of the earth in its orbit; and the duration of the month, as marked by the revolution of the moon round the earth, and its return to conjunction with the sun, were determined by the concurrent observation of all nations. The determination was, however, deficient in accuracy, from a want of a scientific knowledge of astronomy, of scientific instruments for observation, (including a good instrument for the measurement of time,) and of recorded observations of the heavenly bodies.

Owing to the disagreement in details which this want of precision engendered, each Hellenic State had a civil or religious calendar of its own; and the years of different national calendars began at different periods.

Each State, moreover, had a peculiar nomenclature of months; and as the commencement of the year differed according to the country, each national series of months commenced its course from a different day. The calendar, as is stated in a passage already cited from Geminus, was regulated with reference to sacrificial observances; and hence the names of the months were, as a rule, derived from a god, or from some public festival or other sacred celebration.([82]) In later times,

---

Zwölf-Tafel-Fragmente, p. 283. Ten months is treated as the time of pregnancy in Dig. xxviii. 2, 29; Cod. vi. 29, 4. See likewise Varro ap. Non. in spissum, p. 266, ed. Gerlach et Roth (vol. i. p. 293, ed. Bipont); Virgil, Ecl. iv. 61; Ovid, Heroid. xi. 45; Fast. i. 33; ii. 175, 445: iii. 124; v. 534; Met. ii. 453; viii. 500; ix. 286; x. 296, 479, 512.

Tertullian, de Animâ, c. 37: Legitima nativitas ferme decimi mensis ingressus est, who discovers in this number a connexion with the Decalogue. He likewise states, ibid., that the Roman goddesses Nona and Decima derived their names from the months in which the birth usually occurred. Ausonius, Eclog. de Ratione Puerperii, v. 39, speaks of nine months as the period of pregnancy. Macrobius, Comm. Somn. Scip. i. 6, § 66, gives the same period.

([82]) The derivations of the names of the Attic months will throw

when flattery had introduced the system of apotheosis, the names of powerful rulers were given to certain months. Thus the month of Munychion was called by the Athenians Demetrion, in honour of Demetrius Poliorcetes.(83) The Roman month Quintilis was called Julius, in honour of Julius Cæsar,(84) and the month Sextilis afterwards received the name of Augustus, in honour of Augustus Cæsar.(85) Some of the subse-

---

light on this subject. Hecatombæon took its name from the festival of Hecatombæa; Metageitnion from Apollo Μεταγείτνιος; Boëdromion from the festival of Boëdromia; Pyanepsion from the festival of Pyanepsia; Mæmacterion from the festival of Jupiter Mæmactes; Poseideon from the god; Gamelion, probably from a festival of Gamelia; Anthesterion from a festival of Anthesteria; Elaphebolion from the festival of Elaphebolia; Munychion, apparently from an epithet of Diana; Thargelion from the festival of Thargelia; Scirophorion from the umbrella carried in a sacred procession. See K. F. Hermann, Gottesdienst. Alt. der Gr., § 54—61. Clinton, F. H., vol. ii. p. 324.

The number of months which derived their names from gods attests their origin. Thus there was a month Ἀπέλλαιος at Delphi and other places; a month Ἀπολλώνιος in Elis, Ἄρειος in Bithynia, Ἀρτεμίσιος in many States; Ἀφροδίσιος in Bithynia and Cyprus, Δημήτριος in Bithynia and Bœotia, Διονύσιος in Naupactus, Chalcedon, and Tauromenium; Δῖος in Macedonia, Διόσκουρος in Crete, Ἕρμαιος in Bœotia, Ἡράκλειος in Bithynia and elsewhere; Ἥραιος in Crete, Bithynia, and Delphi; Ἡφαίστιος in a town of Asia Minor; Ποσειδέων in Athens and other places. Κρόνιος or Κρονίων was the original name of the Attic Hecatombæon. In other cases the reference to a festival or god is clear; as Ἀλαλκομένιος, Ἀπατουρίων, Βουφονίων, Γεραίστιος, Δάλιος, Δελφίνιος, Διόσθυος, Εὐαγγέλιος, Θεσμοφόριος, Ἱπποδρόμιος, Ἰώνιος, Ξανθικὸς, Προστατήριος. Concerning the sacred month Carneus, see Thuc. v. 54.

Galen, vol. xvii. P. 1, p. 21, Kühn. ἐν ἀρχῇ τοῦ κατ' ἐνιαυτὸν πέρατος μηνός. See Clinton, F. H., vol. iii. p. 350, who emends Περιτίου. Peritius is the third month from Dius; i.e. there are two between. See K. F. Hermann, Griech. Monatskunde, p. 101. The text in Clinton has καθ' ἑαυτοῦ. Perhaps we should read κατ' ἐνιαυτοῦ πέρας Περιτίου μηνός. This gives a meaning to τοῦτο γὰρ σημαίνει, &c. Peritius was the fourth month of the Macedonian year, and was not therefore at its end; but as it fell about the time of the winter solstice, Galen may consider it as the end of the year. It nearly corresponded to the Roman January. It is probable that Περίτιος was not derived from πέρας, but took its name from a Macedonian festival. See Hermann, p. 74, who derives the name from περιιέναι, and thinks that it was equivalent with the Roman Ambarvalia.

(83) Plut. Demetr. 12. The last day of the month—the ἕνη καὶ νέα—was also called Demetrias, and the festival of Dionysia was called Demetria.

(84) See Censorin. c. 22; Macrob. Sat. i. 12, § 34; Dio Cass. xliv. 5; Appian, B.C. ii. 106; Suet. Cæsar, 76; Plut. Num. 19. This month was selected because his birthday fell in it. The change took place in 44 B.C.

(85) Suet. Aug. 31; Dio Cass. 55, 6; Macrob. i. 12, § 35; Plut. ib. The date of this change was 8 B.C.

quent Roman emperors made similar attempts, but not with the same enduring success.(86) In Rome the care of the calendar was considered a religious function, and it had from the earliest times been placed in the hands of the pontifices.(87)

That a civil calendar of months, having a fixed position in the year, was not recognised in common by the Greeks, even in the second century after Christ, is clearly expressed in the following passage of Galen.(88) This writer, in a commentary upon a Hippocratean treatise, takes occasion to remark that Hippocrates defines the season of the year by a direct reference to astronomical facts.

'Let me mention, once for all, with reference to the reduction of the time of year into months, that if all nations had the same months, Hippocrates would not have spoken of Arcturus, the Pleiades, and the Dog-star, or of equinoxes and solstices; he would have been satisfied with saying that at the commencement of the Macedonian month of Dius, for example,(89)

---

(86) Thus Tiberius declined a proposal to call September Tiberius, and October Livius, Suet. Tib. 26. Caligula gave September the name of Germanicus, in honour of his father, Suet. Calig. 15. Nero ordered April to be called after his name, Neroneus, Suet. Ner. 55; Tac. Ann. xv. 74, xvi. 12. Domitian changed the name of September into Germanicus, and of October into Domitianus. Suet. Domit. 13; Macrob. Sat. i. 12, 36; Martial, ix. 2; Plut. Num. 19. Upon this last unsuccessful attempt, Macrobius says :—' Sed ubi infaustum vocabulum ex omni ære vel saxo placuit eradi, menses quoque usurpationis tyrannicæ appellationis exuti sunt: cautio postea principum cæterorum diri ominis infausta vitantium mensibus a Septembri usque ad Decembrem prisca nomina reservavit.' Commodus, however, made a similar attempt at a later date on a larger scale. Herodian, i. 14; Dio Cass. 67, 4; 72, 15; Suid. in Κόμοδος. Lamprid. in vit. c. 11. The Senate decreed that the months of September and October were to be called Antoninus and Faustinus, in honour of Antoninus Pius; but he declined the honour: Jul. Capit. in vit. Anton. Pii, c. 10. The Emperor Tacitus ordered the month of September to be called by his name, Vopisc. in vit. Tac. c. 13. See Censorin. ib. ad fin., and compare Græv. Thes. Ant. Rom. vol. viii. p. 305.

(87) See Becker, vol. iv. p. 233-438.

(88) Galen was born about 130, and died about 200 A.D. The lifetime of Hippocrates extended, according to Clinton, from 460 to 357 B.C. He therefore preceded Galen by about 500 years.

(89) Dius was the first month of the Macedonian year, which began at the autumnal equinox. It nearly corresponds with October. The proper form of the name is ὁ Δῖος: it is printed τὸ Δίον in the text of

the temperature was in a particular state. But inasmuch as the month of Dius is intelligible only to the Macedonians, but is unintelligible to the Athenians and the rest of mankind, and as Hippocrates wished to write what was useful to men of all nations, it was best for him to mention the equinox alone, without naming any month; for the equinox is a physical event, affecting the whole world; whereas months are local, and differ in each nation.'(90)

§ 6 Not only were the Greek States without any common calendar of time, by reference to which days could be determined, but they had no common chronological era for the designation of years. No trace of any such era appears either in Herodotus or Thucydides. Both of them denote a series of years by stating the interval between the event in question and their own lifetime: or by stating the interval between the event in question and some other previous event, the time of which is assumed to be known. Thus Herodotus informs us that the Trojan war preceded his own time by 800, the poets Hesiod and Homer by 400 years.(91) The same historian reports the statements of the Egyptian priests that 900 years had elapsed between the death of King Moeris and their own time;(92) and that 17,000 years had intervened between the time of Hercules and the reign of Amasis;(93) also the statement of the Tyrian priests, that 2300 years had elapsed since the foundation of Tyre.(94)

In like manner, Thucydides fixes the date of the passage of the Siceli into Sicily, by stating that it happened 300 years

---

Galen, but the correct reading is τὸ Δίου; i.e. τὸ ὄνομα τοῦ Δίου. See Clinton, Fast. Hell. vol. iii. p. 350.

(90) See Hippocrat. Epid. 1, vol. xvii. part 1, p. 19, ed. Kühn. The Achæans held their annual elections at the rising of the Pleiads, Polyb. iv. 37.

(91) ii. 53, 145. (92) ii. 13. (93) ii. 43.

(94) ii. 44. On the chronology of Herodotus, see Rawlinson's Herodotus, vol. i. p. 111. Mr. Rawlinson, though desirous of doing justice to Herodotus, seems to me not to make sufficient allowance for the helpless position of a historian writing at a time when there was no recognised chronological era, and no certain chronological data for distant periods.

before the foundation of Naxos, which preceded his own birth by about 550 years:([95]) he states that the interval between the legislation of Lycurgus and the end of the Peloponnesian war was a little above 400 years.([96]) He speaks of Melos as having been founded 700 years before its capture. He likewise reckons the interval from the building of triremes for the Samians by Ameinocles of Corinth, and from the sea-fight of the Corinthians and Corcyræans, to the end of the same war respectively at 300 and 260 years.([97]) For the Peloponnesian war, the main subject of his work, he counts from the beginning of the war, and makes the first year of the war his era, to which the successive years are referred.

Xenophon adopts a mode of chronological notation similar to that employed by Thucydides.([98])

An era or epoch is a conventional point of past time, determined by counting back from the present time, and used as a fixed point of reference for chronological purposes. The reference may be made by reckoning the years after it; as when a date is fixed by the years after the era of Nabonassar, after the foundation of Rome, after the birth of Christ, after the Hegira: or by reckoning the years before it, as when a date is fixed by the years before the birth of Christ. This simple and convenient contrivance, by which alone a clear idea can be formed of the chronological relations of past events, was unknown to the ancients, until the time when the Greeks attempted to establish an era by counting from the first Olympiad, and when the Romans, at a later period, reckoned by the years from the foundation of their city. Timæus, who died about 256 B.C., is stated to have been the earliest historian who made a systematic use of the Olympic era:([99]) it was always

---

(95) vi. 2.
(96) i. 18. The date of the last year of the Peloponnesian War is 404 B.C.
(97) v. 112, i. 13.
(98) See Schneider ad Xen. Hell. i. 2. 1.
(99) See Polyb. xii. 11.

confined to historical and literary writers, and never passed into use in civil and political life.(¹⁰⁰)

The era of Nabonassar appears to have been used only for astronomical purposes.

The Greeks frequently reckoned dates from the Trojan war, assuming it to be a historical event, determined chronologically. Thus Thucydides states that the Bœotians migrated from Arne to Bœotia sixty years after the capture of Troy, and that the Dorians conquered the Peloponnese eighty years after the same epoch.(¹⁰¹) All the subsequent Greek writers use the Trojan era occasionally, as a point of reference. But neither the Trojan war nor the first Olympiad ever became a fixed era for all Greeks, as the birth of Christ has become to Christian nations, and even as the Hegira has become to Mahometan nations.

In like manner the Romans reckoned from the foundation of the city; and sometimes from the expulsion of the kings or the Gallic conflagration. In Rome, however, the official mode of designating the year was by the name of the annual consuls; a similar mode of designation was generally adopted in the Greek states—thus, at Athens, an annual archon, at Sparta an ephor was *eponymous*, that is, he gave his name to the year.(¹⁰²) In many cases a priest or priestess, or other sacred officer, performed this function;(¹⁰³) doubtless on account of the connexion of the civil calendar with periodical rites of the national religion, to which reference has been already made.

The reckoning of the year by the names of the annual ma-

---

(100) See Krause's Olympia, p. 58.

(101) i. 12. Concerning the Trojan era, see Clinton, F. H., vol. i. p. 123; Fischer's unfinished Griechische Zeittafeln, p. 3.

(102) See Dodwell de Cyclis, p. 320; Manso Sparta, vol. ii. p. 379; Müller, Dor. iii. 7, 7.

(103) See the enumeration of eponymous sacerdotal officers in K. F. Hermann, Gottesdienstl. Alt. der Griechen, § 44, n. 10. Thucyd. ii. 2, fixes the beginning of the Peloponnesian war by saying that it fell in the archonship of Pythodorus at Athens, the ephoralty of Ænesius at Sparta, and the 48th year of the priesthood of Chrysis at Argos. The Hieromnemon was the eponymous magistrate at Byzantium, Demosth. Coron. p. 255; Polyb. iv. 52. Compare Müller, Dor. iii. 9. 10.

gistrates implied an era, because the registration of the names was continuous up to a certain point. But the conversion of the names into a numerical date must have often required reference or calculation, and could only have been obvious to the memory for recent years. Yet the Romans appear to have used this mode of chronological notation not only for civil, but also for domestic purposes; thus they reckoned the year of their birth by the names of the consuls; and they marked their wine-casks, in order to denote the year of the vintage, by the names of the same officers.([104])

The official or legal practice of designating the year by the king's reign—still practised in England—is similar in its nature. It exactly corresponds to the determination by the year of an officer who was not annual: as when Thucydides states that the commencement of the Peloponnesian War fell in the forty-eighth year of the Argive priestess, Chrysis.([105])

([104]) Compare the following passages:
   'Hic dies, anno redeunte festus,
   Corticem astrictum pice dimovebit
   Amphoræ fumum bibere institutæ
      Consule Tullo.'—Horat. Carm. iii. 8.
The ode to the Amphora begins:
   'O nata mecum consule Manlio.'—iii. 21.
L. Manlius Torquatus, consul in 689 u.c. is meant. He is again alluded to in Epod. xiii. 6: 'Tu vina Torquato move, consule pressa meo.'
   'Defluat, et lento splendescat turbida limo
   Amphora centeno consule facta minor.'—Martial, viii. 45.
   'Ipse capillato diffusam consule potat
   Calcatamque tenet bellis socialibus uvam.'—Juv. v. 30-1.
Where Ruperti says: 'Nam cadis inscribebantur nomina consulum, quibus vinum condebatur.' A person dated his birth by the consuls; as Horace above, and Cic. Brut. 43. Horace dates his youth by a consul. Carm. iii. 14. Ulpian, Dig. ii. 13, 1, uses the expression 'dies et consul' several times for the day and year. He lays it down that when the plaintiff gives written notice of action to the defendant, he is not to add the day and year: 'editiones sine die et consule fieri debent.' By 'the day' is meant the day of the month. A complete Roman date expressed the day of the month and the names of the consuls: for example, the day of Cæsar's assassination would have been dated 'the Ides of March, in the consulship of Cæsar and Antonius.'

([105]) The twenty-sixth year of a fabulous Argive priestess, Alcyone (or Alcinoe), who lived three generations before the Trojan war, was mentioned by Hellanicus, Dion. Hal. i. 22. A work upon the series of these priestesses was written by Hellanicus, Fragm. Hist. Gr. vol. i. p. xxvii.

There is no visible phenomenon to mark the beginning or end of a solar year, as there is to mark the beginning and end of a lunar month. The sun has always the same appearance: unlike the moon, it has no phases.([106]) The commencement of the year might therefore be placed at any point: wherever it was taken, it would necessarily include the four seasons. The Attic year began at the summer solstice; and the same rule was observed in several Ionic states. Sparta, with the other Peloponnesian States, and Macedon, began their year with the autumnal equinox: the Bœotian, Delphian, and Bithynian years began at the winter solstice.([107]) The commencement of the Roman year appears likewise to have been regulated by the last-mentioned standard. January was the first month after that which included the winter solstice.([108]) March, however, continued to be the beginning of the year for many legal and domestic purposes among the Romans, after the civil year began with January.([109])

In modern states, it is not unusual, notwithstanding the prevalence of the established calendar, to make use of a year which does not begin on the first of January. Thus in England the year for many municipal and parochial purposes is reckoned from Lady-day or Michaelmas-day. The tenure of land is generally computed by the same periods. In Scotland, the period in contracts of landlord and tenant is often dated from Lammas or Candlemas. As according to this system, the year is reckoned between two days determined by the established

---

(106) See the description of the moon's phases in the verses of Sophocles, Fragm. 713, Dindorf.

(107) K. F. Hermann, Griech. Monatskunde, p. 122-9.

(108) See the remarks of Plut. Quæst. Rom. 19. Simplicius, ad Aristot. Phys. p. 400, b. ed. Brandis, says that different nations begin their year at different periods: that the Athenian year begins at the summer solstice, the year of the States of Asia Minor at the autumnal equinox; the Roman year at the winter solstice; the year of the Arabians and Damascenes at the vernal equinox. The word Μίνωες is corrupt, and I am unable to emend it. The small town of Minoa, in Crete [Μινῷαι] cannot be meant.

(109) Macrob. Sat. i. 12, § 7.

calendar, no confusion or uncertainty is created by the departure from the ordinary computation.

In the middle ages, the system, derived from the Roman calendar, of beginning the year with the first of January was, to a great extent, abandoned; and years commencing at different festivals of the Church were introduced.([110]) The most prevalent of these was the year commencing on the festival of the Annunciation of the Virgin, or Lady-day, March 25, which was generally used in England from the fifteenth century, till the abolition of the old style in 1752.([111])

§ 7 All credible testimony, and all antecedent probability, lead to the result that a solar year, containing twelve lunar months—determined within certain limits of error—has been generally recognised by the nations adjoining the Mediterranean, from a remote antiquity. Some statements of ancient writers, and some conjectures of modern critics, are, however, inconsistent with this conclusion, and point to the existence of divergent years, wholly independent of the course of the sun through the zodiac, according to the view of the ancients, or of the periodic time of the earth in its orbit, according to the Copernican system.

Thus the Arcadians are reported to have used a year of three months, or four months (for its duration is variously reported);([112]) and the Carians and Acarnanians a year of six months.([113]) This short year is given as a reason for the fabulous epithet of the Arcadians, which designated them as *Prælunarians*:([114]) it is stated that they had a year before it was regu-

---

([110]) See Ideler, Chron., vol. ii. p. 325-343. Arago, Pop. Astr. b. 33, c. 21.

([111]) See the short Dissertation 'On the Ancient Manner of Dating the Beginning of the Year,' in the Annual Register for 1759, vol. ii. p. 410.

([112]) Censorin. c. 19, Plin. vii. 48, Macrob. Sat. i. 12, 9, and Solin. i. § 34, state the Arcadian year at three months. Plut. Num. 18 states it at four months.

([113]) Censorin. 19; Augustin, C. D. xv. 12; Plut. Num. ib.; Macrob. ib.: Solin. ib.

([114]) προσέληνοι. See Schol. Apollon. Rhod. iv. 264. Eudoxus is here

lated in Greece according to the moon's course;(115) but this reason is insufficient, for if their year consisted of months, it must have been regulated by the moon. It was the solar year from which this trimestrial or quadrimestrial period deviated.

A different, though not more authentic, origin for this epithet of the Arcadians was assigned by Aristotle, in the chapter of his Collection of Constitutions which related to the Tegeates, an Arcadian community. The great philosopher thought fit to inform his readers that Arcadia was originally inhabited by barbarians, who were expelled by the Arcadians in consequence of a battle fought before the rising of the moon.(116) Mnaseas, who wrote about the beginning of the second century B.C. resorted to another explanation: he converted the epithet into a man, and derived it from Proselenus, an ancient king of the Arcadians.(117) In the feeble treatise on Astronomy ascribed to Lucian, the epithet is explained by the supposition that the Arcadians despised astronomy, and were ignorant of the moon.(118)

These abnormal years are designated by Censorinus as 'involved in the darkness of remote antiquity;'(119) and the short years attributed to the Arcadians, Carians, and Acarnanians, are probably not less unreal than the long year of thirteen

---

quoted as an authority for the fable that the Arcadians were anterior to the moon. Compare Ovid, Fast. i. 469—
    Orta prior luna, de se si creditur ipsi,
    A magno tellus Arcade nomen habet.
Fast. ii. 287—
    Ante Jovem genitum terras habuisse feruntur
    Arcades, et lunâ gens prior illa fuit.
According to Steph. Byz. in 'Αρκὰς, Hippys of Rhegium first gave the epithet of προσέληνοι to the Arcadians. See Frag. Hist. Gr. vol. ii. p. 12. He appears to have been earlier than Herodotus.
Compare Lycophr. 482.
(115) Censorin. ib.
(116) Ap. Schol. Apollon. ib.; Fragm. Hist. Gr. vol. ii. p. 133. Compare the dissertation of Heyne, de Arcadibus lunâ antiquioribus, Opuscula, vol. ii. p. 332-353, who is not able to throw much light upon the origin of the epithet.
(117) Ap. Schol. Apollon. ib; Fragm. Hist. Gr. vol. iii. p. 150.
(118) De Astrol. 26.
(119) Caligine jam profundæ vetustatis obducti, c. 20.

months and 374 days which is attributed to the Lavinians.([120])

§ 8 There are likewise numerous statements respecting a short Egyptian year, which we shall have occasion to examine hereafter, in connexion with the general subject of Egyptian astronomy.([121])

Of the testimonies respecting a short Egyptian year, the only one entitled to much consideration is that of Eudoxus, who (according to the report of Proclus, a writer of the fifth century after Christ) stated that the Egyptians designated a *month* by the appellation of a *year*.([122]) Now it is impossible to suppose that, in the time of Eudoxus, the Egyptians were unacquainted with the common solar year. The Egyptian astronomy is, both by ancients and moderns, referred to a remote date; the Egyptian astronomers and geometers are supposed to have been the teachers of the Greeks, as early as the time of Thales and Pythagoras. Herodotus states that the Egyptians were the first of all mankind who invented the year, and divided it into twelve parts.([123]) Macrobius affirms that while other nations had erroneous years of different lengths, the Egyptian year was always accurate.([124]) We are even informed that the Egyptian year was changed from 360 to 365 days, in the year of the World 3716. The statement of Eudoxus cannot therefore be supposed to mean that the chronological period of a solar year, as received by other nations, was, in the fourth century before Christ, about thirty years before the foundation of Alexandria, unknown to the Egyptians.

If therefore the statement of Eudoxus referred to his own time, the designation of a lunar month by the term *year* must

---

([120]) Solin. § 34; Augustin, C. D. xv. 2.

([121]) Below, ch. v. § 4, 7.

([122]) εἰ δὲ καὶ ὅ φησιν Εὔδοξος ἀληθές, ὅτι Αἰγύπτιοι τὸν μῆνα ἐνιαυτὸν ἐκάλουν, οὐκ ἂν ἡ τῶν πολλῶν τούτων ἐνιαυτῶν ἀπαρίθμησις ἔχοι τι θαυμαστόν. Proclus in Plat. Tim. p. 31, F., referring to the passage p. 22, B., ὦ Σόλων, Σόλων, Ἕλληνες ἀεὶ παῖδές ἐστε.

([123]) ii. 4. By δυώδεκα μέρεα, Herodotus must mean months.

([124]) Sat. i. 12, § 2.

have been grounded on some peculiar phraseology, or have been derived from some peculiar origin, as to the nature of which we are uninformed. If, on the other hand, it referred to the remote antiquity of Egypt, the statement cannot be regarded as deserving of credit; and it may be put on the same footing as the many wild and extravagant stories concerning the fabulous ages of that country, which Herodotus had heard from the Egyptian priests about half a century earlier.

The statements of later writers, that the original Egyptian year consisted of four months, or of three months—the division being in one case determined by three seasons, and in the other by four seasons—are apparently of late fabrication. They are wanting both in external attestation and internal probability; they are likewise accompanied with an etymological legend, which represents the division of the *seasons* (ὧραι) as the work of the Egyptian god and king Horus([125])—a legend which is clearly of Greek origin.

The ancient Egyptians are, moreover, stated by some of the late chronographers to have given the appellation of *year* even to a day.([126]) This statement appears to have been devised merely as a contrivance for reducing the enormous periods of time assigned to the fabulous antiquity of Egypt. Bailly uses the same hypothesis for reducing the long periods cited for the astronomical observations of the Chaldeans.([127]) It is difficult to understand the meaning of this hypothesis, or to give it any rational construction. The ancient Egyptians must have been conscious of that period, consisting of twenty-four hours of light and darkness, which we call a day, and which the later Greeks denoted by the more precise name of

---

([125]) Censor. 19. Orus, the son of Osiris, is described by Herodotus as having been King of Egypt, ii. 144. Diodorus calls him Horus, and says that he was the last of the gods who reigned over Egypt, i. 25.

([126]) See the Anonymi Chronologica prefixed to Malalas, p. 21, and Malalas, lib. ii. p. 23, ed. Bonn. Suidas in ἥλιος. Chron. Pasch. vol. i. p. 81, ed. Bonn.

([127]) Astron. Anc. p. 295, 373, 377.

νυχθήμερον.([128]) The word which expressed this simple and necessary idea must have been appropriated to it. If they had not carried their observations of the sun's course so far as to form an idea of a year, they could have had no name for this important measure of time. As soon as they formed an idea of a year, they would naturally give it a name; but this name must have been different from that by which they signified a day.([129])

§ 9 The most celebrated, however, of the abnormal years is the ancient Roman year, attributed to the institution of Romulus.

The belief in an original year of ten months was prevalent among the antiquarian and historical writers of Rome. The most ancient authority cited for it is M. Fulvius Nobilior,([130]) a contemporary of Cato the Elder, who was consul in 189 B.C.: he had a taste for literature; he had devoted a peculiar attention to chronology; and had set up some fasti in a temple which he erected to Hercules and the Muses.([131]) Junius Gracchanus, a contemporary of the Gracchi, who flourished about the year 124 B.C., and wrote on legal and constitutional antiquities, is likewise quoted as a witness to the same fact.([132]) Varro, the great antiquarian, whose lifetime extended from 116 to 28 B.C., is also stated to have accepted the decimestrial year of Romulus: and the same current article of national faith was adopted by Ovid, in his Fasti; by Suetonius, who wrote a treatise on the Roman year; and by Macrobius, Gellius, and others.

([128]) It occurs in 2 Cor. xi. 25.

([129]) Curtius, viii. 33, states that the Indians used months of fifteen days. A month of fifteen days, like a year of three or four months, is doubtless fabulous.

([130]) Censorin. de D. N. c. 20, 22. The work of Censorinus was composed in 238 A.D.

([131]) He is likewise mentioned as treating of the etymologies of Maius and Junius, and is cited as a testimony to another chronological point, in Macrob. Sat. i. 12, § 16; i. 13, § 21.

([132]) Censorinus, ib. See Mercklin, De Junio Gracchano Comm. i. ii. Dorpat, 1840, 1841.

The accounts of the decimestrial year of the Romans, delivered to us by the ancient writers, differ not only in details, but also (as we shall see presently) in fundamental points; the following, however, may be considered as an outline of the most generally received version of the story.

Romulus, at the foundation of the Roman state, instituted a year of ten months, or 304 days, which he borrowed from Alba, the mother-city of Rome, and birthplace of his parent Ilia. He appointed Martius and Aprilis to be the two first months of this year, in honour of his father Mars, and of Venus, the progenitress of the Ænean race; the name *Aprilis* being referred to ἀφρὸς, or *Aphrodite*.[133] Having given the place of honour to the gods of his own family, he named the two next months from the people whom he was about to govern; namely, Maius, from the *majores*, or elders, and Junius, from the *juniores*. The remaining six months received no allusive appellations, but were named by him numerically, in their order, from Quintilis to December. Of the ten Romulean months, April, June, Sextilis, September, November, and December had thirty days; and March, May, Quintilis, and October had thirty-one days: thus making the sum of 304 days (180 + 124 = 304).[134]

---

(133) The derivation of Aprilis from Ἀφροδίτη is attributed to Fulvius and Gracchanus by Varro, L. L. vi. § 33. Censorin. c. 22. Ovid is copious upon this courtly theme, Fast. iv. 19—40:—

    Si qua tamen pars te de Fastis tangere debet,
      Cæsar, in Aprili quod tuearis habes.
    Hic ad te magnâ descendit imagine mensis,
      Et fit adoptivâ nobilitate tuus.

See also iv. 61—4.

    Qui dies mensem Veneris marinæ
      Findit Aprilem.
          Horat. Carm. iv. 11.

where the epithet marina alludes to the derivation from ἀφρός.

(134) See Censorin. de D. N. 20, 22; Macrob. Sat. i. 12, § 3—5, 38. Gell. iii. 16. Festus in Martius, p. 150, ed. Müller. Serv. Georg. i. 43. Plut. Num. 18, 19; Ovid, Fast. i. 27—42, iii. 75—100, 119—122, 149—50; Solin. i. 35; Lydus de Mens. i. 14, 16.

Ausonius, in two epigrams among his Eclogæ, entitled Monosticha and Disticha de Mensibus, has these verses:—

    Martius antiqui primordia protulit anni.
    Martius et generis Romani præsul et anni
    Prima dabas Latiis tempora consulibus.

All authorities seem to have agreed in making the year of Romulus commence with the month of March; though the place of that month in the solar year must have been extremely variable, if the civil year, consisting of only 304 days, was uncorrected by intercalation. They likewise were unanimous in supposing, that the numerical names of the six months from Quintilis to December originated in this fact.([135a]) But there was much discrepancy of opinion as to the origins of the names of the other four months. Some said that Martius was made the first month of the year, and obtained its name, on account of the warlike character of the people.([135b]) Others thought that Romulus had a deeper meaning in the names which he gave to the first two months. They supposed that he named the first from the god of slaughter: and the second from the goddess of generation, by whom the evils flowing from the deeds of man were repaired. Other explanations were likewise given for the names of these two months: one found in them a reference to the zodiac([136]); another denied that the name of Venus was known at Rome under the kings, and derived Aprilis from *aperire,* because April, the commencement of spring, opened the life of the year, after the torpor of winter.([137]) The derivation

---

([135a]) Thus Solinus says:—Maxime hunc mensem principem testatur fuisse, quod qui ab hoc quintus erat, Quinctilis dictus est, deinde numero decurrente December solemnem circuitum finiebat intra diem trecentesimum quartum, i. 35.

([135b]) See Festus in Martius. So Ovid, in explaining the origin of the name of this month, says:—
  Mars Latio venerandus erat, quia præsidet armis;
  Arma feræ genti remque decusque dabant.
  Fast. iii. 85, 6.

([136]) Aries was assigned to Mars, and Taurus to Venus. Scorpius was likewise divided between Mars and Venus. The sting was assigned to Mars, and the anterior part to Venus—the libra of the Latins and ζυγὸς of the Greeks—because this goddess united persons in the *yoke* of matrimony.

([137]) In this explanation of the name Aprilis, as given by Macrobius, Cincius the antiquarian, and Varro, L.L. vi. § 33, concurred. It is alluded to by Ovid, Fast. iii. 85—90, whose loyalty to the Julian family is, however, shocked at the idea of rejecting the derivation of April from Venus, ib. 115—24. Greswell, Origines Kalendariæ Italicæ, vol. i. p. 161, thinks that the month Martius derived its name from the planet Mars, and that the month Aprilis was named from *aperire,* in the sense of the opening

of *Aprilis* from *aperire* overlooks the fact that with a year of 304 days, April would not always have been a spring month. The second year of Rome would have begun on October the 31st, so that April would have fallen in December, and have been a winter month.

The origins assigned for the name Maius are even more various. The month was said to have been borrowed from Tusculum, where the chief god was named Maius. Cincius derived Maius from Maia, the wife of Vulcan. According to other etiologists, it was derived from Maia, the mother of Mercury, or from Maia, goddess of the earth, and Bona Dea, who had numerous titles and attributes. Ovid is perplexed as to the true origin of the name Maius, and professes himself as unable to choose among the many explanations proposed.[138] He mentions three as equally probable. 1. *Majestas*, or sovereign power. 2. *Majores natu*, corresponding with *Junius*, from *Juvenes*. 3. Maia, one of the Pleiades.[139]

He is likewise in doubt as to the origin of the name *Junius*, and traces it, 1. To Juno, referring to other examples of the same name.[140] 2. To Hebe, (*Juventas*) wife of Hercules. 3. To *Juniores*. 4. To the *junction* of Romulus and Tatius.[141] Others supposed the month Junius to have taken its name from Junius Brutus.[142]

The original year instituted by Romulus is stated to have been reformed by his successor Numa. The reform of Numa is thus described.[143] To the year of Romulus, which consisted

---

month. But the month Martius was doubtless so called by the Italian nations long before they had received the names of the planets from the Greeks.

[138] Fast. v. 1—6. [139] Ib. 11—110.
[140] Fast. vi. 59—63. There was a month Ἡραῖος in Crete, Bithynia, and Delphi.
[141] Fast. vi. 1—100.
[142] See Macrob. Sat. i. 12; Plut. Num. 19; Serv. Georg. i. 43. Concerning the names of the Roman months, compare Merkel ad Ovid. Fast. p. lxxix.
[143] See Macrob. i. 12, 13; Censorin. 20. Ovid, Fast. i. 43, ii. 47, iii. 151, states that Numa prefixed January and February to the ancient

of 304 days, he added fifty-one days; thus making a year of 355 days. These 355 days he distributed between the ten existing months and two new months, January and February, which he prefixed to the year; so that March now became the third month; and the months beginning with Quintilis, which were founded upon a numerical nomenclature, were all misnumbered by two. The month Januarius was named from the god Janus: the month Februarius was named from the god Februus, who presided over ceremonies of purification. Numa gave them the place of honour, at the beginning of the year, and assigned them a priority over the month of Mars, in his character of a pacific and philosophical king.

As a year of 355 days was shorter than the solar year by 10¼ days, Numa sought to bring it into harmony with the sun by intercalation. He is stated to have effected this purpose by intercalating a month of twenty-two or twenty-three days in alternate years; which was borrowed from the Greek system of intercalation, but, being inaccurately applied, made the year too long.

The Greek system of intercalation to which allusion is made, is the octaëteric cycle, attributed to Cleostratus. This system was based upon a year consisting of 12 alternate

---

ten months. Florus, i. 2, and Eutrop. i. 3, say that Numa distributed the year into twelve months. Victor, de Vir. Ill. 3, states him to have distributed the year into twelve months, and to have added to it January and February. According to Lydus, de Mens. i. 16, iii. 4, Numa made a year of twelve months, and reconciled the courses of the sun and moon. He also prefixed January and February to the year: making it begin in the middle of Capricorn; at which time the day has lengthened by half an hour since the winter solstice. (In our latitude, the increase of the day in the ten days from Dec. 21 to Jan. 1, is sixteen minutes.)

The popular belief that the year of Romulus consisted of ten months, and that it began with March, was referred to at Rome in interpretation of a Sibylline prophecy, in the time of Belisarius, 537 A.D. Procop. de Bell. Goth. i. 24.

Serv. Georg. i. 43, states that the original Roman year consisted of ten months, and that March was its first month. Tibullus, iii. 1, 1, says that March was originally the first month of the year.

Suidas, in Νουμᾶς Πομπίλιος, says of Numa: ἐνιαυτόν τε πρῶτος εὕρατο, εἰς δώδεκα μῆνας τὴν ἡλιακὴν κατανείμας περίοδον, χύδην τε καὶ ἀκατανοήτως παντάπασι πρὸ αὐτοῦ παρὰ Ῥωμαίοις φερομένην.

months of 29 and 30 days, and therefore containing 354 days. The difference between this year and the solar year, being 11¼ days, was multiplied by 8, in order to avoid fractional parts of a day, and to produce a number of days convenient for intercalation. The intercalation obtained by this method for an octennial period was 90 days, which was distributed into 3 months, and these were intercalated at convenient intervals. According to the system ascribed to Numa, an equal number of days was intercalated in a period of eight years, namely, two months of 22 and two months of 23 days. But as his year consisted of 355 instead of 354 days, there was an excess of 8 days in 8 years.

On the other hand, the intercalation ascribed to Numa is more accurate than the most ancient method of intercalation which Censorinus attributes to the Greeks. His statement is that the Greeks, observing that the moon was sometimes at the full twelve times and sometimes thirteen times in a solar year, thought that lunar and solar time could be reconciled by intercalating a month every other year. This period was, by reckoning the two extreme years inclusively, called a trieteris.[144] But the intercalation in question produced an error of 7½ days in excess of the true time for each biennial period; and deviated further from the truth than the intercalation attributed to Numa. Unless, therefore, we make the comparison with the improved octaëteric system of Greek intercalation, the method of Numa was less defective than the primitive Greek method.

A somewhat similar account of the intercalation of Numa is given by Plutarch; but he arrives at it by an entirely different road. Plutarch mentions, indeed, the tradition which assigns ten months to the year of Romulus; but he adopts another version of the story, according to which the original Roman year contained 360 days, distributed into months of

---

[144] c. 18. Compare Ideler, vol. i. p. 270. The difference is between 738 and 730½ days.

very unequal length, some having less than twenty days, some having thirty-five days and even more: January and February were the two last months of this year. Plutarch says that before Numa the Romans were ignorant of the difference between the lunar and solar years, and only sought to bring the year to 360 days; but that Numa knew the true lengths of these years to be respectively 354 and 365 days, with a difference of eleven days, and that he introduced an intercalary month, Mercedinus, of twenty-two days, in every alternate year, by which the calendar was kept in harmony with the sun. Numa likewise prefixed the months January and February to the reformed year, probably, as Plutarch thinks, on account of their vicinity to the winter solstice.([145]) The difference between the account of Censorinus and Macrobius on the one hand, and that of Plutarch on the other, with respect to the operation of Numa, is that the former supposes the year to which the intercalation is applied to consist of 355 days, and the intercalation to be of a month alternately of twenty-two and twenty-three days; whereas the latter supposes the year to consist of 354 days, and the intercalary month to consist uniformly of twenty-two days.

According to Plutarch's supposition, the Roman calendar, as reformed by Numa, would have exactly resembled the Egyptian calendar, as assumed by the Sothiac period. With the correction of the biennial intercalary month, it took the solar year at 365 days, and therefore lost one day in four years. This is a nearer approximation to the truth than the method of intercalation ascribed to Numa by Macrobius and Censorinus, according to which there was an annual error of a day.

Valerius Antias, the author of a voluminous history of Rome, who lived at the time of Sylla, declared Numa to have been the author of the system of intercalation in the Roman

---

[145] Num. 18, 19; Quæst. Rom. 19. The passage at the end of c. 18, in the Life of Numa, is repeated by Zonaras, vii. 5; vol. ii. p. 21, ed. Bonn.

calendar.(146) Cicero speaks of the system of intercalation having been established by Numa on a scientific basis, but as having been subsequently maladministered by the Pontifices.(147)

Livy's statement is, that Numa first divided the year, according to the moon, into twelve months; and that, in order to bring it into harmony with the sun, he arranged intercalary months, so that the days coincided after a period of twenty-four years.(148) This cycle of intercalation is not mentioned by any other ancient writer with reference to Numa. Macrobius, however, states that it was introduced as a correction to Numa's calendar. His account (as we have already stated) is that the intercalation of Numa was founded on the Greek octaëteric system, by which ninety days were intercalated in eight years of 354 days. But as the year of Numa contained 355 days, there was an error of eight days in excess at the end of each octennial period. In order to correct this error, the intercalation in each third octennial period was afterwards reduced from ninety to sixty-six days; so that at the end of twenty-four years the error was adjusted by the deduction of twenty-four days.(149) The motive for committing an error in order to correct it subsequently, seems to have been the desire of producing an inter-

---

(146) Antias libro secundo Numam Pompilium sacrorum causâ id invenisse contendit. Macrob. Sat. i. 13, § 20.

(147) De Leg. ii. 12.

(148) i. 19. Livy does not mention the year of Romulus. Dionysius mentions neither the year of Romulus nor the reform of Numa.

Livy seems to be ignorant of the meaning of solstitium, and to make solstitialis in this passage equivalent to solaris. A similar misapplication of the word is cited from Servius, ad Æn. iv. 653. The word solstitium is commonly used by the Latin writers to denote the summer solstice: thus Ovid—

Longa dies citius brumali sidere, noxque
Tardior hibernâ solstitialis erit.
Ep. ex Pont. ii. 4, 25-6.

The phrase exortus solstitialis is used by Pliny, N. H. ii. 46, to denote the north-east. Solstitialis orientis meta is used by Gell. ii. 22, in the same sense.

(149) Sat. i. 13, § 9—13.

calary period which admitted of convenient distribution into months. According to the plan described, three months of thirty days apiece were intercalated in each of the two first octaëterids; and two months of thirty-three days each, or two months of thirty days each, and six additional days, were intercalated in the third octaëterid: whereas if eighty-two days (which would have been the correct number) had been intercalated in each octaëterid, it would have been difficult to digest this number into months. If the Roman year ever consisted of 355 days, it is not unlikely that this mode of intercalation was applied to it; but the conflicting accounts of Livy and Macrobius do not point to any certain historical result.([150])

An entirely different and peculiar account of the measures consequent upon the original reform of Numa, by which the year had been lengthened to 355 days, is given by Solinus. He states that when it was perceived that the solar year consists of 365¼ days, they added 10¼ days to the reformed year.([151]) This method, he says, was approved throughout the world, and various schemes of intercalation were devised on account of the odd quarter of a day. The Greeks introduced the intercalary cycle of eight years, which the Romans at first adopted; but the method was after a time disused, as the regulation of the calendar was transferred to the priests, who administered it corruptly, in order to suit the interests of the farmers of the revenue; and it fell into confusion, until it was reformed by Julius Cæsar.([152])

([150]) Siccama de Vet. anno Romuli et Numæ, Græv. Thes. vol. viii. p. 83, supposes the Romulean year to have consisted of 354 days, and twelve months, of which March was the first, and January and February the last. He supposes further that Numa altered this year by adding a day, and by prefixing January and February to the beginning.

([151]) Solinus supposes this to have been done for the purpose of preserving the odd number of days, in obedience to the precept of Pythagoras: whom he therefore connects with Numa. See below, n. 173.

([152]) i. 39—44. Compare the commentary of Salmasius upon this passage, in his Plinianæ Exercitationes. Salmasius points out the improbability of the statement of Solinus, that the Roman year before the reform of Cæsar consisted of 365¼ days.

The version of the original institution of the year by Romulus was not universally received by the Roman historians. Licinius Macer, a contemporary of Cicero, who wrote a history of Rome from the foundation of the city,([153]) and is cited by Livy as having made researches into authentic documents,([154]) stated the primitive Roman year to have consisted of twelve months;([155]) and Fenestella, who lived about half a century later, and likewise wrote a history of his country in at least twenty-two books, shared this opinion.([156])

Plutarch likewise rejects the common account of the decimestrial year of Romulus. He supposes the primitive Roman year to have consisted of twelve months and 360 days, but to have begun with March, and to have ended with January and February.

The utmost discordance of testimony likewise exists with respect to the introduction of intercalation into the Roman calendar. Licinius Macer, who held the primitive Roman year to consist of twelve months, stated that Romulus was the author of intercalation. Junius Gracchanus, who assumed a primitive decimestrial year, supposed this corrective to have been introduced by Servius Tullius.([157]) Cassius Hemina and Sempronius Tuditanus, two historians who lived in the second century B.C.,([158]) stated it to have been first established by the Decemvirs, in 451—49 B.C.; while Fulvius Nobilior placed its introduction even as late as the consulship of Manius Acilius Glabrio, 191 B.C., only two years before his own consulship.([159])

([153]) See Krause, Vit. et Fragm. Hist. Rom. p. 234.
([154]) Livy, iv. 7, 20, 23.   ([155]) Censorin. c. 20.
([156]) Censorin. ibid.
([157]) See above, p. 34.
([158]) Concerning Hemina and Tuditanus, see the author's work on the Credibility of the Early Roman History, vol. i. pp. 29, 30. Dionysius includes Sempronius among the λογιώτατοι τῶν Ῥωμαίων συγγραφέων, i. 11.
([159]) These statements respecting the origin of intercalation are in Macrob. Sat. i. 13, § 20. Varro confuted the statement of Fulvius, by citing an ancient law engraved on a brass tablet, of the consulship of Pinarius and Furius (472 B.C.), which was dated by a reference to intercalation. This was also prior to the Decemvirate.

Ovid likewise, in his Fasti, ascribes the introduction of intercalation to the Decemvirs.([160]) All that he assigns to Numa, is, the conversion of the decimestrial year into a year of twelve months, with January and February at its head.

This discordance of testimony, or rather of statement, (for none of these writers, except Fulvius Nobilior, can be regarded as witnesses) shows that no authentic account of the introduction of intercalation into the Roman calendar had descended to the times of the historians. The earliest of these writers were separated from the supposed reign of Romulus by 550, and from that of Numa by 500 years. There is, moreover, no reason for supposing that Gracchanus had access to better information than Fulvius Nobilior respecting the introduction of intercalation; or that either of them knew more of the year of Romulus than Licinius Macer.

The details of the intercalatory operation of Numa are likewise described by the different writers with irreconcilable diversity. Censorinus and Macrobius differ from Livy, and all three differ from Plutarch; while Solinus gives an account different from all four. Whatever may have been the system of the early Roman calendar, it is incredible that the regulators of it could have possessed a scientific method of intercalation, implying a tolerably accurate knowledge of the length of the tropical year, in the eighth century before Christ, long before the time when similar methods began to be applied to the Greek calendars.

If it is admitted that the intercalation—which is an essential part of the operation ascribed to Numa—is unsupported by credible evidence, all that remains is the decimestrial year of Romulus, beginning with March; and the addition of January and February at the commencement of this year, by

---

([160]) Postmodo creduntur spatio distantia longo
Tempora bis quini continuasse viri.
Fast. ii. 53-4.
The meaning of this obscure couplet seems to be, that the Decemvirs harmonized, or equalized, the years which were of different lengths.

Numa. Now this part of the story has all the appearance of a legend invented partly on verbal or etymological, and partly on religious grounds. The numerical names of the months from Quintilis to December, which were manifestly counted from March, naturally suggested the notion of a decimestrial year. It was thought that if the primitive year had contained twelve months, the nomenclator would not have stopped at December, but would have denoted the two remaining months by the names of *Undecember* and *Duodecember*. But there is no good reason why this system of numeration (which seems to have been peculiar to the Roman calendar) should have been carried consistently through all the months. The numerical naming was capricious, it began at the fifth month; why should it not stop at the tenth? January and February may have been the last, instead of the first months of the year; as indeed Plutarch states them to have been. Cicero plainly declares that February had in ancient times been the last month of the year.([161]) A fragment of Varro also alludes to some ancient time when March was the first and February the last month of the year.([162]) Ovid declares that February was the last month of the ancient Roman year. ([163]) The Terminalia, which were considered the last in the series of festivals, were celebrated on the 23rd of February; and after this day the intercalary month was inserted in years of intercalation; ([164]) that is

---

([161]) Sed mensem, credo, extremum anni, ut veteres Februarium, sic hic Decembrem sequebatur. Cic. de Leg. ii. 21. The rest of the passage is mutilated, but the meaning of these words is clear and certain.

([162]) Varro, Epist. Quæst. ap. Serv. Georg. i. 43. Other fragments from the same work are in Varro, vol. i. p. 194, ed. Bipont.

([163])   Sed tamen, antiqui ne nescius ordinis erres,
   Primus, ut est, Jani mensis et ante fuit,
Qui sequitur Janum, veteris fuit ultimus anni:
   Tu quoque sacrorum, Termine, finis eras.
Primus enim Jani mensis, quia janua prima est:
   Qui sacer est imis manibus, imus erat.
                              Ovid, Fast. ii. 47—52.

([164]) The intercalation was after the 23rd of February, between the Terminalia and the Regifugium, Macrob. i. 13, § 15; Censorin. 20.

Terminalia, quod is dies anni extremus constitutus; duodecimus enim

to say, it was added near the end of the final month of the year. Macrobius remarks that the custom of making the intercalation in February proves it to have been the last month.([165]) There was nothing which necessarily fixed a month of a certain name to a particular place in the series. Ovid mentions the place of March in various Italian States; it was third, fourth, fifth, sixth, and tenth in different calendars.([166])

Starting from this explanation of the names of the numbered months, the historical etiologist would naturally assign the origin of the decimestrial year to the founder of the State; and he would complete the scheme by ascribing the reform of this rude and primitive calendar to Numa, the founder of sacred rites and institutes, because the calendar was considered a matter of religious concern, intimately connected with the times of sacrifices and festivals, and consequently placed under the control of the priests.

The whole story may therefore be considered as an explanatory legend, invented to account for the places in the calendar occupied respectively by January and February, and by the six months bearing numerical names. The notion of the founder, Romulus, a warlike youth of eighteen, establishing a calendar, belongs to fable, not to history; the month Martius, which Romulus is supposed to have named from Mars, and to have placed, in honour of his divine parent, at the head of the year, was not peculiar to Rome. There was a month of the same name in the calendars of Alba, Falerii, Aricia, Tusculum,

---

mensis fuit Februarius; et quum intercalatur, inferiores quinque dies duodecimo demuntur mense. Varro, L. L. vi. 13.

According to Augustin, C. D. vii. 7, January was dedicated to Janus, and February to Terminus, 'propter initia et fines.' He seems to consider February as having at one time been the last month of the Roman year.

([165]) Sat. i. 13, § 14.

Philargyrius ad Georg. iii. 304, likewise considers February as the last month in the year, of which March was the first month. He makes this statement in explanation of the passage of Virgil, in which he speaks of Aquarius as marking the end of the year. This constellation sets in February.

([166]) Fast. iii. 87—96.

Laurentum, Cures, the Æquicoli, the Hernici, the Peligni, and the Sabines.([167]) The Bithynian year likewise had a month Ἄρειος.([168]) Moreover, even if the derivation of Aprilis from Aphrodite were true, it would not prove any connexion with the Æneadæ and the Alban race of kings, for a month Aphrodisius was extant in several Greek calendars.([169]) The existence of Venus as a Roman deity in early times was likewise denied by the antiquarians, Cincius and Varro.([170]) Both the original institution of Romulus, and the reform by his successor Numa, are, like the other measures attributed to the agency of these mythical kings, destitute of authentic attestation. The story is, moreover, delivered to us with the wide and insoluble discordance which is the characteristic of legendary tradition. In particular, the introduction of intercalation, by which the defects of the rude primitive year are supposed to have been rectified, is related with every variety of circumstance. It was attributed to Servius Tullius as well as to Numa; it was also ascribed to the Decemvirs; and the earliest and apparently the best authority placed it as late as 191 B.C.([171]) Even those who supposed a primitive year of ten months were not unanimous in assigning the introduction of the twelve month year to Numa; for Gracchanus considered this important change to have been made by Tarquin.([172]) With respect to the method of intercalation which Numa is stated by Macrobius to have borrowed from the Greek system,

---

(167) Ovid, Fast. iii. 89—98.
(168) Hermann, Griech. Monatskunde, p. 47.
(169) Ibid, p. 48.
(170) See Varro, L. L. vi. § 33; Macrob. Sat. i. 12, § 13.
(171) A singular story respecting the origin of the short duration of February is told by Dio Cassius, and is repeated by later writers. The accuser of Camillus is stated to have borne the name of Februarius, and when Camillus returned to Rome after his exile, he curtailed the month which bore his accuser's name, in revenge for the injury done him. See Dio Cass. Fragm. 27, ed. Bekker; Suid. in Φεβρουάριος and Βρῆννος, Malalas, p. 183, ed. Bonn; Cedren. vol; i. p. 203, ed. Bonn.
(172) Censorin. c. 90, who does not specify which Tarquin is meant. It is probably Tarquinius Priscus.

but to have applied incorrectly, it may be observed that the invention of this system is attributed to Cleostratus, a Greek astronomer who was posterior to 500 B.C., whereas the reign of Numa is placed by chronologers at 715-673 B.C., at least a century and a half earlier. The derivation of the intercalation of Numa from the octaëteric system of Cleostratus involves, therefore, an equal anachronism with the derivation of his philosophy from that of Pythagoras.([173])

The inferiority of the Romans to the Greeks in astronomy, as well as in other departments of physical science, is fully recognised by the Latin writers. Virgil expressly includes astronomy among the subjects in which he proclaims the preeminence of the Greeks, while he vindicates to Rome the mastery in the art of government.([174]) The Romans were the tardy pupils of the Greeks in astronomical and mathematical science; and it may be safely assumed that while the astronomical science of the Greeks was in its infancy, that of the Romans had scarcely an existence.

The account of Numa's intercalation, whether it be in the form of a month of twenty-two days in alternate years, or of sixty-six days in an octaëteris, assumes that his year consisted of 355 days. There is, however, a record of a statue of Janus, dedicated by Numa, which was furnished with symbols of 365 days, representing the year.([175]) Whatever may have been the date of this statue, the inventor of the story of its dedication

---

([173]) See the author's Inquiry into the Cred. of the Early Rom. Hist. vol. i. p. 451; Schwegler, R. G. vol. i. p. 560.

([174]) Æn. vi. 848. Compare Ovid, Fast. iii. 101—120, which passage contains a singular pun upon the word *signum*, which signified both a constellation and a military standard. The comparative recency of astronomy among the Romans is stated by Seneca:—' Nondum sunt anni mille quingenti ex quo Græcia *stellis numeros et nomina fecit*. Multæque hodie sunt gentes, quæ tantum facie noverint cœlum, quæ nondum sciant, cur luna deficiat, quare obumbretur. Hoc apud nos quoque nuper ratio ad certum deduxit.' Nat. Quæst. vii. 25.

([175]) Plin. N. H. xxxiv. 16. The statue of Janus, with its allusive symbols, but without the mention of Numa, is described by Macrob. Sat. i. 9, § 10; Lydus, de Mens. iv. 1; Suidas, in Ἰανουάριος. Ideler, Chron. vol. ii. p. 34, thinks that the right reading in Pliny is 355.

by Numa could not have recognised Numa's year of 355 days and the method of intercalation by which it was supposed to have been rectified.

The improbability of a year differing from the true solar year by 61¼ days is felt by Macrobius, who supposes that even in the reign of Romulus, and before the reform of Numa, a number of days, undigested into months, was added to the year of 304 days, so as to bring it into harmony with the sun and the course of the seasons.[176] This correction, however, would have been only one mode of intercalation; and if it was constantly applied, there could be no reason why the sixty-one days should not be distributed into two months, even if no provision was made for the fraction of a day; and such a provision could easily be made if the true length of the solar year was known at Rome at the time attributed to Romulus. This supposition of Macrobius, while it betrays his consciousness of the improbability of the Romulean year, entirely destroys the essence and meaning of the traditional account, because it supposes the year of ten months to have been virtually a year of twelve months. Ovid adheres to the spirit of the legend, and makes the ancient Roman year really consist of only ten months. He thus describes its length:

> Ergo animi indociles et adhuc ratione carentes
>   Mensibus egerunt lustra minora decem.
> Annus erat, decimum cum luna receperat orbem.
> Fast. iii. 119—21.

If the lustrum, or period of five years, was ten months too

---

(176) Sed cum is numerus [viz. 304 days] neque solis cursui neque lunæ rationibus conveniret, nonnunquam usu veniebat ut frigus anni æstivis mensibus et contra calor hiemalibus proveniret: quod ubi contigisset, tantum dierum sine ullo mensis nomine patiebantur absumi quantum ad id anni tempus adduceret quo cœli habitus instanti mensi aptus inveniretur. Sat. i. 12, § 39. It should be observed that, in a year of 304 days, there could be neither winter nor summer months without intercalation. For example, if, in the first year of Romulus, the year began at the same time as our March, Quintilis or July would have been a summer, and December a winter month: but, in the third year, Quintilis would have held the same place as our March, and December the same place as our August.

short, each year must have consisted only of ten months, and there could not have been any intercalation of undigested days. On the other hand, the later chronologers, who assigned a definite number of years to the reign of Romulus, and connected them with the series of years reckoned from the foundation of the city,([177]) must have considered the years of Romulus as ordinary solar years, and have disregarded the tradition of the decimestrial year.

The account of the natal theme of Rome, cast by L. Tarutius Firmanus, an eminent astrologer, contemporary with Varro and Cicero, likewise implies that the years of the reign of Romulus were ordinary solar years, like those of the succeeding kings. Tarutius fixed the conception of Rhea Silvia at the first year of the second Olympiad (772 B.C.), the third hour of the twenty-third day of the Egyptian month Choeak, and the birth of Romulus at sunrise on the twenty-first of Thoth. By adding eighteen years, the supposed age of Romulus at the time of the foundation of Rome, we obtain exactly Olymp. VI. 3 = 753 B.C., the Varronian era for the foundation of Rome, from which the received reckoning proceeds uninterruptedly in solar years.([178])

The Romulean year of ten months and 304 days is rejected by Scaliger as incredible.([179]) The year of 304 days is likewise rejected by Dodwell, but he retains the ten months, by supposing that they were not regulated according to the moon, and were of sufficient length to make up an ordinary solar

---

([177]) The reign of Romulus is reckoned by some ancient authors at thirty-seven, by others at thirty-eight years: see Schwegler's Röm. Gesch. vol. i. p. 517. Cicero, Livy, Dionysius, and Plutarch agree in the former number. The seven kings of Rome are stated to have reigned 244, or, in round numbers, 240 years. See the author's Inquiry into the Early Roman History, vol. i. p. 527.

([178]) See Plut. Rom. 12; Solin. i. 18; Cic. de Div. ii. 47; Lydus, de Mens. i. 14. Compare the author's work on the Cred. of the Early Rom. Hist. vol. i. p. 393. The 23rd of Choeak corresponds to Dec. 20, and the 21st of Thoth to Sept. 18, in the Julian Calendar, according to the table in Ideler, vol. ii. p. 143. The pregnancy of Rhea was therefore reckoned by Tarutius at nearly nine calendar months.

([179]) De Emend. Temp. lib. ii. p. 172, ed. 1629.

year.([180]) This last hypothesis, it may be observed, destroys the essence of the traditionary account, which supposes the ten months to be ordinary lunar months, and connects them indissolubly with the number of 304 days, wanting sixty days, or two ordinary months, to complete the number 364. Merkel, in his Prolegomena to Ovid's Fasti, considers the story of a decimestrial year ending with December, and of the subsequent addition to it of the months January and February, as apocryphal.([181])

On the other hand, Ideler, though he abstains from dogmatizing upon details, holds that there is no valid ground for doubting of the existence of a decimestrial year in early Rome;([182]) and Mommsen, whose authority upon a subject of Roman antiquity is of much weight, considers the primitive Roman year of ten months and 304 days as a reality and no fiction, though he admits that the accounts of it may have descended to us in an inaccurate and imperfect form.([183])

Mommsen calls to his aid the decimal argument,([184]) to which Ovid and Lydus had previously had recourse, for the purpose of explaining the number of months in the Romulean year.([185]) This, however, is a merely fanciful and arbitrary analogy, and proves nothing. It would be as reasonable to use the duodecimal division of the Roman As, or of the day, as an argument

---

(180) De Cyclis, x. § 108, p. 672.
Lord Macaulay, Hist. of Engl. c. xiv. vol. iii. p. 461, says of Dodwell: 'He had perused innumerable volumes in various languages, and had acquired more learning than his slender faculties were able to bear. The small intellectual spark which he possessed was put out by the fuel.' Dodwell was a man of great learning, and great logical acuteness, but was destitute of judgment and good sense. 'Felicis memoriæ, expectans judicium.' I question whether his intellect was oppressed by his knowledge. On the contrary, his opinions would probably have been still more extravagant, if he had known less of positive facts, and of the opinions of others.
(181) Præf. ad Fast. p. lxxvii. (Berlin, 1841.)
(182) Handbuch der Chron. vol. ii. p. 27.
(183) Die Römische Chronologie bis auf Cæsar (Berlin, 1858), p. 47.
(184) Ib. p. 50. Upon the universal use of the denary scale of arithmetic in antiquity, see Aristot. Problem. xv. 3.
(185) Ovid, Fast. i. 31—6, iii. 121—6; Lydus, de Mens. i. 15.

in favour of a year of twelve months. More weight may appear to be due to the legal and customary periods of ten months, which were in use at Rome during the historical age, and which the defenders of the decimestrial year regard as a relic of the ancient calendar. Thus the mourning of widows for their husbands was fixed by usage at ten months;[186] and women are stated to have mourned a similar period for their nearest relations. The rule with respect to children is stated to have been that there was no mourning for children under three years of age; that the mourning for children between three and ten years of age was of as many months as the child was years old; and that the mourning for children above ten years of age was of ten months.[187] The same period of ten months is likewise stated by Plutarch to have applied to the mourning for a father and a brother.[188] The rule laid down by Paulus, who flourished about 200 A.D., is that the mourning for parents and children above six years of age is a year; for children below six years of age a month; for a husband ten months; and for kindred of less degree eight months.[189]

The ancient customary period of ten months for mourning cannot, however, be considered as proving the existence of a primitive decimestrial year. It may more probably have arisen from the period of pregnancy, which was commonly assumed as equal to ten months, and it may have been extended from the mourning for husbands to the mourning for other near relations. It appears that by the ancient Roman law a widow was

---

(186) Cicero, pro Cluent. 12; Ovid, Fast. i. 35, iii. 134; Seneca ad Helv. c. xvi. § 1. Majores decem mensium spatium lugentibus viros dederunt. Dionysius speaks of the Roman matrons mourning a year (ἐνιαύσιος χρόνος) for Brutus and Publicola, as for their nearest relations, v. 48. Plutarch, Publ. 23, makes the same statement respecting Publicola. Plutarch, Num. 12, says that Numa fixed the longest period of mourning at ten months, and that this was the period for widows. He repeats this statement in the Life of Coriolanus, Cor. 39. Dionysius states that the matrons mourned a year for Coriolanus, viii. 62. Seneca, Epist. 63, § 11, states that the time of mourning anciently fixed for women was a year; but that men did not mourn.

(187) Plut. Num. 12. (188) Plut. Cor. 39.
(189) Sent. Recept. lib. i. tit. 21, § 13.

prohibited from marrying within the ten months next after her husband's death.(190) This prohibition was extended to a year by a law of the Emperors Gratian, Valentinian, and Theodosius, in the year 381 A.D.(191) Custom seems to have likewise extended the time of mourning for widows to a year in later times; but the writers who speak of the period of a year for mourning in early times merely use round numbers, and do not allude to a decimestrial year. It should be observed that mourning was limited to women at Rome: it was not customary for men to mourn.(192) The period of mourning in antiquity seems to have been short; in Sparta it was twelve days; in Athens a month; and in other Greek states it did not exceed four or five months.(193) The ten months' credit given for oil and wine, according to Cato, cannot be regarded as evidence of an ancient decimestrial year;(194) and the same may be said of the term of ten months after the father's death, within which the furniture settled on a daughter was to be supplied.(195) The hypothesis of Niebuhr that the twenty years' truce between Rome and Veii consisted of decimestrial years is destitute of foundation.(196)

An attempt has been made by some modern writers to prop up the year of 304 days by a hypothetical substratum which

---

(190) See Plut. Num. 12. Antony married Octavia within ten months of the death of her first husband Marcellus; but for this purpose it was necessary for him to obtain a decree of the Senate dispensing with the law. Plut. Anton. 31. Compare Drumann, Geschichte Roms, vol. i. p. 424.

(191) Cod. Theod. iii. 8, 1; and see Cod. Just. v. 9, 1.

(192) Sen. Ep. 63, § 11. At the death of Augustus, a decree of the senate commanded that men should mourn a few days, and women a year. Dio Cass, lvi. 43. The attendants at a Roman funeral were dressed in black, Horat. Ep. i. 7, 6.

(193) Hermann, Privat-alterthümer der Griechen, § 39, n. 34—6.

(194) De R. R. c. 146—8.

(195) Polyb. xxxii. 13. The dos was payable within three years: there is no ground for the supposition of Niebuhr, Rom. Hist. vol. iii. n. 107, that cyclic years are meant; though it is adopted by Rein, Römisches Privatrecht, p. 194. By 'cyclic years' he means years of ten months.

(196) See the author's work on the Credibility of the Early Roman History, vol. ii. p. 287.

requires examination. According to Censorinus, Macrobius, and Solinus, the Romulean year consisted of four full, and six hollow months, the former of which had thirty-one and the latter thirty days (124 + 180 = 304); and at the reform of the calendar this number was increased to 355 days. As the additional fifty-one days were not sufficient to make two new months, a day was taken from each of the six hollow months, and the fifty-seven days thus obtained were assigned to the two new months; January having twenty-nine, and February twenty-eight days.([197]) The account given by the only three Roman writers who describe the Romulean year in detail—the two former of whom are learned and intelligent antiquarians—therefore supposes that it was founded on a division of months, by which the number of its days was determined.

The hypothesis above alluded to proceeds from a rejection of the constitution of the Romulean year, as described by Macrobius and Censorinus, and finds an origin for it in the nundinal, or eight-day period of Rome. The author of this conjecture was Puteanus, a Dutch philologist of the sixteenth century, who pointed out that 304 is a multiple of 8 (38 × 8 = 304). ([198]) Pontedera, an Italian writer of the eighteenth century, added the remark that six years of 304 years nearly coincide with five years of 365 days (1824 and 1825 days). ([199]) This conjectural explanation of the Romulean year has been developed by Niebuhr, and has been presented by him in a more elaborate form. This historian adopts the derivation of the Romulean year from thirty-eight nundinal weeks, which measure of time he supposes to have been borrowed from the Etruscans; he likewise applies to the sæculum of 110 years the supposed Romulean lustrum of six years, corresponding

---

([197]) Censorin. c. 20. Macrob. Sat. i. 12, § 3, 38; i. 13, § 1—7. Solin. i. 37—8.

([198]) See his Dissertation de Nundinis Romanis, c. 5, 7, in Græv. Thes. Ant. Rom. vol. viii. p. 653, 655. Concerning Puteanus, who lived from 1574 to 1646, see Bayle, Dic. in v.; Biog. Univ. art. Dupuy (Henri).

([199]) See Ideler, vol. ii. p. 25. Pontedera lived from 1688 to 1757.

with the ordinary lustrum of five years. Hence he obtains a sæculum of 132 Romulean years. He next assumes that a month of twenty-four days, or three nundinal weeks, was twice intercalated during a sæculum of 132 Romulean years.([200]) The result is, that if the solar years be taken at 365¼ days, as in the Julian calendar, a sæculum of 110 Julian years is less exact than a sæculum of 132 Romulean years; for the latter makes the solar year too short by 8m. 23s.: while the former makes it too long by 11m. 15s.([201]) Hence Niebuhr thinks it probable that the Etruscans had fixed the tropical year at 365d. 5m. 40s.; that the Etruscans, in the primitive ages of Rome, were profoundly versed in astronomical science; and that the results, though not the scientific foundations, of this science were handed down by tradition among the Etruscan priests. ([202]) Dr. Greswell, in his learned work, printed at the Oxford University press, entitled Origines Kalendariæ Italicæ, has likewise attempted to prove, by copious argument, that the Romulean year was founded upon a nundinal cycle. He conceives that a nundinal year consisted of nine months of thirty-two days, and of a tenth month of sixteen days; but adverting to the numerical coincidence pointed out by former writers, that six years of 304 days are only one day less than five years of 365 days, he holds that the nundinal and solar years could be adjusted to one another. ([203])

Upon this conjectural scheme of Niebuhr, Ideler remarks, that 360 is not less a multiple of 8 than 304; and that 110

([200]) Niebuhr, with his usual confidence in his own conjectures, lays it down that the Roman antiquarians were mistaken in two of their suppositions, viz., that the calendar of ten months was originally the only one in use, and that it was afterwards entirely abandoned. Hist. of Rome, vol. i. p. 282. Engl. Tr.

([201]) Niebuhr assumes the sæculum of 110 Julian years, upon Scaliger's calculation, at 40,177 days. A sæculum of 132 Romulean years, with 48 intercalary days, makes 40,176 days (132×304+48=40,176).

([202]) Hist. of Rome, vol. i. p. 275—9, Engl. Tr. The hypothesis of Niebuhr is adopted by Ukert, Geogr. der Griechen und Römer, vol. i. part 2, p. 163.

([203]) Vol. i. p. 136, 142, 153.

years of 360 days, together with twenty-four intercalary months of twenty-four days, equally make up the sum of 40,176 days.(204) Müller likewise, in his work on the Etruscans, regards the hypothesis as too uncertain for adoption into a statement of historical facts.(205) It may be added that nothing is gained by the gratuitous assumption of a sæculum composed of 132 Romulean years; and that if the solar year of 365¼, or even of 360, days was known to the primitive Romans, there could be no adequate inducement for forming so capricious and inconvenient a period as a year of 304 days, even although 304 days was a multiple of 8. The profound astronomical science supposed by Niebuhr to have been possessed by the Etruscans in the eighth century B.C., is moreover opposed both to evidence and probability. The Etruscans never made any advance in physical and mathematical science, or even in history and literature. Their nearest approach to astronomy consisted in their observation of lightning for purposes of divination.(206)

We may therefore conclude that the Romulean year of 304 days derives no real support from the modern hypotheses devised as props to the tottering structure; that it probably never had any real existence, and was merely a fiction contrived to account for the numerical names of the Roman months.

Dionysius describes Romulus as the author of the nundinæ;(207) which was a market held at Rome every eighth day, or, as the Romans expressed it,(208) every ninth day: reckoning

---

(204) Chron. vol. ii. p. 27.

(205) Etrusker, vol. ii. p. 329.

(206) See Müller's chapter on the Calendar and Chronology of the Etruscans, b. 4, c. 7, and his necessarily meagre chapter, b. 4, c. 8, upon their science and mental cultivation. Concerning the fulguratory discipline of the Etruscans, see Müller, vol. ii. p. 31, 41, 162-178. Compare the author's work on the Cred. of the Early Roman Hist. vol. i. p. 200.

Seven names of Etruscan months, with their Roman equivalents, have been preserved. See Mommsen, Röm. Chron. p. 278.

(207) Dion. Hal. ii. 28.

(208) See Dion. Hal. vii. 8, where he says that seven complete days intervened between each two successive nundinæ.

both days inclusively, in the same manner that the French call a week eight days, and a fortnight fifteen days. Cassius Sempronius Tuditanus, a well-informed historian of his country, who flourished in the second century before Christ, referred the origin of this institution to the arrangement made by Romulus when he formed the joint kingdom with Tatius; and did not therefore include it among his original institutions at the foundation of the city.[209] Cassius Hemina, a historian of the second century B.C., and Varro ascribed its establishment to Servius Tullius.[210] Other writers considered the observation of the nundinæ as a market-day to be subsequent to the expulsion of the kings; and to have originated in the honours paid by the plebs to the manes of Servius, the plebeian king, on every eighth day.[211] It is clear that those Roman antiquarians who attributed the origin of the nundinæ to Servius, or to the beginning of the Commonwealth, could not have connected this eight-day period with the constitution of the Romulean year.

The market-day at Rome seems to have been fixed at intervals of eight days from motives of convenience.[212] There is no valid reason for supposing that the nundinal period had any connexion with the determination of the civil year, and of the months into which the year was divided. The week, on the other hand, as observed by several oriental nations, and as ultimately adopted from them by the Romans, was an astronomical division, being the fourth part of a periodic lunar month.[213]

---

[209] Ap. Macrob. Sat. i. 16, § 32.

[210] Ib. § 33, and c. 13, § 20.

[211] Ib. § 33. Macrobius, who attributes this opinion to Geminus, appears to mean Geminus Tanusius, who was contemporary with Augustus.

[212] The nundinæ was a market-day, and was sacred to Saturn. Plut. Quæst. Rom. 42; Cor. 19.
Compare Virgil's Moretum, v. 79-81:—
Nonisque diebus
Venales olerum fasces portabat in urbem,
Inde domum cervice levis, gravis ære, redibat.

There is a short title de Nundinis in the Digest. L. 11.

[213] Macrobius, Comm. in. Somn. Scip. i. 6, § 52, conceives the lunar

§ 10 Besides the determination of the courses of the sun and moon for the regulation of the calendar, astronomical observation in Greece was, at an early period, directed to the fixed stars. Homer, in a passage which occurs, with slight variations, both in the Iliad and Odyssey, specifies by name the Pleiades, the Hyades, Orion, Boötes, and the Bear, which last, he adds, is also called the Wain; alone, of all the constellations, it remains constantly in sight, being never submerged in the waves of the ocean, and it keeps watch upon Orion.[214] No reasonable doubt can exist that the constellations thus designated by Homer were the same as those known by the same names in later times. In particular, the Great Bear is clearly defined by its relation with Orion, by its double name of the Bear and the Wain, and by its never sinking below the horizon.

Those who, in the days of Greek literature and erudition, occupied themselves with the interpretation of Homer, sought to find a satisfactory reason why the poet should have said that the Bear *alone* was never bathed in the ocean. Aristotle thinks that he thus characterizes it, as being the best known among the stars which never set.[215] The solution of Strabo is similar: he supposes that by 'the Bear,' Homer designates all

---

month of twenty-eight days as founded upon the septenary number. See Bailly, Astr. Anc. p. 32; Ideler, vol. i. p. 60, 87, 480-2, vol. ii. 175; Arago, Pop. Astron. b. 33, c. 3; Winer, Bibl. Real Wörterbuch, art. Woche. According to Porphyr. de Abstin. iv. 7, the seasons of purification observed by the Egyptian priests were sometimes of forty-two days [6×7], and never less than seven days. The division of the week is attributed by Garcilasso to the ancient Peruvians, but erroneously. See Humboldt, Vues des Cordillères (Paris, 1816, 8vo.), vol. i. p. 340.

(214) Iliad, xviii. 485-9; Od. v. 272-5. The difference consists in the first verse, which in the Iliad is—Πληιάδας θ' Ὑάδας τε τό τε σθένος Ὠρίωνος, and in the Odyssey is—Πληιάδας τ' ἐσορῶντι καὶ ὀψὲ δύοντα Βοώτην.

The last verse is translated by Virgil, but is applied to both the Bears:

'Arctos, Oceani metuentes æquore tingi.'—Georg. i. 246.

The passage in the Odyssey is also imitated by Virgil, Æn. iii. 516.

As to the meaning of the words Ὠρίωνα δοκεύει, see Ideler, Sternnamen, p. 298. Concerning the origin of the constellations, see Buttmann, über die Entstehung der Sternbilder, Berl. Trans. 1826.

(215) Καὶ τὸ 'οἴη δ' ἄμμορος' κατὰ μεταφοράν· τὸ γὰρ γνωριμώτατον μόνον. Poet. c. 25.

the arctic portion of the heavens.([216]) One grammarian altered the reading on conjecture, in order to save the poet's astronomical reputation;([217]) while another thought it better to admit that Homer erred through ignorance.([218]) The most probable supposition seems to be that the Great Bear was the only portion of the arctic sky which, in Homer's time, had been reduced into the form of a constellation.([219])

The epithet 'tardily-setting,' applied to Boötes, alludes to the fact that his disappearance, inasmuch as the constellation is in a perpendicular position, occupies some time; whereas, as Aratus signifies, his rising is rapid, being effected in a horizontal position.([220])

Homer likewise designates Sirius, as being called the 'Dog of Orion,' as appearing in the early autumn, as being a star of peculiar brightness; and as being of evil omen, because it brings fever to mankind.([221])

([216]) i. 1, § 6.

([217]) Crates read οἷος δ' ἄμμορός ἐστι. Ap. Strab. ubi sup. His emendation is mentioned by Apollonius, but his text is corrupt.

([218]) Apoll. Lex. Hom. in ἄμμορον. ὁ δὲ Ἡλιόδωρός φησι βέλτιον λέγειν ὅτι ἠγνόει. For Ἡλιόδωρος Heyne reads Ἡρόδωρος; but the name Heliodorus is correct. The Scholia on v. 489, suggest another explanation; viz., that the Bear is the only constellation mentioned in these verses which never sets. Compare Heyne, Hom. vol. vii. p. 525. Nitzsch, Od. vol. ii. p. 41. Schaubach, Geschichte der griechischen Astronomie, p. 15-23; Delambre, Hist. de l'Astr. Anc. vol. i. p. 341; Ideler, Untersuchungen über die Sternnamen (Berlin, 1809), p. 4-32.

([219]) 'Apparemment il compte pour rien le Dragon et la petite Ourse, parce que ces étoiles n'avaient pas de nom,' Delambre, ib. p. 342.

([220]) The same peculiarity is alluded to by Catullus, lxvi. 67, where the Coma Berenices says:—

  Vertor in occasum tardum dux ante Boöten,
  Qui vix sero alto mergitur Oceano.

Ovid, Fast. iii. 405, and Juv. v. 23, call him 'piger Boötes.' Ovid, Met. ii. 178, calls him 'tardus.' Claudian, de Rapt. Pros. ii. 190, piger. See Ideler, Sternnamen, p. 49.

  Quum jam
  Flectant Icarii sidera tarda boves.
         Prop. ii. 23, 24.
  Cur serus versare boves et plaustra Bootes.
         Prop. iii. 5, 35.

On the sudden rising of Boötes, see Arat. 609, et Schol.

([221]) Iliad, xxii. 26-31. Compare v. 5, where the same star is alluded to. An οὔλιος ἀστήρ is mentioned xi. 62. The obscure phrase νυκτὸς ἀμολγῷ

Homer describes Ulysses as steering his course by the stars in the open sea; and therefore the pole round which the starry heaven turns must have been before his time identified with the north: but he does not refer to the stars as indicating the season of the year. Hesiod, on the other hand, in his Didactic poem, frequently refers to them for the latter purpose. He appears to be ignorant of any calendar of months by which the time of year can be described. Thus he advises the husbandman to begin cutting his corn at the rising of the Pleiads, and to plough at their setting. He says that they are invisible for forty days and nights, but reappear at the time of harvest.[222] He dates the early spring by the rising of Arcturus, sixty days after the winter solstice (Feb. 19); which is soon followed by the appearance of the swallow:[223] and he designates this as the proper season for trimming the vines: but he adds that as soon as the snail emerges from the ground, and the Pleiads have risen, the care of the vines must cease.[224] He speaks of Sirius as characterizing the greatest heat of summer.[225] He lays it down that the rising of Orion is the season for threshing: he points to the time when Orion and Sirius are in the mid heaven, and Arcturus is rising, as proper for the vintage: and he directs the husbandmen to plough at the setting of the Pleiads, Hyads, and Orion.[226] He warns the navigator to avoid the dangers of the sea at the time when the Pleiads, flying from Orion, are lost in the waves.[227] The seasonable time for navigation is, he says, sixty days after the

---

appears to be correctly rendered by 'the dead of night,' by Buttmann in his Lexilogus.

The passage in the twenty-second book of the Iliad is imitated by Virgil, Æn. x. 274.

[222] Op. et Di. 381—5.

[223] v. 562.

[224] v. 569.

[225] v. 585. Imitated by Alcæus, Fragm. 39, Bergk. Sirius is twice mentioned in the Hesiodean poem of the Shield of Hercules, as characteristic of the heat of summer, Scut. 153, 397.

[226] v. 596, 607-615.   [227] v. 616-27.

summer solstice, when the heat of summer has come to an end (Aug. 20).

There is reason to believe that the heliacal rising of Sirius was observed as an epoch at an early period by the Egyptians. It had a peculiar importance in Egypt on account of its coincidence with the annual inundation of the Nile.(228) The Egyptian year began with the month of Thoth, which fell at the time of the rising of Sirius; and this circumstance gave a name to the celebrated Canicular period, of which we shall have occasion to speak hereafter.

It is clear that the annual motion of the stars, with respect to their risings and settings, was known in the time of Homer and Hesiod. Their nocturnal motion could not fail to be observed almost as soon as the motion of the sun during the day. It is alluded to by Homer as affording a measure of time during the night.(229)

The ancients not unfrequently speak of the 'dances of the stars.'(230) This comparison is drawn from the circular dances

---

(228) See Ideler, vol. i. p. 125, 129, 328. Σείριος, or σειρός, is an adjective signifying hot, scorching; hence σείριος ἀστήρ is used for the sun by Hesiod, Op. 415. σείριος alone is used by Archilochus and others for the sun, fr. 42, Gaisford, with Jacob's note, Hesych. in v. Σείριος signifies the dog-star in Hesiod. Æschylus, Ag. 967, and Soph. fragm. 941, have σείριος κύων. Compare Eurip. Hec. 1102; Iph. Aul. 7. Virgil, Æn. x. 273, has the expression 'Sirius ardor,' where he imitates the Greek in making *Sirius* an adjective. Comp. Porphyrius de Antro Nympharum, c. 24.

(229) Iliad, x. 251-3 :—

μάλα γὰρ νὺξ ἄνεται, ἐγγύθι δ' ἠώς,
ἄστρα τε δὴ προβέβηκε, παρῴχηκεν δὲ πλέων νὺξ
τῶν δύο μοιράων, τριτάτη δ' ἔτι μοῖρα λέλειπται.

Od. xiv. 483 :—

ἀλλ' ὅτε δὴ τρίχα νυκτὸς ἔην, μετὰ δ' ἄστρα βεβήκει.

The setting of the moon and the Pleiads is used to mark the middle of the night in the fragment of Sappho, fr. 58, Bergk. Compare Anacreont. Od. iii.

(230) χορεῖαι of the stars, Plat. Tim. p. 40. χορεῖον of the stars, Censorin. c. 13. ἄστρων τ' αἰθέριοι χοροί, Eurip. El. 467. Dances of the planets are mentioned in Manil. i. 669. Chorus astrorum, Stat. Achill. i. 643.

The following couplet occurs in Darwin :—

Onward the kindred Bears, with footsteps rude,
Dance round the pole, pursuing and pursued.

Econ. of Veget. canto i. v. 521.

of the Greeks, and alludes principally to the motion of the circumpolar stars.

The planets, on account of their motion among the other stars, and the consequent irregularity of their risings and settings, were not observed in early times. No planet is alluded to by Hesiod, who uses the stars merely as a calendar.

Venus, the most brilliant among the heavenly bodies after the moon, is mentioned by Homer under the names Hesperus, as the fairest star in the heaven;(231) and also of Eosphorus, or the morning star.(232) Hesperus is likewise alluded to by Sappho, in some extant verses, as putting an end to the toils of the day.(233) The early Greeks did not identify the morning with the evening star. The discovery of their identity is variously attributed to Pythagoras, Ibycus, and Parmenides.(234)

§ 11 The religion and mythology of the early Greeks had scarcely any reference to astronomy, or to an adoration of the heavenly bodies. The worship of the powers of Nature, which some modern historians have attributed to the Pelasgian, or Prehellenic, period of Greece,(235) may have once existed; but its existence rests upon hypothesis, not upon testimony. It was not till a comparatively late period that the god Helios was identified with Apollo, or that Diana became the goddess of the

---

(231) Iliad, xxii. 318.

(232) Iliad, xxiii. 226; Od. xiii. 93. In the latter passage the morning star is called ἀστὴρ φαάντατος. According to Varro, the star of Venus guided Æneas from Troy to Latium, Serv. Æn. ii. 801.

(233) Fragm. 96, Bergk.

(234) Diog. Laert. viii. 14, ix. 23; Stob. Ecl. Phys. i. 24; Plin. ii. 8. Pliny dates the discovery by Pythagoras in Olymp. 42, 142 U. C. (=612 B.C.). Achill. Tat. c. 17, assigns the identification to Ibycus, who is placed about 540 B.C. Compare Karsten, Phil. Gr. Rel. i. 2, p. 249, 255. The authority cited by Stobæus for its being a Pythagorean doctrine is Apollodorus in his treatise περὶ θεῶν; see Fragm. Hist. Gr. vol. i. p. 428.

(235) See Müller's Hist. of the Lit. of Anc. Greece, c. 2, § 3. Plato, Cratyl. § 31, p. 397, conjectures that the ancient Greeks, like many of the barbarians, worshipped the sun, the moon, the earth, the stars, and the heaven; hence he derives θεοὶ from θεῖν. But this is not a historical testimony. Herod. iv. 188, says that the nomad Africans sacrifice only to the sun and moon.

moon.(²³⁶) The sun and moon, though they were conceived as celestial beings, driving their chariots across the vault of heaven,(²³⁷) held quite a subordinate rank in the primitive Pantheon, and were not included among the Olympian gods. When certain stars had been formed into a constellation, and designated by a certain name, the fancy of the mythologist supplied a story to illustrate the name. But Otfried Müller, in his Treatise on Mythology, has shown clearly that the astronomical mythi of the Greeks formed an unimportant part even of their mythology, and were for the most part wholly unconnected with their religion.(²³⁸)

---

(236) Müller, Dor. ii. 5, 5. Æschylus identifies Apollo with the sun, Sept. Theb. 859, and Diana with the moon, Xantr. fragm. 158. Dindorf.

In the fragments of the Phaethon of Euripides, Apollo is described by Merope as the esoteric name of the sun:—

ὦ καλλιφεγγὲς Ἥλι', ὥς μ' ἀπώλεσας,
καὶ τόνδ'. Ἀπόλλω δ' ἐν βροτοῖς σ' ὀρθῶς καλεῖ,
ὅστις τὰ σιγῶντ' ὀνόματ' οἶδε δαιμόνων.

This is a play upon the word Ἀπόλλων, on account of its resemblance in sound to ἀπολλύναι. But in the story of Phaethon, Helius, not Apollo, is treated as his father, and his sisters are called the Heliades.

Later writers identify Apollo with the sun, and Diana with the moon. See Cic. de N. D. ii. 27; Plut. de Orac. def. 7; Macrob. Sat. i. 17; Serv. Æn. iii. 73; Augustin, C. D. vii. 16; Nonn. xl. 401. Ovid, Met. ii. 25, makes Phœbus not only the god of the sun, but also of the seasons and of the calendar.

(237) Neither the sun nor the moon drives a chariot in Homer. But Eos or Aurora has a chariot in a passage of the Odyssey, xxiii. 244-6, and the names of her horses, viz., Lampus and Phaethon, are given.

The chariot of the sun is mentioned by Soph. Aj. 845; Eurip. Ion. 1148; Plut. Sept. Sap. Conv. 12. It is also an essential part of the story of Phaethon, which is as early as Hesiod. The god of the sun is called ἱππονώμας, Aristoph. Nub. 572. Æschylus speaks of the λευκόπωλος ἡμέρα, Pers. 386. The chariot of night occurs in Eurip. Ion. 1150, and Andromed. fragm. 1; Dind.: see likewise Theocrit. ii. 163, where in v. 166, ἄντυξ νυκτὸς means the 'vault of night,' not the chariot, as it is incorrectly rendered by the Scholiast. In Mosch. Id. ii. 38, ἄντυξ ἡμίτομος is the semicircular disc of the moon; its shape when it is in dichotomy.

The Roman poets are more frequent than the Greek poets in their allusions to the chariot of the moon. See Ovid, Fast. ii. 110, iv. 374, v. 16; Rem. Am. 258; Manil. i. 667; Claudian. Rapt. Pros. iii. 403; Auson. Ep. v. 3. According to Manilius, the chariot of the sun has four horses, that of the moon only two, v. 3. Nonnus, vii. 234, mentions the chariot of the moon as drawn by mules.

(238) Proleg. zu einer Wissenschaftlichen Mythologie, p. 191 (p. 130 Engl. Tr.).

The nature of the astronomical mythology of the Greeks will best appear from a few examples.

The constellation of the Great Bear seems to have received its name from its resemblance to the shape of an animal; but when it had acquired that appellation, it was connected with the story of Callisto, a nymph loved by Jupiter, to whom she bore Arcas, the mythical progenitor of the Arcadians. She was first metamorphosed into a bear by Diana, as a punishment for her breach of chastity, or by Juno from jealousy; and afterwards being pursued by hunters, was saved by Jupiter, and transferred to the skies as an asterism near the north pole.[239] The mythologists accounted for the constant appearance of the Great Bear above the horizon, by saying that the ocean deities prevented their pure waters from being polluted by the presence of the concubine of Jupiter.[240]

A different explanation is given by Aratus: he supposes both Bears to have been transferred from Crete to the heavens by Jupiter, during the year when he was concealed by the Curetes in the Idæan cave.[241]

---

[239] See Hesiod, Fragm. ed. Marckscheffel, p. 353; Ovid, Met. ii. 401-530; Callimachus ap. Schol. Il. xviii. 487.

[240] The following are the words of Juno to her former nurse Tethys, in Ovid, Met. ii. 527-30:—

> At vos si læsæ tangit contemptus alumnæ,
> Gurgite cœruleo septem prohibete Triones,
> Sideraque in cœlum stupri mercede receptâ
> Pellite, ne puro tingatur in æquore pellex.

The explanation of Hyginus, Poet. Astr. ii. 1, is similar. 'Hoc signum, ut complures dixerunt, non occidit. Qui volunt aliquâ de causâ esse institutum, negant Tethyn Oceani uxorem id recipere, cum reliqua sidera eo perveniant in occasum, quod Tethys Junonis sit nutrix, cui Callisto succubuerit ut pellex.' Hygin. fab. 177, cites some verses referring to the same point:—

> Tuque Lycaoniæ mutatæ semine nymphæ,
> Quam gelido raptam de vertice Nonacrino
> Oceano prohibet semper se tingere Tethys,
> Ausa suæ quia sit quondam succumbere alumnæ.

Succumbere here means to *supplant*, in a sense derived from *succuba*.

[241] v. 30-4. Compare Diod. iv. 79, 80. This fable is inconsistent with the natural history of the island; for the ancients testify that Crete never contained any bears or other noxious animals. See Aristot. Mir.

The Pleiads evidently took their name from the verb πλεῖν,([242]) because their rising was synchronous with the beginning of the season fitted for navigation in the Greek waters, while their setting marked its termination. They were supposed to be the daughters of Atlas, and the sisters of the Hyades, and to have been changed into stars.([243]) The imagination of poets, by playing upon the name, conceived this constellation as a flight of doves (πελεῖαι) :([244]) Æschylus, employing a privative epithet to circumscribe and explain his metaphor, called them doves, but doves devoid of wings.([245])

The constellation of the Pleiads consists, according to Ideler,([246]) of one star of the third magnitude, three of the fifth, two of the sixth, and many smaller stars. It is therefore scarcely possible for the best eye to discern more than six.([247]) The ancients, from some numerical fancy, thought fit to desig-

---

Ausc. c. 84; Diod. iv. 17; Ælian, N. A. v. 2; Plin. viii. 58; Solinus, c. 11, § 11; Plut. de Cap. ex Host. util. c. 1. Compare Pashley's Travels in Crete, vol. ii. p. 261. The Daunian bear in Iamblich. Pyth. 60, Porphyr. Pyth. 23, is a fiction, for Italy never contained wild bears. The 'Caledonius ursus' in Martial, de Spect. 7, is a merely poetical epithet.

(242) See Müller, Mythol. p. 191 (131, E. T.). Other fanciful etymologies are given by Schol. Iliad, xviii. 486; Schol. Arat. 254.

(243) Hesiod gives them the epithet of 'Ατλαγενεῖς, Op. 383. Hence Virgil, Georg. i. 221, calls them Atlantides. Compare Schol. Il. xviii. 486; Hygin. Poet. Astr. ii. 21.
The Homeric and Hesiodean form is Πληιάδες, derived from πλέω, which verb is never contracted in Homer. Its ancient form seems to have been πλέϝω; the trace of the digamma is preserved in the future πλεύσω.

(244) The substitution of Πελειάδες for Πλειάδες occurred in the astronomical poem attributed to Hesiod. See fragm. p. 352, ed. Marckscheffel. The same lengthened form is used by Pindar, Nem. ii. 16, and numerous instances of it are cited by Athen. xi. c. 79, 80, including one from Simonides.

(245)   αἱ δ' ἔπτ' Ἄτλαντος παῖδες ὠνομασμέναι
πατρὸς μέγιστον ἆθλον οὐρανοστεγῆ
κλαίεσκον, ἔνθα νυκτέρων φαντασμάτων
ἔχουσι μορφὰς ἄπτεροι Πελειάδες.—Ap. Athen. p. 491 A.

(See Blomf. ad Æsch. Ag. 81, Gloss.) This passage is alluded to by Schol. Il. xviii. 486. Æschylus assigns their grief for their father's labour in carrying the world, as the cause of their metamorphosis.

(246) Sternnamen, p. 145.

(247) Hipparch. ad Arat. Phæn. i. 14, affirms that seven stars can really be perceived in a clear moonless night.

F

nate the Pleiads as a constellation of seven stars;[248] and they gave a fictitious reason for the fact that only six were visible. Some said that the seventh Pleiad had been struck by lightning: others that it had been removed into the tail of the Great Bear:[249] the reasons assigned by Ovid[250] are either that one of the seven daughters of Atlas was dishonoured by a marriage with a mortal, whereas each of the six others ascended the bed of a god; or that Electra, the wife of Dardanus, was so overwhelmed by the capture and destruction of Troy, that she veiled her head in token of grief.[251] This constellation was called Vergiliæ by the Romans, because it rose after the spring.[252]

As the Pleiads derived their name from *navigation,* so the Hyades derived their name from *rain* (ὕειν).[253] The Hyades, like the Pleiads, were the daughters of Atlas: they were supposed to have been metamorphosed into stars on account of their grief for the death of their brother Hyas.[254] Some de-

---

[248] Called ἑπταπόρος by Eurip. Iph. Aul. 6; ἑπτάστερος, by Eratosth. c. 14, 23. Quæ septem dici, sex tamen esse solent, Ovid, Fast. iv. 170; Arat. 257; Germanicus, Arat. 259.

[249] Schol. Arat. 254, 257.

[250] Fast. iv. 171-8. The former of those reasons appears in Eratosth. c. 23; Hygin. Poet. Astr. 21; Schol. Il. xviii. 486.

[251] It was a sign of grief among the Greeks to muffle the head, Hom. Od. viii. 85; Eur. Hec. 432; Orest. 42, 280; Suppl. 122. Weeping seems to have been considered as of bad omen, and was therefore concealed.

[252] Eas stellas Vergilias nostri appellaverunt, quod post ver exoriuntur, Hygin. P. A. c. 21. The rising of the Pleiads was in the middle of May, Festus, p. 372. Vergiliæ dictæ quod earum ortu ver finitur, æstas incipit.

[253] Ora micant Tauri septem radiantia flammis,
 Navita quas Hyadas Graius ab imbre vocat.
Ovid, Fast. v. 165-6.
Virgil, Æn. i. 744, iii. 516, calls them 'pluviæ Hyades;' Horace, Carm. i. 3, 'tristes.'
Ὑάδες τε ναυτίλοις
σαφέστατον σημεῖον.—Eurip. Ion. 1156.
Where Ὑάς lengthens the first syllable, which Homer and other poets shorten.

[254] Hygin. fab. 192; P. A. ii. 21; Ovid, ubi sup. The names of five Hyades were enumerated in the astronomical poem attributed to Hesiod, p. 353, ed. Marckscheffel. Their number was variously stated.

rived the name of the constellation from the letter Y, to which its form bears a resemblance.(255) The Romans called it Suculæ; which Cicero considers an etymological-blunder, founded on a confusion with ὓς.(256) It is, however, more probable that this was a native Latin word, formed from *succus*, and that, like the Greek ὕας, it alluded to the moisture which accompanied the rising of this constellation.(257)

Orion was conceived as a warrior and hunter, who was transmuted into a constellation. From his position in the heavens, he was said to fly from the Pleiads,(258) as he, in his turn, was said to be watched by the Great Bear. Various fables were connected with his translation into the sky.(259)

One of the stories relative to the Dog-star derived its name from the dog of Orion.(260) The time of its rising connected it, at an early period, with the idea of intense summer heat; and Müller conjectures that it was so called on account of the prevalence of canine madness at this season. The ceremonies of the Greek and Roman worship, in several places, alluded to the connexion between the dog and the star in question.(261)

There were likewise similar fabulous stories explanatory of the name of the constellation Boötes, also called Arctophylax and Arcturus.(262) Arcturus, as he declares of himself in the prologue to the Rudens of Plautus, was, on account of the

---

(255) Schol. Il. xviii. 486; Schol. Arat. 173; Hygin. fab. 192.

(256) Cic. de N. D. ii. 43. Nostri imperite suculas; quasi a suibus essent, non ab imbribus nominatæ. This remark recurs in a passage cited by Gell. xiii. 9. Compare Plin. ii. 39.

(257) This etymon is given in Serv. Æn. i. 744.

(258) Hesiod. Op. 619; Pind. Nem. ii. 11.

(259) See Meineke, Analecta Alexandrina, p. 133; Hygin. Astr. ii. 34. Orion is called a stormy constellation, in allusion to its setting in the late autumn, Virg. Æn. i. 535, iv. 52; Horat. Ep. xv. 7; Ideler, Sternnamen, p. 219.

(260) Hygin. fab. 130; Astr. ii. 4, 36; Eratosth. 33. Another story connected the star with the dog of Icarius.

(261) See Ideler, Sternnamen, p. 238; Notes and Queries, vol. vii. p. 431.

(262) See Ideler, Sternnamen, p. 47.

times of his rising and setting, considered a stormy constellation.([263])

The constellations of the heavenly sphere seem to have been gradually formed by the Greeks. Those which are mentioned by Homer and Hesiod are doubtless the most ancient.([264]) It was not till astronomers had observed the oblique path of the sun, in its annual progress, that the zodiac was marked out. The division into twelve parts took its origin in the number of lunations, which nearly corresponded with the solar year. When the ecliptic had been divided into twelve parts, astronomers perceived that the motions of the moon, and of the planets visible to the naked eye, were all within a certain distance of the ecliptic. They included this space within parallel circles of the sphere; and adopted the constellation which coincided with each twelfth part as a sign. Hence as these constellations were represented by ζῳδια([265]) or figures, on the celestial sphere of the ancients, this circular band was called the *zodiac*, and the constellations which encircled it were called the signs of the zodiac. Aristotle is the earliest extant writer who mentions the zodiacal circle; it is likewise mentioned by Autolycus.([266]) About a century later, Aratus, in his astronomical

---

(263) Increpui hibernum et fluctus movi maritimos.
Nam Arcturus signum sum omnium acerrumum,
Vehemens sum exoriens; cum occido, vehementior.
Prol. v. 69-71.

(264) Grimm, Deutsche Myth. p. 416, thinks that the Great Bear, Orion, and the Pleiads, were the three constellations first formed.

(265) The word ζῳδιον is used by Herod. i. 70, to denote sculptured figures of animals upon a goblet. The κύκλος ζῳδιακὸς means the circle of figures represented in painting or sculpture. Among the figures of the zodiac four were not animals, and the German *Thierkreis* is an inaccurate translation of the word. No diminutive of the form ίδιον is ever used in epic or tragic poetry, and therefore it may be inferred that the name of the zodiac had not a poetical origin. Diminutives of this form do not belong to the early language, see Buttmann, Ausführl. Gr. Gramm. vol. ii. p. 442. The word κῴδιον became *corium*, which is an old Latin word, used by Cato and Plautus.

(266) ὁ κύκλος τῶν ζῳδίων, Aristot. Meteor. i. 6. 6. i. 8. 3. ὁ τῶν ζῳδίων, ib. i. 8, 14. ὁ διὰ τῶν μέσων τῶν ζῳδίων (κύκλος), Metaph. xi. 8. Concerning Autolycus, see Delambre, Hist. d'Astr. Anc. vol. i. p. 29. According to Eudemus, Œnopides was the discoverer of the zodiac.

poem, names the zodiacal circle, and enumerates its twelve signs. He states, moreover, that six of the signs rise, and six set, every night.[267] The zodiac was doubtless known to Eudoxus, whose Phenomena was versified by Aratus.

On the whole, the mythology of the Greeks has little connexion with the heavenly bodies, and the mythical stories which have an astronomical reference are of comparatively recent origin. Sometimes a narrative received from later writers an astronomical application, which it wanted in its primitive form. Thus the story of the sun turning back his chariot, and concealing his light, at the sight of the crime of Atreus, who had feasted Thyestes with the flesh of his own murdered children,[268] was softened into the rationalized version, that Atreus had discovered the proper motion of the sun, from west to east, contrary to the general motion of the heavens;[269] or that

---

[267] Arat. 541-58.

[268] For a full description of the disappearance of the sun, because he is shocked at the unnatural crime of Atreus, see Sen. Thyest. 783-884. Seneca supposes that when the sun disappears during the day, the moon and stars do not become visible. He conceives the latter heavenly bodies as peculiar to the night, and as not being ready to shine before their usual time.

> Ipse insueto novus hospitio
> Sol auroram videt occiduus,
> Tenebrasque jubet surgere, nondum
> Nocte paratâ. Non succedunt
> Astra, nec ullo micat igne polus,
> Nec luna graves digerit umbras.—v. 820-5.

Where Lipsius says: 'Ecce autem sol sese condens noctem inducebat, et surgere jubebat ante tempus, ideoque imparatam.' It does not seem to have occurred to the poet that the moon and stars were in the heaven during the day as well as during the night; and that they were only rendered invisible by the superior light of the sun. *Digero* in this passage means to dissipate, to scatter, as in Georg. iii. 357 :

> Tum sol pallentes haud unquam discutit umbras.

[269] Polybius, ap. Strab. i. 2, 15 (p. 1120, ed. Bekker), says that the benefactors of mankind and the authors of useful inventions became kings and gods. Among these he instances Atreus, who first taught that the course of the sun was contrary to that of the heaven. Lucian, Astrol. 12, has the same statement respecting Atreus. Euripides, ap. Achill. Tat. Isag. c. 1, spoke of Atreus as discovering the proper motion of the sun:

> δείξας γὰρ ἄστρων τὴν ἐναντίαν ὁδὸν
> δήμους τ' ἔσωσα, καὶ τύραννος ἱζόμην.—Cress. fragm. 14.

This passage means δείξας τὴν ὁδὸν τοῦ ἡλίου ἐναντίαν οὖσαν τῆς τῶν

he had discovered the method of predicting a solar eclipse.([270]) Plato gives a different turn to the fable. He supposes Jupiter, in his horror at the crime of Atreus, to have changed the quarter where the sun rises from east to west.([271]) Another fabulous interpretation traced the erratic motion of the planets to the same cause.([272]) These explanations are evidently philosophical figments, foreign to the supernatural character of the old legend.

§ 12 As the religion and mythology, so the divination of the early Greeks had little connexion with the heavenly bodies. The Greeks sought to penetrate into the future by means of oracles and dreams, by the entrails of victims, and by the flight of birds.([273]) To these the Romans added the interpretation of lightning, which they borrowed from Etruria.([274]) Both also drew prognostics from prodigies; that is to say, from rare natural appearances;([275]) among which comets, meteors, and eclipses held an important place.

ἄστρων. It is misunderstood by Achilles Tatius, p. 73, 82, 95, ed. Petav. The conversion of the sun's course from west to east, in consequence of the crime of Atreus, is mentioned by Euripides, Orest. 1000; Iph. T. 816. Manilius, iii. 18, says that the sun turned back at the sight of the impious feast, and that darkness ensued.

(270) Quæ sol ne videndo pollueretur, radios suos ab eâ civitate detorsit. Sed veritatis hoc est: Atreum apud Mycenas primum solis eclipsin invenisse, cui invidens frater urbe discessit tempore quo ejus probata sunt dicta. Serv. Æn. i. 568. Repeated in Hygin. fab. 258. The former explanation refers to the sun turning back in his course; the latter to the darkness caused by the sun's disappearance. Compare Grote, Hist. of Gr. vol. i. p. 220.

(271) Politic. p. 268.

(272) Achill. Tat. p. 96, ed. Petav.

(273) See Æsch. Prom. 492-508. Compare the enumeration in Hermann, Gottesdienstlichen Alterthümer der Griechen, § 37, 38, and in the elaborate article, Divinatio, in Pauly's Real-Encyclopädie.

(274) See Müller's Etrusker, b. 3, c. 1, 2. The φλογωπὰ σήματα in Æsch. Prom. 507, do not refer to lightning. Lightning was considered by the Greeks as the weapon with which Jupiter punished the wicked. See Aristoph. Nub. 397.

(275) διὸ καὶ οὐδ' οἱ μάντεις εἰώθασι τοῦτο κρίνειν ὡς τέρας· τὸ γὰρ εἰωθὸς οὐ τέρας. Theophrast. de Plant. v. 3. Compare Hermann, ib. § 38, n. 16.

Herod. ii. 82, says of the Egyptians: τέρατα δὲ πλέω σφι ἀνεύρηται ἢ τοῖσι ἄλλοισι ἅπασι ἀνθρώποισι· γενομένου γὰρ τέρατος φυλάσσουσι γραφόμενοι τἀποβαῖνον, καὶ ἤν κοτε ὕστερον παραπλήσιον τούτῳ γένηται, κατὰ τὠυτὸ νομί-

But although the Greeks and Romans agreed with the less civilized nations of antiquity in viewing eclipses, comets, meteors, and other extraordinary appearances in the heaven, with alarm, as being marks of the divine displeasure; yet they drew no prognostics from the ordinary movements of the heavenly bodies: they had no astrological science, or system of astrological prediction.([276])

For this deceitful and seductive art the Greeks were, as it appears, indebted to the Chaldeans, who introduced it among them after the time of Alexander.

It was never cultivated with fondness by the Greeks, though they made it an adjunct to their scientific astronomy, and admitted it into that strange mixture of mystical philosophies which prevailed in the third and fourth centuries after Christ. We shall have occasion to consider the Chaldean astrology, in a future chapter, in connexion with the astronomy of the Babylonians and Egyptians.([277])

ζουσι ἀποβήσεσθαι. The method here described is one of strictly scientific induction; if it had been rigorously observed, the Egyptians would either have discovered the vanity of divination from prodigies, or they would have founded it on a scientific basis. Concerning the meaning of τέρας in Homer, see Nägelsbach Homerische Theologie, p. 146. The ordinary Roman prodigies are enumerated by Claudian, Eutrop. i. 1-7, ii. 40-4.

([276]) The later Greeks, who discovered the germs of all arts and sciences in Homer, found even traces of his astrological science. See Heyne ad Il. xviii. 251, vol. vii. p. 467; Marsham, Can. Crit. p. 480.

Euripides describes Hippo, the daughter of Chiron, as predicting the future from the risings of stars:

ἣ πρῶτα μὲν τὰ θεῖα προυμαντεύσατο
χρησμοῖσι σαφέσιν ἀστέρων ἐπ' ἀντολαῖς.
Melanipp. Sap. Fr. iii. Dindorf.

She was supposed to have learnt astronomy from her father Chiron. See below, ch. ii. § 1. Archdeacon Hare, Phil. Mus. vol. i. p. 25, thinks that the predictions of Hippo intended by Euripides, 'relate only to the art of foretelling the weather from the heliacal rising of the stars.'

([277]) See below, ch. v. § 10.

Chapter II.

PHILOSOPHICAL ASTRONOMY OF THE GREEKS FROM THE TIME OF THALES TO THAT OF DEMOCRITUS.

§ 1. THERE is, as we have seen in the preceding chapter, no trace of any scientific astronomical knowledge in the remains of early Greek literature, or in the authentic accounts of the primitive times of Greece and Rome.

Some of the stories of the Greek mythology, indeed, as reduced and rationalized by the later historians, attributed the origin of astronomy to the fabulous ages of Greece. Thus Atlas is said to have discovered the doctrine of the sphere, and even to have been the author of astronomy; he is further reported to have communicated this knowledge to Hercules, who imparted it to the Greeks.[1] According to another story, the sons of Helius in the island of Rhodes were distinguished by their knowledge of astronomy; they also made improvements in navigation, and arranged the seasons. Actis, one of the Heliadæ, emigrated to Egypt, and taught astronomy to the Egyptians. The knowledge of astronomy was afterwards

---

[1] Diod. iii. 66, iv. 27; Plin. N. H. vii. 57; Herodorus ap. Clem. Alex. Strom. i. 15, § 73 (Fragm. Hist. Gr. vol. ii. p. 34); Serv. Æn. i. 741; Heraclit. de Incred. 4. Diogenes Laertius, Procem. 1, makes him a philosopher.

Diodorus likewise makes Uranus, king of the Atlanteans, the first astronomer, and describes him as the ancestor of the sun and moon, iii. 56, 57. The fable of Atlas bearing the heaven on his shoulders (of which the story in the writers above cited is a rationalized version) is of great antiquity. See Hom. Od. i. 52; Hesiod, Theog. 517; Æsch. Prom. 347, 430; Aristot. de Mot. Animal. 3. Hesperus is fabled to have been a son of Atlas, remarkable for his piety and justice. He used to observe the stars on the top of Mount Atlas, but was carried away by a violent wind, and never reappeared. His subjects gave him divine honours, and called the most beautiful star in the heavens by his name.—Diod. iii. 60.

obliterated in Greece by a deluge, which destroyed nearly all the inhabitants; and the Egyptians thus obtained the credit of having been the authors of this science.[2]

A third story represented Hyperion as having been the first systematic observer of the movements of the sun, moon, and stars, and of the seasons determined by their course, and as having taught this knowledge to mankind; whence he was called the father of the sun and moon.[3]

Uranus, again, King of the Atlanteans, was called the first astronomer, and the sun and moon were said to be his descendants.[4]

Sophocles considers Palamedes in the light of a civilizer, and as the author of numerous inventions, among which he includes a knowledge of the movements of the stars, and of their use in navigation.[5] The treatise on astronomy ascribed to Lucian speaks of Orpheus as having taught that science to the Greeks.[6] The fable of Prometheus chained to the rock, on Caucasus, was likewise interpreted by late writers to contain a concealed reference to the observations of a solitary astronomer upon lofty mountains.[7]

Newton, in his treatise on Ancient Chronology, has assigned the origin of astronomical science to the heroic ages of Greece. His argument rests upon three hypotheses:—

1. That Chiron delineated the constellations, and was a

---

[2] Diod. v. 57.

[3] Diod. v. 67. Compare Hesiod, Theog. 371-4.

[4] Diod. iii. 56, 57.

[5] Fragm. 379, ed. Dindorf. The third line appears to be spurious. Palamedes is described as the inventor of dice in Paus. ii. 20, § 3; x. 31, § 1. Philostratus, Heroic. c. 11, makes him the inventor of the divisions of time, money, weights and measures, arithmetic, and alphabetical writing.

[6] c. 10.

[7] Nec vero Atlas sustinere cœlum, nec Prometheus affixus Caucaso, nec stellatus Cepheus cum uxore, genero, filiâ traderetur, nisi cœlestium divina cognitio nomen eorum ad errorem fabulæ traduxisset. Cic. Tusc. Disp. v. 3. Concerning Prometheus as an astronomer, see below, ch. v. § 1. Cepheus, his wife Cassiopea, his daughter Andromeda, and his son-in-law Perseus, were converted into stars.

práctical astronomer. 2. That Musæus, the son of Eumolpus, and master of Orpheus, made a sphere; and that he was reputed to have been the first among the Greeks who made one, the time at which it was made coinciding with that of the Argonautic Expedition. 3. That the people of Corcyra attributed the invention of the sphere to Nausicaa, daughter of Alcinous, King of the Phæacians; and that she appears to have obtained her knowledge of it from the Argonauts, who, on their return home, sailed to that island, and made some stay there with her father.([8])

Chiron, a centaur, was the son of Saturn and the sea-nymph Philyra;([9]) and thus, as Xenophon remarks, he was the brother of Jupiter.([10]) Saturn was fabled to have changed himself into a horse in order to conceal his amour with Philyra from his wife Rhea; hence the mixed form of Chiron. Homer calls him 'the most just of the centaurs,' and designates him as the teacher of Achilles, in reference to his knowledge of the remedies for a wound.([11]) Pindar, likewise, describes Chiron as having given instruction in medicine to Jason and Æsculapius.([12]) A family which had an hereditary knowledge of the medicinal virtues of plants, and which was said to be descended from Chiron, dwelt near Mount Pelion in late times.([13]) The Mag-

---

([8]) Chronology of Ancient Kingdoms amended, Horsley's Newton, vol. v. p. 63-6. Compare p. 17.

([9]) Hesiod, Theog. 1001; Pindar, Pyth. vi. 22; Apollod. i. 2, 4; Apollon. Rhod. i. 554, ii. 1231-41, and Pherecydes ap. Schol. Virg. Georg. iii. 93. Xenoph. de Ven. i. 4, describes Philyra generally as a Naid. Scriptor Gigantomach. ap. Schol. Apoll. Rh. i. 554. Ovid, Met. vi. 226.

([10]) Xen. ib. § 3. Xenophon points out that Chiron was long-lived, for that, while he was the brother of Jupiter, he lived long enough to be the teacher of Achilles.

([11]) Il. xi. 830-2. Compare Ælian, Nat. An. ii. 18. According to Hesiod, Theog. 1001, Chiron reared Medeus, the son of Jason and Medea.

([12]) Nem. iii. 53. Compare Schol. Apoll. Rh. i. 554; Schol. Arat. 436.

([13]) Dicæarch. de Pelio, § 12; Fragm. Hist. Gr. vol. ii. p. 263. The spear of Achilles was made of an ash from Pelion, and was given by Chiron to Peleus, Il. xvi. 143, xix. 390. Chiron was supposed to inhabit Pelion, and there was a cave on this mountain called the Chironian cave. Pelion was the scene of the amour of Saturn and Philyra.

nesians also offered sacrifices to Chiron as the author of medicine.([14])

An epic poem which bore the name of Hesiod contained the precepts which Chiron imparted to Achilles.([15]) They were regarded as of a religious and ethical character: Pindar particularly specifies the worship of Jupiter as the first of Chiron's lessons to his celebrated pupil.([16]) Xenophon says that Apollo and Diana selected Chiron, on account of his justice, for instruction in the art of hunting; and that he afterwards taught this art to numerous heroes.([17]) Besides these subjects, Chiron is described as instructing Achilles in music and the use of the lyre.([18])

In the Achilleid of Statius, Achilles gives a detailed account of his instruction by Chiron. The subjects are hunting, military exercises, playing on the lyre, treatment of wounds, and medicine. He likewise receives lessons of justice. No allusion is made to any branch of astronomical science.([19]) According to Ovid, divination was not among the arts taught by Chiron.([20])

The ancients do not in general mention astronomy, or any

([14]) Plut. Symp. iii. 1, 3. His medical skill is alluded to by Virgil, Georg. iii. 548, and by Ovid, Fast. v. 401, 410. He cured the eyes of Phœnix, Propert. ii. 1, 60. Hyginus, fab. 274, says that he invented the art of surgery from herbs. Before Æsculapius, Chiron was acquainted with the medicinal virtues of herbs, according to Galen. Introd. vol. xiv. p. 675. Kühn. Some herbs used in medicine were named Chironion and Centaureum, from Chiron, Plin. xxv. 13, 14; Theophrast. Hist. Plant. ix. 11, 1. The Chironion grew upon Pelion, Dioscorid. iii. 57. The Schol. Iliad, iv. 219, makes him the inventor of medicine.

([15]) Hesiod, fragm. p. 175, 370, ed. Marckscheffel.

([16]) Pyth. vi. 21; Hesiod, fr. 206. ib. Apollo is described as seeking counsel from Chiron in Pyth. ix. 29. Compare Eurip. Iph. Aul. 709.

([17]) Cyneg. c. i. § 1-4.

([18]) See Ovid, Fast. v. 385; Philostrat. Heroic. c. 10; Imag. ii. 2. According to Plutarch de Mus. 40, ὁ σοφώτατος Χείρων was μουσικῆς τε ἅμα καὶ δικαιοσύνης καὶ ἰατρικῆς διδάσκαλος. Pindar, Nem. iii. 53, calls him βαθυμήτης. His instruction in music is alluded to by Horace, Epod. xiii. 11. Ausonius, Id. iv. 20, speaks of the gentleness of Chiron's discipline.

([19]) ii. 382—453. Compare i. 38.

([20]) Met. ii. 638. Concerning Chiron, see the essay of Welcker, 'Chiron der Phillyride,' Kleine Schriften, vol. iii. p. 3; Fabric. Bibl. Gr. vol. i. p. 13, ed. Harles.

kindred science, among the subjects which were supposed to have been included in the lessons of Chiron. An anecdote is preserved respecting Diogenes, that, being reproached for his ignorance of geometry, he replied that it was permitted to be ignorant of a subject which not even Chiron taught to Achilles.([21]) The evidence of Chiron's astronomical knowledge is discovered by Newton, in a fragment of the Cyclic Titanomachia; the interpretation of which is dubious, and which probably has no reference to astronomy.([22]) But, whatever may be the true meaning of the verses in question, it is certain that

---

([21]) Parallela of Joannes Damascenus, ad calc. Stob. Phys. p. 740. ed. Gaisford. ad calc. Stob. Anth. vol. iv. p. 206, ed. Meineke.

([22]) ὁ δὲ Βηρύτιος Ἕρμιππος Χείρωνα τὸν Κένταυρον σοφὸν καλεῖ, ἐφ' οὗ καὶ ὁ τὴν Τιτανομαχίαν γράψας φησὶν ὡς πρῶτος οὗτος

εἴς τε δικαιοσύνην θνητῶν γένος ἤγαγε, δείξας
ὅρκους καὶ θυσίας ἱλαρὰς καὶ σχήματ' Ὀλύμπου.
Clem. Alex. Strom. i. 15, § 73, p. 132, ed. Sylburg.

By σχήματ' Ὀλύμπου, Newton understands the 'constellations of heaven.' Ὄλυμπος in the Greek writers seems never to signify the heaven in a physical sense; though it may express the aggregate of the Olympic gods. In the Latin writers Olympus is sometimes equivalent to *coelum*: as in Virg. Georg. i. 450, iii. 223. It has been proposed to read σήματ' Ὀλύμπου, in the sense of 'signa divinæ voluntatis.'

Welcker, Ep. Cyclus, vol. ii. p. 411, thinks that the ancient musician Olympus is referred to, and conceives that the dances invented by Olympus are intended. It is clear from the context that something considered by the Greeks as having an ethical influence is meant. Chiron, it is said, first turned mankind to justice, by teaching them the use of oaths and sacrifices. The third subject of his instruction can hardly be the forms of the constellations, which have no connexion with morality. It is possible that 'the dances of the gods,' as connected with sacrifices, are signified. Σχῆμα was used for a figure in dancing. See Aristoph. Vesp. 1485.

It must at the same time be admitted that Clemens appears to have understood the verses of the Titanomachia in the same manner as Newton: for he proceeds to say that Hippo, the daughter of Chiron, having become the wife of Æolus, instructed him in physics, which she had learnt from her father.

The Titanomachia was ascribed to Arctinus or Eumelus. See Welcker, ib. p. 556.

Letronne, Analyse critique des Représentations zodiacales de Dendéra et d'Esné, Mém. de l'Acad. des Inscript. tom. xvi. part 2, p. 103 (1850), understands the dances, or movements, of the stars.

The word σχήματα was used by late writers with reference to the positions of the stars for astrological purposes. See Gothofred ad Cod. Theod. vol. iii. p. 144.

the Centaur Chiron is a purely fabulous being, and that the fragment of the cyclic poem cannot be regarded as a historical testimony.

Staphylus, in his work on Thessaly, related that Chiron, being a wise man and skilled in astronomy, desired to make Peleus celebrated; that for this purpose he sent for Philomela, the daughter of Actor the Myrmidon, and circulated a report that Peleus was about to marry Thetis, with the consent of Jupiter, and that the gods would attend the wedding in a storm of wind and rain. Having waited for a season when there was about to be tempestuous weather, Chiron married Philomela to Peleus; whence the belief arose that Thetis was his wife.[23] The age of Staphylus is not known; but this attempt to rationalize the marriage of Peleus with the goddess Thetis is probably of late date. He is quoted by no writer earlier than Strabo.[24] In this story, the astronomical skill of Chiron is supposed to be evinced in his power of predicting the weather. The Latin Scholiast to the Aratea of Germanicus, reports that Chiron instructed Æsculapius in medicine, Achilles in music, and Hercules in astronomy.[25] The principal connexion of Chiron with astronomy is that he is reputed to have been converted by Jupiter into a constellation, and to have become one of the signs of the zodiac.[26]

The statement of Diogenes Laertius, in the second century after Christ, that Musæus was the first who composed a theogony, and invented a sphere,[27] is undeserving of serious attention. Musæus, like Orpheus, belongs exclusively to mythology.

---

(23) Schol. Apoll. Rhod. i. 558; iv. 816; Frag. Hist. Gr. vol. iv. p. 505.

(24) Daimachus, who lived about 270 B.C., stated that Peleus married Philomela, Frag. Hist. Gr. vol. iv. p. 442.

(25) Ap. Arat. vol. ii. p. 87, ed. Buhle.

(26) Eratosth. Catasterism. 40; Hygin. Astr. ii. 38; Ovid, Fast. v. 381-414; Schol. Arat. 436; Hermippus, ap. Frag. Hist. Gr. vol. iii. p. 54.

(27) Prœm. 3. Compare Suid. in Μουσαῖος.

The astronomical knowledge ascribed by Newton to Nausicaa rests on a singular error. It seems that a female grammarian of Corcyra, named Agallis, or Anagallis, attributed the invention of the ball to Nausicaa, the daughter of Alcinous.[28] This patriotic figment was founded upon a passage of the Odyssey, where Nausicaa is described as playing at ball with her maidens.[29] The Greek language borrowed its technical terms from common life, and the same word denoted both *ball* and *sphere:* hence the confusion in question.[30]

§ 2 The earliest historical name with which we can connect the scientific pursuit of astronomy in Greece, upon satisfactory evidence, is that of Thales. Thales, the founder of the primitive Ionic school of philosophy,[31] was a citizen of Miletus. By Herodotus his family is stated to have been of Phœnician origin;[32] by others he was reported to have descended from an indigenous noble stock.[33] His lifetime is said to have extended from 639 to 546 B.C., according to which determination he died at the age of ninety-three years,[34] and survived the usurpation

---

(28) Athen. i. p. 14 D. Suidas in Ἀναγαλλὶς, ὄρχησις, et σφαῖρα.

(29) Od. vi. 115. The identity of Phæacia and Corcyra was assumed by the later Greeks, on the authority of Thuc. i. 25. Sophocles introduced Nausicaa playing at ball in a tragedy, Fragm. 389, ed. Dindorf.

(30) See the arguments of Schaubach against Newton's proof of the antiquity of the Greek astronomy, Gesch. der Griech. Astr. p. 362. For a refutation of the arguments of Newton respecting the epoch of Chiron, see Fréret, Œuvres, vol. x. p. 132. He cites largely from a work of Whiston against Newton's chronology.

(31) Speaking of the early speculations concerning Nature, Aristotle says: ἀλλὰ Θαλῆς μὲν ὁ τῆς τοιαύτης ἀρχηγὸς φιλοσοφίας, &c. Metaph. i. 3. δοκεῖ δὲ ὁ ἀνὴρ οὗτος ἄρξαι τῆς φιλοσοφίας, καὶ ἀπ' αὐτοῦ ἡ Ἰωνικὴ αἵρεσις προσηγορεύθη, Plut. Plac. Phil. i. § 3. A similar statement in Galen, Hist. Phil. c. 2.

Origenes, Philosoph. i. prooem. p. 5, ed. Oxon. says: λέγεται Θαλῆν τὸν Μιλήσιον ἕνα τῶν ἑπτὰ σοφῶν πρῶτον ἐπικεχειρηκέναι φιλοσοφίαν φυσικήν.

Thales Milesius, qui primus de talibus rebus quæsivit, aquam dixit esse initium rerum. Cic. de N. D. i. 10.

(32) i. 170. The Phœnician family to which Thales belonged was said to have been named the Thelidæ. Diog. Laert. i. § 22.

(33) Diog. Laert. ib.

(34) He is said to have died, of heat, thirst, and weakness, while he was present, as a spectator, at gymnastic games, Diog. Laert. i. 39. It may be observed, however, that there is a disposition to attribute extraordinary longevity to the early Greek philosophers. Aristotle stated that

of Pisistratus. His life nearly coincides with the reigns of Ancus, Tarquinius Priscus, and Servius, in the received Roman chronology.

Various particulars mentioned respecting Thales agree with the statements of the ancient chronologers as to the time of his life. Thus he is said to have predicted an eclipse of the sun, which put an end to the battle between the Medes and Lydians under Cyaxares and Alyattes respectively. The reign of Cyaxares is placed from 634 to 594 B.C.; that of Alyattes from 617 to 560 B.C.; and the solar eclipse in question has been fixed by modern astronomers at various dates from 625 to 585 B.C. He is reported to have lived in the time of Thrasybulus of Miletus,(35) who was contemporary with Sadyattes and Alyattes, and also with Periander (about 625—585 B.C.) There is a universal agreement that he was one of the Seven Wise Men;(36) and he is said to have been renowned for his philosophy in the archonship of Damasius, 586 B.C.(37) According to Herodotus, he advised the Ionians to form a federal council, and to establish its seat at Teos, before the subjugation of Ionia.(38) He is reported to have suggested to Crœsus a contrivance for placing his army beyond the Halys in the last year of his reign,(39) which lasted from 560 to 546 B.C.; and to have advised the Milesians not to make an alliance with Crœsus: a policy which saved their city when Crœsus was conquered by Cyrus.(40) There is likewise a

---

Empedocles died at the age of 60; but others made him live to 109, Diog. Laert. viii. 74.

(35) Diog. Laert. i. § 27. A certain Minyes, of whom nothing is known, is cited as the authority for this statement.

(36) Cic. de Leg. ii. 1 says Thales, qui sapientissimus in septem fuit. According to Plut. Solon, 3, he was the only one of the Seven Wise Men who explored the region of Physics.

(37) See Clinton ad Ann.

(38) i. 170. This advice is alluded to by Ælian, V. H. iii. 17. Mr. Grote, vol. iii. p. 346, supposes this advice to have been given before the conquest of Ionia by Crœsus.

(39) i. 75. The story about the Halys was affirmed by the Greeks, but was disbelieved by Herodotus. It is repeated by Diog. Laert. i. § 38, and Lucian, Hippias, c. 2.

(40) Diog. Laert. i. 25. This statement is inconsistent with Herod. i. 141.

story, repeated with many variations, of a gold cup given by Crœsus as a prize to the wisest man, which came into the hands of Thales.(⁴¹) A philosophical rivalry is reported to have existed between Thales and Pherecydes; the latter of whom seems to have flourished about 544 B.C.(⁴²) The opinions of Thales were controverted by Xenophanes,(⁴³) who flourished 540—500 B.C. On the whole, though all dates for Grecian history at this period are uncertain, the active part of the life of Thales may be referred with confidence to the first half of the sixth century B.C.

He is said to have made a visit to Egypt, and to have derived his scientific knowledge, both geometrical and astronomical, from the lessons of the Egyptian priests.(⁴⁴) Hieronymus, a disciple of Aristotle, reported him to have measured the pyramids by the length of their shadows.(⁴⁵) We are told, moreover, by several writers, that he speculated concerning the annual inundation of the Nile, and attributed it to the resistance of the Etesian winds.(⁴⁶) With respect to his astronomical knowledge, the following notices have been preserved.

---

and is probably unfounded. It assumes that Miletus was not reduced by Crœsus, which seems to be inconsistent with the fact.

(41) Diog. Laert. i. 29; Diod. ix. 7. Bekker, Phoenix ap. Athen. xi. p. 495 D.; Plut. Solon, 4; Val. Max. iv. 1. ext. 7; Schol. Aristoph. Plut. 9.

(42) Diog. Laert. i. 122; ii. 46. Suid. in Φερεκύδης. Diogenes recites letters between Pherecydes and Thales, i. 43, 122. Concerning the lifetime of Pherecydes, see C. Müller, Fragm. Hist. Gr. vol. i. p. xxxiv.

(43) Diog. Laert. ix. 18.

(44) The spurious letter from Thales to Pherecydes in Diog. Laert. i. 44, speaks of his intention to visit Egypt, in order to confer with the priests and astronomers of that country. Josephus, contr. Apion, i. § 2, states it to be universally admitted that the earliest Greek speculators upon astronomical subjects, such as Pherecydes of Syros, Pythagoras, and Thales, were scholars of the Egyptians and Chaldæans, and left little in writing. Plut. Sept. Sap. conv. 2, mentions his visit to Egypt. Plut. Plac. Phil. i. 3 says that Thales, having studied philosophy in Egypt, migrated in his old age to Miletus. Clem. Alex. Strom. i. 15, § 66, p. 130, Sylb., states that he had conferences with Egyptian priests. Compare Euseb. Præp. Ev. x. 4.

Pamphila declared that he learnt geometry from the Egyptians, Diog. Laert. i. 24. Pamphila was a learned Egyptian lady, who lived in the time of Nero. Phot. Biblioth. Cod. 175.

(45) Diog. Laert. i. 27. The story is repeated by Pliny, xxxvi. 17. It is differently told by Plut. Sept. Sap. conviv. 2.

(46) Athen. ii. 87, ed. Dindorf.; Diod. i. 38; Diog. Laert. i. 37; Seneca, Quæst. Nat. iv. 2, 31; Galen, ib. 23; Plut. Plac. Phil. iv. 1.

He is stated by Herodotus to have predicted the eclipse of the sun which separated the Median and Lydian armies, under Cyaxares and Alyattes. He is reported to have discovered the seasons, and to have fixed the year at 365 days, having learned this fact from the Egyptian priests:(⁴⁷) to have predicted the solstices, and to have determined the course of the sun from solstice to solstice;(⁴⁸) also to have divided the heaven into five parallel zones, with a meridian cutting them from north to south, and with an oblique zodiac passing through the three interior zones;(⁴⁹) and to have first called the last day of the month the *triacad*, or thirtieth.(⁵⁰)

He is further said to have held that the sun, the moon, and the stars were all of an earthy, or solid, substance, but that the stars were likewise of a fiery nature; and that the moon derived its light from the sun. He attributed an eclipse of the moon to the interposition of the earth between the sun and the moon; and an eclipse of the sun to the interposition of the moon between the sun and the earth.(⁵¹) In accordance with his tenet that water is the principle of all things,(⁵²) he taught that the

---

(47) Diog. Laert. i. 27.   (48) Diog. Laert. i. 23, 24.
(49) Plut. Plac. Phil. ii. 12; Galen, Hist. Phil. c. 12 (vol. xix. ed. Kühn); Stob. Ecl. Phys. i. 23, p. 196, ed. Gaisford. Eudemus, in his History of Astronomy (ap. Theon. Smyrn. c. 40), stated that Thales explained a solar eclipse, and showed that the circuit of the sun through the solstices is not always equal.
(50) Diog. Laert. i. 24.
(51) Plut. Plac. Phil. ii. 13, 24, 28; Galen, Phil. Hist. 13, 15; Stob. Ecl. Phys. i. 24, 25, 26. p. 214, 216. Stobæus, p. 205, has the following passage: Θαλῆς γεοειδῆ [read γεώδη] τὸν ἥλιον· ἐκλείπειν δὲ αὐτὸν τῆς σελήνης ὑπερχομένης κατὰ κάθετον, οὔσης φύσεως γεώδους· βλέπεσθαι δὲ τοῦτο κατοπτρικῶς ὑποτιθέμενον τῷ δίσκῳ. In the corresponding passages, Galen, c. 14, and Plutarch, ii. 24, have ὑποτιθεμένῳ. Wyttenbach ad Plut. corrects αὐτὴν for τοῦτο and ὑποτιθεμένην. The meaning of the passage thus emended would be, ' and the moon is seen by reflexion, being in a direct line with the sun's disc:' which is not intelligible. On the meaning of the word κάτοπτρον, see Galen, ib. c. 25; Plut. ib. iv. 14.

The words in Stobæus appear to be right, and the sense to be: that the shadow of the moon's disc is projected on the sun. According to Cleomedes, ii. 5 p. 134, the most ancient of the physical philosophers and astronomers of Greece knew that the moon derives its light from the sun.

(52) See Brandis, ib. p. 113.

fire of the sun and stars is fed by aqueous exhalations.(53) He, moreover, determined the magnitude of the moon as the 720th part of the sun.(54)

Thales, likewise, in his old age, determined the ratio of the sun's diameter to its apparent orbit. He communicated this discovery to Mandraytus of Priene, who offered him any reward which he might demand. Thales requested only that when Mandraytus published the discovery to others, he would attribute the merit of it to its true author.(55)

Thales supposed the earth to float upon the water, like a plank of wood, or a ship: he is even stated to have explained earthquakes by the fluctuations of the underlying water.(56) Aristotle remarks that Thales conceives the earth to be supported by water; but that he does not explain how the water is supported. Hence it is apparent that the doctrine of the sphericity of the earth is erroneously ascribed to Thales.(57) The doctrine that the earth floats upon water is stated to have been brought from Egypt by Thales.(58)

He is reputed to have taught his countrymen to imitate

---

(53) Plut. Plac. Phil. i. 3.

(54) In the passage concerning the phases and eclipses of the moon, in Stob. p. 217, where Thales is mentioned, his name is omitted in the corresponding passages of Plut. ii. 29, and Galen, c. 15.

(55) This anecdote is related by Apul. Flor. iv. 18, 6.

(56) οἱ δ' ἐφ' ὕδατος κεῖσθαι· τοῦτον γὰρ ἀρχαιότατον παρειλήφαμεν τὸν λόγον, ὅν φασιν εἰπεῖν Θαλῆν τὸν Μιλήσιον, ὡς διὰ τὸ πλωτὴν εἶναι μένουσαν ὥσπερ ξύλον ἤ τι τοιοῦτον ἕτερον. Aristot. de Cœl. ii. 13, 13. Quæ sequitur, Thaletis inepta sententia est. Ait enim terrarum orbem aquâ sustineri, et vehi more navigii, mobilitateque ejus fluctuare, tunc quum dicitur tremere. Seneca, Nat. Quæst. iii. 13. Thales Milesius totam terram subjecto judicat humore portari et innatare: sive illud oceanum vocas, sive magnum mare, sive alterius naturæ simplicem adhuc aquam et humidum elementum. Hac, inquit, undâ sustinetur orbis, velut aliquod grande navigium grave his aquis quas premit, ib. vi. 6. He argued that the earth was too heavy to be supported by air. This notion is mentioned in Schol. Iliad, xiii. 125.

(57) See Plut. Plac. iii. 10; Galen, c. 21. In Plut. iii. 11, it is stated that the followers of Thales placed the earth in the middle of the universe.

(58) Simplicius, Schol. ad Aristot. p. 506 b. ed. Brandis.

the Phœnicians in steering by the Little instead of the Great Bear.(⁵⁹)

The anecdote of Thales falling into the well, with the consequent saying of the Thracian female slave, that in trying to discover things in heaven, he overlooked those beneath his feet, which is related by Plato, proves his popular reputation as an astronomer and stargazer.(⁶⁰) Aristophanes, in the Clouds, names Thales as the abstract type of the geometer and astronomer:(⁶¹) and Timon the Sillographer, about 280 B.C. relaxed the customary severity of his satirical poetry in order to commend the astronomical attainments of Thales.(⁶²)

According to Pamphila, he first solved the problem of inscribing a right-angled triangle in a circle:(⁶³) and he is described in general terms as the founder of geometrical science in Greece.(⁶⁴)

The death of Thales preceded the manhood of Herodotus by about a century, and his birth preceded it by nearly two centuries. He left nothing in writing; a work on 'Nautical Astronomy,' attributed to him, was considered the production of Phocus or Phocas, a Samian.(⁶⁵) Hence the accounts both

---

(59) Callimach. Fragm. 94; Diog. Laert. i. 23; Schol. Il. xviii. 487; Hygin. Poet. Astr. ii. 2. See below, ch. viii. § 1.

(60) Plat. Theætet. § 79, p. 174, repeated by Diog. Laert. i, 34, Origen. Philos. p. 5, and alluded to by Tertullian de Anim. c. 6.
Mariana, xiii. 20, alludes to the anecdote of Thales in speaking of Alfonso, king of Castille, the author of the Alphonsine Tables. 'Erat Alfonso sublime ingenium, sed incautum, superbæ aures, lingua petulans, literis potius quam civilibus actis instructus, dumque cœlum considerat, terram amisit.' Compare Bayle, Dict. art. Castille, note G.

(61) Nub. 180, where the Scholiast speaks of Thales as τὰ περὶ τὸν οὐρανὸν πρῶτος ἐξευρών. Av. 1009.

(62) Diog. Laert. i. 34. An elegiac inscription under his statue, in which his astronomical fame is commemorated, is recited by Diogenes. Ibid.

(63) Diog. Laert. i. 24.

(64) Apul. Flor. iv. 18, 5. Compare Brandis, ib. p. 110.

(65) Ναυτικὴ Ἀστρολογία, Diog. Laert. i. 23. The name Phocus occurs in Plut. Solon, 14. It is possible that the 200 hexameter verses ascribed by Lobo the Argive to Thales (Diog. Laert. i. 34) refer to this production. Compare Brandis, Griech. Röm. Phil. vol. i. p. 111. He was likewise stated to have written a treatise concerning the solstices and equinoxes,

of his life and doctrines which reached the earliest historians, were confused and inaccurate, or alloyed with fable.

It is difficult to draw any certain conclusions, respecting the astronomical science of Thales, from such loose and incoherent notices as have descended to us. His visit to Egypt, like other journeys to foreign countries, attributed to Greek philosophers and lawgivers, is probably apocryphal; and if Thales profited by the lessons of the Egyptian priests in geometry, it is not likely that he should have taught them a mode of measuring the height of the pyramids. The opinion that the annual inundation of the Nile is produced by the Etesian winds is mentioned by Herodotus, but is not ascribed to Thales.[66]

Even if Thales had expressed this opinion respecting the cause of the annual inundation of the Nile, it would not prove that he had visited Egypt. This phenomenon had at an early period roused the curiosity of the Greeks, and had become a favourite subject of speculation among their writers. There were two circumstances in it which excited the wonder of the Greeks. 1. Their rivers swelled in winter, and were nearly dry in summer:[67] whereas the Nile rose in the greatest heat of summer, at the rising of the Dog-star. 2. The Greek rivers, being torrents, were destructive by their inundation; whereas the productiveness of Egypt depended on the inundation of the Nile.[68] Accordingly, many of their philosophers and historians advanced theories upon it.[69]

Diog. Laert. i. 23. Plutarch speaks with doubt as to the genuineness of the treatise on astronomy ascribed to Thales, De Pyth. Orac. c. 18.

(66) Herod. ii. 20.

(67) See Kruse's Hellas, vol. ii. p. 297; Grote's Hist. of Gr. vol. ii. p. 286. Lucan says of the Nile:
'Inde etiam leges aliarum nescit aquarum;
Nec tumet hibernus, cum longe sole remoto
Officiis caret unda suis: dare jussus iniquo
Temperiem cœlo, mediis æstatibus exit.'—x. 228-231.

(68) See Pauly, Real-Lex. vol. v. p. 645. The Romans considered the inundations of rivers as unfavourable omens, Virg. Georg. i. 481; Horat. Carm. i. 2.

(69) See Herod. ii. 19; Diod. i. 38; Athen. ii. c. 87, ed. Dindorf;

If the statement, attributed to Eudemus, that Thales predicted the solstices,(70) means that he determined the exact times of their recurrence, he must have known the true length of the tropical year, without being indebted for this information to the Egyptian priests. The same conclusion would likewise follow from his determination of the sun's path in the zodiac. Whether he approximated so nearly to the true length of the year as to fix it at 365 days, is uncertain; it is still more doubtful whether he departed so far from the original notion as to conceive the earth to be an immovable sphere in the centre of the universe. The statement that he 'discovered the seasons,' is absurd: it supposes the Greeks of the sixth century B.C. to be in a state of childish ignorance.

§ 3 The alleged prediction of the solar eclipse by Thales has given rise to a great variety of opinions. The account of Herodotus (71) is that while Cyaxares was king of Media, and Alyattes was king of Lydia, these two neighbouring nations were engaged in a war, which had lasted, with alternate fortune, for five years: and that in the sixth year a battle took place, during which the day suddenly became night. Herodotus designates the battle as a 'night battle;' and he adds that this change was foretold to the Ionians by Thales, who fixed the year of the battle as the period within which the eclipse would occur. The narrative proceeds to relate that when the Medes and Lydians saw that night had taken the place of day, they desisted from the combat, and were eager to

---

Schaubach, Anaxagor. p. 179; Mullach, Democrit. p. 395; Strab. xvii. 1, § 5; Schol. Apoll. Rhod. iv. 269; Seneca, Nat. Quæst. iv. 2; Lucan, x. 219; Galen, Phil. Hist. c. 23; Plut. Plac. Phil. iv. 1.

(70) Diogenes professes to cite it from the History of Astronomy by Eudemus, i. 23.

(71) i. 74. Alluded to by Diog. Laert. i. 23. Mr. Rawlinson holds that authentic Median history begins with Cyaxares, Herod. vol. i. p. 410, 416. Respecting the series of Lydian kings, see Rawlinson, ib. p. 353; Clinton, F. H. vol. ii. p. 266. Respecting the series of Median kings, Clinton, vol. i. p. 257, Strabo, xvii. 1, § 18, states that Cyaxares was contemporary with Psammetichus. According to the Egyptian chronology, Psammetichus reigned from 670 to 616 B.C. This synchronism does not agree with the calculated dates of the eclipse.

make peace with one another: that peace was made through the mediation of Syennesis the Cilician and Labynetus the Babylonian, and that it was cemented by the marriage of Aryenis the daughter of Alyattes, with Astyages the son of Cyaxares.

The statement that this eclipse was predicted by Thales was repeated by Eudemus in his History of Astronomy.([72]) It recurs in Cicero,([73]) Pliny,([74]) and Themistius.([75])

Historical testimony to this occurrence may be considered as resolving itself into the account of Herodotus. It is highly improbable that Eudemus, who wrote about 300 B.C., should have had any independent information on the subject; and the other later writers doubtless followed the Herodotean narrative, either at first or second hand. The Lydian origin of the Etruscans, reported by Herodotus, is in like manner repeated by numerous ancient authors, and it even became an article of national faith; but their repetition, as Schwegler has remarked, adds nothing to the authority of the original statement.([76])

Now Herodotus was born in 484 B.C., and may be supposed to have begun the collection of the materials for his history about 455 B.C. If we suppose the eclipse to have taken place in 585 B.C., the interval would be 130 years: if we suppose it to have taken place in 610 B.C., the interval would be 155 years. This period is within the possibility of accurate oral transmission: but the chances against the preservation of the exact truth during so long a time without contemporary registration are preponderant.

The account of Herodotus consists of two parts: 1. The

---

(72) Ap. Clem. Alex. Strom. i. 14, § 65, p. 130, Sylb.; referred to by Diog. Laert. i. 23; and see above, note 49.

(73) De Div. i. 49, who names Astyages instead of Cyaxares, by an error of memory. In Rep. i. 16, he states that the true nature of eclipses was understood by Thales.

(74) N. H. ii. 9. He fixes the date at Olymp. 48. 4, 170 U.C.=585-4 B.C. in the reign of Alyattes.

(75) Orat. xv.

(76) Röm. Gesch. vol. i. p. 253.

occurrence of the eclipse, and of its historical accompaniments. 2. The prediction of the eclipse by Thales.

The time of this eclipse has been variously determined both by ancients and moderns; but the limits of divergence are not wide. Pliny fixes it at the year 585 B.C.(77) Clemens of Alexandria places it about the fiftieth Olympiad, 580 B.C.(78) Eusebius assigns it to Olymp. 48. 2, 583 B.C. Among the moderns, Petavius and others place this eclipse in 597 B.C.;(79) while Oltmanns and Baily, with the approbation of Ideler, fix it at Sept. 30, 610 B.C.(80) The Astronomer Royal, Mr. Airy, in a recent investigation has determined its date at May 28, 585 B.C.;(81) which date nearly agrees with those of Pliny, Clemens, and Eusebius, and had been previously adopted by Scaliger, Des Vignoles, and others.

A solar eclipse, if the sun's disc were partly visible, and the event happened during a battle, would be sufficient to terrify both armies, and to cause them to believe that the gods signified their anger at the conflict. But unless the eclipse is total, the mere diminution of light is not sufficient to create alarm; and the totality of a solar eclipse is of short duration. It can never exceed four minutes. In the eclipse of the sun which occurred on 18th July, 1860, and which was total in the north of Spain, the totality lasted three minutes, and it was only during this short period that darkness prevailed.(82) The expression of Herodotus, 'a night battle,' which implies that the darkness was of some duration, must be inaccurate.

An event so striking as the occurrence of a visible eclipse of

---

(77) ii. 12. He fixes the time by two dates, viz., Olymp. 48. 4, and 170, U.C.

(78) Strom. i. 14, § 65.

(79) Doct. Temp. x. i.

(80) Berlin Transactions, 1812, and Philosophical Transactions, 1811. Compare Ideler, Chron. vol. i. p. 209.

(81) On the eclipses of Agathocles, Thales, and Xerxes, Phil. Trans. 1853, p. 179. Compare Costard, on the Eclipse of Thales, Phil. Trans. Abridg. vol. x. p. 310 (1758).

(82) Compare Delambre, Hist. d'Astron. Moderne, vol. i. p. 601.

the sun during a battle, (especially if it were total, or nearly total,) could not fail to make a profound impression at the period in question; and its preservation by oral tradition until the time of Herodotus is not inconceivable, or even improbable. At seasons of war and danger the minds of men are peculiarly susceptible of superstitious fears;([83]) as we may perceive in the measures of Nicias respecting the eclipse of the moon at the siege of Syracuse.

The connexion of Thales with the eclipse is subject to greater doubts than the occurrence of the eclipse itself.([84])

Herodotus states that Thales merely predicted the *year* within which the eclipse was to occur. Now if he was able to predict the eclipse at all, it seems incredible that he should not have been able to predict it within narrower limits than a year. He is likewise reported to have predicted it to the *Ionians*. If he had predicted it to the Lydians, in whose country the eclipse was to be total, his conduct would be intelligible: but it seems strange that he should have predicted it to the Ionians, who had no direct interest in the event. Other predictions of physical occurrences, which exceed the powers even of modern science, and which are certainly fabulous, were also ascribed by ancient authors to Thales. Thus he is related to have known from his astronomical science in winter, that there would be a large crop of olives in the ensuing year; and, having been taunted with the inutility of his philosophy, to have hired all the oil-presses in Miletus and Chios([85]) at a low rent, and to have made a large profit by letting them when the olive-crop

---

([83]) Livy says of the first year of the Second Punic War: 'Romæ aut circa urbem multa eâ hieme prodigia facta; aut (quod evenire solet, motis semel in religionem animis) multa nunciata et temere credita sunt.' xxi. 62.

([84]) Martin, Timée de Platon, tom. ii. p. 109, thinks that Thales pretended to have predicted the eclipse, after it had occurred, or that the story of his prediction was invented after his death. The prediction of Thales is likewise doubted by Lalande, Astron. § 184, 201.

([85]) It is difficult to understand why Chios should be named. Samos is nearer than Chios to Miletus; but neither island belonged to Miletus in the time of Thales. The words καὶ Χίῳ appear to be corrupt.

was gathered.(86) In a fabulous account of the preservation of Crœsus from death by fire, in the remains of Nicolaus of Damascus, the storm which extinguished the flames is declared to have been foretold by Thales.(87) The fall of a large aërolite at Ægospotamos in Thrace is, in like manner, stated to have been predicted by Anaxagoras;(88) who is also said to have foretold the fall of a house, and the occurrence of a storm of rain during the celebration of the games at Olympia.(89) Democritus saved a portion of the harvest of his brother Damasus by the timely prediction of a violent storm of rain.(90)

It is, moreover, stated that Anaximander predicted an earthquake at Sparta; which shortly afterwards occurred, and was accompanied by the fall of a part of Mount Taygetus upon the town. Pherecydes of Syros is likewise reported to have predicted an earthquake from a draught of water from a well.(91) Other marvellous powers over nature are attributed to the early philosophers. Thus Empedocles is stated to have

---

(86) Aristot. Pol. i. 11. Aristotle does not indeed give entire credit to this story; he says that it is a contrivance of general application which is attributed to Thales on account of his wisdom. The story was repeated by Hieronymus, in his ὑπομνήματα, about 250 B.C. Diog. Laert. i. 26 (see Fragm. Hist. Gr. vol. ii. p. 450), and is also shortly narrated by Cic. de Div. i. 49. Pliny, N. H. xviii. 68, tells it of Democritus: he says that Democritus predicted the abundant olive-crop from the rising of the Pleiads, and that he bought up all the olives.

(87) Fragm. Hist. Gr. vol. iii. p. 409. Nicolaus lived in the age of Augustus.

(88) See Schaubach, Anaxag. Fragm. p. 41. The prediction of this occurrence is expressly mentioned by Plut. Lysand. 12; Diog. Laert. ii. 10; Plin. N. H. ii. 59, who doubts the possibility of the prediction, as being beyond the reach of the human mind. Ammian. Marcell. xxii. 8, § 5.

(89) Philostrat. vit. Apollon. i. 2, § 2. Philostratus also mentions the prediction of an eclipse, as well as of the fall of stones at Ægospotamos. For a story of Simonides receiving a divine warning of the fall of a house, see Cic. de Orat. ii. 86.

(90) Plin. xviii. 78; Clem. Alex. Strom. vi. 13, § 32.

(91) Plin. N. H. ii. 81. Plutarch mentions that in the earthquake which occurred at Sparta, in the fourth year of the King Archidamus II. (464 B.C.) some pinnacles of Taygetus were shaken down, Cimon, 16. The same earthquake is alluded to by Thuc. i. 101. The prediction of the earthquake from putealis aqua is attributed to Anaxagoras by Ammian. Marcell. xxii. 16.

moderated the destructive violence of the Etesian winds by forming a screen of asses' skins on the hills; whence he obtained the epithet of *Wind-averter* (κωλυσανέμας).([92])

§ 4 The legislation of Solon is placed in 594 B.C., and he was one of the Seven Wise Men whose flourishing period is referred to the year 586 B.C. He may be regarded as the contemporary of Thales.([93]) He is stated to have reformed the Athenian calendar, by bringing the length of the lunar month into harmony with the sun; and by calling the thirtieth day of the month ἕνη καὶ νέα, because it belonged partly to the preceding and partly to the succeeding month.([94]) The meaning of this reform appears to be, that, whereas the Athenian year had previously consisted of twelve months, of thirty days each, the result of which was that the months did not coincide with the moon, Solon made the year consist of alternate months of twenty-nine and thirty days; so that the year consisted of 354 days, and thus coincided, within about nine hours, with the true lunar year; and each two successive months were equal to two periodical months within about 1½ hour.([95]) The reform of Solon, as reported to us, was limited to the month; it was intended only to make the months harmonize with the moon. It is not stated by what method of intercalation Solon brought a year of 354 days into agreement with the sun:

---

(92) Diog. Laert. viii. 60; Suid. in Ἐμπεδοκλῆς, Ἀμύκλαι et ἄπνους, Plutarch de Curios. 1, cont. Colot. 32 ; Hesych. in κωλυσανέμας. Clem. Alex. Strom. vi. 3, § 60, p. 267, Sylb.; Porphyrius and Iamblichus, in their lives of Pythagoras, state that he was called ἀλεξάνεμος; Iamb. § 136; Porph. § 29. Compare Sturz, Empedocles, p. 48; Karsten, Phil. Gr. Rel. vol. ii. p. 21. Martin, Etudes sur le Timée de Platon, tom. ii. p. 109, remarks: 'Les écrivains grecs les plus graves ont répété des contes populaires où ces premiers philosophes étaient considérés comme des espèces de sorciers.'

(93) Clinton conjectures that the life of Solon may have extended from 638 to 558 B.C., F. H. vol ii. p. 301; but the exact times both of his birth and death are uncertain. A spurious epistle from Thales to Solon is in Diog. Laert. i. 44.

(94) Plut. Sol. 25; Diog. Laert. i. 57; Proclus in Tim. i. p. 25; Schol. Aristoph. Nub. 1131; Varro, L. L. vi. 10, on the expression ἕνη καὶ νέα as applied to the moon, see Plat. Cratyl. 24, p. 409.

(95) See Ideler, Chron. vol. i. p. 266.

though it cannot be doubted that some method, more or less precise, was employed for this purpose at the period in question. As we shall see below, the earliest system of intercalation practised by the Greeks in order to rectify the year of twelve lunations was the insertion of an intercalary month in every second year.(96)

The reform of Solon rendered the year less accurate than it had previously been. He reduced it from 360 to 354 days, and thus, finding it 5¼ days too short, he made it six days shorter. His object was not to make the year coincide with the sun, but to make the months coincide with the moon. It appears not to have been attained in practice, owing to the imperfect knowledge of astronomy at this period.(97) Hence the reform of Solon, while it failed in adjusting the civil months to the moon—an object of no real importance in a calendar—caused the civil year to deviate more widely from the natural year as measured by the sun.

§ 5 The next after Thales in the series of Ionic philosophers was Anaximander of Miletus. His birth is placed at 610, and his death at 547 B.C.(98) He is reported to have been the leader of the Milesian colony which founded Apollonia, on the Euxine Sea:(99) but this statement is inconsistent with the date assigned by the ancients for the foundation of Apollonia, namely, 609 B.C.(100)

He is called both the disciple and companion of Thales,(101) and was his junior by about thirty years. He left a statement

---

(96) Compare Clinton, F. H. vol. ii. p. 336; Ideler, ib. p. 270.

(97) See Clinton, F. H. vol. ii. p. 336, note e, and below, ch. iv. § 5; Thuc. ii. 28, uses the expression νουμηνία κατὰ σελήνην, in order to designate the time of a solar eclipse. This marks that the νουμηνία of the calendar did not always agree with the real new moon.

(98) Apollodorus, ap. Diog. Laert. ii. 8. Plin. N. H. ii. 8, agrees as to the date.

(99) Ælian, V. H. iii. 17.

(100) Clinton, ad Ann. Compare Clinton in Philol. Mus. i. p. 89; Müller, Hist. of Gr. Lit. vol. i. p. 321.

(101) Brandis, ib. p. 123.

in prose of his philosophical doctrines; having been the first Greek who resorted to this method of communication.([102]) He is said to have written on geography, and on the fixed stars, and to have framed a celestial sphere, or representation of the starry heaven, as well as a map of the earth.([103]) He is likewise stated to have invented the gnomon, or sun-dial, and to have set up one at Sparta, which showed the time, the seasons, the solstices, and the equinoxes.([104])

Anaximander held that the stars were attached to moveable spheres, whence their motions were derived:([105]) likewise that the fiery ether of the heaven, by the velocity of its circular motion, carried up stones from the earth, and converted them into stars.([106]) He conceived, moreover, that the stars were circular bodies of condensed air, containing fire, which escaped through certain apertures: that the sun occupied the highest place in the universe; that after him was the moon; and that the fixed stars and planets came next in order.([107])

---

(102) Brandis, ib. p. 124.

(103) Suid. in 'Αναξίμανδρος; Strab. i. 1, 24; Diog. Laert. ii. 2; Agathemer. i. 1.

(104) Diog. Laert. ii. 1; Euseb. Præp. Ev. x. 14; Plin. vii. 56. Pliny attributes the invention of the gnomon, and the establishment of the sun-dial at Sparta to Anaximenes, ii. 76, apparently confounding the names. Suid. in 'Αναξίμανδρος, ἡλιοτρόπιον, and γνώμων.

(105) Stob. Phys. i. 24, p. 201, Gaisf.; Plut. Plac. Phil. ii. 16; Galen, c. 13. The words ὑφ' ὧν ἕκαστος βέβηκε appear to imply the existence of more than one sphere.

(106) Galen de Phil. Hist. c. 13.

(107) Plut. Plac. ii. 15; Galen, Phil. Hist. c. 13. 'Αναξίμανδρος καὶ Μητρόδωρος ὁ Χῖος καὶ Κράτης ἀνωτάτω μὲν πάντων τὸν ἥλιον τετάχθαι, μετ' αὐτὸν δὲ τὴν σελήνην, ὑπὸ δ' αὐτοὺς τὰ ἀπλανῆ τῶν ἄστρων καὶ τοὺς πλανήτας. Stob. Phys. i. 24, agrees with respect to Anaximander, Plut. ib. 20. 'Αναξίμανδρος κύκλον [τοῦ ἡλίου] εἶναι ὀκτωκαιεικοσαπλασίονα τῆς γῆς ἁρματείου τροχοῦ τὴν ἁψῖδα παραπλήσιον ἔχοντα κοίλην, πλήρη πυρός. ἧς κατά τι μέρος ἐκφαίνειν διὰ στομίου τὸ πῦρ, ὥσπερ διὰ πρηστῆρος αὐτοῦ, καὶ τοῦτ' εἶναι τὸν ἥλιον. Differently in 21. Origen, Phil. p. 11. εἶναι δὲ τὸν κύκλον τοῦ ἡλίου ἑπτακαιεικοσιπλασίονα τῆς σελήνης, καὶ ἀνωτάτω μὲν εἶναι τὸν ἥλιον, κατωτάτω δὲ τοὺς τῶν ἀπλανῶν ἀστέρων κύκλους. In the passage of Origen, Röth, Gesch. der Phil. n. 139, emends τοῦ τῆς σελήνης, and ἀνωτέρω μὲν εἶναι τὸν ἥλιον, ἀνωτάτω δὲ τὸν τῶν ἀπλανῶν ἀστέρων κύκλον. The alteration is inadmissible, as the same statement occurs in Plutarch and Stobæus; but error in the reporter may be suspected.

Metrodorus of Chios was a disciple of Democritus, and lived about 330 B.C.

According to Eudemus, in his History of Astronomy, Anaximander was the first to treat of the magnitudes and distances of the planets.([108]) He compared the sun to a wheel; he held that its rays were emitted from a cavity, and that they diverged from this centre like spokes.([109]) His doctrine respecting the sun was, that it is of circular form, with an opaque annular band on its exterior, the circumference of which is twenty-eight times that of the earth; that within this annular band is a fiery central portion, equal in size to the earth; that the movement of the sun is due to its opaque ring; and that an eclipse of the sun takes place when the central aperture is closed.([110]) His hypothesis regarding the moon was similar: he supposed the luminous centre to be seen through a tube, like the mouth of a bellows; that the eclipses (or phases) are caused by the revolutions of the opaque ring; and that the outer circumference of the ring is twenty-nine times that of the earth.([111]) Hence Anaximander held that the moon shone by her own light, though that light was faint and thin.([112])

Anaximander is further declared to have held that the earth is a cylinder, whose length is three times its thickness, freely suspended in space,([113]) and immovable in the centre of the uni-

---

(108) Simplicius ad Aristot. de Cœl. 497 a, Brandis.

(109) Achill. Tat. c. 19, p. 81.

(110) Stob. Phys. i. 25, p. 203; Plut. Plac. Phil. ii. 20, 21, 24; Galen, Hist. Phil. 14, who states the ratio at twenty-seven to one. Origen, Philosoph. p. 11.

(111) Stob. Phys. i. 26, p. 213, 216; Plut. ib. ii. 25; Galen, c. 15; Origen, Phil. p. 11. In Plut. ii. 29, Ἀναξιμένης τοῦ στομίου τοῦ περὶ τὸν τροχὸν ἐπιφραττομένου, read Ἀναξίμανδρος, from Stobæus and Galen. For an example of πρηστήρ in the sense of *bellows*, see Apollon. Rhod. iv. 777.

(112) Stob. ib. p. 215; Plut. ii. 28; Galen, c. 15. On the other hand, Diog. Laert. ii. 1, states him to have held that the moon is not of a luminous nature, and that her light is derived from the sun.

(113) Plut. ap. Euseb. Præp. Ev. i. 8; Plut. Plac. iii. 10, where Brandis properly reads λιθίνῳ κίονι, Galen, ib. c. 21; Aristot. de Cœl. ii. 13; Origen, Phil. p. 11, has the following clause: τὴν δὲ γῆν εἶναι μετέωρον ἐπ' οὐδένος κρατουμένην, μένουσαν διὰ τὴν ὁμοίαν πάντων ἀπόστασιν. τὸ δὲ σχῆμα αὐτῆς ὑγρόν, στρογγύλον, χίονι λίθῳ παραπλήσιον. The word ὑγρόν seems redundant; and for χίονι λίθῳ read κίονι λιθίνῳ, as in Plutarch. Martin, Etudes sur le Timée de Platon, tom. ii. p. 127, is of the same opinion. Röth, Gesch. der Phil. ii. 2, note 133, reads τροχὸν for ὑγρόν; but this reading is

verse.[114] Others, apparently with less accuracy, stated him to have held that the earth was not cylindrical, but spherical in form.[115] He is moreover stated to have discovered the obliquity of the zodiac.[116]

He accounted for the suspension of the earth in the centre of the universe by saying that, being equidistant from the containing heaven in every direction, there was no reason why it should move in one direction rather than in another.[117]

As Anaximander reduced his physical doctrines to writing, there is more probability of their having been correctly preserved than those of Thales. We may likewise believe that he was among the first Greeks who attempted to delineate the celestial sphere, to make a map of the world, and to measure time by a sun-dial. It was in 500 B.C. that Aristagoras of Miletus exhibited to the Spartans a brazen plate on which a map of the world was engraved.[118] This was at that time a novelty to the Spartans; and Aristagoras had probably derived it, directly or indirectly, from the labours of his fellow-townsman, Anaximander.

§ 6 Anaximenes of Miletus was the third in the series of Ionic philosophers. He is called the disciple, companion, and successor of Anaximander,[119] which supposes him to have been born about 475 B.C. He is likewise stated to have been the teacher of Anaxagoras, who was born in 500 B.C.[120] If both these statements are true, it would follow that Anaximenes

---

inconsistent with the context, according to which the shape of the earth is cylindrical.

(114) In an extract of the History of Astronomy by Eudemus, cited from Dercylides by Theo Smyrnæus, c. 40, p. 323, ed. Martin (compare Fabr. Bibl. Gr. ed. Harless, vol. iii. p. 462) the following clause occurs: Ἀναξίμανδρος δὲ ὅτι ἐστὶν ἡ γῆ μετέωρος, καὶ κινεῖται περὶ τὸ τοῦ κόσμου μέσον. Montucla reads κεῖται for κινεῖται, with the approbation of Ukert, Geogr. ii. 1. p. 20. Bœckh agrees with Ideler that Anaximander did not hold the doctrine of the rotation of the earth on its axis. Philolaos, p. 122, note.

(115) Diog. Laert. ii. 1.   (116) Plin. ii. 8.
(117) Aristot. de Cœl. ii. 13, § 25.
(118) Herod. v. 49.
(119) Brandis, ib. p. 141; Clinton, F. H., vol. ii. ad ann. 548.
(120) Clinton, ib. ad ann. 500, 480.

reached an unusual age. The active period of his life may be placed in the last half of the sixth century B.C., which would make him nearly the contemporary of Xenophanes the philosopher, Hecatæus the historian, and Anacreon the poet.

The following cosmical and astronomical tenets of Anaximenes are reported to us: that the extreme part of the heavenly sphere is earthy, or solid;([121]) that the stars are of an igneous nature, but that there are certain solid substances, invisible to our eyes, which are carried round with them; that the stars are fixed, like nails or studs, in the crystalline sphere; that the summer and winter seasons are denoted exclusively by the sun, and that they are not also marked by the moon and stars, as was held by other philosophers; that the stars revolve not under but round the earth; that the sun does not descend beneath the earth, but that its light is intercepted by lofty mountains, when its distance is great;([122]) that the substance of the sun is fire, and that in shape it is flat, like a leaf; that the stars, being impelled by the resistance of the condensed air, cause the solstitial movements of the sun;([123]) that the nature of the moon is igneous, and therefore that she shines with her own light;([124]) that the earth is a flat trapezium; and that on account of this form it is supported by the air without sinking.([125])

([121]) Stob. Phys. i. 23; Plut. Plac. ii. 11; Galen c. 12. The correct version of this sentence appears to be that of Galen: Ἀναξιμένης τὴν περιφορὰν τὴν ἐξωτάτην γηίνην εἶναι, with which Plutarch nearly agrees. The version of Stobæus gives no clear sense: Ἀναξιμένης καὶ Παρμενίδης τὴν περιφορὰν τὴν ἐξωτάτω τῆς γῆς εἶναι τὸν οὐρανόν.

([122]) Stob. i. 24, p. 199; Plutarch, ii. 14, 19; Galen, c. 13; Origen, Philos. p. 12. According to the latter writer the stars turn round the earth, ὡσπερεὶ περὶ τὴν ἡμετέρην κεφαλὴν στρέφεται τὸ πιλίον.

The opinions attributed to Anaximenes in the latter passages are very obscure. Compare Röth, Gesch. der Abendl. Philosophie, vol. ii. part 2, note 298.

([123]) Stob. Phys. i. 25, p. 203; Plut. Plac. ii. 22, 23; Galen, c. 14. The meaning of the latter clause seems to be that the sun, when it is furthest from the equator, approaches the stars at the poles, and squeezes the air against them: and that their resistance to this pressure turns the sun back in its course.

([124]) Stob. i. 26, p. 213.

([125]) Plut. Plac. iii. 10; Plut. ap. Euseb. Præp. Ev. i. 8; Galen, c. 21; Aristot. de Cœl. ii. 13, § 16.

He applied the same doctrine to explain the suspension of the sun and moon in space.([126])

§ 7 Heraclitus of Ephesus, who may be placed in the line of the Ionic philosophers, is stated to have flourished about 504 B.C.([127]) The active part of his life probably belonged to the last part of the sixth and the first part of the fifth century B.C. He may be considered as nearly contemporary with Æschylus. The obscurity of the written style in which he expressed his philosophical opinions became proverbial.

The following doctrines upon celestial matters are ascribed to him :—

That the heaven is of an igneous nature; ([128]) that the stars are formed of compressed fire, and are fed by exhalations from the earth; ([129]) that day and night, summer and winter, are caused by the prevalence of bright or obscure, warm or moist, exhalations in the sun; ([130]) that the shape of the sun is that of a bowl, or hollow hemisphere; that a solar eclipse takes place when the luminous convexity is turned upwards, and the dark concavity is turned towards the earth; that the real magnitude of the sun is not greater than its apparent magnitude, and that it is no greater than the width of a man's foot: ([131])

([126]) Origen, Philosoph. p. 12. τὴν δὲ γῆν πλατεῖαν εἶναι ἐπ᾽ ἀέρος ὀχουμένην, ὁμοίως δὲ καὶ ἥλιον καὶ σελήνην καὶ τὰ ἄλλα ἄστρα· πάντα γὰρ πύρινα ὄντα ἐποχεῖσθαι τῷ ἀέρι διὰ πλάτος.

According to Eudemus ap. Theon. Smyrn. 40, p. 324, ed. Martin, Anaximenes held that the moon receives her light from the sun, and explained lunar eclipses.

([127]) He mentioned Pythagoras, Xenophanes and Hecatæus, Diog. Laert. ix. 1.

For an account of the philosophy of Heraclitus, and for a collection of his fragments, see Schleiermacher in the Museum der Alterthums-wissenschaft (Berlin, 1807) vol. i. p. 315—533.

([128]) Stob. Phys. i. 23.

([129]) Stob. i, 24; Galen, c. 13; Plut. Plac. ii. 17.

([130]) Diog. Laert. ix. 11. He held that there were exhalations from the land and sea, some pure and bright, and some dark, Diog. Laert. ix. 9.

([131]) Stob. i. 25, p. 204; Plut. ii. 21, 22, 24; Galen, c. 14; Diog. Laert. ix. 7.

Stobæus ascribes to Heraclitus and Hecatæus the tenet that the sun

that the moon, like the sun, is bowl-shaped: ([132]) that the eclipses, and also the phases, of the moon are caused in the same manner as the eclipses of the sun—namely, by the reversion of the hollow hemisphere, so that the luminous side is turned away from the earth, either completely, or at an angle: ([133]) that the stars have the same configuration: that the superior luminousness of the sun is owing to its position in a clear atmosphere, remote from the earth, and the inferior luminousness of the moon is owing to its position in a turbid atmosphere, nearer the earth.([134]) He laid it down that the north, or the Great Bear, is the boundary between the east and the west; and that opposite the north is the province of Jupiter, who presides over the clear sky.([135])

---

is an ἄναμμα νοερὸν τὸ ἐκ θαλάττης. Galen attributes to Heraclitus the tenet that the sun is an ἄναμμα ἐν μὲν ταῖς ἀνατολαῖς τὴν ἔξαψιν ἔχον, τὴν δὲ σβέσιν ἐν ταῖς δυσμαῖς. Stobæus likewise ascribes to Cleanthes the tenet that the sun is an ἄναμμα νοερὸν τὸ ἐκ θαλάττης, p. 206. Plutarch, ii. 20, attributes it to the Stoics generally, and not to Heraclitus. According to Diog. Laert. vii. 145, Zeno held that the sun was fed ἐκ τῆς μεγάλης θαλάττης νοερὸν ὄντα ἄναμμα. It seems that there is some confusion here, and that the opinion in question does not belong to Heraclitus. Compare Schleiermacher, ib. p. 399.

(132) Stob. i. 26, p. 213, 216; Plut. ii. 27, 28; Galen, c. 15. Plutarch, ii. 25, and Galen have the clause: Ἡράκλειτος γῆν ὁμίχλῃ περιειλημμένην, concerning the moon. Stobæus has: Ἡρακλείδης καὶ Ὄκελλος γῆν ὁμίχλῃ περιεχομένην. Considering the opinion of Heraclitus respecting the bowl-shaped figure of the moon, it seems certain that the reading of Stobæus is correct, and that this tenet belongs to Heraclides.

(133) Diog. Laert. ix. 9, 10. Stobæus, i. 26, p. 216, has the following clause: Ἀλκμαίων, Ἡράκλειτος, Ἀντίφαντος κατὰ τὴν τοῦ σκαφοειδοῦς στροφὴν, καὶ τὰς περικλίσεις. In Galen, c. 15, and Plutarch, ii. 29, it stands thus: Ἡράκλειτος κατὰ τὴν τοῦ σκαφοειδοῦς συστροφήν. Ἀντίφαντος is a corrupt form, for which Ἀντιφῶν is restored by Heeren. στροφὴν in Stobæus is correct: συστροφὴν in Galen and Plutarch is an error, and does not give the required sense. The word περίκλισις is not in the dictionaries, but its meaning is clear. Compare Achill. Tat. c. 21, p. 83. κατὰ μῆνα δὲ ἐκλείπει [ἡ σελήνη], ὡς μὲν Ἡράκλειτός φησιν, ὁμοίως τῷ ἡλίῳ τοῦ φωτοειδοῦς σχήματος ἀναστράφεντος.

(134) Stob. ib. p. 216; Plut. ii. 28 (whose version is the best); Galen, c. 15.

(135) Stob. i. 1, § 6. Ζεὺς αἴθριος.
Compare the verse of Theocritus:

Χὡ Ζεὺς ἄλλοκα μὲν πέλει αἴθριος, ἄλλοκα δ᾽ ὕει.
Id. iv. 43.

Jupiter was supposed to preside peculiarly over rain and fine

Heraclitus is likewise stated to have formed a long period of 18,000 solar years;(¹³⁶) but with what view, whether astronomical or chronological, does not appear.

§ 8 The accounts as to the lifetime of Xenophanes of Colophon, the founder of the Eleatic School, are conflicting, but his flourishing period may be placed with probability from 540 to 500 B.C.(¹³⁷)

Some of his cosmological tenets have been preserved. He held that the stars consisted of fiery clouds; that they are extinguished every day, and are lighted again at night, like coals: and that this alternation of flame is the cause of their risings and settings.(¹³⁸) He believed the sun to be formed in the same manner as the stars; and to be renewed every day. He conceived that a solar eclipse is caused by the sun's extinction; and he related that an eclipse of the sun had once lasted an entire month, during which time there had been no day.(¹³⁹) He likewise held that there are many suns and moons, appropriated to different zones of the earth: and that the sun sometimes deviates into a portion of the earth not inhabited by us, in which case it disappears, as it were, in a cavity, and becomes eclipsed to our portion of mankind. He explained the apparent rotation of the sun round the earth by the vast distance to which it is carried.(¹⁴⁰) His doctrine concerning the moon was that it is formed of compressed cloud; that it shines

---

weather. See Aristoph. Nub. 368—73. Jupiter was conceived as dwelling in the pure ether. Thus Euripides says:

ὄμνυμι δ' ἱερὸν αἰθέρ' οἴκησιν Διός.

Melanipp. Soph. Frag. 7. Dind. and Aristoph. Ran. 100, speak of αἰθέρα Διὸς δωμάτιον.

(136) Stob. i. 8, p. 97. Compare Schleiermacher, ib. p. 396.

(137) Karsten, Phil. Græc. Rel. vol. i. part i. p. 10, places the life of Xenophanes from 600 to 500 B.C.

(138) Stob. Phys. i. 24, p. 199; Plut. Plac. ii. 13; Galen, c. 13; Achill. Tat. c. 11, p. 79. ἄνθραξ means *charcoal*, which is the original sense of our word *coal*, as well as the present sense of the German *kohle*.

(139) Stob. Phys. i. 24, 25; Plut. ii. 20; Galen, c. 14. These notices are not quite consistent with one another.

(140) Stob. Phys. i. 25; Plut. ii. 24; Galen, c. 14.

with its own light; and that its phases are owing to the extinction of its light. He remarked that the sun is necessary for the production and preservation of the world and of the living things which it contains, but that the moon is a superfluity.(141) He likewise held that comets and meteors are composed of ignited clouds.(142) His doctrine concerning the earth was that its foundations are infinite in depth.(143) He therefore held it to be motionless, and not of a spherical figure.

§ 9 Parmenides of Elea, who likewise belonged to the Eleatic School, appears to have been born about 520 B.C., and to have flourished about 460 B.C.:(144) he was considered in antiquity to have been the disciple of Xenophanes, although Aristotle does not speak with confidence as to this fact.(145)

He is reported to have held the following doctrines: that the universe is composed of three circular bands, surrounded by a solid firmament, like a wall. That the highest of these bands is of air; that the next, or the heaven, is of fire; and that the third is the terrestrial circle.(146) That the air is secreted from the earth; that the sun is an evaporation from the fiery band; and that the moon is a mixture of fire and air;(147) that the stars are formed of condensed flame;(148) that the Morning

---

(141) Stob. Phys. i. 26; Plut. ii. 25; Galen, c. 15.
(142) Stob. Phys. i. 29; Plut. iii. 2; Galen, c. 18.
(143) Aristot. de Cœl. ii. 13, § 12. The verses in which he laid down this doctrine are preserved in Achill. Tat. c. 4.

γαίης μέν τόδε πείρας ἄνω πὰρ ποσσὶν ὁρᾶται
αἰθέρι προσπλάζον, τὰ κάτω δ' ἐς ἄπειρον ἱκάνει.

According to the emendation of Karsten, ib. p. 49. Προσπλάζω is an Homeric word: it occurs both in the Iliad and the Odyssey. The astronomical doctrines of Xenophanes are copiously illustrated by Karsten, ib. p. 164—183.

(144) The date of Parmenides is examined by Karsten, Phil. Gr. Rel. vol. i. part 2, p. 3—9.

(145) Metaph. i. 5. Compare Diog. Laert. ix. 21; Clem. Alex. Strom. i. 15, § 64; Sext. Emp. p. 213, ed. Bekker. Suidas in Παρμενίδης.

(146) Stob. Ecl. i. 22, 23; Plut. Plac. ii. 7; Galen, c. 11; Euseb. Præp. Evang. xv. 38; Cic. N. D. i. 11.

(147) Stob. Ecl. ib.

(148) Stob. Phys. i. 24. Concerning an opinion ascribed to Anaximenes and Parmenides, see above, p. 95, n. 121.

star, which he identified with the Evening star, occupies the highest place in the ether, and that next in order comes the sun; and that under the sun are the stars, in the fiery heaven:([149]) that the sun and moon are equal in size, but that the moon receives its light from the sun :([150]) that the phases of the moon are caused by the mixture of fiery and opaque elements in its composition, whence he called the moon *pseudo-luminous*. (ψευδοφανής).([151]) He is stated to have been the first who taught that the earth is spherical, and is situated in the centre of the universe.([152]) It is consistent with this account that the first division of the earth into five zones was ascribed to him:([153]) for this division naturally implies a spherical or at least a hemispherical figure.

§ 10 Xenophanes and Parmenides selected hexameter verse as the vehicle of their physical speculations, and gave the first examples of the philosophical didactic poem. In this respect they served as a model to Empedocles of Agrigentum, who flourished about 455 — 444 B.C.,([154]) and who is stated to have been the disciple of Parmenides.([155])

([149]) Stob. i. 24. That Parmenides was the first to identify the Morning and Evening stars was stated by Favorinus in the fifth book of his Commentaries. Diog. Laert. ix. 23. Favorinus lived at the time of Hadrian.

([150]) Stob. i. 25, 26; Plut. ii. 26.

([151]) Stob. Ecl. i. 26. This paragraph does not occur in Plutarch and Galen, and as the same opinion is previously attributed by Stobæus in nearly the same words to Anaxagoras, there is probably some error.

([152]) Diog. Laert. ix. 21.

([153]) By Posidonius, ap. Strab. ii. 2, 2. The cosmology of Parmenides is illustrated by Karsten, ib. vol. i. part 2, p. 240—256.

([154]) Karsten, Phil. Gr. Rel. vol. ii. p. 12, supposes Empedocles to have lived from 492 to 432 B.C.

According to Aristotle, cited by Eratosthenes in his work on the Olympic conquerors, the grandfather of Empedocles was victor in Olymp. 71 (496 B.C.) Diog. Laert. viii. 51, 52.

Empedocles was subsequent to the expedition of Xerxes against Greece, upon which subject he left an unfinished poem. Diog. Laert. viii. 57.

Some extant verses of Empedocles appear to refer to Pythagoras, Karsten, Phil. Gr. Rel. vol. ii. p. 150, 297. The person in question is spoken of as no longer alive.

([155]) Diog. Laert. viii. 55, 56. The poem of Empedocles was doubtless

Empedocles may be regarded as the contemporary of Herodotus.(156)

He is related to have held the following doctrines on the heavenly bodies. That the heaven is a solid firmament, formed of air condensed by fire, so as to assume the substance of ice; that the matters of fire and air are included within its two hemispheres;(157) that the stars are of a fiery nature, being formed of particles of fire separated from the air; that the fixed stars are fastened to the crystalline vault, but that the planets are free;(158) that the circuit of the sun is coincident with the boundary of the universe:(159) that there are two suns; one, formed of fire, placed in the lower hemisphere, and invisible to us; the other, in the upper hemisphere, opposed to the invisible sun, and sharing in its motion; reflecting its own light from the fiery air upon the earth: that the retrograde movement of the sun at the solstices is produced by the resistance of the containing sphere; that the sun of the upper hemisphere is equal in magnitude to the earth; and that it suffers an eclipse when the moon intervenes between it and the earth;(160) that the substance of the moon is mist congealed or hardened by fire, and that its shape is that of a disc: that it is lighted by the sun; and that its distance from the sun is twice as great as its distance from the earth;(161) that the universe is

---

the model which Lucretius had before his eyes. See the eulogy of him in i. 727—34.

(156) Both Herodotus and Empedocles went as colonists to Thurii, which was founded in 443 B.C. Diog. Laert. viii. 52.

(157) Stob. Ecl. i. 23; Plut. Plac. ii. 11; Galen. c. 12; Achill. Tat. Introd. in Arat. c. 5. Compare Sturz, Emped. p. 321. The word κρύσταλλος always meant *ice* in the earlier writers, though the later writers use it for *crystal*. The expression ψυχρὸς κύκλος appears to be used in the sense of crystalline orb by Alexander Ætolus in the verses cited by Theo Smyrnæus, c. 15, p. 186, ed. Martin. *Crystallus* always signifies crystal in the Latin writers.

(158) Stob. Ecl. i. 24; Plut. ii. 13; Galen, c. 13; Sturz, ib. p. 334.

(159) Stob. Ecl. i. 21; Plut. Plac. ii. 1.

(160) Stob. Ecl. i. 25; Plut. Plac. ii. 20, 23; Galen, c. 14.

(161) Stob. Ecl. i. 26; Plut. ii. 27; de facie in orbe lunæ, c. 16; Euseb. Præp. Ev. i. 8. The verse of Empedocles, cited by Plutarch,

egg-shaped, and that the width of the heaven is greater than its height above our heads.([162]) Empedocles explained the elevation of the northern pole, the depression of the southern pole, and the inclination of the world, by the impulse of the sun and the consequent withdrawal of the air.([163])

He held the earth to be motionless at the centre of the universe; and he attributed its state of rest to the rapid circular movement of the heaven around it. He compared this effect to the water in a cup, which when whirled round rapidly is prevented from falling to the ground.([164]) He conceived the heaven to be an elongated sphere, similar to an egg.([165])

§ 11 Anaxagoras of Clazomenæ belonged to the Ionic School of philosophy, and he is reported to have been the disciple of Anaximenes.([166]) This statement, however, cannot be reconciled with the chronology of the two philosophers; for the birth of Anaxagoras is placed at 499 B.C., and that of Anaximenes at 575 B.C. Anaxagoras removed from Clazomenæ to Athens; and in this city he imparted his opinions on physical science to Pericles, Euripides, Archelaus, and Thucydides the historian.([167]) He was accused of impiety at Athens, and was

---

de fac. in orbe lunæ, c. 2, is correctly given by Wyttenbach: ἥλιος ὀξυβελὴς ἠδ' ἰλάειρα σελήνη. Sturz, p. 332, misled by the false reading, ἠδὲ λάϊνα, says that Empedocles composed the moon of stone. He called the moon ἰλάειρα, because her heat is never scorching, nor her light dazzling. The emendation of Karsten, ἠδ' ἀγλαΐεσσα σελήνη, ib. vol. ii. p. 217, and that of Mullach, Fragm. Phil. Gr. p. 7, are not needed.

([162]) Stob. i. 26; Plut. ii. 31; Galen, c. 15; and Euseb. Præp. Evang. xv. 53. Stobæus represents Empedocles to have held that the distance of the moon from the earth is twice her distance from the sun.

([163]) Plut. Plac. ii. 8; Stob. i. 15; Galen, c. 11; Euseb. Præp. Ev. xv. 39.

([164]) Aristot. de Cœl. ii. 13, § 21; iii. 2, § 3. Empedocles ridiculed the notion of Xenophanes, that the foundations of the earth are infinitely deep, ib. ii. 13, § 12. See above, n. 143.

([165]) Stob. Ecl. Phys. i. 26. Concerning the astronomical and physical doctrines of Empedocles, see Karsten, ib. vol. ii. p. 416—440; Sturz, Emped. p. 321. An extant poem, in 168 iambic verses, entitled Σφαῖρα, is ascribed to Empedocles; but it is borrowed from Aratus, and is of late date. See Fabr. Bib. Gr. vol. i. p. 814, 825, ed. Harl. Sturz, ib. p. 88.

([166]) See Schaubach, Anaxagoræ Fragmenta (Lips. 1827), p. 3.

([167]) Schaubach, ib. p. 17—33. A full account of the influence of the

defended by Pericles. At the close of his life, he migrated to Lampsacus, where he died at about the age of seventy, and therefore about 430 B.C.([168])

Like his predecessors in the Ionic School, the attention of Anaxagoras was exclusively directed to speculations concerning Nature, and especially to celestial phenomena.([169]) He is reported to have said, in answer to a question, that he was born for the contemplation of the sun, the moon, and the heavens.([170]) Mount Mimas, near Miletus, is mentioned as the seat of his astronomical observations.([171]) The epitaph inscribed on his tomb by the Lampsacenes described him as having carried astronomical discovery to the furthest possible limit.([172]) His travels to Egypt (where he was supposed to have acquired

---

teaching of Anaxagoras upon Pericles is given by Plut. Per. 4—6. Pericles learnt μετεωρολογία from Anaxagoras, Plat. Phædr. § 120, p. 270. Pericles conversed with philosophers; first with Pythocleides and Anaxagoras, afterwards with Damon, Plat. Alcib. i. § 30. The instruction of Pericles by Anaxagoras in physics is mentioned by Cic. Brut. 11; his instruction in De Orat. iii. 34. Some verses of Alexander Ætolus describe Euripides as 'Αναξαγόρου τρόφιμος χαίου, Meineke, Anal. Alex. p. 247. See Gell. xv. 20. Euripides was born in 480, and Thucydides in 472 B.C. The exact year of the birth of Pericles is unknown: he was probably born about 495 B.C. Anaxagoras is related to have predicted the fall of the aerolite at Ægospotami, which is referred to 469, 467, and 465 B.C. (Schaubach, p. 42.) The most authentic account of the lifetime of Anaxagoras is that of Aristotle, who says that he was prior to Empedocles in age, but subsequent in the publication of his writings, Metaph. i. 3. Empedocles was probably born about 480 B.C., and belonged to the same generation as the Athenian disciples of Anaxagoras. This statement agrees well with the supposition that Anaxagoras was born about 500 B.C. Clinton fixes the prosecution of Anaxagoras at 432, and his death at 428, B.C. Eusebius states that Anaxagoras flourished Ol. 69.3, 502 B.C., and died Ol. 79.3, 462 B.C.

(168) Alcidamas, ap. Aristot. Rhet. ii. 23, 11, states that the Lampsacenes gave Anaxagoras a public funeral, although he was a foreigner; and that they continued to show him public honours, even in the writer's lifetime. Alcidamas lived in the generation between Anaxagoras and Aristotle.

(169) Aristot. Eth. Nic. vi. 7.
(170) Diog. Laert. ii. 7, 10. See Schaubach, p. 9.
(171) Philostrat. vit. Apollon. ii. 5.
(172) Diog. Laert. ii. 15; Ælian, V. H. viii. 19.

ἐνθάδ' ὁ πλεῖστον ἀληθείας ἐπὶ τέρμα περήσας
οὐρανίου κόσμου κεῖται Ἀναξαγόρας.

Plut. Per. 5, describes Pericles as having learned μετεωρολογία and μεταρσιολεσχία from Anaxagoras.

physical science from the priests), are imperfectly attested, and appear to be a mere figment of late writers.(¹⁷³) He was likewise a geometer, and he is said to have written upon the quadrature of the circle while he was in prison on the charge of impiety.(¹⁷⁴)

The following are the most important astronomical doctrines which are attributed to Anaxagoras. He held, with Anaximenes, that the earth is a plane; and that on account of its flat shape it is buoyed up and sustained by the air. He conceived it as immovable in the centre of the universe.(¹⁷⁵) He supposed the sun and stars to move in circular orbits round the earth; and to pass under it during the invisible part of their course.(¹⁷⁶) With respect to the composition of the heavenly bodies, he conceived them to be stones which had been carried up to heaven by the force of its circular or vortical motion, and had been ignited by the fiery firmament.(¹⁷⁷) He held the sun to be larger than the Peloponnese, and to be a mass of ignited stone.(¹⁷⁸) Anaxagoras is said to have

---

(173) Ammian. Marc. xxii. 16; Theodoret. Affect. Græc. 2; Cedren. Chron. vol. i. p. 165, ed. Bonn. Compare Schaubach, p. 13; Bayle, Dict. Anaxagore. The statement is rejected by Schaubach, ib.; Ritter, Gesch. der Phil. vol. i. p. 290.

(174) Plat. Erast. ad init.; Plut. de Exil. 18. Compare Schaubach, p. 58.

(175) Diog. Laert. ii. 8; Origen. Philos. p. 14. τὴν δὲ γῆν τῷ σχήματι πλατεῖαν εἶναι καὶ μένειν μετέωρον διὰ τὸ μέγεθος καὶ διὰ τὸ μηδὲν εἶναι κενὸν, καὶ διὰ τὸ τὸν ἀέρα ἰσχυρότατον ὄντα φέρειν ἐποχουμένην τὴν γῆν. According to Aristot. de Cœl. ii. 13, § 16, Anaximenes, Anaxagoras, and Democritus considered the breadth of the earth as the cause of its immobility. Simplicius, ad Aristot. de Cœl. p. 491 B, p. 124 A, distinctly states that Anaxagoras held the earth to be motionless at the centre. Compare Schaubach, p. 174-5.

(176) Aristot. Meteor. i. 8; Plut. Plac. iii. 1; Origen, ib. p. 14.

(177) Plut. Plac. ii. 13; Origen. Phil. p. 14; Euseb. Præp. Evang. xv. 30; Plut. Lysand. 12; Diog. Laert. ii. 12.

(178) Plut. Plac. ii. 21; de fac. in orbe lun. 19; Diog. Laert. ii. 8; Stob. Ecl. i. 25, 3; Achilles Tatius, Isag. c. 2, 11. The expression which he applied to the sun was that it is a μύδρος διάπυρος. See Xen. Mem. iv. 7, 7; Plat. Apol. 14; and numerous other passages cited by Schaubach, p. 139—142. μύδρος meant in general a heated mass of iron fresh from the furnace. Sophocles, Ant. 264, applies it to the bars of iron used in the fiery ordeal. Aristotle applies it to the heated stones ejected from

predicted a celebrated aërolite which fell at Ægospotami, in Thrace, about the year 468 B.C.;([179]) and at all events he explained this phenomenon by his hypothesis of stones attracted from the earth, and revolving in the heaven. He considered the moon to be an earth, having in it plains, mountains, and valleys; and likewise containing inhabitants.([180]) He derived its light from the sun.([181]) Anaxagoras explained eclipses of the moon, by the interposition of the earth between the sun and the moon; and he explained eclipses of the sun by the interposition of the moon between the sun and the earth, at the new moon. He likewise supposed that eclipses might be caused by the opaque stony bodies revolving round the earth. He is stated to have been the first who wrote with boldness and free-

---

Ætna. See Schol. Callim. Dian. 49. The word is derived from μυδάω, to melt. Concerning μύδρος διάπυρος, see Bayle's Dict. art. Anaxagore, note B. The tenet of Anaxagoras respecting the sun is held by the ancient commentators to occur in Euripides. Porson, ad Or. 971.

(179) The date of this event is fixed by the Parian Marble, at Ol. 78.1—468 B.C.; by Plin. ii. 58, at Ol. 78.2 (467 B.C.); and by Eusebius at Ol. 79.1 (464 B.C.) As to the connexion of Anaxagoras with it, see Diog. Laert. ii. 10, 12; Plut. Lysand. 12; Plin. ubi sup. Amm. Marc. xxii. 8, § 5. The event was mentioned by Diogenes of Apollonia, Stob. Ecl. Phys. i. 24; Plut. Plac. ii. 13; and is referred to by Aristot. Meteor. i. 7. Plutarch states that the stone was shown in his time, and was worshipped by the Chersonites. Compare Schaubach, p. 41. The fall of the aërolite is made by Plutarch contemporaneous with the end of the Peloponnesian War (405 B.C.); but this date differs from the other statements, and is, moreover, inconsistent with the reference to Anaxagoras, who had been dead more than twenty years. Compare Ideler ad Aristot. Meteor. vol. i. p. 404.

An aërolite, weighing 200 lbs., fell in Upper Alsace in 1492. The Emperor Maximilian I. treated it as a warning from God to men, and caused it to be suspended in a neighbouring church. See Coxe's Hist. of the House of Austria, vol. i. p. 455.

The reality of aërolites was doubted in modern times, and was not established among men of science till the beginning of this century. Masses of meteoric iron, of immense size, have fallen in different parts of the world. The aërolites mentioned in Livy are collected in Steger's Prodigien, p. 64.

(180) Plut. Apol. 14; Plut. Plac. ii. 25; Stob. Ecl. i. 27; Euseb. Præp. Ev. xv. 26; Diog. Laert. ii. 8; Orig. Phil. p. 14. Some Orphic verses describe the moon as inhabited, Procl. Tim. p. 154 A, 283 B. fragm. 9, p. 470, ed. Hermann.

(181) Plat. Cratyl. § 56, p. 409; Plut. de fac. in orbe lun. 16; Plac. ii. 30; Stob. Ecl. i. 27; Origen. Phil. p. 14.

dom concerning the physical causes of eclipses.([182]) It may be remarked that Thucydides knew that an eclipse of the sun takes place at the new moon, and that an eclipse of the moon takes place at the full.([183]) It is not improbable that he may have derived this knowledge from Anaxagoras. It is certain that the physical nature of eclipses was practically unknown to the Athenian army at Syracuse in 413 B.C.([184])

Anaxagoras attributed the solstices of the sun and the phases of the moon to the same cause, namely to the resistance of the cold air, caused by the pressure of the moving body. This resistance the body was unable to overcome, and was therefore turned back in its course.([185]) He knew the difference between the planets and the fixed stars; but he does not seem to have known the number of the former; he appears to have regarded them as meteors, of no definite number, and with no fixed orbits: hence he explained comets to be produced by the concourse of planets, and by their combined splendour.([186])

Anaxagoras seems to have been the first of the Greek speculators upon celestial subjects who was regarded as an atheist.([187]) He incurred this charge by substituting mecha-

---

([182]) Stob. Ecl. i. 27; Origen. Phil. p. 14; Plut. Nic. 23. A similar hypothesis is described by Aristotle, de Cœl. ii. 13, 7. ἐνίοις δὲ δοκεῖ καὶ πλείω σώματα τοιαῦτα ἐνδέχεσθαι φέρεσθαι περὶ τὸ μέσον, ἡμῖν δὲ ἄδηλα διὰ τὴν ἐπιπρόσθησιν τῆς γῆς. Διὸ καὶ τὰς τῆς σελήνης ἐκλείψεις πλείους ἢ τὰς τοῦ ἡλίου γίγνεσθαί φασιν· τῶν γὰρ φερομένων ἕκαστον ἀντιφράττειν αὐτήν, ἀλλ' οὐ μόνον τὴν γῆν. The attempts of Röth to discredit this passage,' Gesch. der Abendl. Phil. ii. 2, note 1300, are futile.

([183]) Thuc. ii. 28, iv. 52, vii. 50. This knowledge had become familiar at a later date. Macrobius in Somn. Scip. i. 15, § ii. remarks: 'Ideo nec sol unquam deficit, nisi cum tricesimus lunæ dies est, et nisi quinto decimo cursus sui die nescit luna defectum.' Geminus, c. 9, remarks that lunar eclipses take place only at the full moon.

([184]) Thucydides speaks contemptuously of Nicias as being too much devoted to divination and similar superstitious devices; but he says that the army were terrified by the omen, and pressed him to stay, and he does not mention any advice of an opposite tendency.

([185]) Stob. Ecl. i. 26; Plut. ii. 23; Origen. Phil. p. 14.

([186]) Aristot. Meteor. i. 8; Plut. Plac. iii. 2; Diog. Laert. ii. 9. Compare Schaubach, pp. 166-8.

([187]) Lucian, Timon. 10, τὸν σοφιστὴν Ἀναξαγόραν, ὃς ἔπειθε τοὺς ὁμιλητὰς μηδὲ ὅλως εἶναί τινας ἡμᾶς τοὺς θεούς. Iren. adv. Hær. ii. 14, Anaxagoras

nical and unprovidential forces for the direct agency of the gods;(188) and by reducing the heavenly bodies, which were believed to be of a divine nature, to a terrestrial standard, and to earthy materials. Similar doctrines had, indeed, been promulgated by preceding philosophers: but he spoke out with greater plainness and courage, and probably carried his physical explanations further than his predecessors.(189) Thus his doctrine that the sun is a mass of ignited stone, gave great offence; the popular belief still being that the sun was a god who drove his chariot across the sky from east to west.(190) His hypothesis of stones revolving in the air—which was probably suggested by the aërolite of Ægospotami—was likewise regarded as atheistic.(191) His degradation of the heavenly moon to a level with the earth—by his supposition that she

---

autem, qui et atheus cognominatus est. Aristides, Orat. 45, vol. ii. p. 80. Dindorf. classes Aristagoras with Diagoras the Melian.

(188) Plutarch, speaking of the eclipse of Nicias, says: οὐ γὰρ ἠνείχοντο τοὺς φυσικοὺς καὶ μετεωρολέσχας τότε καλουμένους, ὡς εἰς αἰτίας ἀλόγους καὶ δυνάμεις ἀπρονοήτους καὶ κατηναγκασμένα πάθη διατρίβοντας τὸ θεῖον, Nic. 23. Xen. Mem. iv. 7, 6, describes Anaxagoras as ὁ μέγιστον φρονήσας ἐπὶ τῷ τὰς τῶν θεῶν μηχανὰς ἐξηγεῖσθαι. Diod. xv. 50, speaking of a comet which appeared in Greece in 372 B.C., says: ἔνιοι δὲ τῶν φυσικῶν τὴν γένεσιν τῆς λαμπάδος εἰς φυσικὰς αἰτίας ἀνέφερον, ἀποφαινόμενοι τὰ τοιαῦτα φαντάσματα κατηναγκασμένως γίνεσθαι χρόνοις ὡρισμένοις.

With the criticisms on the irreligious opinions of Anaxagoras, compare the following verses from the Dunciad (iv. 469):—

> All-seeing in thy mists, we want no guide,
> Mother of arrogance, and source of pride!
> We nobly take the high Priori road,
> And reason downward, till we doubt of God:
> Make Nature still encroach upon his plan,
> And shove him off as far as e'er we can;
> Thrust some mechanic cause into his place;
> Or bind in matter, or diffuse in space.

(189) Plutarch, ib., says that Anaxagoras was the first who wrote with boldness concerning the cause of eclipses; but that at the time of the Athenian expedition to Sicily, his opinions were limited to a few, and were only mentioned in confidence.

(190) Joseph. cont. Apion. ii. 37: Ἀναξαγόρας δὲ Κλαζομένιος ἦν ἀλλ' ὅτε νομιζόντων Ἀθηναίων τὸν ἥλιον εἶναι θεόν, ὁ δ' αὐτὸν ἔφη μύδρον εἶναι διάπυρον, θάνατον αὐτοῦ παρ' ὀλίγας ψήφους κατέγνωσαν.

(191) Plato, Leg. xii. p. 967, speaks of this tenet as an atheistic error which had brought discredit upon astronomy, and had deterred men from the study of it.

had plains, and mountains, and valleys, and might even be inhabited—was also considered as an unauthorized freedom of speculation.(192) His opinions on astronomical subjects, but principally his doctrine concerning the composition of the sun, gave rise to the prosecution for impiety which is said to have been brought against him either by Thucydides, the opponent of Pericles, or by Cleon. He was defended by Pericles, and certainly escaped with his life; he appears to have been thrown into prison, and perhaps to have been sentenced to a fine.(193) It is also stated that a decree was carried at Athens by Diopeithes, about the beginning of the Peloponnesian War, declaring that persons who denied the existence of gods, or who published opinions concerning celestial phenomena, should be prosecuted; this decree was aimed at Pericles on account of his connexion with Anaxagoras.(194)

The prosecution of Anaxagoras is referred by Clinton to the year 432 B.C., about two years before his death, and when he was nearly 70 years old.(195)

Anaxagoras likewise offended the religious sentiments of the people by giving physical explanations of prodigies,(196) and by allegorizing the gods of Homer; for example, by con-

---

(192) See Plat. Apol. 14, where the published doctrines of Anaxagoras, that the sun is made of stone, and that the moon is an earth, are classed together as irreligious opinions, such as Meletus might make the subjects of criminal prosecution.

According to Favorinus, the doctrines of Anaxagoras concerning the sun and moon were not his own, but were borrowed from some previous speculator, Diog. Laert. ix. 34. He did not, however, specify from whom they were derived.

(193) Plut. Per. 32; Nic. 23; Diog. Laert. ii. 12, 13; Diod. xii. 39. Plut. de Superstit. 10: Ἀναξαγόρας δίκην ἔφυγεν ἀσεβείας ἐπὶ τῷ λίθον εἰπεῖν τὸν ἥλιον. Compare Josephus, n. 190.

(194) Plut. Per. 32. On the accusation of Anaxagoras, see Sintenis, ad Plut. Per. 32, p. 220.

(195) Protagoras, in his treatise περὶ θεῶν, professed perfect scepticism concerning the existence of the gods, Sext. Emp. ix. 55; Diog. Laert. ix. 51. Hence he was regarded as an atheist, Cic. N. D. i. 2, 12; and he was banished from Athens. He died in crossing the sea in an open boat. His death took place about 411 B.C., in the interval between the prosecutions of Anaxagoras and Socrates.

(196) Plut. Per. 6.

verting Jupiter into Mind, and Minerva into Art.(197) The latter method of interpretation may have suggested to Thucydides the historical rationalism by which he reduced to probability the Homeric narrative of the Trojan War.

§ 12 Diogenes, a citizen of Apollonia, in Crete, appears to have been a contemporary of Anaxagoras,(198) and to have addicted himself to physical philosophy. Like Anaxagoras, he supposed that the aërolite of Ægospotami was one of the stones which revolved round the earth with the stars.(199) He conceived the sun and other heavenly bodies to be formed of a porous substance, like the pumice-stone, and to receive light and heat from the æther.(200) He attributed the inclination of the earth's axis to a provision of nature, for the purpose of producing a variety of climates.(201)

§ 13 Socrates belonged to the generation next after that of Anaxagoras. He was born in 469 B.C., and he was condemned to death, at the age of seventy, in the year 399 B.C. The comedy of the Clouds was acted in 422 B.C.; but although Socrates was then nearly fifty years old, the character of his teaching and of his philosophical opinions was completely misunderstood by Aristophanes and the Athenian people. Socrates had nothing in common with the philosophers of the Ionic School, from Thales to Anaxagoras. The characteristic peculiarity of his philosophy was, that it avoided all speculation concerning the universe, and the movements of its component parts; that it descended from the celestial sphere to the surface of the earth; from the æther and the region of the gods to the dwellings of men; and that it was occupied exclusively with human feelings, actions, and interests. His teaching was neither cosmological nor astronomical, but ethical and political.(202)

---

(197) Schaubach, p. 37.
(198) Brandis, ib. p. 273. He is called a disciple of Anaximenes.
(199) Stob. Ecl. i. 24; Plut. Plac. ii. 13.
(200) Stob. Ecl. i. 24, 25, 26; Plut. Plac. ii. 13; Galen, c. 13.
(201) Plut. Plac. ii. 8.
(202) Aristotle states that in the time of Socrates —τὸ ζητεῖν τὰ περὶ

Nevertheless, when Socrates had made himself unpopular by his contentious and sceptical method of argument; when he had wounded the vanity of his fellow-citizens by questioning the grounds of their traditionary and apparently intuitive opinions on some of the principal concerns of life; and when he had set himself in hostility to the current feelings of the Athenian public; the great comic poet, who fixed upon him as a fit subject for satirical exhibition upon the stage, did not think it worth while to inquire either of the friends or the argumentative antagonists of Socrates, what were his real opinions. He had not committed his philosophy to writing, and published it to the world; his disciples had not yet begun to act as his exponents, and to write down his conversations. The Athenians were not a reading public, to whom his speculations could percolate through newspapers and reviews. Hence the prejudice which Aristophanes excited against Socrates was not less unfounded, and the picture which he drew of Socrates was not less unlike, than if Bacon had been decried by his contemporaries as an Aristotelian, and Hobbes or Spinosa as an abstruse theo-

---

φύσεως ἔληξε, πρὸς δὲ τὴν χρήσιμον ἀρετὴν καὶ τὴν πολιτικὴν ἀπέκλιναν οἱ φιλοσοφοῦντες. De Part. An. i. 1. Ab antiquâ philosophiâ usque ad Socratem (qui Archelaum, Anaxagoræ discipulum, audierat), numeri motusque tractabantur, et unde omnia orirentur, quove reciderent: studioseque ab his siderum magnitudines, intervalla, cursus anquirebantur, et cuncta cœlestia. Socrates autem primus philosophiam devocavit e cœlo, et in urbibus collocavit, et in domos etiam introduxit, et coegit de vitâ et moribus rebusque bonis et malis quærere. Cic. Tusc. Disp. v. 4. Quo etiam sapientiorem Socratem soleo judicare, qui omnem ejusmodi curam deposuerit; eaque quæ de naturâ quærerentur, aut majora quam hominum ratio consequi posset, aut nihil omnino ad vitam hominum attinere dixerit. Cic. de Rep. i. 10. Socrates mihi videtur, id quod constat inter omnes, primus a rebus occultis et ab ipsâ naturâ involutis, in quibus omnes ante eum philosophi occupati fuerunt, avocavisse philosophiam, et ad vitam communem adduxisse: ut de virtutibus et vitiis, omninoque de bonis rebus et malis quæreret, cœlestia autem vel procul esse a nostrâ cognitione censeret, vel, si maxime cognita essent, nihil tamen ad bene vivendum. Cic. Acad. i. 4.

Socrates is represented in the Phædo, § 40, p. 96, as saying that when he was young, he had a strong bias for physical philosophy, but that he discovered his unfitness for it. ἐγὼ γὰρ, he says, νέος ὢν θαυμαστῶς ὡς ἐπεθύμησα ταύτης τῆς σοφίας ἣν δὴ καλοῦσι περὶ φύσεως ἱστορίαν.—καὶ αὖ τούτων τὰς φθορὰς σκοπῶν, καὶ τὰ περὶ τὸν οὐρανόν τε καὶ τὴν γῆν πάθη, τελευτῶν οὕτως ἐμαυτῷ ἔδοξα πρὸς ταύτην τὴν σκέψιν ἀφυὴς εἶναι, ὡς οὐδὲν χρῆμα.

logian.(203) Near the beginning of the Platonic Apology, Socrates complains that he has two sets of accusers to deal with, the old and the new; and he implies that the old accusers, whose calumnies have taken root in the minds of his judges, are the more difficult for him to meet. The old accusers were, as he says, the poet of the Clouds (whose comedy had been acted twenty-three years previously), and various unnamed persons, who did not come forward on the public theatre, but whispered their calumnies against him in private.

Aristophanes, in the Clouds, instead of painting a portrait of Socrates from nature, reproduced an ideal picture of the Ionic physical philosopher, added some features of the modern sophist and rhetorician, who taught for money, and called the compound Socrates. The pupils in the school of Socrates are represented as studying astronomy, and as poring over geometrical diagrams inscribed upon the floor. Socrates himself is suspended in a basket; withdrawing his mind from the earth, and intent upon celestial phenomena; engaged in contemplating on the sun, and in investigating the motion of the moon; (204) he is even called another Thales.(205) He is, moreover, represented as denying the existence of the recognised gods, and as substituting for them the clouds, the unsubstantial denizens of the airy region in which the thoughts of Socrates are conceived to expatiate.(206)

---

(203) In the Apology of Plato, Socrates describes the young men who frequented his company, and who belonged to the first families in Athens, as imitating his style of interrogation, and as convicting their fellow-citizens of ignorance and presumption. He then proceeds thus: ἐντεῦθεν οὖν οἱ ὑπ' αὐτῶν ἐξεταζόμενοι ἐμοὶ ὀργίζονται, οὐχ αὑτοῖς, καὶ λέγουσιν ὡς Σωκράτης τίς ἐστι μιαρώτατος καὶ διαφθείρει τοὺς νέους. καὶ ἐπειδάν τις αὐτοὺς ἐρωτᾷ ὅτι ποιῶν καὶ ὅτι διδάσκων, ἔχουσι μὲν οὐδὲν εἰπεῖν ἀλλ' ἀγνοοῦσιν, ἵνα δὲ μὴ δοκῶσιν ἀπορεῖν, τὰ κατὰ πάντων τῶν φιλοσοφούντων πρόχειρα ταῦτα λέγουσιν, ὅτι τὰ μετέωρα καὶ τὰ ὑπὸ γῆς, καὶ θεοὺς μὴ νομίζειν καὶ τὸν ἥττω λόγον κρείττω ποιεῖν, c. 10.

(204) Nub. 171, 177, 194, 201, 202, 225, 227.

(205) τί δῆτ' ἐκεῖνον τὸν Θαλῆν θαυμάζομεν, Nub. 180. The astronomer Meton is, in like manner, called a Thales in the Birds, v. 1009.

(206) Nub. 247, 264-6, 365. Socrates is described as occupied with τὰ μετέωρα πράγματα, v. 228. The clouds appear τῷ φροντιστῇ μετέωροι, v. 266. In v. 360, he is included among the μετεωροσοφισταί. According to Xen.

Meletus and Anytus took advantage of the prejudice created by this witty caricature, and inserted in the bill of indictment against Socrates, an allegation that 'he had transgressed the bounds of legitimate knowledge, in investigating things under the earth and things in heaven.'([207]) In refuting this charge, Socrates expressly refers to the words of Aristophanes, and treats the poet's representation as a ridiculous fiction; ([208]) he says that he does not wish to speak of astronomy with disrespect, but he affirms confidently that he never occupied himself with this class of subjects, and he appeals to his many hearers to say whether they ever heard him discourse upon them. If Anaxagoras had been of sufficient importance for Cratinus to ridicule him on the stage, the picture which Aristophanes drew of Socrates would have suited the philosopher who was accused of impiety for having taught that the sun was a mass of heated stone. But for the thinker who first diverted philosophy from the stars to the affairs and homes of men, the accusation was not only not true, but was the very reverse of the truth. It represented him as teaching what he deliberately avoided, for the sake of investigating and teaching other subjects.

The real opinions of Socrates upon the study of geometry and astronomy are set out at length by Xenophon. Socrates disapproved of abstruse geometrical problems; and considered geometry as useful only so far as it could be serviceable for

---

Mem. i. 1, 1, the indictment distinctly charged Socrates with not recognising the gods recognised by the State, and with introducing new deities.

The idea of cloud-worship is the same as that of the saying reported by Madame de Staël, that the French had the empire of the land, the English of the sea, and the Germans of the air.

(207) The accusation against Socrates stands thus in Plat. Apol. 3, Σωκράτης ἀδικεῖ καὶ περιεργάζεται ζητῶν τά τε ὑπὸ γῆς καὶ τὰ ἐπουράνια, καὶ τὸν ἥττω λόγον κρείττω ποιῶν, καὶ ἄλλους ταῦτα διδάσκων.

(208) τοιαῦτα γὰρ ἑωράτε καὶ αὐτοὶ ἐν τῇ Ἀριστοφάνους κωμῳδίᾳ, Σωκράτη τινὰ ἐκεῖ περιφερόμενον, φάσκοντά τε ἀεροβατεῖν, καὶ ἄλλην πολλὴν φλυαρίαν φλυαροῦντα· ὧν ἐγὼ οὐδὲν οὔτε μέγα οὔτε σμικρὸν πέρι ἐπαΐω, ib. 3. The allusion is to Nub. 225, where Socrates, on being asked by Strepsiades what he is doing, answers: ἀεροβατῶ καὶ περιφρονῶ τὸν ἥλιον.

purposes of land-measuring.(209) With regard to astronomy, he considered a knowledge of it desirable to the extent of determining the day of the year or month, and the hour of the night, for journeys by land or sea, and for military watches. He held that the marks of time for this purpose could easily be learnt from night-watchers,(210) pilots, and others whose business it was to know them; but as to learning the courses of the stars, which move in different orbits from our own, to be occupied with the planets, and to inquire about their distances from the earth, and their orbits, and the causes of their motions, he strongly objected to such a waste of valuable time. Not only did he consider such studies useless, but he disapproved of speculations as to the means by which the gods regulated the courses of the celestial bodies. He declared that a person who should attempt such explanations would violate the laws of reason not less than Anaxagoras, who took pride in resolving the agency of the gods into mechanical causes.(211) He dwelt on the contradictions and conflicting opinions of the physical philosophers, and maintained that the questions which they attempted to resolve were beyond the powers of the human intellect, and beyond the reach of human action; and, in fine, he held that the speculators on the universe and on the laws of the heavenly bodies were no better than madmen.(212)

§ 14 But although Socrates condemned the scientific pursuit of astronomy, and to a certain extent shared the prejudices to which he fell a martyr, the accurate observation of the movements of the heavenly bodies made progress during his lifetime. Meton, an Athenian citizen, conversant with practical astronomy, introduced a reform of the calendar in 432 B.C., which implies an accurate determination of the sun's annual course.

---

(209) Xen. Mem. iv. 7, 3, τὸ δὲ μέχρι τῶν δυσξυνέτων διαγραμμάτων γεωμετρίαν μανθάνειν ἀπεδοκίμαζεν.

(210) νυκτοτῆραι, Xen. Mem. iv. 7, 4.

(211) Xen. Mem. iv. 7, 2-7. Compare Grote, Hist. of Gr. vol. viii. p. 571-6.

(212) Xen. Mem. i. 1, 11-15.

Meton appears to have had no connexion with Anaxagoras and with the speculative astronomy of the Ionic School: he is stated, by Theophrastus, to have derived assistance from an observer named Phaëinus, who determined the time of the solstices from observations made on Mount Lycabettus:([213]) he is ridiculed by Aristophanes in the Birds—which was acted in 415 B.C.—as a geometer, who measures the air and offers to lay out the streets of the new city upon mathematical principles.([214]) He is reported to have entertained gloomy forebodings respecting the fate of the expedition against Syracuse before it sailed in 415 B.C.([215]) His lifetime was probably nearly coincident with that of Thucydides.

§ 15 The common year of the Greeks consisted originally of 360 days. The reformed year of Solon consisted of 354 days. In either of these years, the deviation from the sun was so great, that the calendar would soon cease to be a guide to the seasons. The necessity of intercalation, in order to make the year agree with the periodical circuits of the sun, and thus to keep the calendar in harmony with the seasons, must therefore soon have become manifest.

The earliest intercalation appears to have been that called by the Greeks (who, in general, counted both extremities inclusively) the *trieteric;* it consisted in the insertion of an addi-

---

(213) De Sign. Pluv. 3. Theophrastus says that Phaëinus was a resident alien; and Meton a native Athenian. The proper name Φαεινὸς occurs in Schol. Aristoph. Eq. 963, and Φαεινὶς in Thuc. iv. 133.

(214) Av. 992-1019. Ideler remarks that Aristophanes, who treats Meton as a charlatan and an impostor, probably knew as much about him as he knew about Socrates, vol. i. p. 323.

(215) Plut. Alcib. 17; Nic. 13; Ælian, V. H. xiii. 11.
Meton is mentioned in a verse of Phrynichus the comic poet, Mein. Fragm. Com. Gr. vol. ii. p. 589. Phrynichus was a poet of the old comedy, and exhibited at least as early as 429 B.C. (Clinton ad ann. Mein. vol. i. p. 146.) An extant fragment refers to the Hermocopidæ, Mein. vol. i. p. 155, vol. ii, p. 602. Ptolemy, Magn. Synt. iii. 2, vol. i. p. 162. Halma speaks of a summer solstice observed by Meton and Euctemon, on the 21st of Phamenoth, in the Archonship of Apseudes (433 B.C.). See Schol. Av. 997; Ideler, Berl. T. 1814, p. 239; Müller, Gött. Anz. 1822, p. 459; Philochor. fr. 99, ed. Müller.

tional month in every alternate year.(216) Two solar years contain 730½ days; but the addition of a month of 30 days to two years of 360 days would produce 750 days; and the addition of a similar month to two years of 354 days would produce 738 days. In the former case, the excess would be 19½ days; in the latter, it would be 7½ days. In neither case would the remedy be effectual; for the year of 360 days the intercalation would increase, not diminish, the error. Both Geminus and Censorinus attest the intercalation of one month in alternate years; but they do not mention the length of the intercalary month. Herodotus speaks of the intercalation of a month in every second year, on account of the seasons, as the common practice in Greece at his time;(217) and in the colloquy of Crœsus and Solon, he introduces a computation of the number of days contained in 70 years, in which he assumes the year as consisting of 360 days, with an intercalary month of 30 days in each alternate year.(218) It is inconceivable that an intercalation which increased the error which it was intended to remedy could have been in general use;(219) but the text of Herodotus appears to be sound; his calculation is consistent with itself; and we must attribute the absurd result to his ignorance of astronomy. The probability is, that when the intercalation was made in alternate years, an anomalous month, of less than 30 days, was inserted. Thus the Roman month, Mercedonius, which Numa is said to have intercalated every other year, consisted of 22 days; and as his year contained 355 days, there

---

(216) Censorin. c. 19; Geminus, c. 6. Compare Ideler, vol. i. p. 269, 272.

(217) Herod. ii. 4. Ἕλληνες διὰ τρίτου ἔτεος ἐμβόλιμον ἐπεμβάλλουσι τῶν ὡρέων εἵνεκεν.

(218) i. 32. He reckons 70 years as containing 25,200 days, and 35 intercalary months as containing 1050 days. He states the sum correctly at 26,250 days. This passage, like that concerning the census of Servius in the Republic of Cicero, is a remarkable proof of the disposition to find difficulties where none exist. Larcher says that it is one of the most difficult in Herodotus, and that the text is certainly corrupt. He himself gives a translation founded upon a conjectural alteration of the text.

(219) See Ideler, ib. p. 273.

was an error of only 1½ days in a biennial period.(220) It appears that the Roman year was administered according to this system of intercalation before the Julian reform. Great laxity doubtless prevailed in the Greek calendars at the time of Herodotus: there was no recognised scientific authority on astronomical questions: and the priests or public officers who regulated the calendar doubtless intercalated months of unequal length at their discretion, according to the state of the seasons.(221) We learn from Polybius that Timæus compared the calendars of different Greek States, and thereby detected discrepancies of time, sometimes amounting to three months, or a quarter of a year.(222)

The next intercalary cycle was the octaëteris,(223) or octennial period. It was founded on a different basis from the trieteric or biennial intercalation, and assumed a knowledge of the true length of the solar year, within a trifling error.

When we, with many modern writers, attribute a lunar year to the Greeks, we must bear in mind that a year is neither measured by the moon, nor consists naturally of twelve lunations; and that this number of lunations is taken only because it approaches nearest to the solar year of 365¼ days. A year necessarily depends upon the sun; but the ancients, instead of determining it exclusively by the sun (as is done by

---

(220) Above, p. 38.

(221) On the arbitrary manner in which the Greek States tampered with the calendar at the time of the Peloponnesian War, see Grote, Hist. of Gr. vol. vii. p. 90, whose explanation of the difficult passage of Thucydides seems to me correct.

(222) Polyb. xii. 11.

(223) It will be observed that the terms tetraëteris and octaëteris, for a quadrennial and an octennial period, are formed upon a different principle of counting from the term trieteris for a biennial period.

The ὀκταετηρίς was also called ἐννάετηρίς, and the τετραετηρίς πεντάετηρίς, Censorin. 18. Compare Ideler, vol. i. p. 287.

Aristoph. Plut. 583, says that the Olympic Games were celebrated δι' ἔτους πέμπτου.

Concerning the Greek method of counting from one terminus to another, see Dodwell, de Cyclis, p. 50.

As to the true nature of a lunar year, see Dodwell, de Cyclis, p. 163.

the moderns), made a sort of compromise between the sun and the moon, and regulated its length by fixing the civil or calendar months according to the course of the moon, and eking out the aggregate of months thus measured by intercalary days. The essence of the system was, that the year should be formed of that integer number of lunar months which approximates most closely to the solar year; and that the year thus determined should be brought into agreement with the solar year by intercalation.(224)

Starting, then, from the datum that the year is made up of lunar periods, and therefore consists of alternate full and hollow months, of twenty-nine and thirty days, forming in the aggregate 354 days, we have for each year a deficiency of $11\frac{1}{4}$ days under the true solar time. A fractional period being inconvenient for intercalation, it would naturally occur to multiply this quantity by four, and therefore to form a quadriennial cycle. Censorinus, indeed, mentions that the Greeks used a tetraëteric intercalation; but although a quadriennial cycle is ascribed to Eudoxus,(225) and is mentioned in Stobæus,(226) yet there is no historical trace of its actual use in any Greek calendar.

Although forty-five is an integer number, this number of days would form only a month and a half of the ordinary lunar length; and as it would be desired on grounds not only of convenience, but also of religion, to intercalate entire months, this number was doubled by adopting an octennial instead of a quadriennial period. The multiplication of $11\frac{1}{4}$ days by eight gave ninety days, or three months; and hence, if in a period of eight years a month of thirty days was intercalated in the

---

(224) Compare the remarks of Laplace, Système du Monde, tom. ii. p. 316.

(225) Plin. N. H. ii. 48.

(226) Four great years, of 4, 8, 19, and 60 years respectively, are mentioned by Stob. Ecl. Phys. i. 8. The period of 4 years is omitted, and the period of 60 years is stated at 59, in the corresponding passages of Plut. Plac. ii. 32; Galen, c. 16; Euseb. Præp. Ev. xv. 54.

third, fifth, and eighth years,[227] a year of 354 days would be brought into close accordance with the sun.[228]

If this simple and well-attested explanation be correct, it follows that the octaëteric period implies a knowledge that the solar year consists of 365¼ days. It would be important to determine the time of its introduction; since it would indicate the epoch at which the true length of the solar year was known in Greece. But this point cannot be ascertained with certainty. Its introduction is attributed to Eudoxus; but it seems unquestionably to be of an earlier date. Censorinus ascribes it to Cleostratus of Tenedos, who may be referred upon conjecture to the first half of the fifth century B.C. It appears, both from internal evidence and from the testimony of the ancients, to be earlier than the cycle of Meton.[229]

It can scarcely be doubted that the Greek astronomers, from Thales downwards, had determined the times of the equinoxes and solstices; and if this had been done with accuracy, it must have been known that the tropical year consists

---

[227] This is the statement of Geminus. Solinus, i. 42, states that the three months were added at the end of the eighth year. This seems an inconvenient arrangement; the practice probably varied. Macrobius merely says that the Greeks intercalated at the end of the year, and not, like the Romans, in the middle of a month.

[228] This explanation of the origin of the octaëteric intercalary cycle is given by Macrob. Sat. i. 13.
Solinus starts from the false assumption that the established Greek year consisted of 365¼ days, and he supposes that the intercalation was effected by deducting 11¼ days from this year, and so producing a year of 354 days, instead of adding them to an existing year of 354 days. Allowing for this difference, his explanation of the method is identical with that of Macrobius: 'Græci singulis annis undecim dies et quadrantem detrahebant, eosque octies multiplicatos in annum nonum reservabant, ut contractus nonagenarius numerus in tres menses per tricenos dies scinderetur: qui anno nono restituti efficiebant dies quadringentos quadraginta quatuor, quos *embolimos* vel *hyperballontas* nominabant,' i. 42. (354 + 90 = 444.)

[229] See Ideler, vol. i. p. 287, 304-5. Theophrast. de Sign. Pluv. 3, says that Cleostratus determined the solstices from observations on Mount Ida. If he really determined the solstices, he must have known the true length of the tropical year; and if he knew the true length of the tropical year, he could easily have formed the octaëteric period. Ideler, ib. p. 605, accepts the statement that the octaëteric period was introduced by Cleostratus; but he thinks that the agreement of 99 lunations with 8 tropical years was known at an earlier period.

of 365¼ days. The theory of the solstices had, as we have already seen, attracted much attention; and various explanations of the turns of the sun in its course from the north and from the south, had been propounded.(230) We may naturally suppose that, where the theory had attracted so much attention, continuous observations would have been made for fixing the precise intervals between the great periodical phenomena in the annual course of the sun.

Ideler thinks that the octaëteris was the original intercalary cycle, and that the biennial and quadriennial cycles were submultiples of it, formed by subsequent calculation.(231)

But the octaëteris implies a knowledge of the true length of the tropical year; and if it was once introduced, we cannot suppose that recourse would afterwards be had to the imperfect biennial intercalation. Now it is certain from Herodotus that the biennial intercalation was common in Greece in his time. It seems, therefore, that Geminus and Censorinus are right in representing the octaëteric as an improvement on the trieteric system, and as of subsequent introduction.(232)

An attempt to trace the existence of the octaëteric cycle to an early period, long anterior to the time of Cleostratus, has been made by Prof. Boeckh, and has been supported by Ideler and Otfried Müller.(233) But the evidence adduced in proof of this hypothesis is inconclusive. It consists of certain festival cycles, the duration of which was probably determined by

---

(230) According to Aristot. Meteor. ii. 2, some of the earlier speculators accounted for the solstices by saying that the sun was nourished by moisture, and that the same places could not always supply nourishment to the sun.

(231) Chron. vol. ii. p. 607.

(232) Geminus mentions two other intercalary periods, viz. of 16 and 160 years, introduced for the purpose of correcting the departure from the moon in the octaëteric period, c. 6. See Ideler, Chron. vol. i. p. 296; Dodwell, de Cycl. p. 173.

(233) Böckh, Berlin Transactions, 1818-9, p. 98; Mondcyclen der Hellenen (Leipz. 1855), p. 10; Müller, Orchomenos, p. 218-221; Ideler, Chron. ii. p. 607; see also Höck, Kreta, vol. i. p. 247-255; K. F. Hermann, Gottesd. Alt. § 49, n. 11; Dodwell, de Cyclis, p. 213.

grounds wholly unconnected with intercalation. The principal argument is derived from the ennaëteric period of the Bœotian festival of Daphnephoria, and the account, in the Chrestomathia of Proclus,(234) of a curious usage at that festival allusive to the length of the year. An olive-stick was crowned with laurel branches and flowers, at the top was fastened a brazen ball, to which smaller balls were attached; at the middle a ball, less in size than that at the top, was fixed, and purple fillets, 365 in number, were appended to it. The large ball denoted the sun, the ball next in size represented the moon, and the small balls signified the stars. The 365 fillets expressed the sun's annual course. The same octennial period is mentioned by Plutarch in connexion with religious ceremonies belonging to the temple of Delphi.(235) The statement of Apollodorus,(236) that Hercules completed his twelve labours in eight years and one month, has likewise been pressed into the service of the same argument. There is, however, no proof that the ancient festival cycles were intercalary periods, founded upon a knowledge of the true length of the solar year; or that the symbolical rite described by Proclus, in the fifth century, is of high antiquity.

The octaëteric cycle would, if the intercalations were properly made, have produced a calendar sufficiently accurate for all practical purposes.(237) It produced perfect accuracy upon the assumption that the tropical year is $365\frac{1}{4}$ days. The difference between six hours and 5h. 48m. 48s. (namely, 11m. 12s.) is the measure of its annual inaccuracy. But the coincidence of ninety-nine lunations with the octennial cycle is less close. There is a deficiency of nearly $1\frac{1}{2}$ days; which in a century would cause a deviation of more than eighteen days from the

---

(234) Ap. Phot. Bibl. cod. 239, p. 321, ed. Bekker.
(235) De Def. Orac. 15, 21; Quæst. Gr. 12.
(236) ii. 5, § 11.
(237) Geminus remarks that the octaëteris coincides with the sun; but that the assumption of $29\frac{1}{2}$ days for the lunar month makes it $\frac{1}{33}$ of a day too short. See Ideler, vol. i. p. 295.

moon. The cycle of Meton was intended to cure this defect: but while it brought the months into closer coincidence with the moon, it made the year agree less exactly with the sun. His cycle consisted of 6940 days, divided into nineteen years and 235 lunar months.(238) With regard to the moon, the error is only $7\frac{1}{2}$ hours in the nineteen years; but as this cycle assumes the solar year to consist of $365\frac{5}{19}$ days, it is nearly nineteen minutes further from the true length than a year of $365\frac{1}{4}$ days.(239) The cycle of Meton must therefore have been introduced with reference to the moon rather than to the sun; and it was no improvement upon the civil calendar, as regulated by the octaëteric cycle, but rather the reverse.

The introduction of this cycle does not appear to have been dictated by a religious motive. It was not intended to make the sacred festivals harmonize with the moon. As it was no improvement of the civil calendar, it must have originated in purely astronomical considerations.

The Metonic cycle enjoyed great celebrity among the ancient astronomers;(240) and it is stated by Diodorus to have

---

(238) Theophrast. de Sign. Pluv. 3; Gemin. c. 6; Censorin. c. 18; Diod. xii. 36; Schol. Arat. 752; Ælian, V. H. x. 7. The cycle of nineteen years is ascribed by Geminus to Euctemon, Philip, and Callippus. The name of Meton appears to have fallen from the text.
Compare Ideler, vol. i. p. 297-343; Whewell, Hist. of Ind. Sciences, vol. i. p. 128-132.
Concerning Philip, see Smith's Dict. in Philip Medmæus, no. 16, p. 259.
Euctemon was the assistant of Meton, Ideler, vol. i. p. 298.
Euctemon, Meto, and Philip, are mentioned together by Vitruvius, ix. 6, as the authors of astronomical parapegmata.

(239) Geminus remarks that Meton makes the year $\frac{1}{76}$ of a day too long, that is to say, $\frac{5}{19} - \frac{1}{4} = \frac{19}{76} - \frac{18}{76} = \frac{1}{76}$. Compare Ideler, ib. p. 299.
Censorinus states that Meton fixed the year at 365 days and $\frac{5}{19}$ of a day, c. 19.
Ptolemy, Alm. iii. 2, p. 164, ed. Halma, states that the year of Meton and Euctemon contains $365\frac{1}{4}$ days and $\frac{1}{76}$ part of a day.

(240) δοκεῖ δὲ ὁ ἀνὴρ οὗτος ἐν τῇ προρρήσει καὶ προγραφῇ ταύτῃ θαυμαστῶς ἐπιτετευχέναι, Diod. xii. 36.
The celebrity of this cycle is alluded to by the astronomical poet of Greece:—

τὰ γὰρ συναείδεται ἤδη
ἐννεακαίδεκα κύκλα φαεινοῦ ἠελίοιο.—Arat. v. 752-3.

The Scholiast on this passage states that the Metonic cycle was derived

been generally used in Greece. The octaëteric and Metonic cycles prove that the scientific astronomy of the Greeks had attained, not later than the fifth century B.C., to great accuracy in determining the respective lengths of the lunar month and of the solar year.

§ 16 After the lapse of a century, the Metonic cycle was reformed by Callippus.([241]) As Meton had assumed the solar year to consist of $365\frac{5}{19}$ days, and in so doing had made the year longer by $\frac{1}{76}$ of a day than the year of $365\frac{1}{4}$ days, Callippus quadrupled the Metonic period, and thus produced a cycle of seventy-six years. From this cycle he deducted one day, and thereby reduced his solar year to the standard of the more accurate year previously adopted in the octaëteric cycle. The same period of 27,759 days, divided into 940 lunations, made the lunation more exact than in the Metonic cycle. The Callippic cycle agreed therefore more closely than the Metonic cycle both with the sun and the moon, and hence it was used by scientific astronomers.

The first year of the Callippic period was 330 B.C.([242]) There is no reason to suppose that it was ever applied to the regulation of the civil calendars of Greece.

§ 17 Pythagoras and his school exercised an influence upon astronomy, as well by their speculations upon arithmetic and geometry, as by their cosmological hypotheses.

Not only the philosophical teaching, but even the life and date of Pythagoras are involved in doubt and obscurity. The

---

by the Greeks from the Egyptians and Chaldæans. Theophrastus says that Meton obtained his cycle from Phaëinus.

Dr. Whewell conjectures that the length of the month was determined by the observation of eclipses; and he points out that the cycle of Meton is so exact, that it is still used in calculating the new moon for the time of Easter, ib. p. 129, 130.

See Ideler, vol. ii. p. 197.

([241]) Callippus and certain of his astronomical hypotheses are mentioned by Aristot. Metaph. xi. 8, p. 1073, Bekker. On this passage, see Simplicius ad Cœl. p. 500, Brandis.

([242]) Geminus, c. 6; Censorin. 18; Ideler, ib. p. 344; Whewell, ib. p. 130.

statements concerning his lifetime are widely discrepant; the most probable supposition is, that his birth was not earlier than 569, and his death not later than 470 B.C.(243)

Pythagoras himself was exclusively an oral teacher: he made no written record of his philosophical opinions. He is reported to have visited Phœnicia and Egypt, and to have profited by the instruction of the Egyptian priests in geometry and astronomy: he is even stated to have journeyed as far as Babylon, and to have there received information from the Chaldæans. Both Phœnicia and Egypt could easily be reached from his native island of Samos. The extent of his travels, and the nature of his communications with Oriental priests upon scientific subjects, must, however, remain doubtful;(244) for the accounts of his life which have reached us are derived from sources too suspicious to fulfil the conditions of historical credibility.(245)

The earliest of the followers of Pythagoras who committed his opinions to writing and published them to the world was Philolaus, a contemporary of Socrates and Democritus.(246)

---

(243) Röth, Gesch. der Abendl. Philosophie, ii. 1, p. 288, fixes his birth at 569, and his death at 470 B.C. He considers Xenophanes, Anaximenes, and Pythagoras as exact contemporaries, ib. p. 174. Brandis, ib. p. 422-4, does not attempt to fix the dates.

He was mentioned by Xenophanes and Heraclitus.

The writer of the art. Pythagoras in Pauly, places his birth between 580-568 B.C.

Aristoxenus states that Pythagoras, being forty years old, seeing the despotism of Polycrates confirmed, left Samos and migrated to Italy; Porphyr. vit. Pyth. c. 9. A different account is given by Iambl. vit. Pyth. 11: he states that Pythagoras escaped from the despotism of Polycrates in his eighteenth year, and went to Pherecydes, Anaximander, and Thales. The reign of Polycrates lasted from 532 to 522 B.C. See Aristoxen. fragm. 4, 23, Fragm. Hist. Gr. vol. ii.; Inquiry into the Cred. of Early Rom. Hist. vol. i. p. 431.

Meiners, Gesch. der Wissenschaften in Griechenland und Rom. vol. i. p. 362, 368, 370, places the birth of Pythagoras at Ol. 49. 2 (583 B.C.), and his death at Ol. 68.3, or 69.2 (506 or 503 B.C.).

(244) See below, ch. v. § 5. The eastern travels of Pythagoras are doubted by Meiners, ib. p. 376.

(245) See Meiners, ib. vol. i. p. 179-185, where sound criteria are laid down for determining the truth of the traditional accounts of Pythagoras.

(246) Plato, Phædon, p. 61, D.; Diog. Laert. ix. 38; see Boeckh, Phi-

How far these opinions were derived from the great master himself, and how far they were peculiar to the writer, we have no means of judging: but they may be considered as characteristic of the Pythagorean School in the latter part of the fifth century B.C.

A highly important astronomical doctrine, attributed by Stobæus and other late compilers to Philolaus, agrees with that ascribed by Aristotle to the Pythagoreans generally.[247] According to this doctrine, the earth is not motionless in the centre of the universe, but the central place is occupied by a mass of fire, which was designated by the mystical appellation of the 'Hearth of the Universe,' 'the House, or Watch-tower, of Jupiter,' 'the Altar of Nature,' 'the Mother of the Gods.' Round this centre ten bodies moved in circular orbits, in the following order:—At the highest extremity the heaven, containing the fixed stars, next the five planets, then the sun, then the moon, then the earth, and after the earth the antichthon. The region of the fixed stars and planets, and of the sun and moon, was considered as orderly and the subject of science; the

---

lolaos, p. 5. Philolaus was probably born about a century after Pythagoras.

[247] See Boeckh, Philolaos, No. 11. The doctrine is stated in Stob. Ecl. i. 22. Compare Aristot. de Cœl. ii. 13, and the account of Simplicius founded upon the Treatise of Aristotle respecting the Pythagorean tenets, Schol. Aristot. p. 504 b, and 505 a, Brandis. This last important testimony is omitted by Boeckh. See Brandis, Aristoteles, vol. i. p. 85.

Aristotle, in his work on the Pythagorean philosophy, reported a secret dictum of the sect, that there were three sorts of rational beings, gods, men, and Pythagoras, Iamblich. vit. Pyth. § 31.

The Pythagorean scheme is likewise described by Plutarch in his Life of Numa, with reference to the supposed tradition which made Numa a disciple of Pythagoras (c. 11). It is there stated that the Pythagoreans placed fire in the centre of the universe, and called it Vesta and the Monad: that they did not suppose the earth to be immoveable, or in the centre of the moveable orbs, but revolving in a circle round the central fire; and that they did not conceive the earth as first in dignity among the celestial bodies.

Diog. Laert. viii. 85, states that Philolaus was the first to teach that the earth moves in a circle; but that others ascribed the doctrine to Hicetas. We shall see lower down that the doctrine of Hicetas was properly heliocentric, which was not the case with that of Philolaus, although the latter supposed the earth to move in an orbit.

sublunary and earthy region was regarded as disturbed, and the subject of a different intellectual faculty.(248)

The remarkable part of this system is, that it conceived the earth as moving in an orbit through space. But the entire system was formed by an unscientific method. The inventor of it proceeded from certain arbitrary principles, and reasoned deductively from those principles, until he had constructed a scheme of the universe.(249) It was assumed that fire is more worthy than earth; that the more worthy place must be given to the more worthy; and that the extremity is more worthy than the intermediate parts; hence it was inferred that as the centre is an extremity, the place of fire is at the centre of the universe, and that the earth, together with the heavenly bodies, move round the fiery centre.(250) But this was no heliocentric system: the sun moved, like the earth, in a circle round the central fire; and it caused day and night, and the changes of the seasons, but in what manner it is not easy to divine. The orbit of the earth was supposed to be oblique to that of the sun:(251) and by this hypothesis an explanation was apparently found for the phenomena of the year. It was assumed that the inhabited side of the earth was always turned to the sun, and the uninhabited side to the central fire: so that the central fire was never visible from the earth.(252)

It is difficult to understand how the diurnal movement of

---

(248) See Boeckh, ib. p. 95, 167, and Anon. de vit. Pyth. § 10, τὴν δὲ τάξιν μέχρι σελήνης σώζεσθαι, τὰ δὲ ὑπὸ σελήνην οὐκέτι ὁμοίως.

(249) The method is thus described by Aristotle:—οὐ πρὸς τὰ φαινόμενα τοὺς λόγους καὶ τὰς αἰτίας ζητοῦντες, ἀλλὰ πρός τινας λόγους καὶ δόξας αὐτῶν τὰ φαινόμενα προσέλκοντες καὶ πειρώμενοι συγκοσμεῖν. De Cœl. ii. 13, 2. The unscientific character of the Pythagorean astronomy is pointed out by Meiners, ib. p. 555.

(250) Aristot. ib. The Pythagoreans likewise held that the circular motion is the most perfect motion, see Aristot. Probl. xvi. 10.

(251) Boeckh, ib. p. 116.

(252) See Martin, Etudes sur le Timée de Platon, tom. ii. p. 97. It is expressly stated that the antichthon could not be seen from the earth (Boeckh, p. 115), and the antichthon was between the earth and the central fire.

the sun, and the nocturnal movement of the stars, were reconciled with the system of a synchronous revolution of the stars, sun, and earth, in concentric circles; but the apparent motion of the stars by night was as easily explained upon this hypothesis as the apparent motion of the sun by day: and it would be hazardous in the extreme to infer that this Pythagorean system implies a knowledge of the precession of the equinoxes.([253]) The precession is a minute annual inequality, which was discovered by Hipparchus; the Pythagoreans were not close scientific observers of celestial phenomena, and were not likely to have made this discovery for themselves; it is a purely gratuitous assumption that the Egyptians may have been the discoverers, and that they may have communicated their discovery to Pythagoras.([254])

It is further stated that Philolaus supposed the existence of two suns, one of which, formed of glass or crystal, reflected light and heat from the original sun, and afterwards communicated them to the earth.([255]) If this part of the Philolaic system is correctly represented, we must suppose that the source from which the crystalline sun derives its light and heat is the central fire round which the ten bodies revolve.

The nature and functions of the antichthon, in this scheme of the universe, are not clearly determined. It is a distinct body from the earth, and nearer than it to the central fire; it is moreover invisible to the inhabitants of the earth. Practically, it seems equivalent to the hemisphere opposite to that

---

([253]) Œttinger, art. Planetæ, in Pauly's Real Encyclopädie, vol. v. p. 1667, shows that this system fails to explain the phenomena of night and day.

([254]) This conjecture was advanced by Boeckh, Philolaos, p. 118, and was accepted by Martin, ib. p. 98. It is, however, rejected by Ideler, Astron. Beob. der Alten. p. 89, and by Œttinger, ib. p. 1669, and has been abandoned by Boeckh himself, Manetho, p. 54; Kosmisches System des Platon, p. 93.

([255]) See Boeckh, ib. no. 14. In the passages which contain the statement of this doctrine, the original sun is said to be in the heaven or the æther. But unless we suppose it to be the central fire, this doctrine is not consistent with the Pythagorean or Philolaic system of the world.

known to the ancients. Aristotle ascribes to the hypothesis of the antichthon, an origin similar to that which he assigns to the hypothesis of the central fire. He says that the Pythagoreans framed their plan of the universe so as to make it harmonize with their principles respecting the properties and virtues of numbers; that the number ten was held to be a perfect number, and to comprehend the entire notion of numbers; and that, as the apparent celestial bodies (namely, the sphere of the fixed stars, the five planets, the sun, the moon, and the earth) were only nine, they imagined the antichthon, and added it to the rest, in order to make the full decad.(256)

It was stated by Aristotle and Philip of Opus, that, according to the doctrine of certain of the Pythagoreans, an eclipse of the moon might be caused by the interposition either of the earth or of the antichthon.(257) This harmonizes with the Philolaic scheme of the universe just described. Hicetas the Pythagorean likewise held the doctrine of an antichthon.(258)

Another branch of this Pythagorean scheme is preserved by Plutarch, in one of his scientific tracts. It relates to the distances of the revolving bodies, which were determined by numerical ratios. The central fire was set down as unity; and the proportionate distances from the centre to the several orbits were represented numerically by assuming the distance to the antichthon to be three, to the earth nine, to the moon twenty-seven, and so on, tripling each successive number. According to this method, the sun was represented by the number 729,

---

(256) Metaph. i. 5; Simplicius, Schol. Aristot. p. 505 a, states that the Pythagoreans called the moon the antichthon; but his own report of Aristotle's account of the Pythagorean scheme shows that it supposed the moon to be different from the antichthon, and to be next in order beyond the earth, while the antichthon was between the earth and the central fire.

(257) Stob. Ecl. Phys. i. 26; Plut. Plac. ii. 29; Galen, c. 15; Euseb. Præp. Ev. xv. 51. Philip of Opus was a disciple of Plato, and a voluminous writer: see Diog. Laert. iii. 46. In the list of his writings given by Suidas, several treatises on astronomical subjects are included.

(258) Plut. Plac. Phil. iii. 9.

which is both a square and a cube; and therefore the sun was called both square and cube.(259)

It results from this account that the Pythagorean system of the universe, as reported by Philolaus, scarcely deserves the name of a philosophical hypothesis, devised for the explanation of observed phenomena. It is rather a work of the imagination, guided and governed by certain mystical abstractions, and certain principles as to the virtues of numbers. The central fire and the antichthon were invisible; the motion of the earth in a circle concentric with the orbits of the sun, moon, and stars, explained nothing, and failed in reducing the apparent motions of numerous bodies to the real motion of one body. The mythological appellations given to the central fire harmonize with the fanciful character of the entire system. Nevertheless this hypothesis, unscientific and useless as it was, had the undoubted merit of boldness and originality. So far as we know, it was the first speculative system which gave to the earth a motion in space, and made it move in a circular orbit round a centre.

Aristotle thinks that the opinion as to the earth not being the centre of the universe, may have been shared by other philosophers in his time besides the Pythagoreans.(260) Simplicius is unable to discover who these others were; he only knows of a certain Archidemus, subsequent to Aristotle, who entertained the opinion in question.(261)

Although the Philolaic and Copernican systems have been identified by some modern writers,(262) it is plain that they have

---

(259) See Plutarch de Anim. Procreat. e Tim. c. 31. The number 729 is the square of 27 and the cube of 9. On the virtues of the number 27 (which is the cube of 3), according to the Pythagorean doctrine, see Boeckh, Philolaos, p. 77.

(260) πολλοῖς δ᾽ ἂν καὶ ἑτέροις συνδόξειε μὴ δεῖν τῇ γῇ τὴν τοῦ μέσου χώραν ἀποδιδόναι· τὸ πιστὸν οὐκ ἐκ τῶν φαινομένων ἀθροῦσιν ἀλλὰ μᾶλλον ἐκ τῶν λόγων. De Cœl. ii. 13.

(261) Schol. Aristot. p. 505 a. Archidemus, the Stoic philosopher, of whom some notices are preserved, is probably meant; see Dr. Smith's Dict. of Anc. Biog. in v.

(262) See the references in Martin, ib. p. 93.

nothing in common, except the motion of the earth round a centre. The essential features of the latter—the immovability of the sun, and its central position—are wanting in the Philolaic scheme. The extant accounts of the Philolaic hypothesis are indeed meagre, and to some extent conflicting; but its outlines rest upon the certain authority of Aristotle; and the conjecture of Röth, a recent historian of ancient philosophy, that Philolaus supposed the earth to be at the centre, either wholly immovable, or with no other motion than a rotation upon its axis, and the central fire to be within the earth,(263) may be confidently rejected, as inconsistent with the ancient accounts.

The Pythagorean School appears to have varied in its astronomical and cosmological theories. Thus we are informed by Plutarch, that Heraclides of Pontus, and Ecphantus the Pythagorean, supposed the earth to move, but to turn only upon its axis from west to east, and not to have a motion of translation.(264) The time of Ecphantus is unknown, but he appears to have been posterior to Aristotle.

Again, several late writers attribute to Pythagoras himself

---

(263) Geschichte der Abendländischen Philosophie, ii. 2, p. 811, and note 1284, p. 855—860, and notes 1300, 1343. The words of Aristotle, in his treatise De Cœlo, prove beyond a doubt that the Pythagoreans placed the fire at the centre, and conceived the earth to move in a circle round it, like the other heavenly bodies. ἐπὶ μὲν γὰρ τοῦ μέσου πῦρ εἶναί φασι, τὴν δὲ γῆν, ἓν τῶν ἄστρων οὖσαν, κύκλῳ φερομένην περὶ τὸ μέσον, νύκτα τε καὶ ἡμέραν ποιεῖν, ii. 13. Aristotle adds that many others, besides the Pythagoreans, who found their systems upon abstractions, and not upon phenomena, might agree in thinking that the earth is not in the centre of the universe. In Metaph. i. 5, the earth is likewise included in the ten bodies φερόμενα κατὰ τὸν οὐρανόν.

M. Martin, ib. p. 96—8, concurs with Boeckh in holding that the Philolaic system made the earth revolve with the nine other bodies round the central fire. Likewise Ritter, Gesch. der Philosophie, vol. i, p. 408—411.

Röth attempts to weaken the adverse authority of Aristotle by arbitrarily altering the text in the perfectly perspicuous and decisive passage De Cœlo (ἔνιοι for ὅσοι), note 1300. The alterations of text made in note 1353, for the purpose of supporting the writer's own hypothesis, are likewise untenable.

(264) Plut. Plac. iii. 13. The same opinion is ascribed to Ecphantus by Origen, Phil. p. 19, τὴν δὲ γῆν μέσον κόσμου κινεῖσθαι περὶ τὸ αὐτῆς κέντρον ὡς πρὸς ἀνατολήν.

the doctrine that the earth is an immovable sphere, placed at the centre of the universe, round which the sphere of the fixed stars, the five planets, and the sun and moon revolve.([265]) This, which was the common belief of the scientific astronomers of Greece, was undoubtedly shared by some of the earlier Pythagorean philosophers, and probably by Pythagoras himself.

Simplicius enumerates the following philosophers—viz., Empedocles, Anaximander, Anaximenes, Anaxagoras, Democritus, and Plato—as holding that the earth is at the centre of the universe.([266]) The Orphic sect compared the heaven to the white of an egg, and the earth to the yolk: which comparison illustrates, by a sensible image, the idea of the universe commonly held by the Greek philosophers.([267]) Aristophanes in the Clouds likewise speaks of the earth as freely suspended in the air.([268])

A version of this belief, likewise attributed to Pythagoras himself, is more intricate and artificial. According to this scheme there are 12 orbits, which succeed each other in the following order, beginning from the remotest:—1. The sphere of the fixed stars. 2. Saturn. 3. Jupiter. 4. Mars. 5. Venus. 6. Mercury. 7. The Sun. 8. The Moon. 9. The sphere of fire. 10. The sphere of air. 11. The sphere of water. 12. The Earth.([269]) This is a reduction of the Philolaic scheme to the received Greek system; for the earth is evidently taken as the centre, round which the other bodies revolve. According to another version of the Pythagorean doctrine, the earth was the

---

([265]) Diog. Laert. viii. 25 (repeated by Suid. in Πυθαγόρας). Varro, L. L. vii. 17, which latter passage is uncertain. The harmonic intervals of the planets, attributed to Pythagoras in Censorin. c. 13; Plin. N. H. ii. 19, 20, likewise assume that the earth is at the centre.

([266]) Schol. Aristot. p. 505 a.

([267]) Achill. Tat. c. 4. See other illustrations of the immobility of the earth in the same passage.

([268])   ὦ δέσποτ' ἄναξ, ἀμέτρητ' ἀήρ, ὃς ἔχεις τὴν γῆν μετέωρον.
                      Nub. 264.

([269]) Anon. de Vit. Pyth. 10 (in Kiessling's edition of Iamblichus and Porphyrius, vol. ii. p. 108).

centre of the system, the fire was at the centre of the earth, and the moon was the antichthon.([270])

The Pythagoreans are declared to have first laid down the position of the planets;([271]) and to have first taught that the motions of the sun, moon, and five planets are circular and equable.([272]) They are likewise reported to have held that the force which causes the motion of the starry heaven is at the centre of the universe.([273])

The early Pythagoreans further conceived that the heavenly bodies, like other moving bodies, emitted a sound; and they supposed that the sounds of the different spheres, in their respective circular orbits, were combined into a harmonious symphony.([274]) Hence they established an analogy between the intervals of the seven planets, and the intervals of the tones in the musical scale.([275]) These intervals were determined in the following manner, according to Censorinus:([276])

	TONES.
From the Earth to the Moon	1
From the Moon to Mercury	$\frac{1}{2}$
From Mercury to Venus	$\frac{1}{2}$
From Venus to the Sun	$1\frac{1}{2}$
From the Sun to Mars	1
From Mars to Jupiter	$\frac{1}{2}$
From Jupiter to Saturn	$\frac{1}{2}$
From Saturn to the Zodiac	$\frac{1}{2}$
	6

The distribution of Pliny agrees, except that he makes $1\frac{1}{2}$ tones

---

(270) Simplicius ad Aristot. de Cœlo, p. 505, Brandis.

(271) Simplicius, Schol. Aristot., p. 497 a.

(272) Geminus, c. i. p. 2. Sosigenes ascribed this hypothesis to Plato, below ch. iii. § 3.

(273) Simplicius, ib. p. 453 a.

(274) Aristot. de Cœl. ii. 9. Compare Athen. xiv. p. 632. Pseud. Aristot. de Mundo, c. 6.

(275) See Theo Smyrn. c. 15, p. 181, and the verses of Alexander of Ephesus cited by him, Achill. Tat. c. 16; Cic. Somn. Scip. c. 5; Macrob. Comm. in Somn. Scip. ii. 1—4.

(276) c. 13.

from Saturn to the zodiac, and therefore his entire scale contains 7 tones instead of 6.([277])

In this music of the spheres, the moon, as being the lowest in the system, and the tardiest in its movements, represented the grave end of the scale; while the sphere of the fixed stars, as being the highest above the earth, and the most rapid in its circular motion, represented the acute end.

Aristotle informs us that the Pythagoreans answered the obvious objection founded on the inaudibility of the music of the spheres, by saying that constant habit rendered it imperceptible.([278]) The explanation given by Cicero is, that the sound is so loud, as to transcend the capacity of our sense of hearing.([279])

Pythagoras is further stated to have originated the division of the heavenly sphere into five zones, cut obliquely by the zodiac, and to have considered the obliquity of the zodiac as the cause of the retrograde motion of the sun at the solstices. The best authorities, however, attribute to his follower, Œnopides of Chios, the discovery of the obliquity of the sun's course in the zodiac, and its proper motion in reference to the stars. This knowledge Œnopides is stated to have derived from Egypt.([280]) Œnopides appears to have been a contemporary of Anaxagoras.([281])

Alcmæon of Croton, a disciple of Pythagoras, who was at the prime of life when Pythagoras was an old man,([282]) held

---

([277]) ii. 22.

([278]) De Cœl. ib.

([279]) De Rep. vi. 18, repeated by Censorinus, c. 13. Compare Martin, Timée de Platon, tom. ii. p. 36.

([280]) Eudemus, cited below, n. 293; Diod. i. 98; Anon. de Vit. Pyth. 12; Plut. Plac. ii. 12, 23.
Compare Martin, ib. p. 101—8.

([281]) See Plat. Erast. ad init.; Diog. Laert. ix. 37. Two sayings attributed to him are preserved in the Parallela Sacra of Joann. Damasc. ad calc. Stob. Ecl. p. 760, ed. Gaisford. Compare Dr. Smith's Biogr. Dict. in v.

([282]) Aristot. Metaph. i. 5.

that the movement of the planets was from west to east, contrary to that of the fixed stars.(283)

The Pythagoreans are moreover related to have held that comets are planetary stars, which, through lapse of time, have lost their brilliancy and the constancy of their appearances above the horizon; they compared them with the planet Mercury, which seldom rises above the horizon, and only appears at long intervals.(284)

A portion of the Pythagorean sect likewise explained the Milky Way, by supposing that it was composed of stars which had either been disturbed in the catastrophe of Phaethon, or had been burnt up by the sun in its passage through the heavens.(285) This hypothesis, which is seriously refuted by Aristotle, hardly belongs to scientific astronomy. Œnopides is reported to have held that the Milky Way was the original course of the sun; but that when the sun turned back at the Thyestean banquet, it changed its course to the zodiac.(286) Manilius, in enumerating the explanations of the Milky Way, mentions the theory that it had been the ancient course of the sun; he likewise adverts to the mythological explanations of the conflagration of Phaethon and the effusion of Juno's milk; he adds another hypothesis, that it is formed of the souls of illustrious men translated into heaven.(287)

Pythagoras is reported not only to have occupied himself about the mystical properties of numbers, but also to have cultivated geometry, and to have solved some fundamental geometrical problems.(288) He appears to have designated

---

(283) Plut. Plac. ii. 16; Stob. Ecl. i. 24; Galen, c. 13.
(284) Aristot. Meteor. i. 6.
(285) Aristot. Meteor. i. 8. In the words οἷον οὖν διὰ τὸ κεκαῦσθαι τὸν τόπο τοῦτον ἤ τι τοιοῦτον ἄλλο πεπονθέναι πάθος ὑπὸ τῆς φορᾶς αὐτῶν (φασίν), the sense seems to require αὐτοῦ for αὐτῶν. The substance of this passage is repeated in Stob. Ecl. i. 26; Plut. Plac. Phil. iii. 1.
(286) Ap. Achill. Tat. c. 24, p. 86, ed. Petav. See above, p. 96.
(287) i. 701—802.
(288) Plut. Symp. viii. 2, 4. Non posse suav. vivi sec. Epic. 11. In the latter passage Pythagoras is stated to have sacrificed an ox upon the dis-

geometry by the name of ἱστορία, or 'investigation.'([289]) He is nevertheless said to have derived his geometrical science from Egypt.([290])

It may be observed that Aristotle speaks more than once of the Pythagorean School, as 'the philosophers *called* Pythagorean;' by which expression he appears to intimate that the Italian sect known by that name did not in his time represent faithfully the opinions of their founder.([291])

It is impossible, with such information as we possess, to determine the periods of the successive astronomical tenets of the Pythagorean School; neither are we able in general to assign the doctrines to their respective authors, or to distinguish the original dogmas of Pythagoras from the system engrafted upon them by his followers. The later writers did not attempt to discriminate, but attributed every tenet of the Pythagorean sect to its founder. In general, there was doubtless a development of the original idea. Thus the mystical doctrines of the music of the spheres may have originated with Pythagoras himself: but the assumption of the number seven for the planets, and the arrangement of an analogy between their intervals and

---

covery of the relation of the square of the hypotenuse to the squares of the other two sides; or of the problem concerning the area of the parabola, Diog. Laert. i. 25, viii. 12; Callimachus, ap. Diod. x. 11. ed. Bekker.

An epigram referring to this sacrifice is cited in Diog. Laert. viii. 12. Plut. non posse suav. viv. sec. Ep. ubi sup. Athen. x. p. 418 F.

ἡνίκα Πυθαγόρης τὸ περικλεὲς εὕρατο γράμμα
κεῖν' ἐφ' ὅτῳ κλεινὴν ἤγαγε βουθυσίην.

(Anth. Pal. vii. 119.)

This sacrifice is also alluded to by Cic. Nat. D. iii. 36. Ἀπολλόδωρος ὁ λογιστικὸς is quoted as an authority by Diogenes. The name Apollodorus should be restored for Apollodotus in Plutarch.

([289]) Iamblich. de vit. Pyth. § 89. Plato, Phæd. p. 96, applies the word ἱστορία to physical investigation; and the word ἱστορία, used simply, denotes physical knowledge in the fine verses of Euripides, frag. 101, Dindorf.

([290]) Diod. ubi sup.

([291]) In De Cœlo, ii. 13, Aristotle says: οἱ περὶ τὴν Ἰταλίαν, καλούμενοι δὲ Πυθαγόρειοι. In Cœl. ii. 2, § 1. Meteor. i. 8, and Metaph. i. 5, οἱ καλούμενοι Πυθαγόρειοι; Meteor. i. 6 τῶν Ἰταλικῶν τινες καὶ καλουμένων Πυθαγορείων.

the intervals of the tones in the harmonic scale, is unquestionably of posterior date.(292)

§ 18 A cycle of 59 years, made apparently for the adjustment of the calendar, is ascribed to the early Pythagorean School. Œnopides, who has been already mentioned as a contemporary of Anaxagoras, dedicated a brazen tablet at Olympia, containing the details of a cycle of this length.(293) Censorinus states that his year consisted of $365\frac{22}{59}$ days.(294) If this was his assumption, his cycle contained 21,557 days; which is $7\frac{1}{4}$ days in excess of 59 Julian years, and therefore admits of no intercalation. A cycle of 59 years is likewise ascribed to Philolaus, with 21 intercalary months; and he is stated to have fixed the natural year at $364\frac{1}{2}$ days.(295) The obvious interpretation of these statements is, that Philolaus made a cycle of 59 years of $364\frac{1}{2}$ days, and added 21 months. If this be the meaning, the two statements are irreconcilable with each other; for the object of intercalation is to equalize the year of the cycle with the true solar year; and 59 years of $364\frac{1}{2}$ days only differ from 59 years of $365\frac{1}{4}$ days by $44\frac{1}{4}$ days. An intercalation of 21 months, or $619\frac{1}{2}$ days, would therefore produce an enormous error in excess. We may indeed assume, with Boeckh, that the year of $364\frac{1}{2}$ days includes the intercalary months distributed over the cycle:(296) this interpretation is, however, forced; and it still leaves a wide uncorrected depar-

---

(292) Iamblichus, de vit. Pyth. § 31, speaks of Pythagoras as the originator of scientific astronomy in Greece. Comp. Ovid, Met. xv. 71.

(293) Æl. n, V. H. x. 7. Eudemus, in his History of Astronomy, stated that Œnopides εὗρε πρῶτος τὴν τοῦ ζωδιακοῦ διάζωσιν καὶ τὴν τοῦ μεγάλου ἐνιαυτοῦ περίστασιν. (Ap. Theon. Smyrn. c. 40, p. 322, ed. Martin.)

(294) Censorin. 19.

(295) Censorin. 18, 19. The cycle of 59 years is mentioned by Plut. Plac. ii. 32, but is not attributed to any one. Stob. Ecl. i. 9, ascribes it to Pythagoras himself; Galen, c. 16; Euseb. xv. 54. Compare Boeckh, ib. No. 16.

Concerning the cycle of Œnopides, see Dodwell, de Cyclis, p. 262—6, 782.

(296) Philolaos, p. 134.

ture from the true solar year. Some corruption in the numbers may perhaps be suspected.([297])

The cycle of Philolaus would be rendered more intelligible, if we supposed his year to consist of $354\frac{1}{2}$ instead of $364\frac{1}{2}$ days. A cycle of 59 years of $354\frac{1}{2}$ days, together with 21 intercalary months of $29\frac{1}{2}$ days, would make a total of 21,535 days; while 59 Julian years contain $21,549\frac{3}{4}$ days; being a difference of only $14\frac{3}{4}$ days. It is difficult to explain either of these Pythagorean cycles. The number 59 may perhaps, as Boeckh has conjectured,([298]) be the double of the mean lunar month of $29\frac{1}{2}$ days.

§ 19 Leucippus, the founder of the Atomic philosophy, is said to have been a disciple of either Parmenides or Zeno, and was probably somewhat posterior in time to Anaxagoras.([299])

Upon astronomical subjects the following were his doctrines. That the universe is a sphere; that its external coat, which involved it like a membrane, was able, in the whirl of its movement, to attract foreign bodies; that these at first were moist and muddy, but afterwards were hardened; and that being carried round in the rapid circular motion, they ultimately acquired a fiery nature, and became stars. That the stars were thus ignited by the rapidity of their motion, but that the sun received its fiery property from the stars. That the orbit of the sun was furthest from, and that of the moon nearest to, the earth, and that the planets were in the intermediate space; that eclipses of the sun and moon were caused by the inclination of the earth to the south: that the axis of the earth inclines to the south, on account of the rarity of the southern regions, and their consequent inability to resist pressure; whereas the northern regions, being frozen, are hard and solid, and

---

([297]) Ideler, vol. i. p. 303, thinks it incredible that Philolaus should have given the solar year only $364\frac{1}{2}$ days.

([298]) Philolaos, p. 135. The comments of Röth, ii. 2, note 1269, throw no light upon the subject.

([299]) See Brandis, Gesch. der Phil. vol. i. p. 294.

resist the depressing force. That the cause of the comparative frequency of lunar, as compared with solar eclipses, was the inequality of the orbits of the sun and moon. He held that the earth was suspended in space at the centre, being fixed there by the creative vortex; and that its shape was that of a circular plane.[300]

§ 20 Democritus of Abdera, the disciple and associate of Leucippus, belonged to the generation next after Anaxagoras. He was a youth when Anaxagoras was an old man.[301] He may be considered as the contemporary of Socrates. His birth may be referred with probability to the year 459 B.C.; he is stated to have reached a great age, so that he must have lived beyond the middle of the next century.[302] He appears from his own testimony to have travelled to many foreign countries.[303] It is asserted that he visited Egypt for the purpose of learning geometry, and extended his journeys to the Chaldæans at Babylon, to Persia, to the Red Sea, and even to India and Ethiopia, where he had intercourse with the gymnosophists.[304] That he made scientific observations in these countries, and

---

(300) Diog. Laert. ix. 30—33; Plut. Plac. ii. 7, iii. 12; Stob. Ecl. Phys. i. 15, 22; Euseb. Præp. Ev. xv. 38; Origen, Ref. Hær. p. 17; Galen, c. 11, 21.

The passage respecting the earth stands thus in Diogenes, n. 30, τὴν γῆν ὀχεῖσθαι περὶ τὸ μέσον δινουμένην, σχῆμά τε αὐτῆς τυμπανοειδὲς εἶναι. The words περὶ τὸ μέσον δινουμένην appear to mean, 'fixed by the δίνη about the centre;' they may be compared with the celebrated passage respecting the earth in the Timæus. The statement that Leucippus supposed the shape of the earth to resemble that of a tympanum, recurs in Plut. Plac. iii. 10; Euseb. Præp. Ev. xv. 56. The Greek tympanum appears to have been like a tambourine, or to have been hemispherical.

In Stob. Ecl. Phys. i. 22, Λεύκιππος καὶ Δημόκριτος χιτῶνα κύκλῳ καὶ ὑμένα περιτείνουσι τῷ κόσμῳ, διὰ τῶν ἀγκιστροειδῶν ἄστρων ἐμπεπλεγμένον, the sense requires ἀτόμων for ἄστρων, the conjecture of Heeren. The words συμπλέκειν and περιπλέκειν are used, with reference to the Leucippic concourse of atoms, by Origen, ib. p. 17.

(301) This fact rests upon the authority of Democritus himself. He was stated to be 40 years younger than Anaxagoras, Diog. Laert. ix. 34, 41. Aristotle, Meteor. ii. 7, states that Democritus was later than Anaxagoras.

(302) Concerning the lifetime of Democritus, see Democriti Fragmenta, ed. Mullach, p. 2—36.

(303) See Mullach, p. 3.

(304) Mullach, p. 40—49. See Cic. de Fin. v. 19; Strab. xv. i. 38.

held intercourse with learned men, he himself declares; but he adds that not even the land-measurers of Egypt excelled him in geometrical demonstrations; from which it may be inferred that he did not consider himself as having derived much instruction from the Egyptian or other foreign priests. He is reported by Diodorus to have passed five years in Egypt, and to have there obtained much astronomical knowledge.([305]) In the list of the works of Democritus are included treatises on the sacred writings at Babylon and at Meroe, and a discourse on the Chaldæans.([306]) Among his numerous writings are several treatises on astronomical and mathematical subjects;([307]) he, moreover, like other physical philosophers of that period, framed a cycle of intercalation for the government of the calendar. It consisted of eighty-two years, with twenty-eight intercalary months.([308]) This cycle (as Mullach, the collector of the fragments of Democritus, has remarked) evidently assumes a year of 355 days, and intercalary months of thirty days. Upon this assumption, the cycle of Democritus accords with eighty-two years of $365\frac{1}{4}$ days, within only half a day.([309]) The year is nearly the lunar year of 354 days; but as the intercalary months are of thirty days, this cycle does not agree with the moon.

Democritus supposed the earth to be at the centre of the universe, and immovable. He conceived its form to be that of a disc, but hollow in the middle;([310]) and its support to be

---

([305]) Clem. Alex. Strom. i. p. 304; Sylb. Euseb. Præp. Ev. x. 2; Diod. i. 98. The statement that he passed 80 years in Egypt must be erroneous. See Mullach, p. 19. Concerning the word ἁρπεδονάπτης, or *rope-tier*, see Sturz de Dialect. Maced. p. 98.

([306]) Mullach, p. 124—6.

([307]) Mullach, p. 129—131, 142—147.

([308]) Censorin. 18. Compare Mullach, p. 143—5.

([309]) $82 \times 355 + 28 \times 30 = 29110 + 840 = 29950$; and $82 \times 365\frac{1}{4} = 29950\frac{1}{2}$.

([310]) Δημόκριτος δισκοειδῆ μὲν τῷ πλάτει, κοίλην δὲ τὸ μέσον, Plut. Plac. iii. 10. Repeated, with slight verbal variations, in Galen, c. 21; Euseb. Præp. Ev. xv. 56.

derived from the resistance of the subjacent air.[311] He held the shape of the surrounding heaven to be spherical.[312]

He accounted for the immobility of the earth at the centre by saying, that, as it was equidistant on all sides from the containing sphere, there was no reason why it should incline in one direction rather than in another. Hence he explained how the earth remained in a state of rest, though it was subject to vibratory motions, which produced earthquakes.[313] With regard to the inhabited portion of the earth, he first represented it as a rectangle, having its length to its breadth as three to two.[314] He accounted for the inclination of the earth's axis to the south, by the more luxuriant growth of plants in the milder climate of the southern part of the earth, and its consequent preponderance over the northern part.[315] He held the doctrine of a plurality of worlds, at various distances from each other, and of different magnitudes. He supposed the earth which we inhabit to have existed before the stars; he placed the moon nearest to the earth, beyond the moon the sun, and beyond the sun the fixed stars. He conceived the planets as stationed at unequal distances.[316] He perceived that the movement of the stars is from east to west. He likewise held the following tenets:—That the sun is an ignited mass of stone, and that the solstices are produced by its vortical movement;[317] that the moon is a solid body, containing mountains and valleys, which caused the marks apparent upon the moon's face;[318] that the velocity of the motion of the celestial bodies is in proportion to their distance from the earth:

---

[311] Aristot. de Cœl. ii. 13, § 16.
[312] Stob. Ecl. Phys. i. 15.
[313] Plut. Plac. iii. 15; Galen, c. 21.
[314] Agathemer. i. 1.
[315] Plut. Plac. iii. 12; Galen, c. 21.
[316] Origen, Ref. Hær. p. 17. Compare Euseb. Præp. Ev. i. 8, vol. p. 50, ed. Gaisford; Diog. Laert. ix. 44.
[317] Stob. Ecl. i. 25; Galen, c. 14.
[318] Stob. Ecl. i. 26, p. 550, 564; Heeren; Galen, c. 15.

hence that the motion of the fixed stars is more rapid than those of the sun and moon, the moon being the least rapid of all; and that the stars gain upon the sun and moon.(319) With respect to planets, Democritus did not profess to know more than their existence; he did not attempt to name them, or to determine their number and orbits.(320) Like Anaxagoras, he conceived that comets were the result of a concourse of certain planetary stars.(321)

---

(319) See Lucret. v. 619—34, where this singular attempt to account for the proper motion of the sun is clearly expounded.

(320) Democritus, subtilissimus antiquorum omnium, suspicari ait se plures stellas esse quæ currant; sed nec numerum illarum posuit nec nomina, nondum comprehensis quinque siderum cursibus. Seneca, Nat. Quæst. vii. 3.

(321) Aristot. Meteor. i. 6; Galen, c. 17; Stob. Ecl. Phys. i. 27.

Chapter III.

SCIENTIFIC ASTRONOMY OF THE GREEKS FROM PLATO TO ERATOSTHENES.

§ 1 THE writings of Plato do not contain any dialogue which turns principally upon astronomy. Even the Timæus, though it touches upon astronomical questions, is mainly an exposition of cosmological and physical ideas. The great philosopher, however, in his discursive flights of speculation, comes occasionally into contact with astronomy proper; and these passages must occupy a place in the historical survey which we are now taking.

The general view which Plato forms of astronomical science is diametrically opposed to that of Socrates. In the Republic he condemns all close study of the celestial phenomena as immersing the mind in matter, and withdrawing it from the contemplation of abstract truth. He rejects, as a vulgar prejudice, the idea that the observation of the heavenly bodies is sublime; and he approves of astronomical speculation only as an exercise of the pure intellect.(1) In like manner he held that the supposed atheistic tendency of astronomy, owing to its use of mechanical causes, was founded on a gross and material conception of the science; he condemned the doctrines of those who taught that the heavens were full of stones and earth, circulating in space.(2)

Plato appears to have held that the earth is a sphere, suspended in space, and stationary in the midst of the universe.(3)

---

(1) De Rep. vii. p. 529, 530.
(2) De Leg. xii. p. 967.
(3) In the Phædo, c. 132, p. 109, Socrates lays it down that the earth is of circular form, in the midst of the heaven; and that it is kept in its

His language on this subject is not free from ambiguity, and even Aristotle understood him to affirm that the earth has a motion of rotation about the axis of the universe. Nor was this opinion confined to Aristotle among the ancients. It was shared by others, who lived after him, and who, therefore, had time for examining the words in the Timæus, for comparing them with the context, and weighing the arguments of those who took a different view. But the majority of the best modern expositors agree in rejecting this construction.[4] It appears probable that Plato meant merely to describe the world as *wound* or *twined* round the axis of the universe, this word referring to the globular shape of the earth, and suggesting, metaphorically, the idea of a ball of thread.[5]

It was reported by Theophrastus that Plato, in his old age, repented at having placed the earth at the centre of the universe, on the ground that this was not its fitting position.[6]

---

place by its equilibrium, and by its resemblance to the surrounding circular heaven. That the sphericity of the earth was taught by the followers of Socrates is stated by Cleomed. i. 8. Compare Ukert. i. 2, p. 29. Martin, Timée, p. 90, 118.

[4] See Martin, ib. vol. ii. p. 88—91; Boeckh, Kosmisches System des Platon, Berlin, 1852; Prantl, Translation of Aristot. de Cœlo, p. 312. Plato is distinct in making the celestial sphere revolve upon its own axis, in the previous part of the Timæus. διὸ δὴ κατὰ ταὐτὰ ἐν τῷ αὐτῷ καὶ ἐν ἑαυτῷ περιαγαγὼν αὐτὸ ἐποίησε κύκλῳ κινεῖσθαι στρεφόμενον, § 11, p. 34.

[5] γῆν δέ, τροφὸν μὲν ἡμετέραν, εἰλλομένην δὲ περὶ τὸν διὰ παντὸς πόλον τεταμένον, Tim. c. 15, p. 40. For Aristotle's interpretation of these words, see his treatise de Cœlo, ii. 14. Compare Plut. Plat. Quæst. viii.; Cic. Acad. ii. 39. The difficulty of reference to the manuscript books was much greater than it is to printed books; and the ancients were obviously in the habit of citing from memory. Concerning the meaning of εἰλλομένην in this passage, see Boeckh, Kosm. Syst. des Plat. p. 64-5, 67. Letronne, Journ. des Sav. 1841, p. 76, thinks that the word implies the rotundity of the earth, without expressing any rotatory motion. Compare Journ. des Sav. 1819, p. 329. The use of the word πόλος for axis is peculiar to this passage; see notes A. and C., at the end of the chapter.

Cicero reports the theory of Hicetas that the earth is in the centre of the universe, but moves on its axis; that the other heavenly bodies are stationary, and that their apparent motions are due to the real motion of the earth. He then proceeds to say: Atque hoc etiam Platonem in Timæo dicere quidam arbitrantur, sed paullo obscurius, Acad. ii. 39.

[6] Θεόφραστος δὲ καὶ προσιστορεῖ τῷ Πλάτωνι πρεσβυτέρῳ γενομένῳ μεταμελεῖν, ὡς οὐ προσήκουσαν ἀποδόντι τῇ γῇ τὴν μέσην χώραν τοῦ παντός, Plut.

The truth of this statement is questioned by M. Martin; who remarks that Plato was already advanced in years when he composed the Timæus.(⁷) But the testimony of Theophrastus, the disciple of Aristotle, and nearly his contemporary, has great weight on this point. The ground of the opinion alludes to the Pythagorean doctrine mentioned by Aristotle, that the centre is the most dignified place, and that the earth is not the first in dignity among the heavenly bodies. It has no reference to observed phenomena, and is not founded on inductive scientific arguments. It may be added in support of Theophrastus, that Plato is stated, after the death of Socrates, to have studied the Pythagorean philosophy in his visits to Italy and Sicily, to have there conversed with Archytas of Tarentum, and with Timæus the Locrian, both members of the Pythagorean sect, and to have obtained the writings of Philolaus.(⁸) Now the doctrine as to the superior dignity of the central place, and of the impropriety of assigning the most dignified station to the earth, was (as has been already shown) of Pythagorean origin, and was probably combined with the Philolaic cosmology.

Plato alludes to the great circles of the equator and ecliptic in a manner which shows him to have been aware of their inclination to each other.(⁹) He likewise makes mention of the planets in a manner which proves that in his time they had attracted attention in Greece, and that their movements had been explored by the astronomers. Plato speaks of seven

---

Plat. Quæst. viii. 1. ταὐτὰ δὲ καὶ Πλάτωνά φασι πρεσβύτην γενόμενον διανενοῆσθαι περὶ τῆς γῆς ὡς ἐν ἑτέρα χώρα καθεστώσης, τὴν δὲ μέσην καὶ κυριωτάτην ἑτέρῳ τινὶ κρείττονι προσήκουσαν, Plut. Num. 11. Compare Aristot. de Cœl. ii. 13, 3.

(7) Timée, vol. ii. p. 91. Boeckh agrees with Martin, Kosm. Syst. des Plat. p. 149.

(8) See Diog. Laert. iii. 9; Iamblichus, de Vit. Pythag. c. 31, p. 406, ed. Kiessling; Cic. de Rep. i. 10, de Fin. v. 29; Gell. iii. 17. Iamblichus describes the three volumes as not being the compositions of Philolaus himself; but as traditionary writings of the school; and he says that before Philolaus, all the treatises of the Pythagorean sect had been kept secret. Compare Ideler, Mus. der Alterth. vol. ii. p. 405; Martin, Timée, vol. i. p. 42; Theo Smyrn. p. 120, ed. Martin.

(9) Martin, ib. p. 39.

planets; namely, the sun and the moon, together with the five planets visible to the naked eye. To the latter he gives the names by which they appear to have been originally known in Greece. These names are Lucifer and Hesperus, the Morning and Evening star, which he identifies as a single planet; Stilbon, also called the Star of Mercury; Pyroeis, or the fiery, which Plato remarks is so named on account of its red colour; Phaethon, the planet of the slowest course but one; and, lastly, the planet of the slowest course, which by some is called Phænon.(10) He not only knows that Saturn makes the slowest circuit, and that next after Saturn comes Jupiter; but he likewise distinguishes Mercury and Venus from the other three planets, as having courses different from them.(11)

The early Greeks conceived the Morning star to be distinct from the Evening star. They called the former Eosphorus or Phosphorus, and the latter Hesperus.(12) Plato consolidates the two stars into one, under the name of Eosphorus: it had not as yet received the name of the Star of Venus. With regard to the other planets, Plato says that they were first observed and first received names in Egypt and Syria. This priority he attributes to the clearness of the summer sky in those countries, as compared with that of Greece. He designates them, however, by names, which, he says, have been given to them by some persons in Greece,(13) and which, though not at that time popular, were at all events native and unborrowed. The three names, Stilbon, Phaethon, and Phænon, have nearly the same signification, and express nothing distinctive and charac-

---

(10) Tim. p. 38; Epinom. p. 987.

(11) Tim. p. 38. Compare Martin, ib. p. 66. Pyroeis is one of the horses of the sun in Ovid, Met. ii. 193.

(12) See Pauly, art. Hesperus; Martin, ib. p. 63. Plato, however, appears to treat the Morning and Evening stars as distinct in Leg. vii. 22, p. 821. In the epigram ascribed to Plato, Anth. Pal. vii. 670:—

ἀστὴρ πρὶν μὲν ἔλαμπες ἐνὶ ζωοῖσιν ἐῷος,
νῦν δὲ θανὼν λάμπεις ἕσπερος ἐν φθιμένοις,

the Morning and the Evening stars appear to be distinguished.

(13) Epinom. p. 987.

teristic; the name Pyroeis for Mars alludes to the peculiar colour of that planet, which can be perceived by the naked eye.([14]) Plato likewise applies to Stilbon the appellation of the Star of Mercury.

§ 2 We have now reached the period when the Greek astronomy, though still in an imperfect and unformed state, was beginning to assume the character of a science founded on systematic observation. Eudoxus of Cnidos, whose life probably extended from about 406 to 350 B.C., who was junior to Plato, and senior to Aristotle,([15]) may be considered as the father of scientific astronomical observation in Greece. He was a disciple of Plato and Archytas; he resided chiefly at Cyzicus, at the mouth of the Euxine Sea. He is reported to have visited Egypt, and to have there received astronomical instruction from the priests. The account of Strabo is, that Plato and Eudoxus remained thirteen years at Heliopolis, in communication with the priests, for the purpose of extracting their secret doctrine.([16]) This period appears, however, to be exaggerated, nor is it likely that the visits of Plato and Eudoxus were simultaneous or connected with each other. More credit is due to the statement of Sotion, who wrote the History of the Greek Philosophers about 205 B.C.([17]) This statement is, that Eudoxus went to Egypt with Chrysippus, the physician, carrying with him letters of recommendation from Agesilaus to Nectanabis, king of Egypt; that Nectanabis introduced him to the priests; that he remained in Egypt sixteen months, shaving his eyebrows after the native fashion; and that he then composed his

---

(14) M. Arago states that the redness of Mars is greater when the planet is seen by the naked eye, than when it is viewed through a telescope, Popular Astronomy, b. 24, c. 7.

(15) Eudoxus was born about twenty-four years after Plato, and about twenty-two years before Aristotle. For a full account of Eudoxus and his writings, see the dissertation of Ideler, Berlin Transactions, 1828 and 1830. Letronne, on Eudoxus, Journ. des Sav. 1840, p. 744, fixes the birth of Eudoxus at 409, his visit to Egypt at 362, and his death at 356 B.C.

(16) xvii. i. 29.

(17) See Clinton, F. H. vol. iii. p. 526.

cycle of eight years.([18]) The expedition of Agesilaus to Egypt, in which he placed Nectanabis on the throne, occurred in 361 B.C.; ([19]) this year must therefore be the date of the recommendation given by Agesilaus to Eudoxus, and of the visit of the latter to Egypt, if the account of Sotion be true. Eudoxus was at this time more than forty years old, and he died at the age of fifty-three. Diogenes Laertius reports that while Eudoxus was residing with Chonuphis of Heliopolis, the sacred bull Apis licked his garment; whence the priests predicted that he would be celebrated, but short-lived.([20]) This prediction, which supposes Eudoxus to be a young and obscure man, does not agree with the date of his visit to Egypt, according to the account of Sotion. It is mentioned by Plutarch that Eudoxus visited Egypt, and became the disciple of Chonuphis, whom he calls a native, not of Heliopolis, but of Memphis.([21])

Plato, as well as Eudoxus, is stated to have profited by the lessons of Chonuphis;([22]) which belief may have given rise to the story of their joint journey to Egypt. The instruction in astronomy which Eudoxus derived from the priests, during his residence in Egypt, is likewise mentioned by Diodorus.([23]) Eudoxus is reported to have visited Mausolus, king of Caria,([24]) whose reign extended from 377 to 353 B.C., and whose wife, Artemisia, erected the celebrated Mausoleum to his memory.([25]) This fact affords another indication of his lifetime. He is, moreover, related to have visited Dionysius the younger of

---

([18]) Diog. Laert. viii. 87.

([19]) Clinton, F. H. vol. ii. p. 213. Compare Grote, Hist. of Gr. vol. x. p. 499.

([20]) viii. 90.

([21]) De Is. et Osir. 10.

([22]) Plutarch, De Gen. Socrat. 7; Clem. Alex. Strom. i. 15, § 69, says that Sechnuphis was the instructor of Plato, and Conuphis of Eudoxus.

([23]) i. 98. Compare Sen. Q. N. vii. 3. Eudoxus was likewise stated to have published some Dialogues of Dogs, translated from the Egyptian, Diog. Laert. viii. 89.

([24]) Diog. Laert. viii. 87.

([25]) Clinton, F. H. vol. ii. p. 286.

Syracuse;(26) which visit must have taken place between 367 and 356 B.C.

Eudoxus was a geometer and mathematician: when the Delians applied to Plato for a solution of the problem of the duplication of the cube, which the oracle of Apollo had declared to be the condition of the cessation of a pestilence in their island, he referred them to Eudoxus and Helicon, as being abler than himself to afford the desired assistance.(27)

His chief scientific reputation was, however, founded upon his astronomical researches. He was a practical observer of the heavens, and he consigned to writing the results of his observations. According to Ptolemy, he made astronomical observations, not only in Asia Minor, but also in Sicily and Italy.(28) His observatory at Cnidos was extant in the time of Posidonius, the early contemporary of Cicero, who saw from it the star Canopus.(29) He is related to have lived on a high hill, in order to observe the stars.(30) The notion that the stars could best be observed from high places occurs elsewhere among the ancients, and may even be traced in the mythological stories concerning Atlas and Prometheus. As a proof of his ardent desire for astronomical knowledge, a saying of his was cited, that he would willingly suffer the fate of Phaethon, provided he could approach

---

(26) Ælian, Var. Hist. vii. 17.

(27) Plut. de Gen. Soc. 8. Compare Ideler, Berl. Trans. 1828, p. 207; Montucla, Hist. des Mathématiques, vol. i. p. 186.
Helicon wrote a work entitled ἀποτελέσματα, and a treatise περὶ διοσημειῶν, Suidas in v. The latter of these was of an astronomical, but the former of an astrological character. Helicon is likewise stated by Plutarch to have predicted an eclipse to Dionysius of Syracuse, when he was staying at his Court with Plato, and upon the verification of his prophecy to have been presented by Dionysius with a talent of silver, Dion. 19.
Concerning Eudoxus as a geometer, see Ideler, ib. p. 203—212. The curvilinear geometry of Eudoxus is celebrated in an epigram of Eratosthenes, who lived from 276 to 196 B.C. See Bernhardy's Eratosthenica, p. 180.

(28) De Apparentiis, p. 53, ed. Petav.

(29) Strab. ii. 5, 14; xvii. 1, 30.

(30) Petron. Sat. c. 88: Eudoxus quidem in cacumine excelsissimi montis consenuit, ut astrorum cœlique motus deprehenderet.

within such a distance of the sun as would enable its figure and magnitude.(31)

The principal astronomical labour of Eudoxus appea. have been a descriptive map of the heavens, which he executeu in two works, nearly identical with each other, one called the *Enoptron*, or Mirror, the other the *Phænomena*, or Appearances.(32) These works are no longer extant; but the second of them was versified by Aratus, an Alexandrine poet,(33) who wrote about 270 B.C., and therefore about a century after Eudoxus; his poem enjoyed a high reputation in antiquity; it received the honour of being rendered into Latin verse by Cicero and Germanicus Cæsar, the grandson of Augustus; and it has descended to our days, accompanied with a large apparatus of ancient scholia.(34) It continued to be used as a practical manual of sidereal astronomy as late as the sixth century of our era.(35) It appears that Aratus undertook the task of versifying the prose work of Eudoxus, at the suggestion of Antigonus Gonatas, king of Ma-

---

(31) Εὔδοξος δὲ ηὔχετο, παραστὰς τῷ ἡλίῳ, καὶ καταμαθὼν τὸ σχῆμα τοῦ ἄστρου καὶ τὸ μέγεθος καὶ τὸ εἶδος, ὡς ὁ Φαέθων καταφλεγῆναι, Plutarch, non posse suav. viv. sec. Epic. 11.

(32) According to Cic. Rep. i. 14, a solid celestial sphere or globe was first constructed by Thales; and afterwards by Eudoxus. Cicero, however, appears to confound the description of Eudoxus with the material globe.

(33) ἀναφέρεται δὲ εἰς τὸν Εὔδοξον δύο βιβλία περὶ τῶν φαινομένων, σύμφωνα κατὰ πάντα σχέδον ἀλλήλοις, πλὴν ὀλίγων σφόδρα. τὸ μὲν οὖν ἐν αὐτῶν ἐπιγράφεται Ἔνοπτρον, τὸ δὲ ἕτερον Φαινόμενα· πρὸς τὰ Φαινόμενα δὲ τὴν ποίησιν συντέταχεν. Hipparch. ad Phæn. i. 2, p. 98, ed. Petav.

The following passages relating to the head of the Great Bear, cited from the two works of Eudoxus by Hipparchus, will serve to show their relation. From the Phenomena: ὑπὸ δὲ τὸν Περσέα καὶ τὴν Κασσιέπειαν οὐ πολὺ διέχουσά ἐστιν ἡ κεφαλὴ τῆς μεγάλης ἄρκτου, οἱ δὲ μεταξὺ τούτων ἀστέρες εἰσὶν ἀμαυροί. From the Enoptron: ὄπισθεν δὲ τοῦ Περσέως καὶ παρὰ τὰ ἰσχία τῆς Κασσιεπείας οὐ πολὺ διαλείπουσα ἡ κεφαλὴ τῆς μεγάλης ἄρκτου κεῖται, οἱ δὲ μεταξὺ ἀστέρες εἰσὶν ἀμαυροί. In Phæn. i. 12, p. 109. The differences are merely verbal.

(34) The list of ancient commentators on Aratus, printed in Petav. Uranolog. p. 147, contains no less than thirty-six names. It must, however, have been framed by some ignorant compiler, for it contains the names of Thales and Parmenides. The names of the two Aristylli likewise occur twice.

(35) See the treatise of Leontius Mechanicus, περὶ κατασκευῆς Ἀρατείας σφαίρας, in Buhle's Aratus, vol. i. p. 257.

Sorry:
The defect on the previous page was that way in the original book we reproduced.

cedonia.(36) As Aratus was not himself a scientific astronomer,(37) he may be supposed to have faithfully reproduced the descriptions of his guide. A critical commentary upon these works of Eudoxus and Aratus, by the great astronomer Hipparchus—who was about a century posterior to Aratus—is, moreover, extant. We are therefore able to judge of the method pursued by Eudoxus for determining the places of the stars. This method was to conceive the starry heaven as distributed into constellations, with recognised names, and to define them partly by their juxtaposition, partly by their relation to the zodiac, and to the tropical and arctic circles. He did not, like modern astronomers, deal with the stars singly, and define their places by celestial measurement: he laid down neither their right ascension and declination, nor their latitude and longitude; but he gave a sort of geographical description of their territorial position and limits, according to groups distinguished by a common name.(38)

The constellations had been named by the Greeks before the time of Eudoxus; the course of the sun in the ecliptic had been determined; the ecliptic had been divided into twelve parts, each coinciding roughly with a lunation; the celestial band, of sixteen degrees in width, within which the apparent motions of the sun, moon, and visible planets are made,(39) had

---

(36) See the ancient lives of Aratus, in Westermann's Biogr. Gr. p. 53, 59. The ancient biographer states that Aratus versified the κατόπτρον of Eudoxus, as he calls his ἔνοπτρον; but Hipparchus, whose authority is decisive, states that the poem of Aratus was founded upon the φαινόμενα. Antigonus Gonatas reigned, with certain interruptions, from 277 to 239 B.C. For the time of Aratus, see Clinton, F. H. vol. iii. p. 488; and for an account of his writings, see Donaldson's Hist. of Gr. Lit. vol. ii. p. 425.

(37) Constat inter doctos hominem ignarum astrologiæ, Aratum, ornatissimis atque optimis versibus de coelo stellisque dixisse, Cic. de Orat. i. 16.

(38) For an account of the descriptive method of Aratus, see Delambre, Hist. de l'Astron. Anc. tom. i. p. 61—74; Penny Cyclopædia, art. Zodiac.

(39) Ptolemy, Synt. viii. 4, states that the zodiac is bounded by the motions of the planets in latitude. Cleomedes, i. 4, lays it down that the planets never exceed the zodiac. Compare Macrob. in Somn. Scip. i. 15, § 10.

been defined by twelve constellations, and had therefore received the name of the zodiacal circle. The originality of the labours of Eudoxus consisted in his comprehensive view of the heavens, and in his description of the entire starry sphere.

The ancient astronomers had at an early period observed that the sun's annual course is not parallel to the nocturnal motion of the stars, or at right angles to the axis of the world.(40) They defined this course by observing the points at which the sun successively set, and by marking their coincidence with the constellation which appeared at the same point upon the horizon. Having divided the sun's annual course into twelve parts, by a rude agreement with the number of lunations in the solar year, they distinguished each of these twelve parts by a constellation or sign. The zodiac of the Greeks acquired, after a time, a definite width, determined by the greatest latitude of the visible planets. But originally it could have had no defined width; it had been used as a method of measuring the sun's annual course, before the courses of the planets had been observed, and before their distances from the sun's path had been laid down. The ancient astronomers, until Hipparchus, were ignorant of the precession of the equinoxes, and they made no distinction between the sign, or twelfth part, of the zodiac and the constellation. They did not know that the sidereal year is different from the solar year. The sign of Aries, as corresponding with the vernal equinox, was taken as the beginning of the astronomical year; but Ideler has shown that the practice of Eudoxus, and of other ancients, who reckoned the equinoxes and solstices sometimes at the middle and sometimes at other degrees of the sign, affords no basis for the refined inferences which modern astronomers have founded upon it.(41)

---

(40) On the elevation of the north pole, see Ideler, ad Aristot. Meteor. vol. i. p. 505.

(41) Berl. Trans. 1830, p. 55—61. Compare the remarks of Delambre on other inferences founded on the statements of Eudoxus, Astr. Anc.

§ 3 Eudoxus does not appear to have adverted to the planets in his description of the heavens. They are not once mentioned in the comment of Hipparchus; and the manner in which they are noticed by Aratus raises a strong presumption that they were passed over in nearly total silence by his guide.

The following—the only passage in which the planets are mentioned by the versifier of Eudoxus[42]—is thus rendered into English verse by Dr. Lamb, in his translation of the poem of Aratus:—

> Five other stars remain of various size,
> That lawless seem to wander through the skies:
> Hence planets called; yet still they ever run
> Through the twelve signs, the circuit of the sun.
> Thousands of ages come, thousands depart,
> Ere all return and meet where once they start.
> Rash the attempt for artless hand like mine
> To trace their orbits and their bounds define:
> My easier task the circles to rehearse
> Of the fixed stars, and trace Sol's annual course.

But although Eudoxus (whose object it probably was to compose a practical manual for finding the time of night)[43], did not introduce any mention of the planets into his Mirror of the Heavens and his Phænomena; yet he was the first Greek astronomer who devised a systematic theory for explaining the

---

tom. i. p. 122. 'Mais les données d'Eudoxe ne s'accordent pas entr'elles; c'est qu'il n'a point regardé le ciel, qu'il a recueilli les observations grossières faites à vue, peut-être en différens tems et en différens pays. Il n'est pas étonnant qu'avec des élémens aussi imparfaits, il ait donné des discordances énormes; ce qui étonne davantage, c'est la peine inutile que se sont donnée quelques modernes pour expliquer tout cela, en supposant des observations faites à des époques éloignées les unes des autres. Il faudrait autant d'époques différentes qu' Eudoxe a nommé d'étoiles. On s'est accordé à prendre pour idée fondamentale que les observations étaient bonnes. Il était bien plus naturel de les supposer mauvaises; mais alors on n'aurait pu bâtir aucun système.'

(42) Aratus, v. 454-61. The last two verses are correctly explained by the Scholiast thus: οὐκ ἂν εὐθαρσὴς περὶ τῶν πλανήτων εἰπεῖν, ἀρκοῦν δ' ἂν εἴη μοι τὸ περὶ τῶν ἀπλανῶν μάθημα.

(43) Delambre lays it down that the commentary of Hipparchus upon Eudoxus and Aratus, and which has the same character as the works it interprets, "appears to have been composed exclusively for the purpose of facilitating the means of finding the hour during the night," Hist. de l'Astr. Anc. tom. i. p. 172.

periodic motions of the planets. His planetary theory was, with certain corrections and developments, adopted generally by the scientific astronomers of Greece who succeeded him, and it ultimately assumed the form of the Ptolemaic system of the world.

It appears to have originated in a suggestion of Plato, who propounded it as a problem to the astronomers to explain the movements of the planets by a hypothesis of equable and uniform movements.[44]

The heavenly bodies to which the ancients gave the name of *planets*, as having each a peculiar motion, independent of the general and uniform course of the fixed stars, were the five planets visible to the naked eye, and the sun and moon. Sometimes the term was limited to the five erratic stars.

The hypothesis of Eudoxus comprehended the seven planets, and was as follows:—The sun and the moon are each carried by three spheres, one of which is that of the fixed stars, one moves along the zodiacal circle (or ecliptic); and the third moves along a circle oblique to the zodiacal circle; the latter circle being more oblique for the moon than for the sun. Each of the five planets has four spheres, two of which are the spheres of the fixed stars and the zodiacal circle. Of the other two, one has its poles in the zodiacal circle,[45] and the other moves in a direction oblique to this circle.

---

[44] Simplicius, on the authority of Sosigenes, ap. Schol. Aristot. p. 498, Brandis. Geminus attributes the invention of the hypothesis of uniform circular movements to the Pythagoreans: see above, p. 131.

[45] Dr. Whewell, Hist. of the Inductive Sciences, vol. i. p. 166, thinks that Aristotle reported the hypothesis of Eudoxus erroneously, and that instead of saying 'has its poles in the ecliptic,' he ought to have said, 'has its axis perpendicular to the ecliptic.' According to the report of Aristotle, Eudoxus supposed each planet to have four spheres: 1. that of the fixed stars, the poles of which would be identical with those of the earth; 2. a sphere whose motion is along the ecliptic, and whose poles would therefore be perpendicular to the ecliptic; 3. a sphere whose poles were in the ecliptic, and whose motion, therefore, would be perpendicular to the ecliptic; each of the planets had the poles of this sphere in different points of the ecliptic, with the exception of Venus and Mercury, which agreed in having the same poles for this sphere: 4. a sphere whose motion was

Such is the general account of Aristotle;(⁴⁶) from which it results that the total number of revolving spheres supposed by Eudoxus for effecting the complex motions of the sun, moon, and five planets, was twenty-six; namely, six for the two former, and twenty for the five latter.(⁴⁷) Further details respecting his scheme are given by Simplicius, in an elaborate passage of his Commentary upon Aristotle's treatise de Cœlo,(⁴⁸) which it would be foreign to my purpose to repeat, and still more to analyse. Some valuable criticisms upon the exposition of Simplicius are introduced by Ideler in his Dissertation on Eudoxus.(⁴⁹)

The general principle of all these systems of revolving orbs for each movable heavenly body was the same; namely, to resolve each apparent motion into its elements; to decompose it into its compounding directions; and to suppose each of these simple or decomposed movements to be effected by a separate orb. It is difficult to understand how these co-revolving orbs were conceived to harmonize in producing a single resulting motion: but the Greeks, even in the time of Eudoxus, were subtle geometers, though from the want of clocks and telescopes their astronomical knowledge was limited and unprecise; and they doubtless had formed a clear idea as to the solution of a problem which was substantially geometrical. The hypothesis

---

transverse to the ecliptic, and whose poles therefore would likewise be transverse to the ecliptic. The report of Aristotle is consistent with itself, and appears to be correct. If the third sphere had its poles perpendicular to the ecliptic, its motion would be in the direction of the ecliptic, and it would coincide with the second sphere. The subject is copiously explained by Simplicius, ib. p. 499, col. b. Compare Ideler, Berl. Trans. 1830, p. 78.

(46) Metaph. xi. 8.

(47) Ideler, Berl. Tr. 1830, p. 81, by an oversight or error of the press, states the number as 27.

(48) See Schol. Aristot. ed. Brandis, p. 498. In composing this account Simplicius consulted the History of Astronomy by Eudemus, and the Commentary of Sosigenes, the astronomer who guided Julius Cæsar in the reform of the Roman calendar. He likewise refers to a work of Eudoxus on Velocities—περὶ ταχῶν.

(49) Berl. Trans. 1830, p. 73.

of revolving spheres originated in the primitive idea of a solid crystalline firmament, in which the stars were set, and which made a rotatory motion round the earth every twenty-four hours. This idea was entertained by the Ionic philosophers, who spoke of the stars being attached, like nails or studs, to the hollow celestial sphere; ([50]) and even Aristotle conceives the fixed stars as owing their motion exclusively to the sphere in which they are fastened, and by which they are whirled round the earth. This supposition was natural, and not unphilosophical, with respect to the apparent motion of the fixed stars; their movements being simultaneous and uniform, were adequately explained by the simple hypothesis of a revolving sphere to which they were all fastened. It might seem improbable that, if each star had an independent movement, all should move uniformly, and retain constantly the same relative positions to one another. The Copernican hypothesis likewise reduces the diurnal revolution of the fixed stars to a single cause; but it makes that cause the rotation of the earth itself, instead of the rotation of the starry sphere. But when the theory of the diurnal revolution of the starry firmament round the earth, which afforded a simple and satisfactory explanation of the movement of the fixed stars, came to be applied to the sun, moon, and planets, its unsuitability became apparent. It was necessary to multiply the spheres, in order to account for the anomalous tracks of these bodies, as compared with that of the fixed stars; and thus the explanation lost its simplicity, which was its principal recommendation, while it remained limited to the fixed stars, according to the original intention.

The periodic times of the five planets were stated by Eudoxus, as we learn from Simplicius: ([51]) the following is his statement, to which the true times are subjoined, for the sake of comparison:—

([50]) See above, p. 95. The word ἧλος in Homer is used merely to signify an ornamental stud, and not a nail for fastening. It probably had the same meaning in the writings of the ancient Ionian Anaximenes.

([51]) Ib. p. 499 b.

	Statement of Eudoxus.	True time.		
	Y.	Y.	D.	H.
Mercury	1	—	87	23
Venus	1	—	224	16
Mars	2	1	321	23
Jupiter	12	11	315	14
Saturn	30	29	174	1

Upon this determination two remarks may be made. First, the error with respect to Mercury and Venus is considerable; with respect to Mercury, it is, in round numbers, 365 instead of 88 days, more than four times too much. Aristotle remarks that Eudoxus distinguishes Mercury and Venus from the other three planets by giving them one sphere each, with the poles in common. The proximity of Mercury to the sun would render its course difficult to observe and to measure; but the cause of the large error with respect to Venus (130 days), is not apparent.

With respect to the long periodic times of Jupiter and Saturn, the approximation is close; and the accurate determination of an astronomical period of thirty years proves the continuity of observation for a considerable antecedent period: such a fact could not have been ascertained by the observation of a single revolution of Saturn, or within the lifetime of a single observer.

Eudoxus further stated the synodic periods of the several planets ([52]), as is shown in the following table:—

	Statement of Eudoxus.		True time.	
	M.	D.	Y.	D.
Mercury	—	110	—	116
Venus	19	—	1	219
Mars ([53])	8	20	2	49
Jupiter } nearly 13			1	34
Saturn }			1	13

([52]) Simplic. ib.; in p. 499, col. b, line 25, read ἀπὸ φάσεως for ἀποφάσεως.
([53]) Ideler, ib. p. 78, thinks that the numbers in the text of Simplicius

Setting aside Mars (as to which a corruption of the text may be supposed), the numbers given by Eudoxus for the periodic times of the planetary conjunctions approach closely to the true times, as may be more clearly seen in the subjoined comparison:—

	Time of Eudoxus. Days.	True time. Days.
Mercury	110	116
Venus	570	584
Mars	260 [qy. 770]	780
Jupiter }	about 390(54)	399
Saturn }		378

The early Greek astronomers, from Thales to Anaxagoras and Democritus, paid little attention to the planets, which they classed rather with wandering meteors, or comets, than with the fixed stars; the latter, which made an equable motion every night, and always retained the same positions with respect to one another, not only appeared to them as endued with that immutability which was characteristic of the divine nature, but also afforded a better measure of nocturnal time, when the sun-dial could give no assistance. The names of the five planets first occur in the cosmical scheme of Philolaus, who was a contemporary of Socrates: the Pythagorean School of that date had doubtless a general idea of their movements. The same general acquaintance with the number and course of the planets was possessed by Plato. The fuller and more exact knowledge of Eudoxus respecting the planets, and particularly his determination of their periodic and synodic times, appears to have been chiefly derived from the Egyptian priests.([55]) His visit to Egypt

---

for Mars are corrupt, and that the true reading is probably twenty-five months twenty days. The irregular curve line described by the planets was called Εὐδόξου ἱπποπέδη. See Simplic. ib. p. 500, col. a, and Ideler, ib. p. 88.

(54) The month of Eudoxus is reckoned at thirty days.

(55) Speaking of the motions of the five planets, Seneca says: Eudoxus primus ab Ægypto hos motus in Græciam transtulit, Nat. Quæst. vii. 3.

is well attested; and without adopting the extravagant theory of a profound astronomical science, handed down by successive generations of the Egyptian priesthood from a remote antiquity, we may reasonably believe that the Egyptians preceded the Greeks as practical observers of the celestial bodies, and that they had, at the beginning of the fourth century before Christ, accumulated a larger stock of astronomical facts than their more intelligent and more scientific neighbours. We have the distinct testimony of Aristotle as to the astronomical observations of the Egyptians having been carried on for years before his time. In his treatise de Cœlo, after describing an occultation of Mars by the moon, which he had himself observed, he proceeds to state that similar occultations of other stars had been observed by the Egyptians and Babylonians: he remarks that they had observed the heavens for many years, and that the Greeks had received from them many oral reports concerning each of the stars.(56) In his Meteorologics, he appeals to the Egyptians as attesting the fact that some of the fixed stars acquire a tail like a comet; he adds that their testimony is credible though only hearsay.(57) It will be observed that Aristotle does not speak of the Egyptians as having composed any astronomical treatises, or as having communicated to the Greeks any observations in writing.

The knowledge of the planetary bodies began at this time to be regarded by the Greeks as essential to a truly scientific and accomplished astronomer. 'It is necessary (says Plato, or the Platonic author of the Epinomis) that the genuine astronomer should, not like Hesiod and others such as Hesiod, confine himself to a knowledge of the risings and settings of the constellations: he ought to be likewise acquainted with the circuits of the seven planets, and of the eighth celestial sphere.'(58)

The Chaldæans had by this time applied their knowledge

---

(56) Cœl. ii. 12, 3. See below, § 4.
(57) Meteor. i. 6, § 9.
(58) Epinom. § 11, p. 990.

respecting the planets and their movements to astrological purposes. This application became known to Eudoxus, who, with the scientific spirit characteristic of a Greek, condemned it as deceptive.[59] The prognostics to be derived from the movements of the planets respecting human affairs are mentioned by Plato in the Timæus;[60] but he does not connect them with the nativity of the person whose fortunes are in question.

§ 4 Physical speculation may be considered to have originated in Greece with Thales in the sixth century B.C. Ethical and political speculation, in a systematic and scientific form, took its rise with Socrates in the fifth century.[61] So rapid was the progress of intellectual investigation among this highly endowed nation, that in the following century Aristotle attempted to comprehend the entire circle of physical, metaphysical, logical, ethical, and political science, in his philosophy, and believed himself to have accomplished this object.[62] One of his extant treatises, that concerning the Heaven (περὶ οὐρανοῦ),[63] relates to the form and movement of the universe and of its constituent parts, and properly belongs to astronomical science considered in its widest extent. The treatise on Meteorology is confined to the region intermediate between the earth and the region of the stars. It includes inquiries into the nature of meteors, comets, and the Milky Way. A third treatise, of an astronomical character, attributed to Aristotle, and included in the collection of his works, is entitled περὶ κόσμου (Concerning the World, or the Universe), but it is

---

(59) Cic. de Div. ii. 42.

(60) p. 40. φόβους καὶ σημεῖα τῶν μετὰ ταῦτα γενησομένων τοῖς δυναμένοις λογίζεσθαι πέμπουσι.

(61) Aristotle was born eighty-five years after Socrates, and forty-five years after Plato.

(62) See Meteor. i. 1; Eth. Nic. ad fin.; and compare Cic. Fin. i. 4.

(63) For an account of the Commentary of Simplicius on Aristotle's Treatise de Cœlo, see Delambre, Hist. de l'Astr. Anc. tom. i. p. 301. Concerning this Commentary, see likewise the art. Simplicius, by Brandis, in Dr. Smith's Dict. of Anc. Biogr., and Cramer in the Philol. Mus. vol. ii. p. 588.

unquestionably spurious, and was probably composed in the last century B.C. A Latin translation of it is in the works of Apuleius.(64) Aristotle appears likewise to have written a separate treatise on astronomy, now lost, which, according to Diogenes Laertius, was comprised in one book, and was therefore of limited extent.(65)

Aristotle considered astronomy as a science founded on observation of the celestial phenomena, and on mathematical calculation.(66) He describes it as occupied with an essence which is the object of sensation and is eternal; and as being more intimately connected with mathematical science than any other branch of philosophy.(67)

All men, according to Aristotle, whether Greeks or barbarians, have a conception of gods; and all agree in placing the habitation of the gods in the most elevated region of the universe. This region is called heaven or æther; it is imperishable and immortal; and is therefore fitted for the residence of immortal natures.(68)

---

(64) The treatise De Mundo is properly pseudonymous; it is addressed to Alexander the Great. See c. 1: πρέπειν δὲ οἶμαί γε καὶ σοὶ ἡγεμόνων ὄντι ἀρίστῳ κ.τ.λ. The first chapter is adapted by Apuleius, who omits the dedication to Alexander, and inserts an acknowledgment of obligations to Aristotle and Theophrastus. See below, ch. iv. § 2.

(65) Aristotle, de Cœl. ii. 10, refers to his own treatise on astronomy as containing a sufficiently full account of the order and distances of the stars; hence, unfortunately, he omits this subject in his extant treatise de Cœlo. Aristotle's treatises were not divided by himself into books: the divisions in our manuscripts were made in later times by the grammarians: a book of Aristotle, according to their division, may be taken as equal to about twenty or thirty printed octavo pages.

(66) διὸ τὰς μὲν ἀρχὰς τὰς περὶ ἕκαστον ἐμπειρίας ἐστὶ παραδοῦναι. λέγω δ' οἷον τὴν ἀστρολογικὴν μὲν ἐμπειρίαν τῆς ἀστρολογικῆς ἐπιστήμης· ληφθέντων γὰρ ἱκανῶς τῶν φαινομένων, οὕτως εὑρέθησαν αἱ ἀστρολογικαὶ ἀποδείξεις, Anal. Prior. i. 30. καθάπερ οἱ μαθηματικοὶ τὰ περὶ τὴν ἀστρολογίαν δεικνύουσιν, οὕτω δεῖ καὶ τὸν φυσικὸν τὰ φαινόμενα πρῶτον τὰ περὶ τὰ ζῷα θεωρήσαντα καὶ τὰ μέρη τὰ περὶ ἕκαστον, ἔπειθ' οὕτω λέγειν τὸ διὰ τί καὶ τὰς αἰτίας, ἢ ἄλλως πως, De Part. An. i. 1.

(67) Metaph. xi. 8.
For a summary of the astronomical opinions of Aristotle, see Biese, Philosophie des Aristoteles, vol. ii. p. 59—92.

(68) Cœl. i. 3; ii. 1. Speaking of the æther (which with Plat. Crat. 25, he derives from ἀεὶ θεῖν), he remarks thus: ἔοικε δὲ καὶ τοὔνομα παρὰ τῶν ἀρχαίων διαδεδόσθαι μέχρι καὶ τοῦ νῦν χρόνου, τοῦτον τὸν τρόπον ὑπολαμβανόντων

He conceives the heaven to be divine, without beginning or end; and for this reason to be endowed with the circular form, whose nature it is to move perpetually in a circle.([69]) This form is a sphere. The spherical heaven, or universe, is shaped with greater accuracy than is attainable by the work of any human hand;([70]) its motion is likewise equable.([71])

In this spherical heaven the celestial bodies are fixed. Their motion is due to the motion of the spherical orbs to which they are attached. They are not of a fiery nature, but their light and heat are produced by their circular motion, and by their consequent collision with the air.([72])

Aristotle further holds that the form of the heavenly bodies themselves is spherical. His chief ground for this conclusion is, that a sphere, which has no instrument fitted for motion, is the form best suited to bodies which are fixed in a movable sphere, but have no motion of their own. He adds, that the shape of the moon, in her several phases, proves that she is spherical. The crescent-shaped appearance of the sun in an eclipse likewise proves the sphericity of the moon. But, he

---

ὅπερ καὶ ἡμεῖς λέγομεν· οὐ γὰρ ἅπαξ οὐδὲ δὶς ἀλλ' ἀπειράκις δεῖ νομίζειν τὰς αὐτὰς ἀφικνεῖσθαι δόξας εἰς ἡμᾶς, i. 3, § 12. He repeats almost the same words in Meteor. i. 3.

Lucretius thus explains the reason why heaven is supposed to be the seat of the gods (v. 1186—91):—

> In cœloque deum sedes et templa locarunt,
> Per cœlum volvi quia lux et luna videtur,
> Luna, dies, et nox, et noctis signa serena,
> Noctivagæque faces cœli, flammæque volantes,
> Nubila, sol, imbres, nix, venti, fulmina, grando,
> Et rapidi fremitus, et murmura magna minarum.

The word αἰθήρ is derived from αἴθω; but the author of the pseud-Aristotelic treatise de Mundo, c. 2, rejects this etymology, and derives it from ἀεὶ θεῖν, after Plato and Aristotle.

(69) ἀνάγκη τῷ θείῳ κίνησιν ἀΐδιον ὑπάρχειν. ἐπεὶ δ' ὁ οὐρανὸς τοιοῦτος (σῶμα γάρ τι θεῖον), διὰ τοῦτο ἔχει τὸ ἐγκύκλιον σῶμα, ὃ φύσει κινεῖται κύκλῳ ἀεί, Cœl. ii. 3, 2.

(70) Cœl. ii. 4, 13: ὅτι μὲν οὖν σφαιροειδής ἐστιν ὁ κόσμος, δῆλον ἐκ τούτων, καὶ ὅτι κατ' ἀκρίβειαν ἔντορνος οὕτως ὥστε μηθὲν μήτε χειρόκμητον ἔχειν παραπλησίως μήτ' ἄλλο μηθὲν τῶν παρ' ἡμῖν ἐν ὀφθαλμοῖς φαινομένων.

(71) Cœl. ii. 6.   (72) Cœl. ii. 7, 8.

reasons, if one of the heavenly bodies is spherical, the others must have the same form.(73)

Aristotle accounts for the fact that the fixed stars twinkle, and not the planets, by the comparative distance of the former. He supposes that twinkling is an affection of our sight, caused by the strain of the eye to see so remote an object.(74) He likewise argues that the stars have no rotatory motion, from the fact that the moon always turns the same face to the earth.(75)

With regard to the distances of the heavenly bodies, and their order in succession from the centre of the universe, Aristotle refers to the exposition in his separate Treatise on Astronomy, now no longer extant. He lays it down, however, that the circular motion of the external sphere of the fixed stars is the most rapid, that the motion of the spheres nearest to the centre is the slowest; and that the velocity of the motions of the intermediate spheres (each of which has a proper motion contrary to that of the universe) is in the ratio of their distances. The latter fact is, he remarks, demonstrated by the mathematicians.(76)

A difficulty, however, occurs to him with respect to this hypothesis. It would be natural, he observes, that the motions of the bodies nearest the external sphere should be the most simple; whereas the motions of some of the planets are more intricate and complex than the motions of the sun and the moon, although these planets are further than the sun and moon from the centre and nearer the sphere of the fixed stars. As a proof that some of the planets are more distant from the earth than the sun and moon, Aristotle mentions that he had himself observed an occultation of Mars by the moon when half full, the planet immerging under the dark side of the moon, and emerging at the bright side. With regard to the other planets, he refers, for a proof of their position beyond the sun and moon, to ancient observations of the Egyptians and Baby-

---

(73) Cœl. ii. 11.
(75) Ib. § 11.
(74) Ib. 8, § 10.
(76) Ib. § 10.

Ionians, which had become known to the Greeks.(77) Another difficulty which he states is, that the external sphere should contain so many stars as to appear innumerable, whereas the other spheres contain only one heavenly body, endued with a movement peculiar to itself. Of these difficulties Aristotle propounds the following solutions: As to the first, he lays it down that the inconsistency arises from the erroneous assumption that the heavenly bodies are destitute of life. If we assume, as we ought to assume, that they are endowed, not only with life, but with will and power of action, the difficulty vanishes. Each orb accomplishes its circuit according to the best means at its command. The external sphere, as being the most perfect and divine, effects its purpose by a simple and uncomposed motion. The earth, being furthest from the external sphere, has no motion; the bodies nearest to it are unable to move with effect; the middle bodies overcome the obstacles by their energy, but only with complex and irregular movements.(78) As to the second

---

(77) ὁμοίως δὲ καὶ περὶ τοὺς ἄλλους ἀστέρας λέγουσιν οἱ πάλαι τετηρηκότες ἐκ πλείστων ἐτῶν Αἰγύπτιοι καὶ Βαβυλώνιοι, παρ' ὧν πολλὰς πίστεις ἔχομεν περὶ ἑκάστου τῶν ἄστρων, Cœl. ii. 12, § 3.

For πίστεις Buttmann emends πύστεις, which is evidently the true reading (Ideler on the Chaldæan astronomy). Aristotle means to say that the information received by the Greeks from the Egyptians and Babylonians on this subject, was not written, but oral.

Macrob. in Somn. Scip. i. 14: Apud Græcos aster et astron diversa significant: et aster stella est, astron signum stellis coactum, quod nos sidus vocamus.

Schol. Arat. 10: ἰστέον δὲ ὅτι ἀστήρ μέν ἐστιν ὁ καὶ μόνον ἐστὶ καὶ οὐ καθ' αὑτὸν κινεῖται, οἷον Κρόνος, Ζεὺς, καὶ τὰ τοιαῦτα· ἄστρον δὲ τό τε κινούμενον καὶ καὶ τὸ ἐκ πλείστων ἀστέρων σύστημα, οἷον Καρκίνος, Λέων. Read καὶ καθ' αὑτὸν κινεῖται. The latter definition adds a condition which is not in the first. Moreover, if it be strictly interpreted, neither word could be applied to a single fixed star, such as Sirius or Arcturus. A similar definition is given by Galen in Hippocr. Epid. i. vol. 17, part i. p. 16, ed. Kühn. He remarks that a single star is sometimes called ἄστρον, but that a constellation is never called ἀστήρ. The distinction in question may be observed in later times, but it is unknown to the earlier writers. Aristotle, for example, uses the two words interchangeably. Achilles Tatius, c. 14, who draws the same distinction, admits that it was not observed by the earlier writers. The Latin writers are equally inconsistent as to the distinction between *stella* and *sidus*, attributed to them by Macrobius. *Sidus* is sometimes applied to the planets, and sometimes to the sun and moon.

(78) This explanation of the motion of the planets is not very unlike the explanation of the origin of evil, given by Leibnitz in his Théodicée.

objection, he thinks that the external sphere, being the first and most perfect, would naturally contain the largest number of stars; but he adds, that Nature in some degree compensates the superior number of stars in the outward orb, by giving to the interior orbs a greater number of motions.(⁷⁹)

The idea that the stars are living bodies, eminently partaking of the divine nature, occurs elsewhere in the writings of Aristotle.(⁸⁰) He supposed this nature to inhere especially in the bounding sphere of the universe, which approached nearest to the habitation of the Godhead.

§ 5 Aristotle elsewhere expounds the views of some of the mathematicians, to which he subjoins his own, respecting the number and agency of the revolving spheres to which the motion of the heavenly bodies was supposed to be due. He gives this exposition with some doubt; for he exhorts any of his readers who may arrive at a different result, either from his own researches or from those of others, to treat both authorities with respect, but to follow the more accurate.

He first describes the hypothesis of Eudoxus respecting the causation of the planetary movements by a plurality of revolving orbs, to which we have already adverted;(⁸¹) and he proceeds to mention the modification of that theory made by Callippus. Callippus was a native of Cyzicus, where he studied with a certain Polemarchus, a friend of Eudoxus: he went subsequently to Athens, where he resided with Aristotle, occupied in concert with that philosopher in correcting and completing the Eudoxean

---

(79) Cœl. ii. 12.

(80) Thus in Phys. ii. 4, § 6, he points out the inconsistency of those who hold that animals and plants were created by design, but that the heaven and the most divine of all visible objects originated spontaneously: τὸν δ' οὐρανὸν καὶ τὰ θειότατα τῶν φανερῶν ἀπὸ τοῦ αὐτομάτου γενέσθαι. Elsewhere he lays it down that the stars are much more divine than man. Eth. Nic. vi. 7, § 4: εἰ δ' ὅτι βέλτιστον ἄνθρωπος τῶν ἄλλων ζῴων, οὐδὲν διαφέρει· καὶ γὰρ ἀνθρώπου ἄλλα πολὺ θειότερα τὴν φύσιν, οἷον τὰ φανερώτατά γε, ἐξ ὧν ὁ κόσμος συνέστηκεν. Plato, in the Timæus, § 15, p. 40, calls the stars θεοὶ ὁρατοί. The divine nature of the stars was a prevalent belief of the ancients.

(81) Above, p. 152.

hypothesis. This hypothesis accorded with the opinions of Aristotle, because it supposed all the heavenly bodies to move in circles round the earth at the centre of the universe.[82] Aristotle describes the hypothesis of Callippus as agreeing with that of Eudoxus in the distances of the spheres, and also in their number with respect to Jupiter and Saturn. But Callippus gave an additional sphere to each of the other three planets, and two additional spheres to the sun and moon respectively. According to the original scheme of Eudoxus, the total number of spheres was twenty-six; Callippus increased this number to thirty-three.

The hypothesis of Callippus was known only from the history of astronomy by Eudemus, the disciple of Aristotle. Callippus did not publish it to the world in any writing of his own. The reason assigned by Eudemus for the two additional spheres allotted by Callippus to the sun was the anomaly of its annual movement as shown in the unequal intervals between the solstices and equinoxes, discovered by Euctemon and Meton.[83] The reason why Callippus allotted an additional sphere to each of the three planets, Mars, Venus, and Mercury, was succinctly and perspicuously stated by Eudemus; but Simplicius either omitted the statement, or it has fallen from his text.[84]

The scheme of Aristotle is founded on that of Callippus, but adds to it a new element. He assumes, with Callippus, eight spheres for the advancing motions of Saturn and Jupiter, and twenty-five for those of the other three planets, together with the sun and moon. He then assumes a separate set of spheres, for effecting the retrograde motions; these, according to his principle of calculation (which is to deduct one sphere

---

[82] Simplic. ib. p. 398, col. b. The first year of the Callippic cycle was 330 B.C. The lifetime of Callippus may be supposed to have coincided nearly with that of Aristotle (384—322 B.C.). Concerning Callippus, see the article in Dr. Smith's Dict. of Anc. Biogr. Concerning the Callippic cycle, and its scientific character, see above, p. 122.

[83] The date of these observations is 432 B.C.

[84] Simplic. ib. p. 500, col. a.

for each, and to omit the lowest planet altogether), are six for the two highest planets, and sixteen for the three others, together with the sun and moon.([85]) By adding twenty-two to the number of Callippus, he makes the total number of spheres fifty-five.([86])

The spheres of Eudoxus, Callippus, and Aristotle are not mathematical hypotheses, imagined for the sake of solving a mechanical problem. As has been already observed, they are solid though transparent substances, to which the heavenly bodies are firmly attached. The heavenly body itself is devoid of all motive principle: its motion is due to the spheres by which it is borne.

§ 6 Aristotle holds that the heat and light of the sun are

([85]) Metaph. xi. 8. The numbers 6 and 16 are obtained by deducting from the 8 spheres of the 2 superior planets one sphere for each, and by making a similar deduction from the 20 spheres of the two intermediate planets, and of the sun and moon. The following table exhibits the difference between the schemes of Callippus and Aristotle:—

	Callippus.	Aristotle.
Moon	5 spheres	5 spheres
Sun	5	5+4
Mercury	5	5+4
Venus	5	5+4
Mars	5	5+4
Jupiter	4	4+3
Saturn	4	4+3
Total	33	55

Aristotle calls the spheres which give the retrograde motion—contrary to the motion of the spheres to which they are respectively attached—σφαῖραι ἀνελίττουσαι, from ἀνελίττω, to unroll or unwind; 'the reversing or retracting spheres.'

In Simplic. p. 500, col. a, l. 34, ed. Brandis, the sense requires, ὁ δὲ Ἀριστοτέλης μετὰ τὸ ἱστορῆσαι τὴν Καλλίππου δόξαν καὶ τὴν αὑτοῦ περὶ τῶν ἀνελιττουσῶν ἐπήγαγεν. The theory respecting the additional σφαῖραι ἀνελίττουσαι is given by Aristotle as his own, and it is so regarded subsequently by Simplicius.

([86]) Aristotle adds, that if the additional spheres for the sun and moon are omitted, the total number will be only 47. On referring to the table in the previous note, it will be seen that this statement cannot be reconciled with it. If the moon has no unwinding spheres, the omitted spheres would be only 4, and the total number would be 51, instead of 47. Various attempts to explain this apparent inconsistency, which puzzled the ancient astronomers and commentators, may be seen in Simplic. ib. p. 505 b, 808. The difficulty would not arise if we could suppose Aristotle to have regarded Mercury and not the Moon as the lowest planet.

the consequences of the velocity of its motion, and that their influence is great upon the earth, on account of the comparative proximity of the sun. He explains the weakness of this influence in the case of the fixed stars, by their remoteness, though their motion is rapid; and in the case of the moon, by the slowness of her motion, though she is near the earth.[87]

As to the figure of the earth, and its position in the system of the universe, Aristotle is very explicit. He examines and rejects the opinions of certain prior philosophers; as that of Thales, that the earth floats on water; that of Anaximenes, Anaxagoras, and Democritus, that it is supported by the pressure of air, owing to its flat shape; and that of Xenophanes, that its foundations are infinitely deep. He reports an argument in favour of the tenet that the earth has the figure of a tympanum, or tambourine, or circular flat drum; namely, that the section of the sun, at its rising and setting, is straight, and not circular, which it would be if the earth were a sphere. He answers this argument by the distance of the sun and the magnitude of the earth's circumference.[88] He likewise rejects the Pythagorean doctrine of the central fire, and of the ten bodies revolving round it; as well as the doctrine of the rotation of the earth upon its axis, which he attributes to the Timæus of Plato.[89]

Having disposed of these divergent opinions, Aristotle establishes the position, that the earth is at rest in the centre of the universe. He infers this partly from the gravitation of all bodies to the centre of the earth, and partly from other arguments. Thus he remarks that all the heavenly bodies, except the external sphere of the fixed stars, have several movements, and are unable to accomplish their circuit by a simple motion. Hence he argues, that if the earth revolved in an orbit round the centre, or if it turned upon its axis at the centre, it would

---

[87] Meteor. i. 3.
[88] Cœl. ii. 13.
[89] Ib. 13, § 8; 14, § 1.

have a double movement. But if it had a double movement, the fixed stars ought to exhibit deviations and turns in their course; whereas they always rise and set in the same places. Another argument is, that the figures devised by the mathematicians for exhibiting the order of the heavenly bodies and their changes are framed upon the assumption that the earth is the centre of the system; and they agree with the phenomena. That the earth is a sphere, he infers from this being the form which matter gravitating to a centre would naturally assume. He draws the same inference from the eclipses of the moon: for the moon in her phases, he remarks, exhibits every variety of form, being sometimes gibbous, sometimes a half moon, and sometimes a crescent: whereas in her eclipses the outline of the shadow is always circular; and as the eclipse is produced by the interposition of the earth, this result must be owing to its spherical form. He likewise appeals to the phenomena of the fixed stars as proving, not only that the earth is a sphere, but that it is a sphere of moderate size; for, he remarks, with a small change of distance to the north or south, we have a new horizon, and a change in the stars visible in those directions; thus some of the stars which are visible in Egypt and the Island of Cyprus are invisible in the countries to the north, and some of the stars which never set in the countries to the north set further south. Hence he thinks that those who infer the vicinity of the western coast of Africa to India, from the presence of elephants in both regions, cannot be accused of maintaining a paradoxical opinion. He adds that the mathematicians who attempt to calculate the circumference of the earth, reckon it at 400,000 stadia; whence we must infer, not only that the earth is spherical, but that its size is inconsiderable compared with that of the other heavenly bodies.(90) The system of the universe adopted by Aristotle is therefore, with some additions,

---

(90) Cœl. ii. 14. The immobility of the earth is likewise asserted by Aristotle in Meteor. i. 9; and in Meteor. i. 3, he remarks that he has shown in his astronomical writings, that the earth is much smaller than some of the stars.

that which had been promulgated by Eudoxus, which was afterwards accepted by Euclid, Archimedes, and Hipparchus, and generally by the Greek mathematicians, and which received its full development in the comprehensive treatise of Ptolemy.([91])

Comets were the objects of much speculation among the early Greek astronomers; the opinions of Anaxagoras and Democritus, of the Pythagoreans, and of Hippocrates of Chios, and of his disciple Æschylus, respecting them are reported and analysed by Aristotle.([92]) Differing in other respects, they agreed in considering the comets to be planets. Against this general position, Aristotle argues by saying that the planets are always confined within the zodiacal band; whereas many comets have been seen without these limits, and it has often happened that more than one comet has been visible at the same time.([93]) He points out further, that some of the fixed stars have been seen with a tail. For this fact, he refers to the general report of the Egyptian observers: he adds, however, that he had himself seen a star in the leg of the constellation Sirius, with a faint tail. He states that it could scarcely be seen if the vision was fixed directly upon it, but that it was more visible if the sight was turned slightly on one side. Against the theory that comets were a congeries of planets,([94]) he remarks that all those which had been seen in his time disappeared

---

(91) Cleomedes, i. 8, p. 51, states that the doctrine of the sphericity of the earth was held by all the mathematicians.

(92) Meteor. i. 6. In the Eudemian Ethics, vii. 14, Aristotle mentions Hippocrates as a geometer, who was devoid of penetration in worldly affairs, and was cheated of a large sum of money, on account of his simplicity, by the collectors of the 2 per cent. custom duty at Byzantium. Plutarch, Sol. 2, mentions that this Hippocrates was a merchant. He is alluded to by Aristotle as a geometer in Soph. El. 11. See Fab. Bibl. Gr. vol. i. p. 848, Harl. Concerning a Byzantine custom duty in later times, see Polyb. iv. 46, 47.

(93) Pliny notices this opinion, and thinks that Aristotle is mistaken: Aristoteles tradit et plures simul cerni; nemini compertum alteri, quod equidem sciam, N. H. ii. 25.

(94) This opinion is also alluded to by Sen. Nat. Quæst. vii. 11: Quibusdam antiquorum hæc placet ratio: Quum ex stellis errantibus altera se alteri applicuit, confuso in unum duarum lumine, faciem longioris sideris

without setting, while they were still above the horizon: they faded away gradually, and left no trace either of one planet or of several. He adds that the great comet, in the archonship of Asteius (373 B.C.),(95) appeared in the winter, in a clear sky: on the first day it was not visible, because it set before the sun; on the second day, it was seen imperfectly, for it set immediately after the sun in the west; its brightness extended over a third part of the sky: it reached as far as the belt of Orion, and there ceased. Aristotle points out that a concourse of stars does not constitute a comet. The Egyptian astronomers, he says, report that conjunctions of planets, both with one another, and with fixed stars, occur. He himself had observed Jupiter, in the constellation Gemini, on two occasions, coming into conjunction with a star, and occulting it, but without assuming the appearance of a tail.(96) Aristotle himself thinks that comets are in the nature of meteors, and that their range is in the region nearest the earth.(97)

Aristotle further reports three opinions respecting the Milky Way; one, of the Pythagoreans, already mentioned, that the Milky Way was the result of some great catastrophe in the heavens caused by the sun; another, that it was that portion of the heaven which was in the shadow of the earth as the sun passed beneath it; and a third, that it was the reflexion of the sun's rays in the heaven. His own doctrine concerning the Milky Way is that it is of the same nature as comets, but more diffused.(98) The true explanation of the Milky Way had been

---

reddi. Nec hoc tunc tantum evenit, quum stella stellam attigit, sed etiam quum appropinquavit. Intervallum enim, quod inter duas est, illustratur ab utrâque, inflammaturque, et longum ignem efficit.

(95) Aristotle was in this year eleven years old.

(96) Meteor. i. 6.

(97) Meteor. i. 7. The opinions respecting comets are thus reported by Pliny: 'Sunt qui et hæc sidera perpetua esse credant, suoque ambitu ire, sed non nisi relicta a sole cerni. Alii vero qui nasci humore fortuito, et igneâ vi, ideoque solvi,' N. H. ii. 26.

(98) Meteor. i. 8. These opinions recur in Plut. Plac. iii. 1. The second opinion reported by Aristotle is attributed to Anaxagoras by Plutarch. Concerning the Milky Way, see Achill. Tat. c. 24, p. 85, who

given by Democritus, namely, that it is a congeries of small stars, close to one another.

§ 7. The doctrine of the immobility of the earth at the centre of the universe was firmly held by Eudoxus and by Aristotle. The same doctrine was probably held by Plato, though a doubt existed in antiquity, as well as at present, respecting the interpretation of the passage in the Timæus, to which we have already adverted. Some philosophers, however, at this time perceived that the diurnal movement of the sun and the nocturnal movement of the moon and stars might be accounted for on the supposition that they were only apparent, and were produced by the rotation of the earth upon its axis. The earliest Greek to whom this hypothesis is ascribed is Hicetas of Syracuse, a Pythagorean.([99]) His date is not exactly known; but he was anterior to Theophrastus, and may be supposed to have been contemporary with Socrates or Plato.([100])

Heraclides of Pontus was the friend and associate of Plato, and is likewise called his disciple.([101]) He is also said to have been among the disciples of Aristotle.([102]) His lifetime may be placed, upon conjecture, from 410 to 340 B.C. His writings were numerous, and comprehended a great variety of subjects.

---

refers to the spurious catasterisms of Eratosthenes, c. 44, and the collection of opinions on the subject in Macrob. Comm. Somn. Scip. i. 15; Stob. Ecl. i. 27; Plut. Plac. Phil. iii. i. Above, p. 133.

([99]) Hicetas Syracusius, ut ait Theophrastus, cœlum, solem, lunam, stellas, supera denique omnia, stare censet; neque præter terram rem ullam in mundo moveri; quæ cum circa axem se summâ celeritate convertat et torqueat, eadem effici omnia, quasi stante terrâ cœlum moveretur. Atque hoc etiam Platonem in Timæo dicere quidam arbitrantur, sed paullo obscurius, Cic. Acad. ii. 39. Diogenes Laertius states that some consider Philolaus as the originator of the hypothesis, that the earth moves in a circular orbit, and others, Hicetas, viii. 85. See Boeckh, Kosm. Syst. des Plat. p. 122. This statement is founded on a confusion of a revolution on an axis with a revolution in an orbit.

([100]) Ukert, Geogr. der Griechen u. Römer, i. 2, p. 119, makes him contemporary with Eudoxus.

([101]) Simplicius, ad Aristot. Phys. p. 362, ed. Brandis, calls him a companion of Plato.

([102]) Diog. Laert. v. 86.

Without being a scientific astronomer, he speculated upon astronomy; and he promulgated the same doctrine as that ascribed to Hicetas. He laid it down that the earth turns round its axis at the centre of the universe, and that the heaven is at rest.([103]) It is specially mentioned that he did not give it a movement of translation in space; but only a movement round its own axis, from west to east.([104]) The same hypothesis is ascribed to Ecphantus the Pythagorean, who was probably posterior to Heraclides.([105]) The distinct statement, in the cases both of Heraclides and Ecphantus, that they conceived the rotatory movement of the earth to be from west to east, shows them to have perceived that where one body moves and the other is at rest, and the vision is not corrected by a comparison with a third body, the body at rest may appear to be in motion.

This truth was clearly understood by Aristotle, and applied by him to the movement of the heavens. In his Treatise de Cœlo, he examines the question, whether the stars have an

---

([103]) ἐν τῷ κέντρῳ δὲ οὖσαν τὴν γῆν καὶ κύκλῳ κινουμένην, τὸν δὲ οὐρανὸν ἠρεμεῖν, Ἡρακλείδης ὁ Ποντικὸς ὑποθέμενος, σώζειν ᾤετο τὸ φαινόμενον, Simplic. ad Aristot. de Cœl. p. 506. A similar statement recurs in p. 505, 508. In the Commentary on the Physics, p. 348, Simplicius says: διὸ καὶ παρελθών τις, φησὶν Ἡρακλείδης ὁ Ποντικός, ἔλεγεν ὅτι κινουμένης πως τῆς γῆς, τοῦ δ᾽ ἡλίου μένοντός πως, δύναται ἡ περὶ τὸν ἥλιον φαινομένη ἀνωμαλία σώζεσθαι. The latter dictum refers apparently to an objection that the simple rotation of the earth does not explain the annual motion of the sun.

The words παρελθών τις ἔλεγεν mean, 'Some one came forward (in an assembly) and said;' as Boeckh has explained, Kosm. Syst. des Platon, p. 137.

Ἡρακλείδης μὲν οὖν ὁ Ποντικός, οὐ Πλάτωνος ὢν ἀκουστής, ταύτην ἐχέτω τὴν δόξαν, κινῶν κύκλῳ τὴν γῆν. Πλάτων δὲ ἀκίνητον αὐτὴν ἵστησιν, Proclus in Tim. p. 281 E. Proclus is referring to the supposed doctrine of the rotation of the earth in the Timæus, which he attributes to Heraclides.

([104]) Ἡρακλείδης ὁ Ποντικὸς καὶ Ἔκφαντος ὁ Πυθαγόρειος κινοῦσι μὲν τὴν γῆν, οὐ μὴν γε μεταβατικῶς, ἀλλὰ τρεπτικῶς, τροχοῦ δίκην ἐν ἄξονι στρεφομένην ἀπὸ δυσμῶν ἐπ᾽ ἀνατολὰς περὶ τὸ ἴδιον αὐτῆς κέντρον, Euseb. Præp. Evang. xv. 58. The passage is less correctly given in Plut. Plac. Phil. iii. 13; Galen, c. 21.

([105]) The doctrine of the rotation of the earth is likewise attributed to Ecphantus in Orig. Ref. Hær. p. 19. τὴν δὲ γῆν μέσον κόσμου κινεῖσθαι περὶ τὸ αὑτῆς κέντρον ὡς πρὸς ἀνατολήν. The same doctrine has been erroneously ascribed to Cleanthes the Stoic, in consequence of a false reading in Plut. de fac. in orbe lun. 6. See Ukert, ii. 1, p. 129.

independent motion, or whether they are carried round by the sphere in which they are fixed. The entire starry heaven (he remarks) appears to move; and it is evident either that both the heaven and the stars must be at rest, or that both must move, or that one body must move and the other be at rest. Now it is impossible (he proceeds to say) that both should be at rest, unless indeed the earth moves; because the appearances would not be explained. If, then, the earth is supposed to be at rest, it follows that either the heaven and the stars both move, or that one of the bodies moves and the other is at rest.(106) It is manifest from these remarks that Aristotle's mind was familiar with the hypothesis, that the diurnal motion of the stars might be explained by the hypothesis of the rotation of the earth.

The hypothesis of Hicetas, Heraclides, and Ecphantus, was confined to the rotation of the earth upon its axis, or, what was equivalent, upon the axis of the universe.(107) The earth was still supposed by them to retain its central position, round which the heavenly bodies were carried in their respective spheres. It had no motion in space; and hence, as it appears, we do not hear that any of these philosophers were charged with impiety. The only hypothesis which up to this time had given the earth a motion in an orbit, was that of Philolaus.

The Greeks perceived at a remote period that the nocturnal

---

(106) De Cœl. ii. 8. Simplicius on this passage correctly expounds the reasoning of Aristotle. ὡς εἴγε μὴ κινοῖτο ἡ γῆ, ὅπερ μετ' ὀλίγον μὲν ἀποδείξει, νῦν δὲ ὡς ὑπόθεσιν ἔλαβεν, ἀδύνατον τοῦ οὐρανοῦ καὶ τῶν ἄστρων [φαινομένων] σώζεσθαι τὰ φαινόμενα, Schol. Brandis, p. 495. The word φαινομένων appears superfluous.

(107) This hypothesis is also described by Seneca in the following passage: he supposes a simple daily rotation of the earth in the centre of the universe, not a motion of the earth in an orbit: 'Illo quoque pertinebit hoc excussisse, ut sciamus, utrum mundus terrâ stante circumeat, an mundo stante terra vertatur. Fuerunt enim qui dicerent, nos esse quos rerum natura nescientes ferat, nec cœli motu fieri ortus et occasus, ipsos oriri et occidere. Digna res est contemplatione, ut sciamus in quo rerum statu simus; pigerrimam sortiti an velocissimam sedem; circa nos Deus omnia, an nos, agat,' Nat. Quæst. vii. 2. By *ipsos* is meant *ourselves;* Lipsius proposed *ipsos nos.*

motion of the heavens appears to be round an axis, of which only one pole was visible to them, and which, therefore, was inclined to the plane of the horizon. Homer knew that the Great Bear never sets in Greece: many ancient writers speak of the dances of the stars round the pole;[108] and the inclination of the universe was a familiar idea to the early Greek philosophers.[109] The Greek word πόλος originally signified a ball or sphere; and hence it was applied to the cavity of heaven. As the celestial vault was only a hemisphere, the word was afterwards used to denote the basin of a sundial, and at an early period it was applied to the central point of the hemisphere, or the vertex of the axis of the sphere. Even Eudoxus employed the word to denote the star nearest the north pole: it was not, however, till a later age that its modern use was fully established.

Achilles Tatius, in his Introduction to the Phænomena of Aratus, describes the axis of the heaven as terminating in the centres of the arctic and antarctic circles of the celestial sphere, and the heaven as revolving round it in the same manner that the wheel of a chariot revolves round its axle. He adds that Aratus does not specify the material of which the cosmical axis is formed; but that he uses the metaphorical language suited to a poet, and likens it to a spit. Aratus, however, does not, in fact, use this inappropriate similitude: for if the axis of the universe resembled a spit, the earth must be conceived as the piece of meat traversed by it, and would, therefore, partake of the rotatory motion of the heaven, which is contrary to the supposition. Achilles Tatius proceeds to remark, that if the axis of the universe is supposed to be of fire, it will either be consumed in passing through the sphere of fire, or extinguished in passing through the sphere of water; if it is supposed to be of air or water, it will equally be destroyed by one of the three elemental spheres. The geometers, he

---

[108] Above, p. 61.
[109] Compare Plut. Plac. Phil. iii. 12.

adds, conceived it as a mathematical line; while the physical philosophers regarded it as spiritual or immaterial.([110])

Whatever language may have been used for characterizing the cosmical axis, it must have been considered both by the philosophers and the vulgar as unsubstantial, and merely as a metaphysical entity. They were familiar with the idea of a ball spinning on its axis, and their senses showed them how this motion was effected.([111]) They did not suppose that any solid or visible cylinder protruded from the north pole of the earth, and was fixed in the north pole of the heavenly sphere.

§ 8 Two of the most eminent disciples of Aristotle, Theophrastus and Eudemus, composed histories of astronomy. The work of the former was in six books;([112]) of the work of Eudemus, the second book is cited by Simplicius as containing his account of the spheres of Eudoxus and Callippus.([113]) The science of astronomy had, therefore, at the end of the fourth century B.C., made such progress as to admit of its history being written at length in separate works.

§ 9 The received opinion among the Greek astronomers and geometers of this period was, that the earth remained at rest in the centre of the universe, having a motion neither of rotation nor of translation, and that the several heavenly bodies —the sun, the moon, the five planets, and the fixed stars— were carried round it in solid but transparent spheres. The main object of the Greek astronomers was to determine the

---

(110) Isag. in Phæn. c. 28, p. 88. Compare note A at the end of the chapter.

(111) Aristotle, Cœl. ii. 8, 8, states that $κύλισις$ and $δίνησις$ are the two motions proper to a sphere. By $δίνησις$ he means rotation without any change of place.

(112) Diog. Laert. v. 50.

(113) Simplic. p. 498.

Eudemus wrote a treatise on Physics, Schol. Aristot. p. 334, 343, 353, 362, 370, 389, 411, and a history of geometry, ib. p. 327. He likewise composed $ἀξιώματα περὶ τόπου$ jointly with Theophrastus, ib. p. 374, 377, 379. Theophrastus and Eudemus are mentioned together as companions of Aristotle, ib. p. 394. There was a life of Eudemus by Damasus, ib. p. 404. Concerning Eudemus and his writings, see Brandis, Uebersicht über das Aristotelische Lehrgebäude, p. 215—250; Martin, Theo Smyrnæus, p. 60.

relation of the sun's annual course to the nocturnal motion of the fixed stars. This determination had a double object. It served the purpose of a calendar, by dating the annual periods of the rising and setting of the most conspicuous constellations: it served the purpose of a nocturnal clock, which was of essential use when the sundial was the principal instrument used for measuring hours.

Socrates advised his friends to learn astronomy for practical purposes, and in order to furnish them with a measure of time at night.[114]

Plato lays it down, with respect to astronomy, that a knowledge of the times of the year and month is useful not only for agriculture and navigation, but also for military command.[115]

Polybius likewise points out at length the utility of astronomical knowledge to a military commander, by means of which he will know the times of the equinoxes and solstices, and the increase or diminution of the day during the intervening periods. He can likewise tell the hour of the day by the sun's shadow; and the hour of the night by the rising and setting of the zodiacal constellations, and in cloudy nights by the moon.[116]

The importance of a direct reference to the heavenly bodies, as the measures of time among the Greeks, was owing to their want of two appliances which have become so familiar to us from long habit, that it requires a vigorous effort of the imagination to conceive the state of things implied in their absence. These two appliances are almanacs and clocks. For the formation of an almanac, an advanced knowledge of astronomy is requisite; but that knowledge is possessed by the makers of

---

(114) οὐκοῦν καὶ ἐπειδὴ ὁ μὲν ἥλιος φωτεινὸς ὢν τάς τε ὥρας τῆς ἡμέρας ἡμῖν καὶ τἆλλα πάντα σαφηνίζει, ἡ δὲ νὺξ διὰ τὸ σκοτεινὴ εἶναι ἀσαφεστέρα ἐστίν, ἄστρα ἐν τῇ νυκτὶ ἀνέφηναν, ἃ ἡμῖν τὰς ὥρας τῆς νυκτὸς ἐμφανίζει; καὶ διὰ τοῦτο πολλὰ ὧν δεόμεθα πράττομεν. Words of Socrates in Xen. Mem. iv. 3, 4.

(115) De Rep. vii. 9, p. 527.

(116) ix. 15, 16. Euripides describes the military night-watches as determined by the stars, Rhes. 527.

almanacs in all civilized countries; their methods are similar, and their results exactly coincide. The conventional divisions of time—such as months and hours, and the beginning of the year—are likewise the same for all Christian countries of the Western Church, and nearly the same for all Christendom. As an almanac is complete and unerring, it forms a universally recognised guide, and nobody thinks of going to the source from which it is derived. It is founded upon astronomical measurements; but those measurements having once been made, are blindly, though safely, followed in practice. In Greece, in the fourth century B.C., every person had to provide his own almanac: he collected the materials for it from various indications, partly natural, and partly civil or religious; his work was often done on the spur of the occasion, and therefore inaccurately or imperfectly.(117) The mechanism adopted by the Greeks for determining the time of day or night was still more defective and inconvenient than their measures of annual time. Herodotus informs us that the Greeks derived the sundial (consisting of a hollow hemisphere and a gnomon or style), and the twelve parts of the day, from the Babylonians.(118) The reports of the ancients as to the origin of inventions are often fabulous, and, when not manifestly fictitious, are generally liable to well-grounded suspicion. This statement is, however, rendered probable by many circumstances, which Prof. Boeckh has collected in his Meteorological Treatise, and particularly by the relation of the Greek weights to the Babylonian. It is proved by satisfactory evidence that the Greek talent and mina are

---

(117) For specimens of almanacs in the later Roman period, the consequence of a fixed calendar, see Græv. Thes. Ant. Rom. vol. viii.

(118) πόλον καὶ γνώμονα καὶ τὰ δυώδεκα μέρεα τῆς ἡμέρης παρὰ Βαβυλωνίων ἔμαθον οἱ Ἕλληνες, ii. 109. Boeckh thinks that the sundial passed to Greece from Babylon through Phœnicia.

The primitive sundial of the Greeks, ascribed to Berosus, is described by Delambre, Hist. Astr. Anc. vol. ii. p. 510. Two things are, however, certain:—1. That the primitive Greek sundial was not confined to the cabinets of the curious; 2. That the shadow was cast by a gnomon, not by a suspended globule.

weights of Babylonian origin, and that the latter word was borrowed from the Chaldæan language.[119] By the twelve parts of the day must be meant twelve portions of the day, as distinguished from the night, which could be measured by the sundial. The Greeks of later times, as we shall show presently, had a double mode of reckoning the hours of the day. According to the popular method, they divided the period from sunrise to sunset into twelve equal parts. The hours reckoned upon this principle varied in length with the season. According to the more scientific method, the day and night at the equinox were severally divided into twelve equal parts, and each of these was reckoned as an hour. The division of the day into twelve parts, which Herodotus describes the Greeks as having derived from the Babylonians, together with the sundial, was doubtless reckoned according to the former method.

The introduction of the use of the gnomon and dial into Greece, as well as of an instrument for measuring hours, is attributed to Anaximander by Diogenes Laertius, on the authority of the Miscellaneous History of Favorinus.[120] Pliny makes a similar statement with respect to Anaximenes.[121] The obvious interpretation of these statements is, that the sundial was the invention of Anaximander or Anaximenes, and that one of these two philosophers taught the use of it to the Greeks. If they are to be regarded as historical, they cannot be fairly re-

---

[119] Boeckh, Metrologische Untersuchungen (Berlin, 1838), p. 32—42. Ahaz, eleventh king of Juda, whose sundial is mentioned in 2 Kings xx. 8—11, Isaiah xxxviii. 8, is stated to have reigned from 740 to 724 B.C. His reign corresponds with the period of the earliest Greek colonies in Sicily. Compare Winer, Bibl. R. W. in Hiskias.

[120] εὗρε δὲ καὶ γνώμονα πρῶτος καὶ ἔστησεν ἐπὶ τῶν σκιοθήρων ἐν Λακεδαίμονι, καθά φησι Φαβωρῖνος ἐν Παντοδαπῇ Ἱστορίᾳ, τροπάς τε καὶ ἰσημερίας σημαίνοντα, καὶ ὡροσκόπια κατεσκεύασε, ii. 1. Favorinus lived in the time of Hadrian: his testimony on such a question can only be regarded on the assumption that he copied his authorities correctly. The statement is repeated by Euseb. Præp. Evang. x. 14, and by Suidas in Ἀναξίμανδρος, γνώμων, and ἡλιοτρόπιον. See above, p. 92.

[121] Umbrarum hanc rationem et quam vocant gnomonicen invenit Anaximenes Milesius, Anaximandri (de quo diximus) discipulus; primusque horologium, quod appellant sciothericon, Lacedæmone ostendit, N. H. ii. 76.

conciled with the statement of Herodotus. The use of the word ὥρα, or *hour*, for the twelfth part of the illuminated day, was posterior to the age of Alexander;(122) but it can hardly be supposed, with Salmasius, that the sundial and gnomon, when first introduced into Greece, were used merely for determining the equinoxes and solstices, and not for measuring the parts of a day.(123) In the time of Aristophanes, and even in that of Menander, the common mode of denoting the time of day at Athens was by the length of the shadow of the gnomon. A person was invited to dinner by asking him to come when the shadow of the gnomon was so many feet long. This mode of measuring time supposes either that reference was made to some common gnomon, or to a gnomon of recognised length.(124)

According to the statement of Philochorus, Meton, the celebrated astronomer, set up a sundial against the wall of the

---

(122) See Ideler, Chron. vol. i. p. 238.

(123) See Salmas. Plin. Exerc. (1689), p. 445, sq.; Menage, ad Diog. Laert. ii. 1. Pliny says that the 'gentium consensus tacitus' extended to three main points: 1. The use of the Ionic alphabet. 2. Shaving the beard. 3. The use of hours, N. H. vii. 58—60. Compare Casaubon, ad Athen. i. 1.

(124) A. τὴν γῆν δὲ τίς ἔσθ' ὁ γεωργήσων.—B. οἱ δοῦλοι. σοὶ δὲ μελήσει ὅταν ᾖ δεκάπουν τὸ στοιχεῖον, λιπαρῶς χωρεῖν ἐπὶ δεῖπνον.

Aristoph. Eccl. 651, 2, where the Scholiast says: ἡ τοῦ ἡλίου σκιὰ ὅταν ᾖ δέκα ποδῶν.

πόλος τόδ' ἐστίν· εἶτα πόστην ἥλιος τέτραπται;

Aristoph. Gerytad. fragm. 210, Dindorf, where πόλος signifies a sundial; and the question is, What is the time of day?

Menander, Org. 2 (Fragm. Com. Gr. vol. iv. 179, Meineke) tells a ludicrous story of a man who being invited to supper at the time when the sun's shadow was twelve feet long—that is, apparently, when it was of great length, and therefore late in the afternoon—mistook the shadow of the moon for that of the sun, and thinking he was after his time, arrived at daybreak. See the Commentary of Casaubon on Athenæus, lib. vi. c. 10, and Salmasius, ib. p. 455.

The passage of Eubulus, Fragm. Com. Gr. vol. iii. p. 261 (a contemporary of Demosthenes, see Meineke, ib. vol. i. p. 355), is so mutilated that it is in part unintelligible; but its general meaning is, that a person invited to supper when the shadow of the gnomon was twenty feet long, came early in the morning, making excuses for his pretended lateness.

Plutarch, De Discrim. Adul. et Amici, c. 4, speaks of a parasite as measuring the length of the shadow, in order to ascertain the dinner-hour. Where see Wyttenbach's note.

Pnyx at Athens, in the year 433 B.C.(125) There was a similar dial at Achradina, near Syracuse, in the time of Archimedes, a copy of which was placed on the deck of the great ship of Hiero.(126)

The sundials of the ancients were of various construction; but they must all have been contrivances for measuring the length and direction of the shadow of the sun during the day. The common Greek dial seems to have consisted of a gnomon or style, and a hollow basin (called πόλος, or σκαφίον);(127) on which the hours were marked by their numbers up to twelve. Hence the ingenious Greek epigram :—

ἓξ ὧραι μόχθοις ἱκανώταται· αἱ δὲ μετ' αὐτὰς
γράμμασι δεικνύμεναι ΖΗΘΙ λέγουσι βροτοῖς.(128)

A *horologium*, or sundial, for determining the hour of the day, is mentioned in a fragment of the comic poet Baton, who lived about the beginning of the third century B.C.(129)

The day was divided by the Greeks into twelve equal parts or hours; but the *day*, or period of time divided, varied with the season. The instrument of division was the sundial, and the interval between sunrise and sunset, whatever might be its length, was divided into twelve equal parts. Hence it followed that the length of the hour varied with the time of year; it was longer in summer and shorter in winter. These were the hours which

---

(125) ὁ δὲ Φιλόχορος ἐν Κολωνῷ μὲν αὐτὸν οὐδὲν θεῖναι λέγει, ἐπὶ 'Αψεύδους δὲ τοῦ πρὸ Πυθοδώρου ἡλιοτρόπιον ἐν τῇ νῦν οὔσῃ ἐκκλησίᾳ, πρὸς τῷ τείχει τῷ ἐν τῇ Πνυκί. Schol. Aristoph. Av. 997; Frag. Hist. Gr. vol. i. p. 400. Compare Ideler, Chron. vol. i. p. 326. If this date, and the date of Diodorus for the cycle of Meton are both correct, the sundial was prior to the cycle by one year.

(126) κατὰ δὲ τὴν ὀροφὴν πόλον ἐκ τοῦ κατὰ τὴν 'Αχραδίνην ἀπομεμιμημένον ἡλιοτροπίου, Athen. v. p. 207 E.

(127) See note C at the end of the chapter.

(128) Anth. Pal. x. 43.

(129) Fragm. Com. Gr. ed. Meineke, vol. iv. p. 499. He was contemporary with Arcesilaus and Cleanthes, ib. vol. i. p. 480. It appears to me that this passage is misinterpreted by Meineke. The poet means to say that the person addressed carries round his oil-cruse, and scrutinizes it closely as if it were a sundial. He does not mean to imply either that a sundial is portable, or that it resembles an oil-cruse.

were in common use. Thus Achilles Tatius remarks, that in the latitude of Greece the day consists of fifteen hours at the summer solstice, and of nine hours at the winter solstice; but that in the mechanical contrivances for measuring time, the day always consists of twelve hours.([130]) The hours, which were exactly a twenty-fourth part of a day and night, were called equinoctial hours, because an equinoctial day contained exactly twelve hours.([131])

Palladius, in his treatise on agriculture, written in the fourth century after Christ, arranges his work according to the Julian months, and annexes to each a table for finding the hour of the day by the length of the shadow of the gnomon. He divides the interval between sunrise and sunset into twelve equal hours, and states the length of the shadow for the end of each hour; so that he only enumerates eleven hours, the end of the twelfth hour being marked by sunset.

The following tables for June and December will illustrate the method of measuring the time of day by variable, or, as they were called, *temporal*, hours:—

Hours reckoned from sunrise.	Length of the shadow of the gnomon in feet.	
	June.	December.
1	22	29
2	12	19
3	8	15
4	5	12
5	3	10
6	2	9

(130) ὁπότε ἐν τοῖς μηχανικοῖς ὡρολογείοις καὶ ὑδρολογείοις ἀεὶ ἡ ἡμέρα δυώδεκα ὡρῶν φαίνεται, c. 25, p. 87.

(131) Concerning the ancient division of hours, see Aldus, in Schneider, Script. Rei Rust. vol. iii. 2, p. 65; Delambre, Hist. Astr. Anc. tom. ii. p. 511; and Ideler, Chron. vol. i. p. 86. It is stated in Stuart's Antiq. of Athens, vol. i. p. 20, in reference to the dials on the Tower of the Winds at Athens, that 'not only the hours of the day, but the solstices also, and the equinoxes, are projected on these dials; and that the longest as well as the shortest days are divided alike into twelve hours.' The division of the day into twenty-four hours is mentioned by Hipparchus ad Phæn. i. 1, p. 98.

Hours reckoned from sunrise.	Length of the shadow of the gnomon feet.	
	June.	December.
7	3	10
8	5	12
9	8	15
10	12	19
11	22	29

Noon always corresponded with the end of the sixth hour, when the sun was on the meridian, and the shadow of the gnomon was the shortest. The length of the gnomon is not stated.([132])

The division of the day into hours at Rome was subsequent to 450 B.C., the date of the Twelve Tables, in which the only parts of the day mentioned were noon, sunrise, and sunset.([133]) Up to the First Punic War, 264 B.C., it was the custom for the attendant of the consul to call the time of noon, when he saw the sun from the senate-house between the Rostra and the Græcostasis; and to call the last hour, or sunset,([134]) when he saw the shadow of the Columna Mænia touch the Carcer.([135]) It was reported by Fabius Vestalis,([136]) that the first sundial was erected at Rome by L. Papirius Cursor, in the temple of Qui-

---

([132]) Vitruvius, ix. 7, states that at Rome the gnomon was divided into nine parts or degrees, and that its shadow on the equinox was eight of those parts.

([133]) Concerning the introduction of sundials into Rome, see Plin. N. H. vii. 60; Censorin. c. 23; Gell. N. A. xvii. 2. Compare Ernesti, De Solariis, Opuscula Philologica, p. 21; Ideler, Chron. vol. ii. p. 7. Censorinus says: In horas duodecim divisum esse diem, noctemque in totidem, vulgo notum est. Sed hoc credo Romæ post reperta solaria observatum.

([134]) Suprema summum diei, id a superrimo. Hoc tempus xii Tabulæ dicunt occasum esse solis; sed postea lex Plætoria id quoque tempus jubet esse supremum quo præco in comitio supremam pronuntiavit populo. Varro, de L. L. vi. § 5, ed. Müller. The enactment in the Twelve Tables was: 'Sol occasus suprema tempestas esto.' Censorin. c. 24; Dirksen, Zwölf-Tafel-Fragmente, p. 180. The Roman senate probably used no artificial light, and did not sit after sunset.

([135]) The rostra are stated to have received this appellation in 338 B.C., from the beaks of ships taken at Antium in that year. The Columna Mænia was also erected in the same year.

([136]) This writer is mentioned only by Pliny. He probably lived about the end of the Republic. The sister of Terentia, the wife of Cicero, was named Fabia, and she was a vestal virgin. Sallust, Cat. 15; Oros. vi. 3; Plut. Cat. Min. 19.

rinus, twelve years before the war with Pyrrhus, when he dedicated that temple in fulfilment of a vow made by his father; but this statement is considered by Pliny as insufficiently authenticated. The year indicated is 293 B.C.([137]) According to Varro, the earliest public sundial at Rome was that brought from Sicily by the Consul Manius Valerius Messala, after the capture of Catana,([138]) in the First Punic War, 263 B.C., and placed on a pillar near the Rostra. This dial, having been arranged for the latitude of Sicily, did not show the hours accurately for Rome; nevertheless, it remained in use for ninety-nine years. In the year 164 B.C., Q. Marcius Philippus included, among the works of his censorship, a new and accurate sundial in the Forum.([139]) Plautus, who began to write comedies about 224 B.C., introduces a slave as complaining of the novel introduction of sundials and hours, which makes him dependent for his meals upon the sun; whereas, when he was a boy, he used to eat when he was hungry.([140]) It has been supposed that this passage was translated by Plautus from the Greek original of the comedy in which it occurred; but it appears to refer to the recent introduction of sundials at Rome.([141])

A clepsydra, or contrivance for measuring time by allowing water to escape through the orifice of a vessel—similar to a modern hour-glass—was known at Athens in the time of Aristophanes. It was used for regulating the time allowed for speeches of accused persons before the courts of justice.([142]) The sundial had two great defects as a measure of hours: it was un-

---

([137]) The temple of Quirinus was probably vowed by L. Papirius Cursor, the father, after his great victory over the Samnites, in 309 B.C.

([138]) It appears that Catana submitted to the Romans at this time, and was received as an ally, but that it was not besieged and taken. See Eutrop. ii. 10.

([139]) It was noted that the *carbunculus*, or ruby, was first brought to Italy in this year. Plin. N. H. xxvi. 4.

([140]) Ap. Gell. N. A. iii. 3.

([141]) See Salmas. Plin. Exerc. p. 458.

([142]) Acharn. 692; Vesp. 93.

serviceable during cloudy weather,(143) and also during the night. Hence Plato is reported to have made an instrument like a clepsydra of great magnitude, for measuring time at night.(144) Scipio Nasica, while censor in 159 B.C., erected a water-clock at Rome, under a shed, in a public place. This water-clock acquired the name of *Solarium*, or sun-dial, from the habit of measuring hours by the sun.(145) The improved sundial of Philippus had been set up only five years previously. Ctesibius, a celebrated mechanician of Alexandria, constructed a complicated water-clock in that city, about 140 B.C.(146)

The rarity of sundials, the difficulty of using them, and their failure during the night and in cloudy weather, gave rise to the employment of slaves among the Greeks and Romans, whose duty it was to announce the hour.(147)

§ 10 With such imperfect contrivances for determining the seasons of the year and the hours of the night, it was natural that the scientific astronomy of the Greeks should be at first directed principally to an observation of the movements of the fixed stars, and of the revolution of the starry sphere. Such, as we have already seen, was the object of the astronomical trea-

---

(143) The epigram of Antiphilus, in Anth. Pal. vii. 641, speaks of a public water-clock, with twelve divisions, the object of which was to indicate the time of day when the sun was obscured by clouds. Antiphilus lived under the reign of the Emperor Nero.

(144) λέγεται δὲ Πλάτωνα μικράν τινα ἔννοιαν δοῦναι τοῦ ἐπισκευάσματος νυκτερινὸν ποιήσαντα ὡρολόγιον ἐοικὸς τῷ ὑδραυλικῷ, οἷον κλεψύδραν μεγάλην λίαν. Athen. iv. p. 174 c.

(145) Plin. vii. 60; Censorin. c. 23. Cicero, de N. D. ii. 34, opposes a solarium descriptum, a sundial, to one ex aquâ, a water-clock, for measuring hours.

(146) Vitruv. ix. 9; Plin. vii. 37. Compare Ideler, Chron. vol. i. p. 230. A clepsydra is described by Simplicius, Schol. Aristot. p. 380. The division of a water-clock into twelve hours is mentioned in an epigram of Paulus Silentiarius, Anth. Pal. ix. 782. Paulus Silentiarius lived in the time of Justinian I. Concerning water-clocks, both ancient and modern, see Beckmann's Hist. of Inventions, vol. i. p. 135, Eng. tr.

(147) περήτρια, ἡ παραγγέλλουσα τὴν ὥραν ταῖς κεκτημέναις, Hesychius, Photius, and Suidas. The etymology of the word is not apparent. The slave who announced the hour is mentioned by Juv. x. 216, Martial, viii. 67. Artaxerxes is described by Josephus, Ant. xi. 6, § 10, as inquiring the hour of the night from the officer whose duty it was to observe the time.

tises of Eudoxus; and such likewise is the object of those of Autolycus and Euclid, the earliest Greek writings now extant in which Astronomy is treated geometrically.

Autolycus the mathematician, of Pitana, in Æolis, was a fellow-citizen of Arcesilaus the philosopher. Before the latter removed to Athens, he was a disciple of Autolycus, and travelled with him to Sardis.([148]) As the lifetime of Arcesilaus extended from 316 to 241 B.C., Autolycus may be supposed to have flourished about 320—300 B.C.([149])

Two astronomical treatises of Autolycus are extant, but they have never been printed completely in the original Greek; and the incomplete edition of the Greek text which exists is so rare that it can with difficulty be consulted.([150]) No copy of it exists in the library of the British Museum. One is entitled Περὶ Κινουμένης Σφαίρας, 'Upon the Sphere in Motion;' the other is entitled Περὶ Ἐπιτολῶν καὶ Δύσεων, 'On the Risings and Settings of the Stars.' The former consists of twelve, the latter of eighteen propositions.

The treatise on the Movement of the Sphere is intended to illustrate the rotatory motion of the celestial globe. It assumes that the earth is at the centre, and it explains the apparent motion of the starry heaven upon this hypothesis. Delambre (who, in his History of Ancient Astronomy, has given a copious analysis of the two treatises of Autolycus) ([151]) makes the following remarks on the former:—' Out of the twelve propositions (he says), nine are fundamental, and have retained their place in all the elementary books of astronomy; they are always implied when they are not stated formally. They are propositions of pure geometry, and must have been conceived by all who sup-

---

([148]) Diog. Laert. iv. § 29. Bailly, Hist. Astr. Anc. p. 465, confounds Æolis with the Lipari Islands.

([149]) Clinton, F. H. vol. ii. ad ann. 299, and p. 367.

([150]) The Greek text of the propositions of the two treatises of Autolycus, without the demonstrations, is printed in Dasypodius, Sphæricæ Doctrinæ Propositiones, Argentor. 1572. A Latin translation, with notes, was published at Rome in 1588.

([151]) Vol. i. p. 19—48.

posed the spherical movement of the heaven; but they cannot have been reduced into a systematic form when Autolycus composed this little treatise. His work is a monument of the application of geometry to astronomy, but it is only a first step; it contains no trace of spherical trigonometry.'[152]

The treatise on the Risings and Settings of the Stars relates to their true and apparent risings and settings. The apparent or heliacal risings and settings are alone the subjects of observation by the naked eye.[153] The theorems are confined to the fixed stars; and there is a particular reference to the twelve parts of the zodiac, as denoted by constellations. The following are the most important propositions which he lays down on this subject:—

1. The zodiacal sign occupied by the sun neither rises nor sets, but is either concealed by the earth or lost in the sun's rays. The opposite sign neither rises nor sets, but it is visible during the whole night.

2. Of the twelve signs of the zodiac, that which precedes the sign occupied by the sun rises visibly in the morning; that which succeeds the same sign sets visibly in the evening.

3. Eleven signs of the zodiac are seen every night. Six signs are visible, and the five others, not occupied by the sun, afterwards rise.

4. Every star has an interval of five months between its morning and its evening rising, during which time it is visible. It has an interval of at least thirty days between its evening

---

(152) Hist. d'Astr. Anc. vol. i. p. 21. Philoponus, on Aristotle's Physics, remarks that Autolycus has treated the motion of the sphere in a less purely geometrical method than Theodosius, and more according to physical methods. Schol. Aristot. p. 348, ed. Brandis. Theodosius was a mathematician later than the reign of Trajan.

(153) 'Ces deux espèces de levers et de couchers sont les seules qu'on pût réellement observer; ces observations faciles que ne supposent qu'un peu d'attention, de bons yeux et un horizon libre, ont fait longtems toute l'astronomie des anciens, et la matière de leurs calendriers; ces levers et ces couchers ont réglé l'ordre des travaux agricoles et des tems propres à la navigation.' Delambre, ib. p. 23.

setting and its morning rising, during which time it is invisible.

Aratus states, in his astronomical poem, that half the zodiac is visible, and that six signs set every night; in which number he seems to include the sign occupied by the sun. This statement was doubtless copied from Eudoxus.([154])

Autolycus makes no mention of the planets in this treatise. Their irregular movements rendered them unsuited to the practical object which he had in view. Autolycus is, however, stated by Simplicius to have proposed some hypotheses, similar to those of Eudoxus, for explaining the anomalous motions of the planets, and to have failed in his attempt.([155])

§ 11 A treatise of the same character, and probably composed nearly at the same time, though belonging to a more advanced period of astronomical science, is that of the celebrated geometer Euclid, entitled Φαινόμενα, or 'The Appearances of the Heavens,' a title similar to that of the astronomical treatise of Eudoxus, as well as of the astronomical poem of Aratus. According to Proclus, he lived in the time of the First Ptolemy, 323—283 B.C.([156])

The *Phænomena* of Euclid consist of eighteen theorems, with their demonstrations, some of which are long and intricate.([157]) A brief account of the treatise, in which the theorems are extracted, is given by Delambre;([158]) but as the treatise is little known, even to professed students of antiquity, and as the Introduction to it contains a summary of Euclid's astronomical system, expounded with the precision and perspicuity charac-

---

([154]) Arat. 550—68. Compare Schaubach, Gesch. der Astron. p. 334 —5, who points out the importance of the doctrine of the risings and settings of stars to the Greeks, even to those who were not professed astronomers. Without them, he remarks, neither the season of the year nor the hour of the night could be properly determined.

([155]) Schol. Aristot. p. 502 b, Brandis.

([156]) Clinton, ad ann. 306.

([157]) P. 557—597, ed. Gregor. Oxon. 1703. The treatise consists of 40 folio pages, half of which is occupied with the Latin translation.

([158]) Hist. d'Astr. Anc. vol. i. p. 51—58.

teristic of the great geometer, I extract the substance of the latter, which will exhibit the most finished specimen of Greek astronomical science about the year 300 B.C.

The fixed stars rise at the same point, and set at the same point; the same stars always rise together, and set together; and in their course from the east to the west they always preserve the same distances from one another. Now, as these appearances are only consistent with a circular movement, when the eye of the observer is equally distant from the circumference of the circle in every direction (as has been demonstrated in the treatise on Optics), it follows that the stars move in a circle, and are attached to a single body, and that the vision is equally distant from the circumference.

A star is visible between the Bears, not changing its place, but always revolving upon itself. Since this star appears to be equally distant from every part of the circumference of each circle described by the other stars, it must be assumed that all the circles are parallel, so that all the fixed stars move along parallel circles, having this star as their common pole.

Some of these neither rise nor set, on account of their moving in elevated circles, which are called the 'always visible.' They are the stars which extend from the visible pole to the Arctic circle. Those which are nearest the pole describe the smallest circle, and those upon the Arctic circle the largest. The latter appear to graze the horizon.

The stars to the south of this circle all rise and set, on account of their circles being partly above and partly below the earth. The segments above the earth are large, and the segments below the earth are small, in proportion as they approach the Arctic circle, (^{159}) because the motion of the stars nearest this circle above the earth is made in the longest time, and of those below the earth in the shortest. In proportion as the stars recede from this circle, their motion above the earth is made in less time, and that below the earth in greater. Those that are nearest the south are the least time above the earth, and the longest below it.

The stars which are upon the middle circle make their times above and below the earth equal; whence this circle is called the Equinoctial. Those which are upon circles equally distant from the equinoctial, make the alternate segments in equal terms. For example, those above the earth to the north correspond with those below the earth to the south; and those above the earth to the south correspond with those below the earth to the north. The joint times of all the circles, above and below the earth, are equal. The circle of the Milky Way and the zodiacal circle being oblique to the parallel circles, and cutting each other, always have a semicircle above the earth.

(^{159}) In p. 560, l. 9, τῶν δὲ ὑπὲρ γῆς τμημάτων ἑκάστου αὐτῶν μόνον φαίνεται, the sense requires μέγιστον for μόνον.

. Hence it follows that the heaven is spherical. For if it were cylindrical or conical, the stars upon the oblique circles, which cut the equator, would not in the revolution of the heaven always appear to be divided into semi-circles; but the visible segment would sometimes be greater, and sometimes less than a semicircle. For if a cone or a cylinder were cut by a plane not parallel to the base, the section is that of an acute-angled cone, which resembles a shield (an ellipse). It is, therefore, evident that if a figure of this description is cut in the middle both in length and breadth, its segments will be unequal: and, likewise, that if it be cut in the middle by oblique sections, the segments will be unequal. But the appearances of the heaven agree with none of these results. Therefore the heaven must be supposed to be spherical, and to revolve equally round an axis of which one pole above the earth is visible, and the other below the earth is invisible.

The horizon is the plane reaching from our station to the heaven, and bounding the hemisphere visible above the earth. It is a circle; for if a sphere be cut by a plane, the section is a circle.

The meridian is a circle passing through the poles of the sphere, and at right angles to the horizon.

The tropics are circles which touch the zodiacal circle, and have the same poles as the sphere.

The zodiacal and the equinoctial are both great circles, for they bisect one another. For the beginning of Aries and the beginning of the Claws (or Libra) are upon the same diameter; and when they are both upon the equinoctial, they rise and set in conjunction, having between their beginnings six of the twelve signs, and two semicircles of the equinoctial; inasmuch as each beginning, being upon the equinoctial, performs its movement above and below the earth in equal times. If a sphere revolve equally round its axis, all the points on its surface pass through similar arcs of the parallel circles in equal times. Therefore these signs pass through equal arcs of the equinoctial, one above and the other below the earth. Consequently the arcs are equal, and each is a semicircle: for the circuit from east to east and from west to west is an entire circle. Consequently the zodiacal and equinoctial circles bisect one another. But if in a sphere two circles bisect one another, each will be a great circle. Therefore the zodiacal and equinoctial are great circles.

The horizon is likewise a great circle; for it bisects the zodiacal and equinoctial, both great circles. For it always has six of the twelve signs above the earth, as well as a semicircle of the equator. The stars above the horizon which rise and set together reappear in equal times, some moving from east to west, and some from west to east.

These propositions imply a completely geocentric system, in which the earth is at rest, and the starry sphere revolves round it every twenty-four hours. Accordingly, Euclid's first theorem is, that 'the earth is in the middle of the universe, and stands to it in the relation of centre.'

Galen remarks that Euclid, in his Treatise on Phænomena, demonstrates in a few sentences, that the earth is at the centre of the universe, and that its relation to the universe is that of a central point; he adds, that learners are as much convinced of the conclusiveness of the demonstration, as that twice two make four.[160]

§ 12 Although the geocentric theory always retained a firm ground in the general belief of the Greek astronomers and geometers, still certain rival hypotheses, dissenting from the received system of the universe, were from time to time promulgated. We have already seen that a wild and fanciful scheme was devised by Philolaus the Pythagorean, according to which the earth, together with the sun, the moon, and the other heavenly bodies, revolved in circular orbits round the central fire. Hicetas, a Pythagorean philosopher, (who probably was nearly contemporary with Philolaus,) and Heraclides of Pontus, retaining the geocentric system, ventured nevertheless to set the earth in motion. They supposed it, preserving its place in the centre of the universe, to revolve on its axis from west to east; and by this hypothesis they accounted, on scientific grounds, for the diurnal motions of the sun and fixed stars. Aristotle and others of the ancients likewise supposed Plato to have given a rotatory motion to the earth in the Timæus. In the first half of the third century B.C., about one generation after Euclid, Aristarchus of Samos proposed a theory of the world exactly similar to the Copernican. His date is fixed by the testimony of Ptolemy, who states him to have observed the summer solstice of the year 280 B.C.[161] He was likewise, as we shall see presently, contemporary with Cleanthes, who succeeded to the primacy of the Stoical School in 264 B.C.[162] His

---

(160) De Hippocrat. et Plat. Plac. viii. 1. vol. v. p. 654, ed. Kühn.
(161) Mag. Synt. iii. 2. Compare Clinton, F. H. vol. ii. p. 340, note.
(162) See Clinton, ad ann. 279.

lifetime may be placed conjecturally from about 320 to 250 B.C.([163]) The theory of Aristarchus is only known to us from secondary sources; but the chief of these is Archimedes, who could not fail to understand the hypothesis, and to report it correctly. Archimedes was born in 287 B.C.; and the theory of Aristarchus may have been published after he had reached the years of manhood.

Aristarchus is stated by Archimedes to have rejected the geocentric doctrine held by the majority of Greek astronomers, and to have promulgated the following hypothesis in a controversial treatise directed against that doctrine:—That the fixed stars and the sun are immovable; that the earth is carried round the sun in the circumference of a circle of which the sun is the centre; and that the sphere of the fixed stars having the same centre as the sun is of such a magnitude, that the orbit of the earth is to the distance of the fixed stars as the centre of the sphere of the fixed stars is to its surface. Archimedes treats as absurd the hypothesis, that the distance of the fixed stars from the earth is as a point to a surface, and consequently infinite: he therefore supposes Aristarchus to have meant to say that, as the earth is to the sphere of the fixed stars, according to the received geocentric hypothesis, so is the sphere by which he supposes the earth to be carried round to the sphere of the fixed stars. By this interpretation, Archimedes substitutes a proportion between four spheres; and he adopts this interpretation as correctly representing the meaning of Aristarchus, in nis subsequent reasonings and calculations. It will be observed that Archimedes assumes Aristarchus to mean that the earth is carried round the sun in a solid sphere. Archimedes does not inform us whether the hypothesis of Aristarchus included the planets. The express limitation of the hypothesis to the fixed

---

([163]) Martin, Timée, vol. ii. p. 127, cites Pappus as stating that the fame of Aristarchus attracted Apollonius of Perga to Alexandria. I have been unable to verify this quotation. Apollonius, moreover, who was born under Ptolemy Evergetes, 247—222 B.C., and who died under Philopator, 222—205 B.C., seems to belong to a later date.

stars, and the omission of all reference to the planets, shows that the latter were considered of secondary importance by the Greek astronomers.[164]

The compiler of the work on the Opinions of the Philosophers likewise describes Aristarchus as placing the sun amongst the fixed stars, and as supposing the earth to move along the solar circle (or the ecliptic); he adds, that Aristarchus accounted for solar eclipses by the inclination of the ecliptic to the axis of the earth.[165] It appears, moreover, from another notice of the theory of Aristarchus, that he agreed with the Copernican theory in giving the earth a motion of rotation as well as an orbital motion; and that he thus explained both the diurnal and the annual motions of the sun and fixed stars. 'He endeavoured,' says Plutarch, 'to account for the appearances by supposing the heaven to be motionless, and the earth to turn in an oblique circle, revolving at the same time round its axis.'[166] These words show that Aristarchus explained the apparent annual motion of the sun in the ecliptic by supposing the orbit of the earth to be inclined to its axis. Simplicius, likewise, in his Commentary upon Aristotle *de Cœlo*, couples Aristarchus with Heraclides as holding that the appearances would be explained by supposing the starry heaven to be at rest and the earth to make nearly one diurnal revolution from the west, round the poles of the equinoctial. 'The word *nearly* is added,' says Simplicius, 'on account of the sun's daily motion of one degree.'[167] Sextus Empiricus likewise refers to the hypothesis of the earth's rotation, when he includes Aristarchus

---

[164] See note B, at the end of the chapter.

[165] Plut. Plac. Phil. ii. 24; Stob. Ecl. Phys. i. 25; Galen, Phil. Hist. c. 14; Euseb. Præp. Ev. xv. 50. In the two latter writers, read τὴν δὲ γῆν for τὴν δὲ σελήνην, from the two former.

[166] Ἀρίσταρχον ᾤετο δεῖν Κλεάνθης τὸν Σάμιον ἀσεβείας προκαλεῖσθαι τοὺς Ἕλληνας, ὡς κινοῦντα τοῦ κόσμου τὴν ἑστίαν, ὅτι φαινόμενα σώζειν ἀνὴρ ἐπειρᾶτο, μένειν τὸν οὐρανὸν ὑποτιθέμενος, ἐξελίττεσθαι δὲ κατὰ λοξοῦ κύκλου τὴν γῆν, ἅμα καὶ περὶ τὸν αὑτῆς ἄξονα δινουμένην. De fac. in orbe lun. 6. As to the use of ἐξελίττεσθαι, compare Eurip. Herc. Fur. 977, ὁ δ᾽ ἐξελίσσων παῖδα κίονος κύκλῳ, where the verb denotes a circular motion.

[167] Schol. Aristot. p. 495, Brandis.

the mathematician among those who deny the movement of the heaven, and ascribe a movement to the earth.(168) Plutarch, again, in another treatise, speaks of Aristarchus and a certain Seleucus as having held the doctrine of the earth's rotation: he says that the former only advanced it as a hypothesis, whereas the latter demonstrated it.(169) Archimedes mentions the 'Hypotheses' of Aristarchus as containing his heliocentric theory: it is probable that he promulgated the opinion, unaccompanied with any geometrical proof. Seleucus the mathematician is described in the work on the Opinions of the Philosophers as holding the doctrine of the earth's motion.(170) He lived before Hipparchus, by whom he was quoted on a question relating to the tides.(171) He wrote against Crates, who likewise speculated on tides;(172) this circumstance fixes his lifetime to about the middle of the second century B.C. Seleucus likewise held, with Heraclides of Pontus, that the universe is infinite.(173)

We are informed by Plutarch that Cleanthes (who probably at the time was head of the Stoical School at Athens, the most religious of the Greek philosophical sects) declared his opinion that Aristarchus of Samos ought to be prosecuted for impiety, because he taught that the hearth of the universe was movable.(174) By 'the hearth of the universe' he meant the earth; and he employed this mystical appellation as alluding to its central and sacred character.(175) His indignation at the here-

---

(168) οἵ γε μὴν τοῦ κόσμου κίνησιν ἀνελόντες, τὴν δὲ γῆν κινεῖσθαι δοξάσαντες, ὡς οἱ περὶ Ἀρίσταρχον τὸν μαθηματικὸν, οὐ κωλύονται νοεῖν χρόνον. Sext. Emp. adv. Dogmat. iv. § 174, Bekker.

(169) Quæst. Plat. viii. 1.

(170) Plac. Phil. iii. 17; Excerpt. Joan. Damasc. ad calc. Stob. Phys. vol. ii. p. 775, ed. Gaisford. Also in Stob. Anth. vol. iv. p. 244, ed. Meineke.

(171) Strab. i. 1, § 9.

(172) Joan. Damasc. ib.

(173) Stob. Phys. xxi. 3; Plut. Plac. ii. 1. Concerning the birth-place of Seleucus, see Boeckh, Kosm. Syst. des Plat. p. 142.

(174) De fac. in orbe lun. 6.

(175) See Martin, Timée, tom. ii. p. 115.

tical doctrine of Aristarchus implies that he conceived it not merely as making the earth revolve round its axis (which Heraclides and Hicetas had done), but as uprooting it from the tranquil and dignified seat which it had occupied at the centre of the heavenly sphere, and as sending it a wanderer through space. The appeal of Cleanthes, however, met with no response: the general opinion of Greece had become more tolerant of physical speculation than it was in the time of Anaxagoras and Socrates. Cleanthes seems to have been driven to use argumentative, not penal weapons, in defence of the immobility of the earth. We find, at least, in the list of his works, a treatise 'against Aristarchus,'(176) which probably referred to his heliocentric hypothesis.

A treatise of Aristarchus, 'On the Magnitudes and Distances of the Sun and Moon,' is still extant.(177) The geometrical methods by which he attempts to measure the apparent diameters of the sun and moon, and to determine their distances, are considered sound by modern astronomers; but as his observed data were inexact, owing to the defective instruments which he used, his results are erroneous.(178) In this treatise the author makes no allusion to his heliocentric hypothesis.

---

(176) πρὸς 'Αρίσταρχον, Diog. Laert. vii. 174.
Schaubach, Geschichte der Griech. Astron. p. 466—477, has attempted to show that the hypothesis of Aristarchus did not anticipate the Copernican theory; and his arguments have received some countenance from Mr. Donkin, the Savilian Professor of Astronomy at Oxford, in his article *Aristarchus*, in Dr. W. Smith's Dictionary of Ancient Biogr. and Myth. The subject has been fully investigated by Ideler, in his excellent dissertation, Ueber das Verhältniss des Copernicus zum Alterthum, published in the Museum der Alterthumswissenschaft (Berlin, 1810), vol. ii. p. 393 —454; who considers the hypothesis of Aristarchus as agreeing substantially with the Copernican: see p. 429. This view is adopted by Martin, Timée, vol. ii. p. 127. Compare Delambre, Hist. d'Astr. Anc. tom. i. p. 80, 102. The silence of Ptolemy, Mag. Synt. i. 4 and 6, respecting Aristarchus, proves nothing. In the chapters cited, Ptolemy uses general arguments to disprove the motion of the earth, both of rotation and translation, but he mentions no astronomer who held these opinions. He states the hypotheses, and controverts them without attributing them to any one.

(177) Edited by Wallis, Oxon. 1688, and repeated in Wallis, Opera Mathematica (Oxon. 1699, fol.), vol. iii. p. 569.

(178) See Delambre, Hist. de l'Astr. Anc. tom. i. p. 75, and Donkin,

Vitruvius includes Aristarchus among the principal inventors of scientific instruments and dialling;([179]) and describes him as the author of the hemispherical sundial. Aristarchus may have improved it; but this form of sundial certainly existed before his time.([180]) He is likewise reported to have made a great cycle of 2484 years,([181]) which was doubtless founded on some astronomical combination.

§ 13 Archimedes might be assumed on general grounds, even without the evidence afforded by his treatise of the *Arenarius*, to have been acquainted with all the astronomical science of his time. He is characterized as an astronomer by Livy,([182]) who probably had only a general idea of his celebrity in physical science. He is stated by Macrobius to have measured the distances of the moon, sun, and planets;([183]) and he constructed an orrery, in which he exhibited the motions of the sun, moon, planets, and fixed stars round the central earth.([184]) This orrery

---

art. *Aristarchus*, in Dr. Smith's Dictionary of Anc. Biogr. and Myth. Delambre says: 'L'écrit que nous venons d'analyser, s'il ne nous donne que des résultats très inexacts, est du moins recommandable par la finesse des aperçus et une méthode vraiment géométrique,' p. 80. Speaking of the determination of the sun's distance by Aristarchus, he says: 'Quand on compare cette distance d'Aristarque aux idées de ses prédécesseurs, on conçoit qu' Aristarque a dû passer en son tems pour un grand astronome,' p. 79.

(179) Vitruv. i. 1, 17.

(180) Scaphen sive hemisphærium Aristarchus Samius (dicitur invenisse), ix. 9.

(181) Censorin. c. 18.

(182) Unicus spectator cœli siderumque, mirabilior tamen inventor ac machinator bellicorum tormentorum operumque, xxiv. 34. The exposition of the orrery of Archimedes seemed to Cicero to prove, 'plus in illo Siculo ingenii, quam videretur natura humana ferre potuisse,' De Rep. i. 14.

(183) In Somn. Scip. ii. 3.

(184) See Cic. de Rep. i. 14; Nat. D. ii. 35; Tusc. Disp. i. 25, from which passages it appears that the orrery of Archimedes was made of brass; that it exhibited the revolutions of the sun, moon, and five planets, and showed the nature of eclipses; and that it was removed from Syracuse by Marcellus, and deposited in the Temple of Virtue at Rome. The Temple of Virtue was dedicated by Marcellus from the plunder of Syracuse. Plut. Marcell. 28; Val. Max. i. 1, 8.

Ovid, Fast. vi. 271,

> Arte Syracosiâ suspensus in aëre clauso
> Stat globus, immensi parva figura poli.

was doubtless intended to exhibit the system of the world which was generally received among the scientific men of Greece. Nevertheless, his extant works show that his attention and inventive powers were principally directed towards mechanics and geometry. His life extended from 287 to 212 B.C. He appears to have resided at Syracuse, his native town, where he was killed by a Roman soldier, when the town was taken in the Second Punic War, and where his tomb was found by Cicero.([185]) His works are written in the Doric dialect of his fellow-citizens, who had derived it from their mother-city, Corinth.([186])

§ 14 The Museum of Alexandria was the creation of the early Ptolemies. Its vast library was its principal feature; but it was likewise a college or academy of men distinguished in literature and science, who were attracted to Alexandria by the patronage of the Greek kings of Egypt.([187]) Among the sciences cultivated by the Greek Alexandrine School was Astronomy. Aristarchus, to whom reference has been already made, is stated to have resided at Alexandria. Aristyllus and Timocharis, two astronomers who appear to have been contemporary with each other, conducted their operations in the same city. They are classed by Plutarch with Aristarchus and Hipparchus, as authors

---

Claudian, in his epigram 'in sphæram Archimedis (Ep. 18), describes it as being of glass, and particularly mentions that it represented the zodiac, and the moon's motions.

Sextus Empiricus adv. Dogm. iii. § 115, Bekker, speaks of this orrery as made of wood, and as exhibiting the motions of the sun, moon, and other stars.

Lactantius, Div. Inst. ii. 5, following Cicero, is more copious: An Archimedes Siculus concavo ære similitudinem mundi ac figuram potuit machinari, in quo ita solem ac lunam composuit, ut inæquales motus et cœlestibus similes conversionibus singulis quasi diebus efficerent, et non modo accessus solis et recessus, vel incrementa diminutionesque lunæ, verum etiam stellarum vel inerrantium vel vagarum dispares cursus orbis ille dum vertitur exhiberet?

Martianus Capella, vi. § 583, ed. Kopp., describes an orrery, similar to the Archimedean, in which the earth is immovable.

([185]) Tusc. Disp. v. 23.

([186]) See Theocrit. Id. xx. 91.

([187]) See Müller's Hist. of Greek Literature, by Donaldson, vol. ii. p. 418.

of prose treatises on Astronomy.(188) It is stated by Ptolemy that the only observations of the fixed stars to which Hipparchus had access were those made by the two astronomers in question; and that they were neither unambiguous nor complete.(189) In another place, Ptolemy mentions that the observations of Timocharis were roughly taken.(190) Observations of Timocharis are specified by Ptolemy, of which the dates extend from 293 to 272 B.C.(191) His lifetime must, therefore, have nearly coincided with that of Aristarchus.

Conon of Samos, another celebrated astronomer of this period, belonged to the Alexandrine School. He was the friend of Archimedes, who survived him, and who mentions him with admiration in his extant works.(192) He is stated by Ptolemy to have himself made astronomical observations in Italy,(193) and by Seneca to have formed a collection of the solar eclipses observed by the Egyptians.(194) His astronomical fame is like-

---

(188) De Pyth. Orat. 18. It seems that there were two persons named Aristyllus, who were celebrated as astronomers. In the list of writers who composed commentaries on Aratus, in Petav. Uranolog. p. 147, there is first Ἀρίστυλλοι δύο γεωμέτραι, and afterwards Ἀρίστυλλος μέγας and Ἀρίστυλλος μικρός.

(189) Synt. Mag. vii. 1, p. 2, Halma.

(190) vii. 3, p. 15.

(191) An observation of Timocharis, taken in the 47th year of the first Callippic period, or 465 of Nabonassar, is mentioned in Ptol. iii. 3, p. 21; two in the 36th year of the first Callippic period, or 454 of Nabonassar, p. 23, 26; another in the 48th year of the first Callippic period, or 466 of Nabonassar, p. 24; another in the 13th year of Ptolemy Philadelphus, x. 4, p. 205. Timocharis is mentioned in Schol. Arat. 269.

(192) See the Introduction to the Treatises on Helices, and on the Quadrature of the Parabola, where Archimedes mentions with regret the death of his friend Conon, and speaks with admiration of his geometrical genius. Conon had treated conic sections before Apollonius. Procem. ad Apollon. Perg. lib. iv. p. 217, ed. Halley.

(193) De Apparent. p. 53, Petav.

(194) Conon, diligens et ipse inquisitor, defectiones quidem solis servatas ab Ægyptiis collegit, Sen. Nat. Quest. vii. 3, where 'inquisitor' signifies observer. Mr. Donkin, art. *Conon*, in Dr. Smith's Dictionary of Anc. Biog. and Myth., renders this passage, 'he made a collection of the observations of solar eclipses *preserved* by the Egyptians.' Servatas, however, here means *observed;* compare Virgil, Æn. vi. 337—9.

> Ecce gubernator sese Palinurus agebat:
> Qui Libyco nuper cursu, *dum sidera servat*,
> Exciderat puppi mediis effusus in undis.

'Cœli signa servare,' Cic. de Div. i. 19. 'Servare de cœlo' seems to

wise commemorated by Virgil.(195) Upon the safe return of Ptolemy Evergetes from his Syrian expedition in 243 B.C., his queen, Berenice, dedicated a lock of her hair in the Temple of Arsinoe-Aphrodite at Zephyrium, in Lower Egypt.(196) This lock of hair has become immortal; for having disappeared in the temple where it was deposited, it was translated by Conon into the heaven as a constellation,(197) which has retained the name of *Coma Berenices,* and still appears in the celestial sphere. This incident (which was made by Callimachus the subject of an elegiac poem),(198) proves that the Alexandrine astronomers were desirous of marking their gratitude for the patronage which they received from the Greek Court. The name of this constellation has been more permanent than the appellation of *Medicean stars,* which Galileo gave to the satellites of Jupiter, and of *Georgium Sidus,* which Herschel gave to the sixth planet discovered by him. The fame of this lock of hair has likewise been perpetuated in the word *vernice, vernis,* and *varnish,* which alludes to the amber colour of the queen's beautiful tresses.(199)

---

have been the technical phrase for the observations of the flight of birds made by the augurs.

(195)   In medio duo signa, Conon; et quis fuit alter,
Descripsit radio totum qui gentibus orbem?
Tempora quæ messor, quæ curvus arator haberet?
                                Virg. Ecl. iii. 40—42.

The second portrait is conjectured to be that of Eudoxus.

The astronomical reputation of Conon is likewise described in the introductory verses of the Coma Berenices of Catullus (Carm. 66):—

      Omnia qui magni dispexit lumina mundi,
        Qui stellarum ortus comperit atque obitus;
      Flammeus ut rapidi solis nitor obscuretur,
        Ut cedant certis sidera temporibus;
      Ut Triviam furtim sub Latmia saxa relegans
        Dulcis amor gyro devocet aërio.

(196) Strab. xvii. 1, § 16.

(197) Hygin. Poet. Astr. ii. 24; Eratosth. Catasterism. 12; Schol. Arat. 146; Achill. Tat. Isag. c. 14. The constellation lies between the tail of Leo and the star Arcturus.

(198) The few extant fragments of the elegy of Callimachus are in Blomfield's edition, p. 321. The substance of the poem is preserved in the rugged version of Catullus, Carm. 66.

(199) See Eastlake's Materials for a History of Oil Painting (Lond. 1847), p. 230—268. Her golden hair is mentioned by Catullus, v. 62,
            Devotæ flavi verticis exuviæ.

Βερενίκη was used in lower Greek as the name of amber. Compare Du-

Eratosthenes, whose lifetime reached from 276 to 196 B.C., was another distinguished member of the scientific School of Alexandria. Having been invited to that city from Athens by Ptolemy Evergetes, he was placed by him at the head of the library. He was the father of systematic chronology and scientific geography; [200] and he was likewise a proficient in astronomy.[201] His chief astronomical performance was his determination of the circumference of the earth by a method identical with that which would be employed by a modern astronomer. It was known that Syene in Upper Egypt is under the tropic of Cancer; [202] and that the gnomon of the sundial cast no shade there at the summer solstice. It was likewise assumed that Syene and Alexandria had the same longitude. The distance between Syene and Alexandria had been ascertained by the measurers of King Ptolemy to be 5000 stadia. The zenith distance for Alexandria was determined by Eratosthenes at $\frac{1}{50}$th of the circumference of the meridian. Hence the circumference of the earth was fixed at 250,000 stadia. In order to obtain a number divisible by 360 without a remainder, Eratosthenes supposed the number to be 252,000, and thus obtained 700 stadia for a degree.[203]

---

cange, Gloss. inf. Græc. in Βερονίκη, Diez, Romanisches Wörterbuch, p. 368.

[200] This is the view of him which is taken by modern writers; his contemporaries considered him to have obtained a good secondary position in all sciences, but the first in none; hence they called him *Beta*. See Bernhardy, Eratosthenica (Berlin, 1822), p. viii.

[201] It has been conjectured that the erection of the large armillæ, or circular instruments for astronomical observation, at Alexandria was due to his influence; but the conjecture is unsupported by express testimony. Bernhardy, ib. p. xii.

[202] That the sun is perpendicular at Syene at the summer solstice, and shines into a deep well, is stated, with full explanation, by Strabo, xvii. 1. 48. Compare Dr. Smith's Dict. of Anc. Geogr., art. *Syene*. Lucan's ideas on this phenomenon were very indistinct. In one place he implies that the absence of shadows is constant at Syene:—

  Calidâ medius mihi cognitus axis
  Ægypto, atque umbras nusquam flectente Syene.
           ii. 586, 7.

In other passages he indicates a consciousness that the perpendicularity of the sun in the tropic of Cancer is only temporary, viii. 851, x. 234.

[203] See Cleomed. i. 10, where the method of Eratosthenes is fully

The circumference of the earth being determined, its diameter was known. Eratosthenes further determined the diameter of the sun to be twenty-seven times greater than that of the earth.(204) He likewise made the distance of the sun from the earth to be 804,000,000 stadia, and that of the moon 780,000 stadia.(205)

According to a story told by Pliny, Dionysodorus, a celebrated geometer, died in his native island of Melos. A short time after his burial, a letter was found on his sepulchre, addressed by him to the people on earth, stating that he had reached the lowest point of the earth, and that the distance was 42,000 stadia. This was interpreted by geometers to mean that he had reached the centre of the earth, which was a sphere; whence it followed that the circumference of the earth was 252,000 stadia ($= 42,000 \times 6$).(206) This value for the earth's radius was, however, in fact, taken from the estimate of the earth's circumference by Eratosthenes. The date of Dionysodorus is unknown.(207)

There is a work on the constellations which bears the name of Eratosthenes; but it has been proved by Bernhardy to be pseudonymous, and to be the compilation of a late Greek grammarian, closely resembling the Poeticon Astronomicon of Hyginus, and probably derived from a common source.(208) The

---

described. Also Incerti Auctoris excerptum Mathematicum, in Macrobius, ed. Jauus, vol. i. p. 219. See also Plin. ii. 108; Strab. ii. 5, 7; Vitruv. i. 6; Martianus Capella, vi. § 596, ed. Kopp. Compare Delambre, Astr. Anc. vol. i. p. 89, 221; Mr. De Morgan's art. *Eratosthenes*, in Dr. Smith's Dict. of Anc. Biog. and Myth.; De la Nauze, Mém. de l'Acad. des Inscript. tom. xxvi.; Schaubach, Gesch. der Griech. Astr. p. 275; Bernhardy, Eratosthenica, p. 57. Assuming the stadium to be $\frac{1}{8}$th of a mile, the degree of Eratosthenes would be $= 87\frac{1}{2}$ miles.

(204) Macrob. in Somn. Scip. i. 20, § 6. He cites the 'libri dimensionum.'

(205) Stob. Phys. i. 26 (where the word μυριάδων should be expunged, with Bernhardy); Euseb. Præp. Ev. xv. 53; Galen, Hist. Phil. c. 15. The passage of Plut. Phil. Plac. ii. 31, is mutilated. Compare Bernhardy, ib. p. 56. The number expressing the sun's distance is μυριάδες τετρακοσίαι καὶ ὀκτακισμύριαι $=$ 4 millions $+$ 800 millions.

(206) N. H. ii. 109.

(207) See Dr. Smith's Dict. of Biogr. in v.

(208) The καταστερισμοὶ of Eratosthenes were edited with notes and

author and the date of the latter work are likewise undetermined.([209])

Another scientific astronomer of Alexandria, the survivor of Conon, but the contemporary of Archimedes, who dedicated to him his treatises on the Sphere and Cylinder, and on Spirals, was Dositheus. He is one of the authorities whose observations on the fixed stars are followed in the calendars of Geminus and Ptolemy.([210]) In a passage appended to the Calendar of Ptolemy, it is mentioned that his observations were made at Colonia, where the length of the longest day was $14\frac{1}{2}$ hours.([211]) It does not appear what place can be meant. He is mentioned by Censorinus as one of the improvers of the octaëteric cycle ascribed to Eudoxus.([212])

Apollonius of Perga is stated to have been born in the reign of Ptolemy Evergetes,([213]) and to have flourished in the reign of his successor, Ptolemy Philopator.([214]) The latter of these reigns extended from 222 to 205 B.C. He was, therefore, a younger contemporary of Archimedes. Apollonius was one of the principal ornaments of the Alexandrian School: his trea-

---

an *epistola* of Heyne, Gotting. 1795, by Schaubach, the author of the History of Greek Astronomy. The best text is in Westermann's Mythographi, p. 239. Compare Bernhardy, ib. p. 114, sq.

(209) See Dr. Smith's Dict. of Anc. Biog. and Myth. art. *Hyginus*.

(210) P. 36, 42, Petav. He is also mentioned as an astronomical observer by Pliny, N. H. xviii. 74.

(211) P. 53.

(212) C. 18.

(213) Eutocius (Comm. ad Apollon. Perg. Conica, lib. i. p. 8, ed. Halley, Oxon. 1710, fol.) cites Heraclius in the Life of Archimedes, as stating that Apollonius γέγονε μὲν ἐκ Πέργης τῆς ἐν Παμφυλίᾳ ἐν χρόνοις τοῦ Εὐεργέτου Πτολεμαίου. Evergetes reigned from 247 to 222 B.C., and Apollonius is stated to have flourished in the succeeding reign of Philopator, 222—205 B.C., the entire period being 42 years. Apollonius was subsequent to Archimedes, from whom he was unjustly accused of having pirated his demonstrations of the Conic Sections, Eutoc. ib. Conon of Samos (who was a contemporary of Evergetes) preceded Apollonius, Eutoc. ad lib. iv. p. 217. If Apollonius flourished under Philopator, his birth probably preceded the reign of Evergetes by a few years. His lifetime may be conjecturally placed from 250 to 180 B.C.

(214) Ptol. Hephæstion, ap. Phot. Bibl. Cod. 190, p. 151 b, ed. Bekker.

tise on Conic Sections acquired him the title of 'the Great Geometer,' by which he was known to the ancients:([215]) he was likewise celebrated as an astronomer, and he is reported to have obtained the nickname of Epsilon, from his fondness for observing the moon, whose crescent resembled that letter.([216])

It appears that he was the first to discard the theory of revolving spheres, and to substitute that of eccentrics and epicycles, for explaining the movements of the planets.([217]) This ingenious and refined theory could only have been devised by an accomplished geometer. The hypothesis of the eccentric circle approximated closely to that of the ellipse; while the theory of epicycles accounted by circular movements for the stations and retrogradations of the planets. This theory, after having been developed and elaborated by Hipparchus and Ptolemy, maintained its ground in scientific astronomy until it was overthrown, in the sixteenth and seventeenth centuries, by the great observers and mathematicians who established the Copernican System.

---

(215) Geminus, in the 6th book, τῆς τῶν μαθημάτων θεωρίας, cited by Eutocius, ibid.

(216) Ptol. Heph. ib.

(217) Ptol. Synt. ix. 1. An excellent account of the theory of epicycles is given in Drinkwater's Life of Kepler, c. 4.

Note A.—(p. 142.)

Mr. Grote has recently made this passage the subject of a separate investigation, in which all the main arguments bearing upon the question are fully and perspicuously stated. See his dissertation entitled 'Plato's Doctrine respecting the Rotation of the Earth, and Aristotle's Comment upon that Doctrine.' Lond. 1860. 8vo. There were two hypotheses respecting the interpretation of this passage: one that Plato meant the earth to be motionless, and to be attached to the cosmical axis; the other that he represented it as revolving round or upon that axis. Mr. Grote substantially adopts the latter hypothesis: he understands Plato to assign a rotatory motion to the earth, but he conceives the axis to be material and rigid, to be a solid revolving cylinder, and the earth to turn *with* the axis, not *round* or *upon* it. He renders the word εἰλλομένην, 'being packed or fastened close round, squeezing or grasping around.'

This view of Plato's conception of the cosmical axis is expressed in the following passage of Achilles Tatius, p. 95, ed. Petav.:—ὁ ἄξων ἀπὸ ἀρκτικοῦ πόλου μέχρι τοῦ ἀνταρκτικοῦ διήκει, διὰ τοῦ αἰθέρος καὶ τῶν ἄλλων στοιχείων ἱκνούμενος. ἡ γῆ οὖν βαρυτάτη οὖσα ἐνείκται [lege ἐνείρκται] καὶ ἐμπεπερόνηται ἐμπεριειλημμένη ὑπὸ τοῦ ἄξονος, ὡς μὴ δύνασθαι κινεῖσθαι. τὰ δὲ ἄλλα χαυνότερα ὄντα τόπον ἔχοντα στρέφεται· ὥσπερ ἂν εἴ τις ὀβελίσκῳ ξύλον ἀκινήτως ἐμπερονήσῃ ἐμπεριθεὶς τῷ ξύλῳ τροχοὺς κυκλοτερῶς. συμβαίνει γὰρ τοὺς μὲν τροχοὺς κινεῖσθαι, τὸ δὲ ξύλον ἀκινητεῖν ὑπὸ τοῦ κρατοῦντος στενοχωρούμενον.

The best sensible image of this hypothesis is that of a joint of meat fixed upon a spit, and turned by it before the fire, according to the suggestion of Achilles Tatius, c. 28. See above, p. 173.

It is, however, difficult to discover what is gained by this hypothesis. Sir Edmund Head seems to me to have conclusively established against Buttmann, that the idea of *rolling* is original and fundamental in εἴλεω and its cognate forms. (Philological Museum, vol. i. p. 405.) ἰλλομένων ἀρότρων, in Soph. Ant. 341, denotes the zigzag course of the plough, always returning upon itself. The word ἰλλάς, applied to oxen, appears to be derived from their use in ploughing, Soph. Fragm. 80, Dindorf. Cleomedes twice uses εἰλέω to denote a circular motion of revolution round a centre. ὁ τοίνυν οὐρανὸς κύκλῳ εἰλούμενος ὑπὲρ τὸν ἀέρα καὶ γῆν, De Met. i. 3, p. 20. ἕκαστος τῶν ἀπλανῶν ἀστέρων σὺν τῷ κόσμῳ εἰλούμενος περὶ τὸ οἰκεῖον κέντρον κύκλον περιγράφει, i. 5, p. 29.

If we are to suppose that Plato assigns a rotatory motion to the earth, the most natural interpretation would therefore be to render εἰλλομένην by *revolving*, and to understand (as Aristotle appears to have done), that the earth turns round the axis of the world.

The ancients in general undoubtedly conceived the cosmical axis as immaterial; as a geometrical line. If Plato supposed the earth to be turned by a solid revolving cylinder, he must have supposed this cylinder to project from the north pole of the earth, and to be visibly fixed in the north pole of the heaven: an idea of which no trace, so far as I am aware, occurs in any ancient writer.

The following passage from the Commentary of Hipparchus upon the Phænomena of Eudoxus and Aratus, proves that neither Eudoxus nor Hipparchus supposed the cosmical axis to be a solid cylinder attached to the celestial poles:—περὶ μὲν οὖν τοῦ βορείου πόλου Εὔδοξος ἀγνοεῖ λέγων οὕτως· 'ἔστιν δέ τις ἀστὴρ μένων ἀεὶ κατὰ τὸν αὐτὸν τόπον, οὗτος δὲ ὁ ἀστὴρ πόλος ἐστὶ τοῦ κόσμου.' ἐπὶ γὰρ τοῦ πόλου οὐδὲ εἷς ἀστὴρ κεῖται, ἀλλὰ κενὸς ἐστι τόπος, ᾧ παράκεινται τρεῖς ἀστέρες, μεθ' ὧν τὸ σημεῖον τὸ κατὰ τὸν πόλον. τετράγωνον ἔγγιστα σχῆμα περιέχει, καθάπερ καὶ Πυθέας φησὶν ὁ Μασσαλιώτης, lib. i. p. 101. Stella quæ dicitur polus, Vitruv. ix. 6.

The unsubstantial nature of the cosmical axis is marked by Manilius:—

    Nec vero solidus stat robore corporis axis,
    Nec grave pondus habet, quod onus ferat ætheris alti.
    Sed cum aër omnis semper volvatur in orbem,
    Quoque semel cœpit, totus volet undique in ipsum,
    Quodcunque in medio est, circa quod cuncta moventur,
    Usque adeo tenue, ut verti non possit in ipsum,
    Nec jam inclinari, nec se convertere in orbem.
    Hoc dixere axem, quia motum non habet ullum;
    Ipse videt circa volitantia cuncta moveri.
    Summa tenent ejus miseris notissima nautis
    Signa, per immensum cupidos ducentia pontum.
                            i. 292—302.

Aristotle, in his Treatise on the Motion of Animals, c. 3, p. 699, informs us that there were different ways of explaining the revolution of the heaven. Some supposed that the motion of the celestial sphere was produced by the energy of its poles. Others conceived that Atlas symbolized the revolving force of the diameter. Here, however, the axis or diameter is used for the purpose of explaining the movement of the starry sphere; it has no reference to the earth.

The received doctrine of the ancient astronomers respecting the poles is stated by Geminus, c. 3, Macrob. in Somn. Scip. i. 16, § 4. Concerning the use of πόλος, see Schol. Aristoph. Av. 179; Scriptor de Mundo, c. 2; Plin. ii. 15.

Dr. Whewell, in his Platonic Dialogues, vol. iii. p. 372, admits the difficulty of supposing Aristotle to have misunderstood Plato, but rightly urges the improbability that Plato 'should hold the rotation of the earth round an axis, except in order that he might thereby account for the apparent diurnal motion of the heavens.'

Prantl, in his translation of the Treatise of Aristotle de Cœlo (Leipzig, 1857), tries to get over the difficulty by supposing that a tremulous or vibratory motion is meant. He translates the passage in ii. 13 as follows: 'Einige aber behaupten auch, dass die Erde, während sie sich am eigentlichen Mittelpunkte befinde, in einer zitternd schwankenden Bewegung um die durch das All gespannte Axe sei, wie dies im Timäus geschrieben steht,' p. 159. The passage in c. 14 is similarly translated, p. 173. Compare the note, p. 311. But Aristotle's argument clearly shows that he understood a rotatory motion to be meant by Plato.

Simplicius, in commenting on Aristot. de Cœl. ii. 1, says, that Aristotle must have been well acquainted with the doctrines of Plato concerning the cause of the circular motion of the starry heaven, inasmuch as he condescended to compose a summary and abridgment of the Timæus: καὶ πάντων οἶμαι μᾶλλον ὁ Ἀριστοτέλης τὴν ἐν Τιμαίῳ περὶ τούτων τοῦ Πλάτωνος γνώμην ἠπίστατο, ὡς καὶ σύνοψιν καὶ ἐπιτομὴν τοῦ Τιμαίου γράφειν οὐκ ἀπηξίωσε, lib. 491 b, ed. Brandis. The lists of Aristotle's writings, however, make no mention of an abridgment of the Timæus of Plato. Diogenes Laertius, v. 25, has the title of a work in one book, τὰ ἐκ τοῦ Τιμαίου καὶ τῶν Ἀρχυτείων, and the Anonymus has a similar title, see Brandis, *Aristoteles*, vol. i. p. 85; but this was a summary of the opinions of Timæus himself, the Pythagorean philosopher, and of the followers of Archytas (like the extant summary of the opinions of Xenophanes, Zeno, and Gorgias), not of the Platonic Dialogue. It seems therefore probable that Simplicius made a mistake.

Note B.—(p. 191.)

Archimed. Arenarius, p. 319, ed. Torelli. κατέχεις δὲ ὅτι καλεῖται κόσμος ὑπὸ μὲν τῶν πλειόνων ἀστρολόγων ἁ σφαῖρα, ἇς ἐστὶ κέντρον τὸ τᾶς γᾶς κέντρον, ἁ δὲ ἐκ τοῦ κέντρου ἴσα τᾷ εὐθείᾳ τᾷ μεταξὺ τοῦ κέντρου τοῦ ἁλίου καὶ τοῦ κέντρου τᾶς γᾶς. ταῦτα γὰρ ἐν ταῖς γραφομέναις παρὰ τῶν ἀστρολόγων διακρούσας Ἀρίσταρχος ὁ Σάμιος ὑποθεσιῶν ἐξέδωκεν γράψας· ἐν αἷς ἐκ τῶν ὑποκειμένων συμβαίνει τὸν κόσμον πολλαπλάσιον εἶναι τοῦ νῦν εἰρημένου. ὑποτίθεται γὰρ τὰ μὲν ἀπλανῆ τῶν ἄστρων καὶ τὸν ἅλιον μένειν ἀκίνητον· τὰν δὲ γᾶν περιφέρεσθαι περὶ τὸν ἅλιον κατὰ κύκλου περιφέρειαν, ὅς ἐστιν ἐν μέσῳ τῷ δρόμῳ κείμενος· τὰν δὲ τῶν ἀπλανῶν ἄστρων σφαῖραν, περὶ τὸ αὐτὸ κέντρον τῷ ἁλίῳ κειμέναν, τῷ μεγέθει ταλικαύταν εἶμεν, ὥστε τὸν κύκλον καθ' ὃν τὰν γᾶν ὑποτίθεται περιφέρεσθαι τοιαύταν ἔχειν ἀναλογίαν ποτὶ τὰν τῶν ἀπλανῶν ἀποστασίαν οἵαν ἔχει τὸ κέντρον τᾶς σφαίρας ποτὶ τὰν ἐπιφάνειαν. τοῦτο δὲ εὔδηλον ὡς ἀδύνατόν ἐστιν. ἐπεὶ γὰρ τὸ τᾶς σφαίρας κέντρον οὐδὲν ἔχει μέγεθος, οὐδὲ λόγον ἔχειν οὐδένα ποτὶ τὰν ἐπιφάνειαν τὰς σφαίρας ὑπολαπτέον αὐτό. ἐκδεκτέον δὲ τὸν Ἀρίσταρχον διανοεῖσθαι τόδε· ἐπειδὴ τὰν γᾶν ὑπολαμβάνομεν ὥσπερ μὲν τὸ κέντρον τοῦ κόσμου, ὃν ἔχει λόγον ἁ γᾶ ποτὶ τὸν ὑφ' ἁμῶν εἰρημένον κόσμον, τοῦτον ἔχειν τὸν λόγον τὰν σφαῖραν, ἐν ᾇ ἐστιν ὁ κύκλος καθ' ὃν τὰν γᾶν ὑποτίθεται περιέρεσθαι, ποτὶ τὰν τῶν ἀπλανέων ἄστρων σφαῖραν. τὰς γὰρ ἀποδείξιας, τῶν φαινομένων οὕτως ὑποκειμένων, ἐναρμόζει· καὶ μάλιστα φαίνεται τὸ μέγεθος τᾶς σφαίρας ἐν ᾇ ποιεῖται τὰν γᾶν κινουμέναν ἴσον ὑποτίθεσθαι τῷ ὑφ' ἁμῶν εἰρημένῳ κόσμῳ.

As the works of Archimedes are not commonly found, even in good classical libraries, and as this passage contains the earliest distinct statements of the heliocentric hypothesis, I have reprinted it at length. The sentence beginning ταῦτα γὰρ ἐν ταῖς γραφομέναις is ungrammatical and incoherent, and its sense can only be guessed. It is thus translated by Peyrard, Œuvres d'Archimède (Paris, 1808), vol. ii. p. 232: 'Aristarque de Samos rapporte ces choses en les réfutant, dans les propositions qu'il a publiées contre les astronomes.' This version supposes κατὰ to be read for παρά. It is likewise impossible to combine ἐν ταῖς γραφομέναις ὑποθεσίων. The following form of the sentence would make sense, but the use of γραφὰς for βίβλους or βίβλον is inadmissible; and the sentence cannot be restored without the help of better manuscripts: ταῦτα γὰρ ἐν τοῖς γραφομένοις κατὰ τῶν ἀστρολόγων διακρούσας Ἀρίσταρχος ὁ Σάμιος ὑποθεσίων ἐξέδωκε γραφάς.

The Arenarius of Archimedes is printed in Wallis, Opera Mathematica (Oxon. 1699, fol.), vol. iii. p. 513, who by ἐν ταῖς γραφομέναις ὑποθεσίων understands ἐν ταῖς γραφομέναις ὑποθέσεσιν.

It may be observed that the Greek astronomers conceived the earth as being merely a point, and having no magnitude, as compared with the distance of the sphere of the fixed stars. The first theorem of Euclid's Phenomena is, ἡ γῆ ἐν μέσῳ τῷ κόσμῳ ἐστὶ, καὶ κέντρου τάξιν ἐπέχει πρὸς τὸν κόσμον. The same proposition is laid down by Ptolemy, Synt. i. 5 ; Cleomedes, i. 11, p. 70, Bake; Geminus, c. 13, p. 31, and Achilles Tatius, c. 21, p. 83.

Ut docent mathematici concinentes, ambitus terræ totius, quæ nobis videtur immensa, ad magnitudinem universitatis instar brevis obtinet puncti, Ammian. Marcellin. xv. 1, § 4.

Macrobius says that the earth is a point with respect to the sun's orbit. Punctum dixerunt esse geometræ quod ob incomprehensibilem brevitatem sui in partes dividi non possit, nec ipsum pars aliqua sed tantummodo signum esse dicatur. Physici terram ad magnitudinem circi per quem sol volvitur puncti modum obtinere voluerunt. In Somn. Scip. i. 16, § 10.

Lower down, ii. 9, § 9, he makes the same assertion with respect to the orbit of the other celestial bodies. Item quia omnis terra, in quâ et oceanus est, ad quemvis cœlestem circulum quasi centron puncti obtinet locum, necessario de oceano adjecit: *qui tamen tanto nomine quam sit parvus vides.* Nam licet apud nos Atlanticum mare, licet magnum vocetur; de cœlo tamen despicientibus non potest magnum videri, cum ad cœlum sit terra signum, quod dividi non possit in partes.

Compare Martianus Capella, ed. Kopp, p. 491, note.

Aristarchus of Samos lays it down that the earth is as a point and a centre to the moon's sphere: Wallis, Op. Math. vol. iii. p. 569, ed. Oxon. This proposition was denied by Hipparchus and Ptolemy, Pappus ib: p. 570. Ptolemy affirms that the distance of the sphere of the moon from the centre of the earth is not, like the distance of the zodiacal circle, so great that the earth is as a point in comparison with it. Synt. iv. 1, vol. i. p. 212, Halma.

According to the exposition of Archimedes, Aristarchus assumed the orbit of the earth to have no magnitude, compared with the distance of the fixed stars.

Note C.—(p. 179.)

It has been already explained that πόλος signified a hollow hemisphere; and hence it came to signify the basin or bowl of a sundial in which the hour-lines were marked. In this sense it is used in the well-known passage of Herodotus, ii. 109. Lucian, Lexiphan. 4, represents the pedant avoiding the word ὡρολόγιον, as a modernism, and therefore as saying, ὁ γνώμων σκιάζει μέσην τὴν πόλον. Pollux, vi. 110, says, that a round hollow vessel, called χαλκίον and σκάφη, resembled the πόλος that shows the hours. The word πόλος is used for sundial in the epigram of Metrodorus, Anth. Pal. xiv. 139. Martianus Capella, vi. § 597, ed. Kopp., gives a description of a sundial, which will apply to an earlier period. 'Scaphia dicuntur rotunda ex ære vasa, quæ horarum ductus stili in medio fundo siti proceritate discriminant, qui stilus gnomon appellatur.' The gnomon and scaphe of the horologium at Alexandria are mentioned in Cleomedes, i. 10, p. 67, ed. Bake. The human head was called σκάφιον from its rotundity, Pollux, ii. 39. A detailed description of a sundial is given by Macrobius, in Somn. Scip. i. 20, § 26. He calls the 'saxeum vas' a 'hemisphærium,' and the gnomon a 'stilus.' Compare ii. 7, § 15.

The words ὡρολόγειον and *horologium* anciently signified a sundial. Compare the epigram to a ὡρολόγιον, in the Doric dialect, Anth. Pal. ix. 780.

ὡρανὸν ἁ χωρεῦσα σοφὰ λίθος, ἁ διὰ τυτθοῦ
γνώμονος ἀελίῳ παντὶ μερισδομένα.

Geminus, c. 5, p. 15, speaks of the gnomon of the horologium casting a shade. Cleomedes, i. 10, p. 66, says that the gnomon of the horologium casts no shade when the sun is at the zenith; and he speaks elsewhere of the shadow of the gnomon of the horologium, i. 11, p. 74. The transfer of the word in the Romance languages to a clock is, like the instrument itself, modern. Compare Ducange, Gloss. Lat. in v. The word ὡρολόγιον was likewise used by the Byzantine Greeks, to denote a book containing the sacred *hours* or services. See Ducange, Lex. Inf. Gr., and Suicer in v.

For an account of the dials on the Tower of the Winds at Athens, see Stuart's Antiquities of Athens, vol. i. ch. 3, p. 20.

The sundial at Orchomenos (figured in Dodwell's Tour through Greece, vol. i. p. 231) consists of a semicircle with twelve radii, forming eleven compartments. Ten of the radii are denoted with numerals, from A to I, so that the four last are ZHΘI; thus agreeing with the epigram in the text. The twelfth hour is omitted, as in the notation of the gnomon in Palladius, for which see p. 180.

The following couplet, from an inscription on a sundial at Constantinople, in the reign of Justin II. (565—578 A.D.) supposes twelve hours to be marked on it:—

> ὡράων σκοπίαζε σοφὸν σημάντορα χαλκὸν
> αὐτῆς ἐκ μονάδος μέχρι δυωδεκάδος.
> Anth. Pal. ix. 779.

The epigram in the text accounts only for ten hours. The modern Latin verses exhaust the entire twelve hours, in the life of a law student:—

> Sex horas somno, totidem dis legibus æquis,
> Quatuor orabis, des epulisque duas,
> Quod superest, sacris ultro largire Camœnis.
> See Philol. Mus. vol. i. p. 691.

The death of Cicero is said to have been foreshown by the omen of a raven striking off the gnomon of a dial. 'M. Ciceroni mors imminens auspicio prædicta est. Cum enim in villâ Caietanâ esset, corvus in conspectu ejus horologii ferrum loco motum excussit, et protinus ad ipsum tetendit, ac laciniam togæ eo usque morsu tenuit, donec servus ad eum occidendum milites venisse nuntiaret.' Val. Max. i. 5, 5.

Chapter IV.

SCIENTIFIC ASTRONOMY OF THE GREEKS AND ROMANS, FROM HIPPARCHUS TO PTOLEMY, 160 B.C. TO 160 A.D.

§ 1 AFTER the death of Archimedes, in 212 B.C., the Alexandrine Museum became almost the exclusive seat both of astronomical observation and of astronomical science in Greece. Aristarchus of Samos, Aristyllus, Timocharis, Apollonius, Conon, and Eratosthenes, were among the astronomers of this School in the third century before Christ: and about the middle of the second century, Hipparchus, the great luminary of Ancient Astronomy, appeared in Alexandria.

Observations of Hipparchus for dates from 162 to 127 B.C., are recorded by Ptolemy.([1]) His lifetime may be placed on conjecture from 190 to 120 B.C. He was a native of Bithynia. His only extant work is the Commentary upon the treatise of Eudoxus, and the poem of Aratus, already mentioned, which relates to the positions of the fixed stars.([2]) A short Commentary on the Phænomena of Aratus, inscribed with the name of Eratosthenes or Hipparchus, has likewise been preserved.([3]) It may be an abridgment of a work by one of these astronomers; but in its present form it is certainly subsequent to the Augustan age, and is probably a compilation of later date.([4])

([1]) See Clinton, F. H. vol. iii. p. 532.
([2]) In three books; at p. 97 of the Uranologium of Petavius.
([3]) Ib. p. 142.
([4]) In c. 2 it is stated that Orion rises on the 22nd of July, and the Dog-star on the 7th of August. The use of the Roman names for the months, and the substitution of Julius and Augustus for Quintilis and Sextilis had therefore been established in Egypt or Greece when this work was compiled. It is stated, ib., that the popular name for Orion is ἀλετροπόδιον. This is considered a recent word by Ducange, Gloss. Inf. Græc. in v. The term *ecliptic*, for the circle passing through the middle of

Our knowledge of the discoveries of Hipparchus is principally derived from Ptolemy.

The following summary of these discoveries is given by Delambre, in his History of Ancient Astronomy.

'The foundations of astronomy among the Greeks were laid by Hipparchus. He determined the positions of the stars by right ascensions and declinations: he was acquainted with the obliquity of the ecliptic. He also determined the inequality of the sun and the place of its apogee, as well as its mean motion; the mean motion of the moon, of its nodes, and of its apogee; the equation of the moon's centre, and the inclination of its orbit: he likewise detected a second inequality, of which he could not, from his want of proper observations, discover the period and the law: he commenced a more regular course of observations, in order to provide his successors with the means of explaining the motions of the planets. His Commentary on Aratus shows that he had expounded, and given a geometrical demonstration of, the methods necessary for finding the right and oblique ascensions of the points of the ecliptic and of the stars, the east point and the culminating point of the ecliptic, and the angle of the east, which is now called the nonagesimal degree. He was therefore possessed of a spherical trigonometry. His calculations on the eccentricity of the moon prove that he had a rectilineal trigonometry and tables of chords. He had drawn a planisphere according to the stereographic projection. He could calculate eclipses of the moon, and use them for the correction of his lunar tables; he had an approximate knowledge of parallax. His operations imply a complete scientific system. His great want was a more perfect set of astronomical instruments. At this period, an approximation to the truth within a degree was rarely attainable. Hipparchus did not always attain it, especially when the operation was

---

the zodiac, is used in c. 7, p. 146. This term appears to be of later introduction than the time of Ptolemy. Hipparchus himself is cited in c. 6, as stating the number of the stars at 1080. Compare Delambre, Hist. Astr. Anc. tom. i. p. 173; Bernhardy, Eratosthenica, p. 185.

complicated; but his error was sometimes only of a few minutes.'(5)

One of the principal titles of Hipparchus to the name of a great astronomer is derived from his determination of the motions of the sun and moon, involving an accurate definition of the year, and a method of predicting eclipses. Combined with this determination was his theory of epicycles and eccentrics, by which these motions were explained. The theory of Eudoxus, developed by Callippus and Aristotle, was exclusively founded on a combination of spheres rotating on their respective axes. It involved therefore nothing but a combination of simple motions of circles round their proper centres. This theory first underwent a transformation in the hands of the great geometer Apollonius, who substituted for the combined system of spheres the more refined and ingenious hypotheses of the eccentric circle and of the epicycle. The hypothesis of Apollonius was adopted by Hipparchus, who applied it to detailed explanations of the motions of the sun and moon.

The two great stages of progress which the Greek astronomy had hitherto achieved, were the conception of the earth as a solid sphere, and the reduction of the phenomena of the heavenly bodies to uniform movements in circular orbits. The first of these hypotheses, which was firmly held by Plato, Aristotle, Euclid, and Archimedes, implied a signal triumph over the superficial impressions of the senses, at a time when navigation had scarcely extended beyond the Mediterranean, and when no Greek traveller or mariner had penetrated into the southern hemisphere.(6) Since the invention of the compass, and the consequent abandonment of the system of coasting voyages; since the southern hemisphere has been explored, since the ocean has been crossed in every direction, and the earth itself been circumnavigated, the sphericity of the earth has become the subject of ocular demonstration. It is only by

---

(5) Hist. d'Astr. Anc. tom. i. p. 184. Compare Disc. Prél. p. 14.
(6) See below, ch. viii. § 8.

supposing the earth to be a sphere freely suspended in space that the facts which have now become the subjects of repeated observations can be explained. The doctrine of the figure of the earth is therefore now independent of astronomical considerations: it belongs rather to physical geography. But with the ancients it was otherwise. This doctrine with them rested almost exclusively on astronomical grounds; and it was only reached by a gradual advance of astronomical reasoning. The conception of the fixed stars as revolving in a solid sphere round the earth, was perhaps not difficult of attainment; but the reduction of the motions of the sun, moon, and five planets, to circular orbits was far removed from ordinary ideas, and implied deep concentrated thought and scientific abstraction. The theory of composite spheres, devised by Eudoxus and developed by Callippus and Aristotle, was ingenious, and required much geometrical resource; but it was intricate, and it failed in the essential point of explaining all the phenomena.(7) The Apollonian and Hipparchean theory of eccentrics and epicycles proceeded on the same astronomical basis: it was more intricate, but it exhibited more geometrical subtlety, and it accomplished the important end of explaining all the known phenomena.

The value of the hypotheses of the Greek astronomers respecting the orbital motions of the heavenly bodies has been ably discussed by Dr. Whewell, who has shown an appreciation of the importance of erroneous but approximative hypotheses in the progress of discovery, which is sometimes wanting in historians of science. The following passage in his *History of the Inductive Sciences* contains the substance of his observations on this subject:—

'That which is true in the Hipparchian theory, and which no succeeding discoveries have deprived of its value, is the resolution of the apparent motions of the heavenly bodies into

---

(7) Simplicius sums up his account of the theory thus: τοιαύτη τις ἐστὶν ἡ διὰ τῶν ἀνελιττουσῶν σφαιροποιία, μὴ δυνηθεῖσα διασῶσαι τὰ φαινόμενα, ὡς καὶ ὁ Σωσιγένης ἐπισκήπτει λέγων, p. 502 b, Brandis. Concerning Sosigenes, see below, ch. iv. § 6.

an assemblage of circular motions. The test of the truth and reality of this resolution is, that it leads to the construction of theoretical tables of the motions of the luminaries, by which their places are given at any time, agreeing nearly with their places as actually observed. The assumption that these circular motions, thus introduced, are exactly uniform, is the fundamental principle of the whole process. This assumption is, it may be said, false; and we have seen how fantastic some of the arguments were, which were originally urged in its favour. But some assumption is necessary, in order that the motions, at different points of a revolution, may be somewhat connected—that is, in order that we may have any theory of the motions; and no assumption more simple than the one now mentioned can be selected. The merit of the theory is this;—that obtaining the amount of the eccentricity, the place of the apogee, and, it may be, other elements, from a few observations, it deduces from these, results agreeing with all observations, however numerous and distinct. To express an inequality by means of an epicycle, implies not only that there is an inequality, but further;—that the inequality is at its greatest value at a certain known place; diminishes in proceeding from that place by a known law;—continues its diminution for a known portion of the revolution of the luminary;—then increases again, and so on: that is, the introduction of the epicycle represents the inequalities of motion as completely as it can be represented with respect to its quantity.

'We may further illustrate this, by remarking that such a resolution of the unequal motions of the heavenly bodies into equable circular motions is, in fact, equivalent to the most recent and improved processes by which modern astronomers deal with such motions. Their universal method is to resolve all unequal motions into a series of terms, or expressions of partial motions; and these terms involve sines and cosines, that is, certain technical modes of measuring circular motion, the circular motion having some constant relation to the time. And thus the problem of the resolution of the celestial motions into equable circular ones, which was propounded above two thousand years ago in the School of Plato, is still the great object of the study of modern astronomers, whether observers or calculators.

'That Hipparchus should have succeeded in the first great steps of this resolution for the sun and moon, and should have seen its applicability in other cases, is a circumstance which gives him one of the most distinguished places in the roll of great astronomers.'

'The unquestionable evidence of the merit and value of the theory of epicycles is to be found in this circumstance;—that it served to embody all the most exact knowledge then extant, to direct astronomers to the proper methods of making it more exact and complete, to point out new objects of attention and research; and that, after doing this at first, it was able to take in, and preserve, all the new results of the active and persevering labours of a long series of Greek, Latin, Arabian, and modern European astronomers, till a new theory arose which could discharge this office. It may, perhaps, surprise some readers to be told, that the author of this next great step in astronomical theory, Copernicus, adopted the theory of epicycles; that is, he employed that which we have spoken of as its really valuable characteristic.'[8]

Hipparchus further discovered that apparent motion of the fixed stars round the axis of the ecliptic, according to the order of the signs, and contrary to the daily movement of the heavens, which is called the Precession of the Equinoxes. As this motion amounts only to fifty seconds in each year, and requires seventy-two years to make a degree; as the only prior observations of even tolerable accuracy, to which Hipparchus had access, were those of Timocharis and Aristyllus, and as their observations only preceded his time by about 150 years, the detection of this anomaly is a conclusive proof of great penetration

---

[8] Vol. i. p. 181, 185.

combined with close attention to observed phenomena. The discovery of the Precession is an important era in astronomy. It is attributed in distinct terms to Hipparchus by Ptolemy, who had ample means of knowing its true author; and it appears to have been advanced by Hipparchus as an original suggestion, unknown to his predecessors and contemporaries.([9]) When we consider the scarcity of accurate recorded observations in antiquity, there is nothing remarkable in the postponement of the discovery of this minute inequality until the middle of the second century before Christ. We may perhaps rather wonder that Hipparchus should have succeeded in discovering this truth by means of the few and faint indicia which were within his reach. Hipparchus, moreover, not only discovered the Precession, but he approximated closely to its real amount. He said that it was not greater than fifty-nine seconds, and not less than thirty-six seconds.([10])

Eudoxus had constructed a map of the heavens, by enumerating the constellations, and by describing their positions with relation to each other and to the zodiacal band. Hipparchus made a further step in advance: he framed a catalogue of the stars, of which he enumerated 1080, and determined their places with reference to the ecliptic, by their latitudes and longitudes.([11]) Although the number of stars visible to the naked eye is, in fact, not very great, yet they produce the impression of being innumerable; and Pliny speaks with wonder of the achievement of Hipparchus, 'who had ventured to count the stars, a work arduous even for the Deity.'([12])

Hipparchus did not attempt to extend the theory of eccen-

---

(9) See Synt. iii. 2, vii. 2, 3. Compare Delambre, Hist. de l'Astr. Anc. tom. ii. p. 240. Whewell, Hist. of Induct. Sciences, vol. i. p. 186.

(10) See Delambre, ib. vol. ii. p. 249, 254. Autolycus and Euclid were both ignorant of the precession of the equinoxes, Delambre, ib. vol. i. p. 26, 55. The discovery of the precession is attributed to Hipparchus and Ptolemy by Simplicius, Schol. Brandis, p. 496 a. The existence of the precession is unknown to Macrobius, in Somn. Scip. ii. 11, § 10.

(11) Delambre, ib. vol. i. p. 185, vol. ii. p. 261; Whewell, ib. p. 191.

(12) N. H. ii. 26.

trics and epicycles to the five planets. He left no geometrical explanation of their movements, and confined himself to laying a foundation of accurate observations, to be used by his successors. 'Hipparchus,' says Ptolemy, 'displayed his love of truth in confining to the sun and moon his demonstration of circular and uniform motions, and in not extending them to the five planets. Inasmuch as his predecessors had not left him a sufficient number of accurate observations, he judged rightly, with reference to the planets, in attempting nothing beyond a collection of good observations for the use of his successors, and a demonstration, by means of these observations, that the hypotheses of the mathematicians of his time did not agree with the phenomena.' Ptolemy proceeds to say that none of his own predecessors had been able to reduce the planets to uniform and circular motions; and that one of the main difficulties of the problem was owing to the shortness of the time for which there were observations of the planetary movements preserved in writing.[13]

'When,' says Delambre, 'we consider all that Hipparchus invented or perfected, and reflect upon the number of his works, and the mass of calculations which they imply, we must regard him as one of the most astonishing men of antiquity, and as the greatest of all in the sciences which are not purely speculative, and which require a combination of geometrical knowledge with a knowledge of phenomena, to be observed only by diligent attention and refined instruments.'[14]

§ 2 Between the times of Hipparchus and Ptolemy, a period of about 300 years, the science of Astronomy made little or no progress. It continued, however, to be expounded in its improved form, and to be studied by various philosophers; and several works of this period, in which Astronomy was treated, are still extant.

Posidonius, a Stoical philosopher, who was born about 135

---

(13) Synt. ix. 2.   (14) Ib. vol. i. p. 185.

B.C., near the end of the lifetime of Hipparchus, combined, like Eratosthenes, the cultivation of Astronomy with that of Physical Geography.([15]) He constructed an orrery which exhibited the diurnal movements of the sun, moon, and five planets: it is mentioned by Cicero, his contemporary and friend, apparently from personal knowledge.([16]) From the description of this orrery, we may infer that, like that of Archimedes, it was made upon the geocentric principle.

Posidonius calculated the circumference of the earth by a method different from that of Eratosthenes, and made it only 240,000 stadia, or 30,000 miles:([17]) he determined the distance from the earth to the sun at 502,000,000 stadia,([18]) and the sun's diameter at 3,000,000 stadia.([19]) He considered the shape of the inhabited part of the earth to resemble a sling,([20]) the long part apparently lying east and west; but he doubtless supposed the earth to be a sphere.

The extant treatise of Cleomedes, entitled Κυκλικὴ Θεωρία τῶν Μετεώρων, in two books, expounds the received system of the world in a concise and perspicuous manner. 'The doc-

---

(15) Galen, De Hippocrat. et Plat. Plac. viii. 1, vol. v. p. 652, ed. Kühn, says that Posidonius was the most scientific of the Stoics, on account of his geometrical training.

(16) Cic. de Nat. D. ii. 34. The sphere of Billarus, taken from Sinope, was preserved by Lucullus, Strab. xii. 3, § 11. Nero had a round table which turned perpetually like the heaven, Suet. Ner. 31. An extant treatise of Leontius Mechanicus, Περὶ Κατασκευῆς Ἀρατείας Σφαίρας, printed in Buhle's Aratus, vol. i. p. 257, describes the method of constructing a celestial globe. Leontius is conjecturally referred to the sixth century A.D. See Dr. Smith's Dict. of Anc. Biogr. and Myth. art. *Leontius Mechanicus*, vol. ii. p. 758.

The word λαζούριον, for *dark blue*, occurs in this treatise, p. 261. It is the Greek form of *lapis lazuli*, and is of Arabic origin: the Italian *azzurro* is borrowed from it. Compare Ducange, Gloss. Gr. in λαζούριον, Gloss. Lat. in lazur, Diez in azzuro. The word ξάριον likewise occurs, as the name of a light colour, in p. 262. The explanation of Ducange, who derives it from ἐξαέριος, is not satifactory.

(17) Cleomed. i. 10.
(18) Plin. H. N. ii. 21.
(19) Cleomed. ii. 1, p. 98, ed. Bake.
(20) Fragm. Hist. Gr. vol. iii. p. 282, fr. 69. Posidonius likewise gave an explanation of the Milky Way, Macrob. in Somn. Scip. i. 15, § 7.

trines which it contains,' says Cleomedes, 'are not the opinions of the author himself, but have been collected from various writings, ancient and modern. The chief part of the treatise, (he adds) is borrowed from the writings of Posidonius.'([21]) The date of Cleomedes is unknown; but as he mentions Eratosthenes, Hipparchus,([22]) and Posidonius, but never alludes to Ptolemy, it may be inferred that he lived before the latter.([23]) His attack upon the Epicureans likewise implies that the work was anterior to the time when Christianity had extinguished the distinctions of the philosophic sects. Letronne, however, makes him coeval with, or subsequent to, Ptolemy.([24])

As the treatise of Cleomedes professes to be in substance an epitome of the astronomical writings of Posidonius, it will serve to exhibit the state of astronomical science in Greece about the Christian era, and may be studied with that view.

The treatise of Geminus, entitled Εἰσαγωγὴ εἰς τὰ Φαινόμενα, 'An Introduction to the Celestial Appearances,'([25]) resembles that of Cleomedes in its general character, and is probably of about the same date.

Geminus, although his name is Roman, shows no trace of any connexion with Rome or Italy:([26]) he appears to be a Greek, writing for Greeks. The latest writers whom he names in his treatise are Crates the grammarian, and Poly-

---

(21) Ad fin.

(22) He quotes Hipparchus from report, and without having read the writing to which he refers, ii. 1, p. 102, Bake.

(23) The work of Ptolemy is quoted several times in the treatise of Leontius 'On the Construction of an Aratean Globe,' Buhle's Aratus, vol. i. p. 257.

(24) See Letronne's review of Bake's edition of Cleomedes, Journ. des Sav. 1821, p. 707. His determination of the date of Cleomedes is founded upon astronomical arguments, and upon the knowledge of refraction shown in his treatise.

(25) Printed in the Uranologium of Petavius, p. 1. A summary of it is given by Delambre, Hist. Astr. Anc., tom. i. p. 190.

(26) The Greeks write his name with a circumflex on the penult, Γεμῖνος. He mentions Rome in c. 14, p. 33.

bius,[27] the latter of whom was alive in 129 B.C. It appears, however, that he composed an epitome of the Meteorologics of Posidonius;[28] which fact brings him down to a lower date, and makes him at least contemporary with Cicero.[29]

---

(27) c. 13, p. 31. The treatise of Polybius cited by Geminus was entitled, περὶ τῆς περὶ τὸν ἰσημερινὸν οἰκήσεως.

(28) Simplic. ad Aristot. Phys. p. 348, Brandis.

(29) Petavius, Doctr. Temp. ii. 6, p. 52, attempts to determine the time of Geminus by a statement in c. 6, p. 19, of his treatise, relating to the Egyptian year of 365 days. Geminus says that, according to the Egyptians and Eudoxus, the festival of the Isia falls at the winter solstice: which was true 120 years previously, but at the time when he wrote, owing to the omission of the ¼ of a day, it fell a month earlier. The number 120 is the result of calculation, and is obtained by quadrupling the number of days in a month. Geminus cannot mean to imply that the interval between Eudoxus and himself was exactly 120 years. Geminus quotes Eratosthenes as having said that the Isia were once celebrated at the summer solstice—so that they went round the entire year. Achilles Tatius, c. 23, p. 85, says that the Egyptians celebrated the Isia at the time of the winter solstice, and put on mourning for the departing sun. Ideler, Ueber die Sternnamen, p. 175, places Geminus before Cæsar.

Some indication as to the date of Cleomedes and Geminus may be drawn from their use of the term *ecliptic*, in its modern sense. In its original meaning it denoted that which pertains to an eclipse; as in Plut. Rom. 12, σύνοδος ἐκλειπτικὴ τῆς σελήνης πρὸς ἥλιον. The word is similarly used by Manilius, iv. 818, 848. Afterwards it was applied to the apparent path of the sun through the heaven; because in this path eclipses take place. Cleomedes, i. 4, says that the path of the sun is in a circle along the middle of the zodiac; ὅθεν καὶ ἡλιακὸς οὗτος κέκληται, p. 25. But in ii. 5, p. 138, he says of the middle circle of the zodiac, ὃς ἡλιακός τε καὶ ἐκλειπτικὸς καλεῖται. Geminus, c. 9, speaking of the moon, says: ὅταν γὰρ διὰ μέσου τοῦ ἐκλειπτικοῦ ἡ σελήνη τὴν πάροδον ποιεῖται [l. ποιῆται], ὅλη ἐμπίπτει εἰς τὸ σκίασμα τῆς γῆς, ὥστε ἀναγκαῖον καὶ ὅλην αὐτὴν ἐκλείπειν. ὅταν δὲ παραψήηται τοῦ σκιάσματος, μέρος τι τῆς σελήνης ἐκλείπει· ἔστι δὲ τὸ ἐκλειπτικὸν αὐτῆς μοιρῶν δύο, p. 25. In this passage the reference to the original meaning of the word is clear. Achilles Tatius, c. 23, after stating that the path of the sun is in the middle circle of the zodiac, proceeds thus: διὸ καὶ ἡλιακὸς ὑπὸ τῶν ταῦτα δεινῶν προσηγόρευται καὶ ἐκλειπτικός, ἐπειδὴ ἐν αὐτῷ αἱ ἡλιακαὶ ἐκλείψεις γίνονται, p. 84. Macrobius, in Somn. Scip. i. 15, § 10, has the following passage: Quantum igitur spatii lata dimensio porrectis sideribus occupabat, duabus lineis limitatum est: et tertia ducta per medium ecliptica vocatur, quia, cum cursum suum in eâdem lineâ pariter sol et luna conficiunt, alterius eorum necesse est evenire defectum: solis, si ei tunc luna succedat; lunæ, si tunc adversa sit soli. Again, he says that when the moon has traversed half its latitude, it reaches the place 'qui appellatur eclipticus,' i. 6, § 53.

The term *ecliptic* is never used in this sense by Ptolemy, and so far as this word goes, it leads to the inference that both Geminus and Cleomedes are subsequent to him, and wrote about the end of the second or

Another work, which apparently belongs to this period, is the treatise Περὶ Κόσμου, *De Mundo*, falsely ascribed to Aristotle. The work is properly pseudonymous, and has not been accidentally confounded with the writings of the great Stagirite. The author of it dedicates the work to Alexander the Great, and evidently means to assume the person of Aristotle.([30]) A Latin version of it, with a modification of the passage in the preface which contains the dedication to Aristotle, is among the works of Apuleius. Its mention of Ireland, and of the keystone of an arch, are decisive as to the lateness of its date.([31]) The only part of it which directly concerns Astronomy, is the first chapter, which describes the system of the world, according to the received geocentric method: the earth is motionless at the centre of the universe; the fixed stars, the planets, and the sun and moon move round it in circular orbits.

The cultivation of Astronomy at Alexandria seems to have been continued by a succession of geometers and observers after the patronage of the Ptolemies, to which it had owed its origin, was extinguished by the Roman arms, and by the destruction of the independence of Egypt. It was, however, continued under less favourable circumstances, and upon a contracted scale. The Greek philosophic world was at this time mainly divided between two sects, the Stoic and the Epicurean. Of these, the former gave some attention to Astronomy. Posidonius was a Stoic; Cleomedes and the author of the pseud-Aristotelic treatise *De Mundo* appear, moreover, to have belonged to the same sect. The Epicureans, however, adhered to the popular Astronomy founded upon the first uncorrected im-

---

the beginning of the third century. Macrobius was as late as the fifth century.

Geminus, c. 13, p. 13, says that the torrid zone had been explored in his time, and that an account of it had been written, through the influence of the kings at Alexandria. This passage appears to fix the age of Geminus to a period prior to the Roman conquest of Egypt.

(30) The treatise is entitled, Περὶ Κόσμου πρὸς 'Αλεξανδρον. In c. 1 the writer says: πρέπειν δὲ οἶμαί γε καὶ σοὶ ἡγεμόνων ὄντι ἀρίστῳ τὴν τῶν μεγίστων ἱστορίαν μετιέναι. See above, p. 159.

(31) See c. 3, 6.

pressions of the senses. Epicurus, whose lifetime extended from 342 to 270 B.C., entertained opinions upon Astronomy singularly inconsistent with the character of a bold, original, and truth-seeing physical teacher, which Lucretius confers upon him.([32]) The Epicureans held the earth to be in the centre of the heaven,([33]) but they denied its sphericity, and the tendency of all things to a centre.([34]) They likewise taught that the magnitude of the sun is no greater than it appears to our senses; and that its diameter is only a foot.([35]) They extended the same primitive doctrine to the moon.([36]) Epicurus likewise held that the stars are extinguished at their setting, and lighted

---

([32]) Particularly in the following verses:—

Ergo vivida vis animi pervicit, et extra
Processit longe flammantia moenia mundi,
Atque omne immensum peragravit mente animoque.
i. 73—5.

Cicero says of Epicurus: Principio, inquam, in physicis, quibus maxime gloriatur, primum totus est alienus. Democrito adjicit, perpauca mutans; sed ita, ut ea, quæ corrigere vult, mihi quidem depravare videatur, De Fin. i. 6.

The physics of Epicurus were principally borrowed from those of Democritus, Cic. N. D. i. 26.

([33]) Lucret. v. 535.

([34]) Lucret. i. 1051—1081. Cleomedes i. 8, p. 51, says: οἱ δὲ ἡμέτεροι, καὶ ἀπὸ μαθημάτων πάντες, καὶ οἱ πλείους τῶν ἀπὸ τοῦ Σωκρατικοῦ διδασκαλείου σφαιρικὸν εἶναι τὸ σχῆμα τῆς γῆς διεβεβαιώσαντο, where by οἱ ἡμέτεροι the Stoic sect appears to be signified. The Epicurean sect is pointedly excluded.

The absurdity of the hypothesis that bodies gravitate to the centre of the earth is maintained in Plut. de fac. in orbe lun. c. 7.

([35]) Diog. Laert. x. 91; Plut. Plac. Phil. ii. 21; Galen, Phil. Hist. c. 14; Stob. Phys. i. 20.

Nec nimio solis major rota, nec minor ardor,
Esse potest, nostris quam sensibus esse videtur.
Lucret. v. 565, 6.

Civ. Acad. iv. 26; Fin. i. 6.

Cleomedes, ii. 1, treats Epicurus with great contempt on account of this doctrine, which he analyses and refutes at length. He declares that in questions of Astronomy Epicurus is κατὰ πολὺ τῶν ἀσπαλάκων τυφλότερος, p. 106, 'blinder than a mole.' See the note of Bake on Cleomedes, p. 389.

([36]) Lunaque, sive notho fertur loca lumine lustrans,
Sive suo proprio jactat de lumine lucem;
Quicquid id est, nihilo fertur majore figurâ,
Quam nostris oculis, quâ cernimus, esse videtur.
Lucret. v. 575—8.

again at their rising. 'A doctrine (says Cleomedes) not less absurd, than if any one were to say that men were alive when they were in sight, and were dead when they were out of sight.'(37) He even supposed the light of the sun to be quenched in the western waves, and to be re-illuminated every morning: and he gave faith to the report of the Iberians, which Cleomedes calls 'an old woman's tale,' that they heard the hissing of the sun, like heated iron, in the ocean.(38) Some of the Epicureans, however, rejected this infantine belief, and supposed the sun to return from the west to the east by an oblique course above the earth.(39)

It appears from the Epistle of Epicurus, preserved by Diogenes Laertius,(40) that Epicurus aimed at explaining physical phenomena by common sense, rather than by scientific arguments, remote from ordinary comprehension; that he denounced 'the low-minded technicalities of the astronomers,'(41) and that he sought to apply the operation of known and familiar analogies to the celestial bodies.

§ 3 It is an undoubted fact that from the time of Eudoxus to that of Hipparchus, scientific astronomy had made immense progress in Greece. The heavenly bodies had been diligently observed; the sun's annual movement and its relation to the earth and to the fixed stars had been determined with a close approach to accuracy; geometrical reasonings, of great sub-

---

(37) Ib. p. 108. Ptolemy, Synt. i. 2, mentions this doctrine, which he calls absurd, but he nevertheless thinks it worthy of an argumentative refutation.

(38) Ib. p. 109.

(39)   Scis nam fuisse ejusmodi sententiam
Epicureorum; non eum occasu premi,
Nullos subire gurgites, nunquam occuli,
Sed obire mundum, obliqua cœli currere,
Animare terras, alere lucis pabulo
Convexa cuncta, et invicem regionibus
Cerni.   Avienus, Ora Maritima, v. 646—652.

Compare Ukert, Geogr. der Gr. und Röm. i. 2, p. 27, 125.

(40) x. 84.

(41). μὴ φοβούμενος τὰς ἀνδραποδώδεις τῶν ἀστρολόγων τεχνητείας. ib. 94.

tlety and ingenuity, had been applied to the explanation of the celestial phenomena; and a system of the world had been devised, which gave an apparent solution of the leading astronomical problems.

The scientific and speculative genius of the Greeks, as applied to Astronomy, did not, however, bear the practical fruits which it might have been expected to produce. This was partly owing to the want of national patronage and recognition, and to the absence of public observatories. It was only at Alexandria that Astronomy could be said to enjoy a national establishment and endowment. But it was owing still more to the want of accurate astronomical instruments; especially of an instrument, such as our clocks, for the measurement of time; to the imperfection of the Greek arithmetical notation; to the scantiness and defectiveness of the astronomical tables, and of the stores of registered observations; and lastly, to the want of the telescope, or of any optical instrument for magnifying distant objects and for assisting the natural sight. Owing to these disabilities, the Greek astronomers were unable to reach the accuracy of determination to which their scientific methods, if supported by proper mechanical appliances, would have carried them; their practice lagged behind their theory; and they did not succeed in exercising that authority, and in commanding that attention, to which their intellectual eminence entitled them. The discovery of the telescope has given to astronomers a position in relation to the rest of mankind which they never had previously occupied. It created, with respect to the unapproachable objects of the heavens, a new sense. It revealed what never before had been, or could be, known. When a distant object on the surface of the earth is seen through a telescope, it is seen under circumstances which would render it invisible to the naked eye. Those who are near it can, however, see it without a telescope. But when the satellites of Jupiter, and Saturn's ring, were seen through Galileo's telescope, they were seen for the first time since man had been an inhabitant of the earth. The great telescopes of modern times, by which the eye of man has been able to discern objects at distances which his mind almost refuses to conceive, have been the subject of general curiosity and wonder, and have given a new significance to the verses of the ancient poet respecting astronomers :—

> Subjecere oculis distantia sidera nostris,
> Ætheraque ingenio supposuere suo.

§ 4 One of the principal uses of Astronomy, as indeed of all the physical sciences, is, by explaining the true sequence of phenomena and the operation of natural causes, to prevent those superstitious fears, that discovery of ominous portents and of terrific prodigies, which occur naturally to the uninstructed mind.([42]) Anything which appeared to transgress the known laws of nature was considered by the ancients as a prodigy; it was believed to be a mark of the divine displeasure, and to be sent by the gods as a warning to men with respect to their future conduct. Among prodigies, none were more striking or more alarming than eclipses; and as soon as the Greek philosophers began to speculate concerning physics, and to explain the order of the universe, they attempted to detect the cause of eclipses, and to bring them under a natural law. This subject did not escape the attention of the early Ionic philosophers; but Anaxagoras is stated to have been the first to promulgate clear and definite notions concerning their astronomical character. Pericles, who had profited by the physical instruction of Anaxagoras, is reported to have relieved the Athenians from their alarm at an eclipse of the sun, which occurred as the army was about to set sail on an expedition against the coasts of the Peloponnese, in the early part of the Peloponnesian War. The story told by Plutarch is that the eclipse took place at the moment of embarcation, and that Pericles, seeing the terror of the troops, and among them of his own pilot, held a cloak before the eyes of the latter, remarking that the eclipse

---

(42) Montucla, Hist. der Math. vol. i. p. 41, remarks that Astronomy put an end to superstitious terrors at eclipses &c.

was not a sign of calamity more than the cloak, the only difference being that in the eclipse the body which caused the obscuration was the larger of the two.(43) Notwithstanding the lessons respecting eclipses which Pericles, in his authoritative and conspicuous position, may have imparted to his countrymen, the educated Athenians at the time of the expedition to Syracuse, about twenty years afterwards, did not understand their nature; and the Athenian army was sacrificed to a superstitious scruple of Nicias, which, indeed, the general doubtless shared with the men under his command.(44)

Thucydides says that most of the Athenians objected to the departure of the fleet on account of the sinister omen; and that Nicias would not even allow the question of departure to be debated before the thrice nine days appointed by the diviners had elapsed. Plutarch's comment on the deplorable event is curious. It happened, he says, that Nicias had at the moment no experienced diviner, for Stilbides, on whom he had been accustomed to rely, and who moderated the extreme superstitious rigour of exposition, had recently died. 'In fact (he continues), the omen, according to Philochorus, was not unfavourable to those who were meditating escape, but rather propitious; for deeds done with fear need concealment, and light is dangerous to them. Moreover, Autoclides,(45) in his "Treatise on the Interpretation of Omens," lays it down that an expiation of three days is the proper time for phenomena of the sun and moon; whereas Nicias induced the army to wait for another circuit of the moon, not seeing that she was purified as soon as

---

(43) Plutarch, Per. 35. The expedition referred to is that described in Thuc. ii. 56 (430 B.C.) The passages of Cic. Rep. i. 16; Val. Max. viii. 11, ext. 1; Quintil. i. 10, § 47, are probably a different version of the same anecdote. A similar story is told of Pericles respecting a thunderbolt, by Frontin. i. 12, 10.

(44) See Thuc. vii. 50; Diod. xiii. 12; Plut. Nic. 23; Polyb. ix. 19. The date of the eclipse is Aug. 27, 413 B.C. Compare Grote, vol. vii. p. 433; Zech, Astron. Unters. über die wichtigeren Finsternisse des Alterthums (Leipzig, 1853), p. 32.

(45) Nothing is known of this writer.

she had passed the dark region under the shadow of the earth.' Plutarch here meets Nicias on his own ground: he admits that some delay was requisite, but he thinks that Nicias was misled by unskilful diviners, and that the delay which he prescribed was unnecessarily long. The 'circuit of the moon' appears to be the periodic month of $27\frac{1}{2}$ days, and to correspond to the 'thrice nine' days of Thucydides. Diodorus says that upon the occurrence of the eclipse, Nicias convened the diviners, and that they advised the postponement of the departure for 'the accustomed three days.' The moral which Polybius draws from this event is, the necessity of astronomical knowledge to a military commander. If Nicias, he says, had understood the true nature of an eclipse, he would have turned it to his own account; for he would have taken advantage of the fear and astonishment of the enemy, whose ignorance of eclipses was equal to that of the Athenians, to withdraw his army, and to escape in safety.

The comment of Polybius may be illustrated by the well-known anecdote of Columbus, who is said to have terrified the Indians of Jamaica by predicting an eclipse of the moon, and thus to have induced them to furnish the provisions necessary for the maintenance of his crew.[46] It may be added, that the astronomical science of Columbus in 1504, unless he was assisted by some calendar prepared for the use of mariners, could not have been superior to that of the later Greek astronomers, since whose time no scientific improvements had then been effected.

A similar interpretation was put upon an eclipse of the sun by the army of Pelopidas, as it was about to march against Alexander of Pheræ, in 364 B.C. The bulk of the army was intimidated by the evil omen, and the diviners discouraged the expedition. Pelopidas, however, made the attempt with a few volunteers and mercenaries, and lost his life in the rash attack.[47]

---

(46) See Robertson's History of America, b. ii.
(47) Plut. Pelop. 31; Diod. xv. 80. According to the latter, the inter-

On the other hand, Helicon of Cyzicus, a friend of Plato, is said to have predicted an eclipse of the sun at the court of Dionysius of Syracuse, and upon the verification of his prediction to have received a reward of a talent of silver.(48) The date of this eclipse has not been determined. As the reward was a large one, even for a despot's munificence, the ability to predict an eclipse must at this time have been of great rarity. Dion likewise is said to have foreknown a lunar eclipse which occurred soon afterwards during a feast which he gave to his soldiers in the island of Zacynthus at the full moon. Their fears were excited; but Miltas the prophet came forward, and assured them that the event prefigured by the gods was the eclipse of some conspicuous man: that nothing was more conspicuous than the empire of Dionysius, and that its brilliancy would be extinguished soon after their landing in Sicily.(49) It will be observed that Miltas, though a diviner, had been imbued with the philosophical opinions of the Academy.(50) The courage which Dion showed on this occasion, and his persistence in his expedition against Dionysius, notwithstanding the serious omen of the eclipse, is likewise ascribed by Plutarch to the astronomical lessons which he had received from Plato.(51)

An eclipse of the moon, nearly complete, which fell on September 20, 331 B.C., preceded the battle of Arbela by eleven days.(52) Arrian says that Alexander thereupon sacrificed to the sun, moon, and earth, as being the three powers concerned in the production of the phenomenon. He adds that Aristander

---

pretation put upon the departure of the army by the prophets was, that the sun of the state was eclipsed.'

The eclipse in question has not been exactly identified. See Grote, vol. x. p. 424.

(48) Plutarch, Dion. 19.

(49) Plut. Dion. 24. Compare Grote, vol. xi. p. 122. The date of this eclipse is 357 B.C.

(50) Plut. Dion. 23.

(51) Plut. Nic. 23.

(52) Compare Clinton, F. H. vol. ii. ad ann.; Ideler, Chron. vol. i. p. 347; Zech, Astron. Unters. p. 33.

the diviner thought that a calamity to the moon was a good omen for the Greeks.(⁵³) The account of Curtius is more circumstantial and picturesque. He describes the army of Alexander as terrified by the eclipse, as complaining of the long distance from home to which they had been dragged, and as being ready to break into a mutiny; when the Egyptian diviners (who were well aware of the true nature of eclipses), calmed their fears by pronouncing that the sun was the friend of the Greeks, and the moon of the Persians; and that an eclipse of the moon portended defeat to Persia.(⁵⁴) If this response was really given, the diviners of Alexander turned against the Persians the arms of their own soothsayers: for Herodotus informs us, that when Xerxes was marching against Greece in 480 B.C., an eclipse of the sun took place, and that the magi being consulted, declared that it signified an eclipse of the Greek cities; for that the sun was the sign for the Greeks, and the moon for the Persians.(⁵⁵)

The complete eclipse of the sun which occurred during the expedition of Agathocles against Africa is not stated to have been predicted. It must have taken the army by surprise, for it filled them with consternation.(⁵⁶) Agathocles is reported to have assuaged their fears, by assuring them that the omen was favourable. He admitted that an eclipse portended evil to

---

(53) Arrian, Anab. iii. 7, 6.

(54) Curt. iv. 39. This eclipse is simply mentioned by Plut. Alexand. 31. Cicero, de Div. i. 53, says that there was a prophecy that if the moon was eclipsed, a little before the rising of the sun, in the sign of Leo, Darius and the Persians would be defeated by Alexander and the Macedonians.

(55) Herod. vii. 37. The ground of this dictum is not apparent. The crescent was not the symbol of the ancient Persians, as it is of the modern Turks. Strabo describes them as worshipping both sun and moon, xv. 3, 13. Compare Aristoph. Pac. 406—413. The solar eclipse of 480 B.C. cannot be identified. Mr. Rawlinson is probably right in considering the anecdote as apocryphal. Airy, Phil. Trans. 1853, p. 199, thinks that a lunar eclipse of 14th March, 479 B.C. is intended.

(56) τῇ δ' ὑστεραίᾳ τηλικαύτην ἔκλειψιν ἡλίου συνέβη γενέσθαι ὥστε ὁλοσχερῶς φανῆναι νύκτα, θεωρουμένων τῶν ἀστέρων πανταχοῦ· διόπερ οἱ περὶ τὸν Ἀγαθοκλέα νομίσαντες καὶ τὸ θεῖον αὐτοῖς προσημαίνειν τὸ δυσχερὲς ἔτι μᾶλλον ὑπὲρ τοῦ μέλλοντος ἐν ἀγωνίᾳ καθειστήκεισαν, Diod. xx. 5. The authority of Diodorus is good upon any event relating to the later Sicilian history.

some power, and that if it had happened before the expedition sailed, it would have been sinister to the invaders. But as it had occurred while the fleet was at sea, it signified disaster to Carthage.([57]) The eclipse in question, according to Diodorus, fell in 310 B.C.,([58]) and the recent researches of Mr. Airy have fixed it to August 14 of that year.([59])

The earliest authentic mention of an eclipse in Roman history appears to be the notice by Livy of an eclipse of the sun in 190 B.C. during the Apollinarian games.([60])

Another eclipse occurred, in 168 B.C., during the campaign of the Consul Æmilius Paullus against Perseus, king of Macedonia. On the eve of the battle of Pydna, C. Sulpicius Gallus, a tribune of the second legion, who had been prætor in the previous year, is stated to have obtained the consul's permission to assemble the soldiers, for the purpose of addressing them. Gallus then informed them, that on the following night the moon would be eclipsed from the second to the fourth hour. As this phenomenon took place according to natural laws and at stated times, it could, he told them, be known beforehand and predicted: they must not consider the eclipse as a portent and a prodigy. They saw that the moon regularly changed at its successive phases; and they must not be surprised if it should be eclipsed, when it passed under the shadow of the earth. On the night of the 3rd of September (according to the unreformed Roman calendar) the moon was eclipsed at the predicted time. The Roman soldiers extolled the foresight of Gallus as almost superhuman; while the Macedonians and

---

([57]) Justin, xxii. 6, copied by Frontin. Strat. i. 12, § 9.

([58]) He places it in the archonship of Hieromnemon, and in the consulate of Caius Junius (Bubulcus) and Quintus Æmilius (Barbula), xx. 3. Compare Clinton, ad ann. 310; Fischer, ad ann. 311.

([59]) Airy, in the Philos. Transactions, 1853, p. 189. Zech, ubi sup. p. 34, agrees as to the date.

([60]) xxxvii. 4. According to Livy, this eclipse fell on the 11th of Quintilis, or July; whereas the calculated time gives March 14th; the difference between these days affords a measure of the error of the Roman calendar at that time. See Ideler, Chron. vol. ii. p. 92. The words of Livy, xxxviii. 36, do not appear to refer to an eclipse.

their diviners were terrified by the ominous occurrence; and their camp resounded with moans and shrieks of alarm until the moon recovered her proper shape. Such is the account of Livy:[61] the difference which he describes between the two camps illustrates the advantage which, according to Polybius, an ancient general might derive from a knowledge of astronomy. The account of Livy is corroborated by Pliny, who states that Sulpicius Gallus predicted the eclipse, and thus relieved the army from their fears. He adds that Gallus published a writing on the subject.[62] The same account is repeated by Frontinus.[63] On the other hand, Cicero represents Scipio as narrating, in the dialogue De Republicâ, that when he was a youth in his father's camp in Macedonia, the army were terrified by an eclipse of the moon, and that Sulpicius Gallus relieved their fears by explaining its physical cause.[64] A similar account is given by Valerius Maximus[65] and Quintilian.[66] Plutarch's narrative implies that the eclipse was unforeseen; he describes the Romans as propitiating the prodigy, and as assisting the moon in her trouble, by the clatter of brass and the exhibition of torches; but the Macedonians as seized with a silent horror, and as believing that the shadow on the moon denoted the spectre of their king. He proceeds to say that Æmilius, though aware of the true nature of eclipses, was fond of religious observances, and believed in divination, and that he immediately sacrificed eleven calves to the moon.[67] He makes no allusion to Sulpicius Gallus in connexion with the event. A fragment of Polybius describes the Romans as elated and the Macedonians as depressed by a report that the eclipse of the moon portended the death of the Macedonian king, which,

---

[61] xliv. 37. Concerning the true date of the eclipse, 21st of June of the Julian year, see Ideler, Chron. vol. ii. p. 104; Fischer, ad ann. 168; Zech, ubi sup. p. 33.

[62] N. H. ii. 9.
[63] Strat. i. 12, § 8.
[64] De Rep. i. 15.
[65] Val. Max. viii. 11, 1.
[66] Quintilian, i. 10, § 47.
[67] Plut. Æmil. 17.

he observes, furnishes an illustration of the common proverb that 'war has many false alarms.'(⁶⁸)

The confidence which the Romans are described in the latter passage as deriving from the prediction in question, seems inconsistent with the anecdote told of Sulpicius Gallus. Nevertheless the Latin writers can scarcely be mistaken in representing him as playing a part on this occasion. The version of Cicero seems indeed preferable to that of Livy. It is not likely that Gallus should, during a campaign, have had time and means for the exact calculation of an eclipse, even if he possessed the necessary skill; but there is nothing improbable in the supposition that when the eclipse had begun, he should have calmed the fears of the soldiers by explaining to them its true cause. We know from the testimony of Cicero, in other passages, that Gallus was devoted to the study of astronomy, and it is even stated that he was able to predict eclipses.(⁶⁹) Here, however, as elsewhere, we are left to judge by probabilities, in our ignorance of the original testimonies now lost. The death of Sulpicius Gallus preceded the birth of Cicero by nearly half a century.

The legions which mutinied in Pannonia, at the accession of Tiberius, in 14 A.D., are described by Tacitus as surprised by

---

(68) Ap. Suid. in πολλὰ κενὰ τοῦ πολέμου (xxix. 6). For βασιλέως ἔκλειψιν Küster conjectures βασιλείας ἔκλειψιν, which gives a better sense; but compare Plut. Æm. 17, where it is stated that the Macedonians supposed the eclipse to represent βασιλέως τὸ φάσμα. With respect to the proverb, see Diogenian. vii. 80, with the note of the Göttingen editors, and Aristot. Eth. Nic. iii. 11.

(69) Mori pæne videbamus in studio dimetiendi cœli atque terræ C. Gallum, familiarem patris tui, Scipio. Quoties illum lux noctu aliquid describere ingressum, quoties nox oppressit cum mane cœpissit! quam delectabat eum defectiones solis et lunæ multo nobis ante prædicere! Words of Cato Major in Cic. de Sen. 14. The father of Scipio is Æmilius Paullus. The astronomical knowledge of Sulpicius Gallus is likewise described in Cic. de Rep. i. 14, where he is represented as explaining the orrery of Archimedes. He is mentioned as an authority on matters of astronomy in Off. i. 6. He was particularly devoted to the Greek writers (Cic. Brut. 20), from whom he doubtless derived his knowledge of astronomy. C. Sulpicius Gallus was a man of an established political position in 171 B.C. (Livy, xliii. 2), and consul in 166 B.C. He lived to be an old man (Cic. de Amic. 27). His lifetime may be placed from about 215 to 145 B.C.

an unexpected eclipse of the moon, as terrified by the appearance, and as ignorant of its physical cause. They follow the prevalent superstition of relieving the moon's sufferings by the clatter of brass and the noise of horns and trumpets. Drusus, who had been sent by Tiberius to quell the mutiny, is represented as taking advantage of the consternation of the soldiers, to bring them back to their obedience.[70]

The cause of a solar eclipse was known to Ennius, writing about 180 B.C.; he describes it as due to the interposition of the moon. The particular eclipse mentioned by Ennius is referred by Cicero to the year 350 after the building of the city, which corresponds to 404 B.C.[71] Cicero states that the eclipses anterior to that of 404 B.C., which was alluded to by Ennius, and was recorded in the *Annales Maximi*, had been calculated backwards up to the eclipse on the nones of Quintilis, which accompanied the supposed translation of Romulus to heaven. These calculations must have been made before the year when Cicero wrote his dialogue De Republicâ, and they appear to have been entered under their proper years in the *Annales Maximi*.

The ancient belief of the Greeks was that the moon in eclipse was bewitched, and that magic incantations or herbs could draw her down from her course. This power was particularly ascribed to the Thessalian women, who had the reputation of witchcraft.[72] A certain Aglaonice, the daughter of Hegetor

---

[70] Ann. i. 24. The eclipse is likewise mentioned by Dio lvii. 4, who states that the violence of the soldiers was checked by its occurrence. Its date, according to Zech, was Sept. 26, A.D. 14.

Livy, xxvi. 5, speaks of a battle before Capua in 271 B.C., in which 'disposita in muris Campanorum imbellis multitudo tantum cum æris crepitu, *qualis in defectu lunæ silenti nocte cieri solet*, edidit clamorem.' Juvenal compares the clatter of a woman's tongue to the noise made at an eclipse of the moon:—

Jam nemo tubas, nemo æra fatiget:
Una laboranti poterit succurrere lunæ.—vi. 442, 3.

[71] De Rep. i. 16. The words of Ennius are, 'Nonis Juniis soli luna obstitit et nox.' Compare the author's work on the Cred. of Early Rom. Hist. vol. i. p. 159, 430. The interval between this eclipse and the time of Ennius was about 200 years.

[72] See Aristoph. Nub. 749; Plat. Gorg. § 146, p. 513; Menander

or Hegemon, was said to have been able to foretel lunar eclipses by her skill in astronomy, and to have used her knowledge for the purpose of creating a belief that she could drag down the moon from heaven; ([73]) thus employing for a superstitious purpose a superiority which Polybius wished to convert to a military object.

The physical nature of eclipses had been explained by Anaxagoras; a knowledge of it appears even in the history of Thucydides; and it was fully understood by Aristotle.([74]) It is certain that Hipparchus and his successors in the Alexandrine School were able, by the assistance of their lunar and solar tables, to predict eclipses both of the sun and moon, with a close approach to exactitude. Hipparchus, in his Commentary, upon Eudoxus, states that the most skilfully prepared predictions of lunar eclipses did not differ from the truth by more than two digits, and that the error was rarely so great.([75]) Even the Chaldæans are stated by Diodorus to

---

ap. Mein. Fragm. Com. Gr. vol. iv. p. 132; Plutarch, de Pyth. Orac. 12; Apollon. Rhod. iii. 533, cum Schol.; Cleomed. ii. 6, cum Not.; Wyttenbach, ad Plut. Præc. Conj. vol. xii. p. 901; Virgil, Ecl. viii. 69; Horat. Epod. v. 45, xvii. 77; Tibull. i. 2, 41; Ovid. Heroid. vi. 85; Manil. i. 227; Lucan, vi. 420. In Schol. Apoll. ubi sup., it is stated that before the time of Democritus eclipses were from this belief called καθαιρέσεις.

On the verses of Sosiphanes, see Wagner, Poet. Trag. Gr. Fragm. vol. iii. p. 376. Sosiphanes lived in the reign of Philip or Alexander.

The subject of the Thessala of Menander was the contrivances of women for drawing down the moon, Plin. xxx. 1.

([73]) Plut. Conj. Præc. 48; De Def. Orac. 13; Schol. Apollon. Rhod. iv. 59.

([74]) The nature of eclipses was understood by Aristarchus of Samos, and was made by him the subject of geometrical reasoning, see Delambre, Ast. Anc. vol. i. p. 79. Aristarchus remarks, in an extant treatise: ὅταν ὁ ἥλιος ἐκλείπῃ ὅλος, τότε ὁ αὐτὸς κῶνος περιλαμβάνει τόν τε ἥλιον καὶ τὴν σελήνην, τὴν κορυφὴν ἔχων πρὸς τῇ ἡμετερᾳ ὄψει, Ap. Wallis. Op. Math. vol. iii. p. 583, ed. Oxon.

([75]) τούτου δὲ γινομένου, ἔδει τὰς τῆς σελήνης ἐκλείψεις πολὺ διαφωνεῖν πρὸς τὰς συντασσομένας ὑπὸ τῶν ἀστρολόγων προρρήσεις· ὑποτιθεμένων δὴ αὐτῶν ἐν ταῖς πραγματείαις τὸ μέσον τῆς σκιᾶς φέρεσθαι ἐπὶ τοῦ διὰ μέσων τῶν ζῳδίων κύκλου. οὐ διαφωνοῦσι δὲ πλέον ἢ δακτύλοις δυσί,—σπανίως δὲ σφόδρα ποτέ,—πρὸς τὰς χαριέστατα συντεταγμένας πραγματείας, Ad Phæn. i. 21, p. 112. With respect to the method of Hipparchus for predicting lunar eclipses, see Delambre, Hist. Astr. Anc. tom. ii. p. 235; and as to his method of predicting solar eclipses, ib. p. 237; on Ptol. Synt. vi. 9, 10.

have been able to predict lunar, though not solar eclipses.(⁷⁶) Cicero, in treating of divination, says that astronomers predict eclipses many years beforehand; he adds, that they can likewise predetermine the places of the planets in the zodiac, and the risings and settings of the fixed stars.(⁷⁷) Augustine, writing about the year 400 A.D., speaks of predictions of solar and lunar eclipses, which fixed the exact day and hour, being verified by the event.(⁷⁸)

The nature of eclipses, and the difference between the nature of solar and lunar eclipses—owing to the sun being a luminous and the moon being an opaque body—is clearly explained in the treatises of Geminus(⁷⁹) and Cleomedes.(⁸⁰) The latter states, moreover, that all eclipses of the moon were predicted by the practical astronomers of his day.(⁸¹)

---

(76) Such appears to be his meaning, though it is not distinctly expressed, ii. 31.

(77) Solis defectiones itemque lunæ prædicuntur in multos annos ab iis qui siderum cursus et motus numeris persequuntur. Ea enim prædicunt quæ naturæ necessitas perfectura est. Vident ex constantissimo motu lunæ, quando illa, e regione solis facta, incurrat in umbram terræ, quæ est meta noctis, ut eam obscurari necesse sit; quandoque eadem luna, subjecta atque opposita soli, nostris oculis ejus lumen obscuret; quo in signo quæque errantium stellarum, quoque tempore futura sit; qui exortus quoque die signi alicujus, aut qui occasus futurus sit, De Div. ii. 6.

(78) Multa invenerunt et prænunciaverunt ante multos annos defectus luminarium solis et lunæ, quo die, quâ horâ, quantâ ex parte futuri essent, et non eos fefellit numerus, et ita factum est ut prænunciaverunt: et scripserunt regulas indagatas, et leguntur hodie, atque ex eis prænunciatur quo anno et quo mense anni, et quo die mensis, et quâ horâ diei, et quotâ parte luminis sui defectura sit luna vel sol, et ita fiet ut prænunciatur, Confess. v. 3.

(79) Gemin. c. 8 and 9. Compare the remarks of Delambre, Hist. Astr. Anc. tom. i. p. 201.

(80) The difference between lunar and solar eclipses is explained by Cleomedes, ii. 4, p. 129, and the nature of lunar eclipses, ii. 6. Compare Ach. Tat. c. 21.

[Luna] etiam tum subjecta atque opposita soli, radios ejus et lumen obscurat, tum ipsa incidens in umbram terræ, cum est e regione solis, interpositu interjectuque terræ, repente deficit, Cic. Nat. D. ii. 40.

The nature of eclipses of the sun and moon is explained by Amm. Marc. xx. 3; that of lunar eclipses by Hygin. Poet. Astron. iv. 14.

(81) καὶ ἤδη γε προλέγονται πᾶσαι αἱ ἐκλείψεις αὐτῆς [i.e. of the moon]

Among the Greek philosophic sects, the Epicureans seem alone to have doubted as to the causation of eclipses. Epicurus admitted, indeed, that the shadow of the earth might be the cause of a lunar eclipse; and the interposition of the moon the cause of a solar eclipse: but he held that an eclipse might be caused by a partial extinction of the light of the sun or moon, or by the interposition of some foreign body, belonging either to the earth or to the heaven.[82]

Delambre thinks that the ancient astronomers promulgated predictions of eclipses, by appending them to their tables of the risings and settings of stars.[83] But these tables were not periodical: the seasons of the rising and setting of stars were only affected by the small anomaly of the precession, of which the ancients were either ignorant, or took no account. The table of this kind, which bears the name of Ptolemy, is exclusively founded on the authority of ancient observers, such as Euctemon, Callippus, Democritus, Dositheus, Eudoxus. No authority later than Hipparchus is named in it.[84]

The ancients had no scientific calendar, constructed by skilful astronomical calculators, and published under the sanction of the State, such as the Nautical Almanac, and the Connaissance des Temps, which are authorized by the English and French Governments, and which again serve as a foundation for numerous popular almanacs. Hence an army might easily be surprised by the occurrence of an eclipse: and the narrow diffusion of astronomical knowledge, owing to the feebleness of the scientific authority, would cause alarm among the soldiers.

ὑπὸ τῶν κανονικῶν, ii. 6, p. 148. By κανονικοί he appears to mean those who made scientific almanacs by means of tables. See the notes, ib. p. 474.

(82) Diog. Laert. x. 96. Lucretius v. 750—769, does little more than state the problem, without solving it. He lays it down, however, that an eclipse may have several causes:—

Solis item quoque defectus, lunæque latebras,
Pluribus e caussis fieri tibi posse putandum est.

(83) Hist. Astr. Anc. vol. ii. p. 258.

(84) p. 42, Petav. Compare Ideler, über den Kalender des Ptolemäus, Berl. Trans. 1816—7.

Even now, an army would scarcely contain any person who could calculate an eclipse; and if an officer of engineers had the necessary skill, he would have little leisure or means of reference for such a calculation. When the eclipses for each year are systematically calculated beforehand, expectation is directed to their occurrence; but when no almanacs exist, the event would not be anticipated, and a military officer in the field would have no reason for setting about a laborious calculation which was not only quite foreign to his proper duties, but which, when made, would probably lead to no interesting result. For one campaign in which a visible eclipse might occur, there would probably be at least a hundred without an eclipse.

The Greeks had no clear idea of the extent of the powers of accurate astronomical prediction; for on the occasion of a bright comet which appeared in Greece in 372 B.C., a short time before the battle of Leuctra, and which was believed by the vulgar to portend disaster to the Lacedæmonians, the men of science affirmed that the Chaldæans of Babylon and the other astronomers could predict the appearance of comets, because they revolved in fixed periods.[85]

§ 5 The reason why the ancients had no accurate almanacs, prepared under the superintendence of scientific astronomers, was not that there were no scientific astronomers, but that the care of the calendar was considered a religious concern, connected with the regulation of sacrifices and holy festivals, and, therefore, committed to the care of sacerdotal authorities.[86] The religious system of each little commonwealth was peculiar —there was no common calendar or computation of time for different communities; and hence arose a facility for tampering with the calendar for temporary or even for personal objects. A remarkable example of the unscrupulous mode in which the calendar was sometimes dealt with, in order to serve a present

---

[85] Diod. xv. 50.

[86] On the duty of the State to regulate the calendar with reference to the sacred festivals, see Plat. Leg. vii. p. 809 D.

convenience, is afforded by the proceeding of the Argives in 419 B.C., who arbitrarily postponed the commencement of the Carnean month until they had completed their expedition against the Epidaurians; the Carnean month being a season during which, by the maxims of their own religious law, they were bound to abstain from hostilities.([87])

Sometimes, doubtless, the calendar fell into disorder from ignorance, or from the unskilful manner in which the intercalations were performed. A humorous passage in the Clouds of Aristophanes represents the moon as sending a message to the Athenians, in order to complain of the confusion of the calendar, and the consequent errors as to the time of the festivals; so that when the gods were mourning, the Athenians were making merry; and that on the days when the Athenians ought to be sacrificing to the gods, the courts were sitting, and they were transacting business.([88])

Cicero describes the Greeks as being in the habit of adjusting their calendar, when it had fallen into error, by lengthening or shortening a month by one or two days.([89])

In some States, indeed, the public regulation of the calendar seems to have made itself so little felt, that its periods were left to the determination of individual caprice.([90])

The result of this astronomical anarchy was, that the calendars of different communities differed; that there were discordances between the beginnings of the ·current months and years; and that the confusion existed which necessarily followed from the absence of a common standard of time.([91])

([87]) Thuc. v. 54. The interpretation of this passage, which Mr. Grote has given, vol. vii. p. 90, seems to me quite satisfactory.

([88]) Nub. 608—626.

([89]) Verr. ii. 52.

([90]) Hesychius et Prov. Gotting. vol. i. p. 405: ' οὐδεὶς γὰρ οἶδεν ἐν Κέῳ τίς ἡμέρα,' ὅτι οὐχ ἑστᾶσιν αἱ ἡμέραι, ἀλλ' ὡς ἕκαστοι θέλουσιν, ἄγουσιν. ὅθεν λέγεται, ' σαυτῷ νουμηνίαν κηρύσσεις.' Compare Crates, ap. Athen. iii. p. 117 B; and Macho, ap. Athen. viii. p. 349 B.

([91]) Aristox. Harm. ii. p. 37: οἷον ὅταν Κορίνθιοι μὲν δεκάτην ἄγωσιν, Ἀθηναῖοι δὲ πέμπτην, ἕτεροι δέ τινες ὀγδόην. And Plutarch, Aristid. 19, speaking of a discrepancy of days of the month at the time of the battle of

§ 6 The Roman year, at the earliest time at which we have historical accounts of it, consisted of 355 days. The calendar was under the exclusive control of the College of Pontiffs, as being a matter of religious concern.(92) Even the preparation of the *Annales Maximi*, on account of their connexion with the annual notation of time, was confided to the Pontifex Maximus.(93) In order to bring a year of 355 days into accordance with the sun, a system of intercalation was necessary. What this was, we are not exactly informed; but it was administered by the pontiffs, who exercised their power neither scientifically nor honestly. They are stated to have falsified the time in order to favour or to spite particular magistrates or farmers of the public revenue, by unduly lengthening or shortening the term of their office or contract.(94)

When Julius Cæsar was Pontifex Maximus, in the year 46 B.C., he found the Roman calendar in great confusion. According to the theory of the calendar, January, the first month of the year, began soon after the winter solstice. This coincidence, however, no longer took place; and as the festivals and public sacrifices were generally regulated by the day of the month, and as they were often connected with the season of the year, their celebration fell at wrong periods.(95) Cæsar rectified this inconvenient state of things by an important change. The

---

Platæa, says: τὴν δὲ τῶν ἡμερῶν ἀνωμαλίαν οὐ θαυμαστέον, ὅπου καὶ νῦν διηκριβωμένων τῶν ἐν ἀστρολογίᾳ μᾶλλον ἄλλην ἄλλοι μηνὸς ἀρχὴν καὶ τελευτὴν ἄγουσι. Plutarch's lifetime falls in the first century after Christ. Compare K. F. Hermann, Gottesdienstl. Alterth. der Griechen, § 35, where the subject is fully illustrated. Anacreon, fragm. 6, Bergk. describes Posideon as a wet and wintry month, which implies that it occurred constantly at the same season. Compare above, p. 116.

(92) Censorin. 20.

(93) See the author's Inquiry into the Cred. of the Early Rom. Hist. vol. i. p. 155; Göttling, Gesch. der Röm. Staatsverf. p. 179. The days on which the successive events occurred were noted in the Annales Maximi (Serv. Æn. i. 373), which gave to these annals the character of a retrospective almanac.

(94) Censorin. c. 20; Macrob. Sat. i. 14, 1; Amm. Marc. xxvi. 1, 12; Solin. i. 43, 44. Dionysius declares that the Romans in early times regulated their months by the moon, Ant. Rom. x. 59.

(95) Solin. i. 45; Plut. Cæs. 59; Suet. Cæs. 40.

error of the calendar amounted, in his time, to no less than ninety days, or three months, in advance; so that January was an autumn month, and occupied the season of the year which ought to have been occupied by October.([96]) Cæsar inserted the regular intercalary month, or Mercedonius, of twenty-three days, and two additional intercalary months containing together sixty-seven days: these, added to a year of 355 days, made altogether a year of transition containing 445 days. By this extensive intercalation, the month of January was brought back to its proper place. In order to guard against the repetition of error, Cæsar further directed that the old lunar year of 355 days should be abandoned, and that the calendar should follow the solar year of 365¼ days. He established this change by adding one day to the months of April, June, September, and November, and two days to the months of January, Sextilis and December each; making altogether an addition of ten days; by which the regular year was lengthened to 365 days. He further provided for a uniform intercalation of one day in every fourth year; which accounted for the remaining quarter of a day.([97])

It had been the custom under the unreformed system to intercalate the month Mercedonius after the 23rd of February, and to subjoin the remaining five days of February to the intercalary month.([98]) The custom of intercalating between the

---

([96]) It is remarked by Mr. Key, in his art. Roman Calendar (Dict. of Gr. and Rom. Ant.), that Cæsar, in his Civil War, makes an interval of several months between January and the winter, Bell. Civ. iii. 5, 6, 9, 25.

([97]) See Ovid, Fast. iii. 155—166; Suet. Cæsar. 40; Plut. Cæs. 59; Dio Cass. xliii. 26; Plin. xviii. 57; Censorin. 20; Macrob. Sat. i. 14; Amm. Marc. xxvi. 1. Compare Ideler, Chron. vol. ii. p. 117.
Ovid thus versifies the length of the Julian year:—

      Ille moras solis, quibus in sua signa rediret,
        Traditur exactis disposuisse notis.
      Is decies senos tercentum et quinque diebus
        Junxit, et e pleno tempora quarta die.
      Hic anni modus est. In lustrum accedere debet
        Quæ consummatur partibus, una dies.

Lustrum is here used for a period of four years.

([98]) Celsus, Dig. 50, 16, 98, states that the intercalary month consisted

23rd and 24th of February was retained in the Julian calendar. But the period intercalated was only one day, and in order to distinguish it, without altering the denominations of the days of February from the 14th to the 23rd inclusive, the method of duplication was resorted to. The Romans, as is well known, reckoned the days of the last half of the month by counting back from the calends, or first day, of the succeeding month. The 24th day of February was, according to their method of reckoning inclusively, the 6th day before the Calends of March. Hence they called the odd day inserted between the 23rd and 24th *bis-sextus,* or (as it would be expressed in French), *sex-bis;* and the previous eleven days which were counted back from the Calends of March retained their former designations unchanged. The year in which the Julian intercalation takes place has hence been called *bissextile.*([99])

Cæsar appears to have been a student of astronomy. Lucan represents him as saying that, even in the midst of his campaigns, he had always found time for astronomical pursuits; ([100]) and he is related to have written a treatise on the motions of the stars.([101]) He did not, however, undertake the reform of the calendar without competent advice; he was assisted by Sosigenes, a scientific astronomer of the Alexandrine School; ([102])

---

of twenty-eight days. This is explained by its interposition after Feb. 23, which added to it the five last days of that month, the intercalary month being assumed to consist of twenty-three days. Compare Livy, xliii. 11, xlv. 44.

(99) See Ideler, Chron. vol. ii. p. 129; Delambre, Hist. Astr. Mod. tom. i. p. 14.

(100)   Media inter prælia semper
      Stellarum cœlique plagis, superisque vacavi;
      Nec meus Eudoxi vincetur fastibus annus.
                                      x. 185—7.

The word *fasti* had another plural form, *fastus,* of the 4th declension. *Supera* is used in a sense equivalent to τὰ μετέωρα.

(101) Siderum motus, de quibus non indoctos libros reliquit, ab Ægyptiis disciplinis hausit, Macrob. Sat. i. 16. The *Ægyptiæ disciplinæ* here referred to are not those of the native priests, but those of the Greek Alexandrine School.

(102) Plin. N. H. xviii. 57. Plutarch, Cæs. 59, says: Καῖσαρ δὲ τοῖς ἀρίστοις τῶν φιλοσόφων καὶ μαθηματικῶν τὸ πρόβλημα προθείς.

a Roman clerk, named M. Flavius, is also stated to have been useful in arranging the details of the change.([103]) By the reform thus effected, the calendar was for the first time emancipated from pontifical control, and was made a matter of purely civil regulation: all reference to lunar months and to the periods of festivals was discarded; the year was defined exclusively by the sun; ([104]) and intercalation, no longer dependent on the discretion of the pontifices,([105]) was regulated by a constant law. This measure was doubtless considered at the time as a bold and unauthorized interference with the course of time: hence Cicero (who feared and hated Cæsar), on being informed that the constellation Lyra would rise on the following day, answered, 'Yes, if the edict allows it.'([106])

The Julian reform of the calendar was founded upon the science of the Greek Alexandrine School; and the reformed year was partly copied from the Egyptian year of 365 days, which, however, was not kept in harmony with the sun by a quadriennial intercalation of a day.([107]) The Romans had first the practical ability to convert the Greek science to a useful purpose, and to found the civil measurement of time upon an accurate basis.

It seems strange that so simple an intercalation as that of a day in every fourth year, in order to provide for the fractional fourth part omitted in the common year, should have been misapplied by the administrators of the reformed calendar.

---

(103) Macrob. Saturn. i. 14, § 2.

(104) Appian, B. C. ii. 154, says: καὶ τὸν ἐνιαυτὸν ἀνώμαλον ἔτι ὄντα διὰ τοὺς ἔσθ' ὅτε μῆνας ἐμβολίμους (κατὰ γὰρ σελήνην αὐτοῖς ἠριθμεῖτο), ἐς τὸν τοῦ ἡλίου δρόμον μετέβαλεν, ὡς ἦγον Αἰγύπτιοι.

(105) The expressions of Cicero, in some of his letters, show that it was uncertain in what year the intercalation would be made by the pontiffs; ad Att. v. 21, ad Div. vii. 2. On the dishonest intercalation of the pontiffs, see l'Art de vérifier les Dates, vol. i. p. 396.

(106) Plut. Cæsar, ib.

(107) Macrobius, Saturn. i. 14, § 3, says that Cæsar imitated the Egyptian year of 365¼ days. Compare i. 16, § 39. This statement is only true if referred to the year of the scientific School of Alexandria. The civil year of the Egyptians certainly contained at this time only 365 days. See below, ch. v. § 7.

Nevertheless, in the first thirty-six years of the reform, the pontiffs committed the error of intercalating the day every third year. This error was detected by Augustus, who ordained that the intercalation should be suspended for twelve years; by which measure the three days in advance were absorbed.[108] From that time the intercalations proceeded regularly; and the Julian calendar, with the Roman months, has become the calendar of the civilized world. The slight error in the Julian year, of an excess of 11' 12", which in the sixteenth century had grown to ten days, was rectified by Pope Gregory in 1581; and its recurrence was prevented by a provision that three intercalary days should be omitted in every four centuries.[109] With this slight correction, the reformed calendar of Julius Cæsar still continues in use, unimproved by the vast advances which speculative astronomy has made since his time. The inconvenient Roman mode of reckoning the days of the months has been discarded, and the simple mode of numerating them in one series from the beginning to the end of the month has been substituted for it. The intercalary day is still added to February; but the peculiar mode of reckoning which gave rise to the term *bissextile* is no longer necessary, and has become a subject of mere antiquarian interest.

Before the Julian reformation of the calendar, and its adoption by the Roman world, it was customary both for Roman and Greek writers to mark the time of the year by the rising or setting of some known constellation, or by the equinoxes or

---

[108] Macrob. Sat. i. 14, § 13; Pliny, xviii. 57 (who states that Sosigenes wrote three treatises on the successive amendments of the calendar); Solin. i. § 46; Suet. Aug. 31.

[109] See Ideler, Chron. vol. ii. p. 298—304; Delambre, Hist. Astr. Mod. tom. i. p. 1—84. For the different times at which the Gregorian amendment was received in different countries, see Delambre, ib. p. 72. In England, the Gregorian reform was introduced in 1752, by the omission of eleven days between 2 and 14 September. The change was established by Act of Parliament, the Bill being moved by Lord Chesterfield, and seconded by Lord Macclesfield (24 Geo. II. c. 23). See Lord Stanhope's Hist. of England, c. 31, vol. iv. p. 21, 8vo; Lord Campbell's Lives of the Chancellors, c. 122, ad fin.

solstices. This method of annual chronology is followed by Varro and Columella in their practical treatises of agriculture; the former of whom lived a short time before, the latter a short time after, the Christian era. But Palladius, a later writer, who probably lived in the fourth century, arranges his treatise on agriculture under the Roman months, and denotes the season of the year in the same manner as it would be marked by a modern writer on husbandry. As soon as the calendar corresponded with the annual course of the sun, and the month denoted the season of the year, the artificial measure of time was the most simple, precise, and convenient, and all reference to natural standards would speedily be discarded.

§ 7 But although the Romans of the Imperial period possessed a calendar which was practically perfect as a measure of annual time, and which gradually passed into universal use,[110] their measures of diurnal time were unimproved and imperfect. They still measured the day by the sundial, and the night by the water-clock.

The use of the sun's shadow as the ordinary measure of time led to a singular consequence in the habits of common life. Instead of making the hour a constant quantity, and of making the number of hours vary with the length of the day, the ancients (as we have explained in a previous chapter)[111] made the number of hours constant, and made the length of the hour vary with the length of the day. Whatever might be the time of sunrise, and whatever the time of sunset, the illuminated interval was divided into twelve equal parts.[112] Hence, if the

---

[110] I am not aware that the means exist of tracing the extension of the Roman calendar, with its national months, over Greece and Asia Minor, and the rest of the Roman empire. The least civilized parts (such as Gaul and Spain) would probably be the first to adopt it. The Alexandrine astronomers continued to use the Egyptian months in the time of Ptolemy. Josephus, in his History of the Jewish War, uses the Syrian months. For a list of the Roman months, see Anth. Pal. ix. 384.

[111] Above, ch. iii. § 9.

[112] Vitruvius states that, whatever may be the form or variety of the sundial, the day, whether the longest, or the shortest, or the equinoctial, is divided by it into twelve equal hours. Omnium autem figurarum

sun, according to our notation, rose at 5 A.M., and set at 8 P.M., each hour was equal to eighty minutes.

Various improvements upon the common sundial, which had been made before the Augustan age, particularly by the scientific men of the Alexandrine School, are enumerated by Vitruvius. Some *horologia* made by Ctesibius of Alexandria, celebrated for his mechanical skill,(113) were regulated by water, and are stated to have been intended to be used during winter; at which time the prevalence of cloudy weather impeded the use of a sundial, even in the climate of the Mediterranean. Portable sundials, for the use of travellers, so constructed that they could be suspended, are likewise mentioned by Vitruvius.(114)

§ 8. The use of variable hours was retained by the Arabs; it appears in the astronomical work of Albategnius about the year 900.(115) Delambre remarks that this mode of dividing the day was only expelled by the use of clocks.(116) Even, however, after the introduction of a mechanical timekeeper, which was unconscious of the motion of the sun, the prevalence of ancient habits led to its regulation according to the unequal

---

descriptionumque earum effectus unus, uti dies æquinoctialis brumalisque itemque solstitialis in duodecim partes æqualiter sit divisus, ix. 7, 7.

By *bruma* the Romans denoted the winter, by *solstitium* the summer solstice.

(113) Ctesibius lived in the reigns of Philadelphus and Evergetes, and was contemporary with Apollonius of Perga. The ῥυτὸν of Ctesibius, dedicated in the temple of Arsinoe, is celebrated in an epigram of Hedylus (Ath. xi. p. 497 D, Anth. Pal. app. 30), a poet, who lived in the reign of Philadelphus.

(114) Vitruv. ix. 8. Compare the illustrations of Delambre, Hist. Astr. Anc. tom. ii. p. 515. The scientific knowledge of the ancients on Dialling was consigned by Ptolemy to a Treatise de Analemmate; which is extant only in a Latin version of an Arabic translation. A full account of this treatise is given by Delambre, ib. p. 458—503.

The analemma was a raised basis, upon which the sundial, whatever its construction, was placed. Hence the word was applied to sundials generally. See Schneider's note on Vitruvius, vol. iii. p. 172, and Steph. Thes. ed. Didot. ἀνάλημμα. Delambre, ib. p. 458, is mistaken in supposing that the word is equivalent to *lemma* in its geometrical sense.

(115) Delambre, Hist. Astr. du Moyen Age, p. 56.

(116) Hist. Astr. Anc. tom. ii. p. 512.

hours. Bernardus Monachus, in his collection of customs of the monastery of St. Victor, at Paris, written in the eleventh century, gives detailed precepts for regulating the clock, so that the hours should be longer or shorter, according to the season of the year.([117]) The use of variable hours still subsists in the less advanced parts of Italy, where French manners have not penetrated.([118]) It is likewise prevalent in Turkey, so far as the Turks have any division of hours.([119]) Clocks and watches appear indeed to be nearly unknown among the Oriental nations. The construction of railways will tend to introduce their use into India.

The same rude methods for the measurement of diurnal and nocturnal time remained in use for many centuries. Cassiodorus, in the sixth century, presented two timekeepers to a monastery in Languedoc; one a sundial, the other a water-clock. The latter, he remarks, tells the hour at night, and also in the day when the sun is hidden by clouds.([120]) The clock presented by the King of Persia to Charlemagne in 807 A.D., and also that presented by Pope Paul to Pepin, King of France, were water-clocks. The latter is called 'horologium nocturnum.' ([121]) Even water-clocks, however, were costly and scarce, and few

---

([117]) c. 64. Ap. Martene de Ant. Rit. vol. iii. p. 739, cited in Beckmann, Hist. of Inv. vol. i. p. 429.

([118]) See Ideler, Chron. vol. i. p. 83; Delambre, Astronomie Théor. et Prat. tom. iii. p. 688.

([119]) 'Though it is generally better for a traveller to conform to local customs, there is some inconvenience in Turkish time, as the watch cannot be kept correct without daily attention. It would seem, however, to be a natural mode of measuring time, being followed by so many nations. The Turkish method differs from the Italian in dividing the day into two twelves, instead of reckoning to twenty-four; so that sunset is always twelve o'clock. One of the commonest questions which a native of the Levant who wears a watch puts to a Frank is, "At what hour is midday?" This, he asks, that he may set his watch. The peasant without a watch generally asks, "How many hours is it to sunset?" this being obviously the principal question for the labourer. To the Turk also it is important; as the afternoon's prayer is three hours before sunset.'
Leake's Travels in Northern Greece, vol. i. p. 254.

([120]) De Inst. Div. Lit. c. 29, cited by Beckmann, Hist. of Inv. vol. i. p. 422, Engl. Transl. ed. 1817.

([121]) See Beckmann, ib. p. 423, 425.

monasteries were rich enough to possess one. It was necessary for the monks to resort to other means for finding the hour at night. Cardinal Peter Damiani, in the eleventh century, thus defines the duties of the *significator horarum,* in a work De Perfectione Monachorum. 'He is not to listen to stories, or to hold long conversations with any one, nor is he to inquire what is done by persons engaged in secular pursuits. Let him be always intent on his duty, and never relax his observation of the revolving sphere, the movement of the stars, and the lapse of time. Moreover, let him acquire a habit of singing psalms, if he wishes to possess a faculty of distinguishing the hours; for whenever the sun or stars are obscured by clouds, the quantity of psalms which he has sung will furnish him with a sort of clock for measuring the time.'[122]

Even at a later period, after the introduction of clocks which struck the hour, the rules of some monasteries directed the monk charged with the care of the clock to note the wax candles, and the course of the stars, or even of the moon, so as to cause the monks to rise at the proper time.[123]

The introduction of clocks moved by weights and wheels, and provided with an apparatus for striking the hour, did not take place till the eleventh or twelfth century.[124] Dante

---

(122) Non fabulis vacet, non longa cum aliquo misceat; non denique quid a secularibus agatur inquirat; sed commissæ sibi curæ semper intentus, semper providus, semperque solicitus, volubilis sphæræ necessitatem quiescere nescientem, siderum transitum, et elabentis temporis meditetur semper excursum. Porro psallendi sibi faciat consuetudinem, si discernendi horas quotidianam habere desiderat notionem; ut quandocumque solis claritas, sive stellarum varietas nubium densitate non cernitur, illic in quantitate psalmodiæ quam tenuerit, quoddam sibi velut horologium metiatur, c. 17, cited by Beckmann, ib. p. 427.

(123) Ut notet in cereo, et in cursu stellarum vel etiam lunæ, ut fratres surgere faciat ad horam competentem. Ord. Clun. Bern. Mon. p. 1, c. 51, cited in Beckmann, ib. p. 433. It seems that time was measured by observing the diminution of a lighted *cereus,* or wax taper.

(124) See Beckmann, ib. p. 429. The writer whom Beckmann follows ascribes the invention of clocks to the Saracens, on account of a horologium sent by the Sultan of Egypt in 1232 to the Emperor Frederic II. This piece of mechanism, however, seems to have been an orrery rather than a clock, ib. p. 433. A clock for the use of the courts of law is said

speaks of an *orologio* which strikes the hour; (¹²⁵) thus using the word in the modern sense of a clock, not in that of a sundial, which it had previously borne. The pendulum clock was unknown till the seventeenth century, when it was invented by Huyghens.(¹²⁶)

§ 9 It has been remarked that astronomy made little progress between the times of Hipparchus and Ptolemy. Nevertheless, it seems that, at some time during this interval, a doctrine was promulgated which to a certain extent anticipated the Copernican system. The early Greek astronomers had perceived a difference between the three superior and the two inferior planets, particularly with respect to their periodic times. They had a difficulty in determining the true periodic times of Venus and Mercury, on account of their proximity to the sun. Eudoxus supposed them to be nearly equal to the solar year:(¹²⁷) and Cleomedes adheres to this determination,(¹²⁸) widely as it departs from the truth.(¹²⁹) These two planets are stated to have an equal velocity with the sun, by Cicero,(¹³⁰) by the author of the pseud-Aristotelic Treatise de

---

to have been set up near Westminster Hall in 1228, and to have been defrayed out of a fine imposed on the Chief Justice of the King's Bench. See Barrington in Beckmann, ib. p. 443.

(125)   Indi, come orologio che ne chiami
Nell'ora che la sposa di Dio surge
A mattinar lo sposo perchè l'ami,
   Che l'una parte e l'altra tira ed urge,
Tin tin sonando con sì dolce nota
Che 'l ben disposto spirto d'amor turge.
                    Paradiso, cant. x. v. 139—144.

This passage describes a clock waking the faithful, and summoning them to matins, by striking on a bell moved by machinery.

The expression 'tin tin sonando' is borrowed from the Latin word tintinnabulum, which was supplanted in Italian by the words *squilla* and *campana*. See Ducange, Gloss. in Skella, Diez in vv.

(126) See Delambre, Hist. Astr. Mod. vol. ii. p. 551. The date of the invention is 1657.

(127) Above, p. 155.   (128) i. 3.

(129) The periodic time of Mercury is 87d. 23h. 15'; that of Venus, 224d. 16h. 49'.

(130) Somn. Scip. 4 : Hunc [the sun] ut comites consequuntur Veneris alter, alter Mercurii cursus.

Mundo,(131) by the spurious Timæus De Animâ Mundi,(132) by Geminus,(133) and by Achilles Tatius.(134)

The earliest Greek speculators on astronomy did not attempt to fix the order of the planets. The Pythagoreans were stated by Eudemus, in his History of Astronomy, to have first determined their place in the celestial system.(135) The Pythagorean doctrine on this head is variously reported; but it is probable that the ancient Pythagoreans assumed the place of the five planets to be between the orb of the fixed stars on the one hand, and the sun and moon on the other.(136) This was the doctrine of Plato, Eudoxus, and Aristotle.(137)

As soon, however, as geometry came to be applied to the celestial movements, and greater exactitude was thus introduced into astronomical science, the difference between the motions of the three superior and the two inferior planets was perceived, and the sun was placed in the midst, between them; so that the seven movable heavenly bodies were made to succeed one another in the following order: 1. Saturn; 2. Jupiter; 3. Mars; 4. The sun; 5. Venus; 6. Mercury; 7. The moon. This order was adopted by Archimedes,(138) and after him generally by the mathematical school of astronomers.(139) It was assumed, as the received creed, by Cicero,(140) Mani-

---

(131) c. 6.   (132) § 4, p. 96 E.
(133) c. 1, p. 3.   (134) c. 18, p. 81.
(135) Ap. Simplic. ad Aristot. de Cœl. p. 497 a, Brandis.
(136) See Martin, Timée de Platon, tom. ii. p. 105. Above, p. 131.
(137) Concerning Plato, see Martin, ib. p. 64. Concerning Eudoxus and Aristotle, see Proclus in Tim. 257 F.
(138) Macrob. in Somn. Scip. i. 19, § 2.
(139) τῶν μαθηματικῶν τινὲς μὲν ὡς Πλάτων, τίνες δὲ μέσον πάντων τὸν ἥλιον, Plut. Plac. Phil. ii. 15; Galen, c. 13; Stob. Phys. i. 24. This passage is correctly explained by Martin, Timée, tom. ii. p. 103, 113, 128. Theo Smyrnæus, c. 15, states that some of the mathematicians placed the planets in the following order: 1. the Moon; 2. the Sun; 3. Mercury; 4. Venus. Some in this order: 1. the Moon; 2. the Sun; 3. Venus; 4. Mercury. On these and other variations in the order of the planets, see Ach. Tat. c. 16.
(140) Somn. Scip. 4. Compare Macrob. i. 19. In de Div. ii. 43, Cicero gives this order of the seven planets according to the 'ratio mathematicorum.'

lius,(141) and Pliny;(142) and it appears in the astronomical treatises of Geminus,(143) and Cleomedes;(144) in a metrical passage of Alexander of Ephesus;(145) and in the Poetical Astronomy of Hyginus.(146) It is established in the work of Ptolemy, who states that the early mathematicians were unanimous in placing Venus and Mercury below the sun, but that some of the later mathematicians had placed the sun below these planets, because they are never seen to pass over the sun's disk. This reason is considered inconclusive by Ptolemy, who remarks that the planets in question may never be in a plane between our eyes and the sun; he accordingly supposes Venus and Mercury to intervene between the sun and moon.(147) The same middle station of the sun appears likewise to have been adopted in the late astronomy of the native Egyptians.(148) According to Macrobius, however, this order of the seven planets, though recognised by the Chaldæans, was not followed by the Egyptians. He states that the Egyptians placed the five planets above the sun and moon; but that, on account of their astronomical skill, they were able to explain those phenomena of Venus and Mercury which had led to the formation of the other hypothesis.(149) When the scientific theory of the Greeks had reached the point of distinguishing between the superior and inferior planets, as to their position in the universe with respect to the sun, a new hypothesis was devised, that Venus and Mercury are satellites of the sun, and that they move

---

(141) i. 803—6. (142) H. N. ii. 6. (143) c. 1.

(144) i. 3. See also Achill. Tat. 16. Both orders of the planets are attributed to Pythagoras; see Martin, ib. p. 105; Theo Smyrn. c. 15, attributes to the Pythagoreans the order last mentioned in the text.

(145) Ap. Theon. Smyrn. c. 15; Meineke, Anal. Alex. p. 372.

(146) Poet. Astr. iv. 14.

(147) Synt. ix. 1.

(148) See Dio Cass. xxxvii. 19; Achill. Tat. c. 17.

(149) In Somn. Scip. i. 19, § 5. This passage, which had been misunderstood by previous writers, is correctly explained by Martin, Timée, tom. ii. p. 130.

round the sun, while all three move round the earth. This hypothesis appears in the work of Vitruvius, who wrote in the Augustan age;([150]) and is repeated by Martianus Capella, who is supposed to have lived in the fifth century of our era.([151]) It occurs, likewise, in a form substantially identical, in the astronomical treatise of Theon of Smyrna, who lived about the middle of the second century.([152]) No mention of it is made by Ptolemy. Its late appearance indicates that it was of late origin. It would doubtless have been noticed by Aristotle and his predecessors, if it had been known to them. It coincides to a certain extent with the theory of the world propounded at the beginning of the seventeenth century by Tycho Brahe. His theory was that the earth is immovable and at the centre of the universe; and that the sun, the moon, and the sphere of the fixed stars revolve round it; but that the five planets revolve round the sun, the orbits of Saturn, Jupiter, and Mars alone surrounding the earth, and those of Venus and Mercury lying between the earth and the sun.([153])

The Greeks became aware before this period that the stars are effaced by the light of the sun. Galen remarks that the stars can be seen from deep wells, especially when the sun is not on the meridian.([154]) Pliny had previously stated that the stars can be seen in the daytime from deep wells, and during an eclipse of the sun.([155])

§ 10 The life of Claudius Ptolemæus probably extended from about 100 to 170 A.D. He was a native of Egypt, and resided at Alexandria. His work, in thirteen books, entitled 'The Mathematical System,'([156]) and generally known by its

---

([150]) ix. 4.
([151]) viii. § 854, 857. Compare Delambre, Hist. Astr. Anc. vol. i. p. 312.
([152]) c. 33; with Martin's comment, p. 119. Compare Boeckh, Kosm. System des Plat. p. 138.
([153]) See Delambre, Hist. Astron. Mod. tom. i. p. 219.
([154]) De Usu Partium, x. 3, vol. iii. p. 776, ed. Kühn.
([155]) ii. 14. See above, p. 226, n. 56.
([156]) Μαθηματικὴ Σύνταξις.

Arabic title of *Almagest*, is the most complete and advanced representation of the Greek astronomy, both practical and scientific. Ptolemy had access at Alexandria to a set of the writings of Hipparchus, which he studied carefully, and used as the foundation of his own symmetrical edifice. He was a skilful geometer and calculator; he was acquainted with all the authentic observations of his predecessors; and he was likewise himself an observer, though to what extent cannot be determined. With respect to the solar and lunar theories, he principally followed Hipparchus; but his theory of the planetary movements was elaborated by himself, from the principles of the eccentric and epicycle, which Hipparchus had applied to the sun and moon. The mathematical methods and reasonings of Ptolemy, in this scientific system of astronomy, have been copiously expounded by Delambre, himself an experienced and accomplished astronomer, in his 'History of Ancient Astronomy.'[157]

It will be sufficient for our purpose to say that Ptolemy, as to the outlines of his system, treads closely in the footsteps of the mathematical school of Greek astronomers, of Eudoxus, Euclid, Archimedes, Apollonius, and Hipparchus. He holds that the heaven is spherical, and that it revolves upon its axis; that the earth is a sphere; that it is situated within the celestial sphere, and nearly at its centre; that it is a mere point, in relation to the distance and magnitude of the sphere of the fixed stars; and that it has no motion, either of translation or of rotation.[158]

That the earth has no magnitude in reference to the fixed stars, he infers from the absence of parallax in the stars, and from the fact that the plane of the visible horizon cuts the

---

[157] See tom. ii. p. 67—410. Delambre begins his analysis of the work by saying: 'L'astronomie des Grecs est toute entière dans la Syntaxe mathématique de Ptolémée.' See likewise Mr. De Morgan's art. *Ptolemæus*, in Dr. Smith's Dict. of Anc. Biog. and Myth. vol. iii. p. 569; and Delambre's art. *Ptolémée*, in the Biographie Universelle.

[158] See i. 1.

celestial sphere into two equal parts.(159) Ptolemy, therefore, with Euclid, conceives the diameter of the earth to be an inappreciable quantity when compared with the distance of the fixed stars.

But his arguments with respect to the central position of the earth, imply that he did not conceive that distance to be so great as to be virtually infinite with reference to its displacement from the centre of the starry sphere. If, he says, the earth is not at the centre of the celestial sphere, it must be either off the axis of the celestial sphere, and equally distant from each of the poles; or upon the axis, and at unequal distances from the poles; or off the axis, and at unequal distances from the poles. He proceeds to show that neither of these hypotheses can be reconciled with the appearances. The arguments by which this demonstration is effected all proceed upon the assumption that the appearances would be different from what they are, if the earth was further removed from one part of the hollow sphere or shell in which the stars are set than from another. They therefore necessarily imply that the line joining the supposed place of the earth and the centre of the celestial sphere is not infinitely small, but is an appreciable distance, with reference to the distance of the fixed stars. For example, he argues that if the earth were upon the axis of the celestial sphere, but were nearer one pole than the other, the horizon would cut the visible heaven into unequal parts in every latitude where the sphere is oblique.(160)

His refutation of the hypothesis of the earth having a motion of translation, is exclusively founded on his refutation of the hypothesis of the earth being motionless at a point not coincident with the centre of the celestial sphere. It is not at all directed against the movement of the earth: it is confined to the proof that the earth cannot be at any place other than the centre of the heaven; which must happen, if it has a motion of translation. Ptolemy simply mentions the hypo-

---

(159) i. 5.   (160) i. 4.

thesis, without specifying the nature of the supposed movement, and without attributing the hypothesis to Aristarchus or any other astronomer.

He proceeds to remark, that some persons admit the force of these objections to the hypothesis of a movement of translation, but conceive that they propound a theory, both probable and free from objection, if they suppose the heaven to be immovable, and the earth to revolve round the celestial axis from west to east, making nearly one revolution in each day; or if they suppose both the heaven and the earth to revolve, but about the same axis, and conformably with their mutual appearances.([161]) Ptolemy admits that, with respect to the stars alone, this is the most simple hypothesis; but he conceives that the objections derived from the phenomena of bodies within our atmosphere are insuperable. The clouds, and birds in their flight, and projectiles, would not accompany the earth in its rotation, but would drop to the west. If the atmosphere shared the rotatory movement of the earth, and these bodies partook of its velocity, they would be carried round with it, and would appear to be at rest, instead of having movements of their own.([162])

According to the first of these hypotheses, the earth makes a revolution of 359 degrees in a day, allowing one degree for the proper motion of the sun.([163]) According to the second hypothesis, both the celestial sphere and the earth are supposed to move, and their reciprocal movements are adjusted so as to account for the appearances, according to some arrangement which is not explained.

With regard to the order of the planets, he adopts that of the ancient mathematicians, who placed Saturn, Jupiter, and

---

([161]) It may be worth remarking that this passage is mistranslated by Halma, and its sense entirely destroyed. He is not aware that κινῶ is an active verb.

([162]) i. 6.

([163]) This hypothesis is alluded to by Simplicius, ad Aristot. de Cœl. p. 495 a, Brandis.

Mars next under the sphere of the fixed stars, then the sun above Venus and Mercury, and lastly, the moon next to the earth. He states, however, this order with doubt, because the planets have no parallax, by which alone their true distance can be determined.([164])

§ 11 The Copernican system had not arisen, even in the form of a hypothesis, in the age of Aristotle. This philosopher, indeed, describes the Pythagorean system, according to which the five planets visible to the naked eye, the earth, the sun, and the moon, all revolved in circular orbits around the central fire. This was not the Copernican system, because it supposed the sun to move; though it agreed with the Copernican system in making the earth move in a circular orbit round the centre of the universe, as one of the planets.([165])

Aristotle further attributes to Plato the hypothesis that the earth, situated at the centre, has a rotatory motion round the axis of the universe.([166]) The same hypothesis is expressly ascribed to Heraclides of Pontus and Ecphantus.

Aristotle and his contemporaries had therefore conceived two ideas essential to the Copernican system. 1. That the earth revolves in an orbit round the centre of the planetary system; 2. That it turns upon its own axis. No speculator had hitherto conceived the heliocentric hypothesis, or had supposed that the sun is at the centre of the universe.

This hypothesis was, however, propounded by Aristarchus about 260 B.C., in propositions published against the astronomers, among whom the received belief, according to Archimedes, was that the earth is at the centre, and that the sun revolves round it in an orbit. The Copernican system had therefore been suggested to the minds of Archimedes and Hipparchus, by whom it was rejected. The double motion of the earth upon its axis and in an orbit had likewise been suggested

---

(164) ix. 1. Compare Delambre, ib. p. 308.
(165) ἐν τῶν ἄστρων, Cœl. ii. 13, § 1, 14, § 1.
(166) Cœl. ii. 13, § 4, 14, § 1.

to Eudoxus, Aristotle, and Euclid, who deliberately preferred the geocentric solution of the phenomena.

There is no doubt that the theory of the planetary motions originally devised by Eudoxus, and brought by gradual developments to the system of Ptolemy, was in the highest degree intricate. Its complexity was such as not unnaturally to suggest the remark of King Alfonso, that if he had been consulted before the creation, he could have recommended a less complex scheme.[167] But it must be confessed that the Copernican hypothesis was startling, not only to the apparent evidence of the senses, but even to the reason, until the united efforts of minds of the highest scientific and mathematical genius, combined with the invention of the telescope, and with the improvement of the astronomical instruments and of measures of time, had established it as a demonstrated truth.

Besides the immutable position of the fixed stars, [168] there is another circumstance which must have tended to lead away men's minds from the notion that the earth revolves round the sun. When we suppose one body revolving round another, we naturally conceive the axis of the revolving body as influenced

---

[167] See Bayle, Dict. art. Castille, note H.; Delambre, Hist. Astr. du Moyen Age, p. 248.

Mariana, De Reb. Hisp. xiv. 5, says of Alfonso X., 'Emanuel sane patruus, suo et aliorum procerum nomine, Alphonsum publicâ sententiâ in conventu pronunciatâ, regno privavit; eâ calamitate dignum quod divinæ providentiæ opera, et humani corporis fabricam, insigni linguæ procacitate ingeniique confidentiâ accusare ausus fuerit; uti vulgo hominum opinio est, ab antiquo ducta per manus. Vocis stoliditatem numen justissime vindicavit. Id fore astra memorant portendisse ejus artis non ignaro: si ars est, et non potius inane mortalium ludibrium, quod a prudentibus semper accusabitur, et semper tamen patronos habebit.'

It will be shown below, ch. v. § 12, that the Church always condemned astrology.

[168] 'It was not until the revival of letters that the annual motion of the earth was admitted. Its apparent stability and repose were until then universally maintained. An opinion so long and so deeply rooted must have had some natural and intelligible grounds. These grounds, undoubtedly, are to be found only in the general impression that, if the globe moved, and especially if its motion had so enormous a velocity as must be imputed to it, on the supposition that it moves annually round the sun, we must in some way or other be sensible of such movement.'—Lardner's Handbook of Astronomy, by Dunkin (1860), § 135.

by the predominant tendency, and as directed constantly towards the centre. We should be inclined to suppose that if the earth made a circuit round the sun, any point on its surface would likewise make a circuit round the heavens; that when the earth had accomplished a fourth part of its orbit, the points of the compass would all have changed ninety degrees in reference to the stars; that assuming the poles of the earth's axis to be north and south, the west would become north, the south would become west, the east would become south, and the north would become east, and so on, with the two successive quarters, until at the end of the fourth the original directions were restored. But as the axis of the earth always remains parallel to itself, notwithstanding the motion of the earth in its orbit, the same star always marks the north, and the pole of the earth is unchanged with reference to the starry sphere, although the earth makes a circuit in space. Copernicus, in fact, supposed the axis of the earth to be always turned towards the sun.[169] It was reserved to Kepler to propound the hypothesis of the constant parallelism of the earth's axis to itself.

The Copernican system of the universe, and its subsequent completion by the Newtonian theory of Universal Gravitation, have had a purely scientific value, and have exercised scarcely any practical influence upon the affairs of mankind. The solar year was fixed with a close approach to accuracy by the Julian calendar, in the year 46 B.C. The reform of the Julian calendar, under the auspices of Pope Gregory, in the year 1581, was only a short time subsequent to the publication of the hypothesis of Copernicus, and was promoted by astronomers who held the Ptolemaic system. This reform of an error amounting only to 11′ 12″ in a year brought the calendar to perfection; the annual measure of time has received no improvement since the modern astronomical revolution. With regard to the determination of a ship's place at sea by astronomical methods, the invention of chronometers has been far more important than any

---

(169) See Delambre, Hist. Astr. Mod. vol. i. p. 96.

improvement in astronomical theory. If the ancients had known the telescope and the clock, their scientific methods would have sufficed for nearly all practical purposes, although they might have held to the geocentric hypothesis.

Astronomy, as it has been developed by Copernicus, Kepler, and Newton, and their modern successors, has been treated by mathematical methods, requiring the highest stretch of the reasoning faculty, and has furnished materials for sublime contemplation. But it is a science of pure curiosity; it is directed exclusively to the extension of knowledge in a field which human interests can never enter. An attempt has been made by some astronomers to distinguish between the solar system and sidereal astronomy; but the distinction rests on no solid foundation. The periodic time of Uranus, the nature of Saturn's ring, and the occultations of Jupiter's satellites, are as far removed from the concerns of mankind as the heliacal rising of Sirius, or the northern position of the Great Bear.

Science ought, indeed, to be pursued for its own sake; and the human mind can be worthily occupied in the acquisition of knowledge which can never lead to any practical result. But if the astronomical science of the ancients was less exact and comprehensive than that of the moderns, it had a closer bearing upon human affairs, and it nearly exhausted those departments which are useful to mankind.

Chapter V.

ASTRONOMY OF THE BABYLONIANS AND EGYPTIANS.

§ 1 THE Greeks, especially the writers of the later literary period, were in the habit of attributing the invention and original cultivation of Astronomy either to the Babylonians or to the Egyptians; and they represented the earliest scientific Greek astronomers as having derived their knowledge from Babylonian or from Egyptian priests. They did not, indeed, adhere quite consistently to this supposition; for some of the mythological stories gave to the Greek astronomy an indigenous origin, as we have shown in a previous chapter.(¹)

The author of the Platonic Epinomis says that Egypt and Syria produced the earliest astronomers, on account of the clear sky which those countries possessed in summer, and the consequent facility of observing the stars.(²) By Syria, in this passage, Assyria is doubtless meant.(³) The same writer speaks of their astronomical observations having been carried on for an infinite series of years. Macrobius also describes the Egyptians as having been the earliest astronomers, and attributes this priority to the clearness of their sky.(⁴)

The priests of Thebes, in Egypt, asserted that they were the originators of exact astronomical observation: they attributed this superiority in part to the clearness of their climate,

---

(1) ch. ii. § 1. (2) § 9, p. 987.

(3) Concerning the use of Syria for Assyria, see Herod. vii. 63; Strab. xvi. 1, § 2.

(4) Ægyptiorum retro majores, quos constat primos omnium cœlum scrutari et metiri ausos, postquam perpetuæ apud se serenitatis obsequio cœlum semper suspectu libero intuentes deprehenderunt, &c., Comm. in Somn. Scip. i. 21, § 9.

which enabled them to discern the risings and settings of the stars.(5) Cicero concurs with the Epinomis in ascribing the origin of astronomical observation to the Assyrians; he derives it, however, not from the clearness of their sky, but from the wide extent of the plains which they inhabited.(6) According to Diodorus, the Chaldæans used the temple of Belus, which stood in the centre of Babylon, for their astronomical observations, on account of its immense height.(7) It may be remarked, that the ancients considered lofty eminences or large plains as best fitted for astronomical observation. The reason was that, as they principally observed the risings and settings of the fixed stars, they wished to have a clear horizon.(8)

Tatian, a Christian writer of the second century, moderates the pride of the Greeks, by suggesting to them the many important inventions for which they were indebted to *barbarians*, or foreigners. He reminds them that the Babylonians were the authors of astronomy, and the Egyptians of geometry.(9) Clement of Alexandria, in the early part of the third century, following up the same argument, mentions the Egyptians, and also the Chaldæans, as the originators of astronomy.(10)

Lactantius, a Latin Christian writer, of the beginning of the fourth century, traces the invention of astronomy in Egypt to a religious origin. He says that the Egyptians were the first to worship the heavenly bodies,(11) and as they lived in the open

---

(5) Diod. i. 50.

(6) Principio Assyrii, ut ab ultimis auctoritatem repetam, propter planitiem magnitudinemque regionum, quas incolebant, cum coelum ex omni parte patens atque apertum intuerentur, trajectiones motusque stellarum observitaverunt, De Div. i. 1.

(7) ii. 9.

(8) On the importance of the risings and the settings of stars in the ancient astronomy, see the letter of Mr. Wales, in Vincent's Comm. and Navig. of the Ancients, vol. i. p. 545, ed. 2.

(9) Orat. ad Græc. c. 1.

(10) Strom. i. 16, § 74.

(11) This worship is assigned to the Phœnicians, not to the Egyptians, by Euseb. Præp. Ev. i. 9, 5, vol. i. p. 59, ed. Gaisford. Lucian, De Deâ Syr. c. 2, says that the Egyptians were the first to introduce a belief in the

air, and enjoyed a cloudless sky, they observed the movements of the stars, and the eclipses of the sun and moon—the beings which they venerated as gods.([12])

According to Pliny, the invention of astronomy was assigned by some to Atlas, the son of Libya; by some to the Egyptians; by others to the Assyrians.([13]) In another place, the same writer speaks of Jupiter Belus as the inventor of this science.([14]) Achilles Tatius declares that the Egyptians were reported to have been the first who measured the heaven, as well as the earth, and to have inscribed their discovery upon stone pillars, for the benefit of other nations; but he adds, that the Chaldæans likewise claimed the invention of astronomy, attributing it to Belus.([15])

Manilius, in his astrological poem, attributes the origin of astronomy to the nations which dwelt upon the Euphrates and the Nile; and he describes the science as having been first cultivated by kings;([16]) alluding apparently to Belus, the mythical king of Babylon.

Phœnix of Colophon, who wrote poetry about 300 B.C. mentions it as a proof of the indolence of the ancient Assyrian king Ninus, that he entirely neglected astronomy: which im-

---

gods, and rites for their worship; and that the Assyrians borrowed theological belief and religious worship from the Egyptians.

([12]) Sed omnium primi qui Ægyptum occupaverunt, cœlestia suspicere atque adorare cœperunt. Et quia neque domiciliis tegebantur propter aëris qualitatem, nec enim ullis in eâ regione nubibus subtexitur cœlum; cursus siderum et defectus notaverunt, dum ea sæpe venerantes curiosius atque liberius intuerentur, Div. Inst. ii. 13.

([13]) N. H. vii. 56.

([14]) N. H. vi. 26. Durat adhuc ibi Jovis Beli templum. Inventor hic fuit sideralis scientiæ. This statement is repeated by Solinus, in the following sentence: Beli ibi Jovis templum, quem inventorem cœlestis disciplinæ tradidit etiam ipsa religio, quæ Deum credit, c. 56, § 3. Seneca, N. Q. iii. 29, speaks of Berosus as the translator of Belus; which implies that Belus was believed to be the author of writings on the primitive history and astronomy of Assyria.

([15]) Isag. c. i. p. 73, ed. Petav.

([16]) i. 40—45.

plies a belief that the cultivation of that science was habitual among the early Assyrians.([17])

A rationalized version of the mythus of Prometheus, preserved in Servius, attributes the origin of astronomy to the Assyrians. According to this interpretation of the fable, Prometheus was a philosopher, who observed the stars on Mount Caucasus, and taught astronomy to the neighbouring Assyrian people. The vulture which consumed his liver was an emblem of the care and solicitude with which he watched the movements of the heavenly bodies.([18])

Diodorus and Diogenes Laertius state that the Egyptians asserted themselves to be the inventors of geometry and astronomy; to which the latter adds arithmetic.([19])

According to Isidorus, the Egyptians were the authors of astronomy, while the Chaldæans invented astrology and the observation of nativities. The Greeks, however, attributed the discovery of astronomy to Atlas, whence he was fabled to have supported the heaven.([20])

A native fable is related by Socrates in the Phædrus of Plato, according to which Theuth, an ancient Egyptian god, invented numbers and calculation, geometry and arithmetic, as well as the games of dice and draughts.([21]) Another story, cited by Diogenes Laertius, refers the origin of geometry to Moeris,

---

(17)   ὃς οὐκ ἴδ' ἀστέρ', οὐδ' ἰδὼν ἐδίζητο.
                                    Ap. Athen. xii. p. 530 E.

Naeke, Choerilus, p. 229, says in explanation of this verse: 'Hoc dicit poeta: Ninum neque vidisse astrum, nec si quando videret, ut fieri non potuit, quin aliquoties cœlum adspexerit, explorasse. Quæ apud populum astrorum cultorem extrema negligentia.'

(18) Ad Ecl. vi. 42. It is added that his theft of fire from heaven was only typical of his explanation of the nature of lightning. Cicero, De Div. i. 19, speaks of the Babylonians, 'Qui e Caucaso cœli signa servantes, numeris et motus stellarum cursusque persequuntur.' Why the Babylonians should observe from Caucasus, does not appear. See above, p. 73.

(19) Diod. i. 69; Diog. Laert. procem. § 11.

(20) Orig. iii. 24, 1.

(21) Phædr. § 134, p. 274. As to the claim of the Egyptians to be the inventors of letters, see Tac. Ann. xi. 24.

King of Egypt.(22) Diodorus, on the other hand, represents Osiris, with the help of Hermes, as having originated language, writing, music, the worship of the gods, and astronomy.(23) The same author informs us that the earliest Egyptian lawgiver was Mneues; and that the second was Sasychis, a man distinguished for his wisdom: that the latter made additional regulations concerning the worship of the gods, that he was the inventor of geometry, and that he taught astronomical science and observation to the natives.(24)

Herodotus attributes the origin of geometry, in its literal sense of *land-measuring*, to Egypt; this art having been rendered necessary by the alterations in the size and boundaries of fields, which the annual inundations of the Nile produced.(25) The same practical origin of geometry is assigned by Strabo, who remarks that the commercial wants of the Phœnicians caused them to invent arithmetic and astronomy.(26) A similar explanation of the origin of geometry in Egypt is given by Diodorus, who, however, couples with it the science of arithmetic.(27) Aristotle rationalizes the supposed Egyptian origin

---

(22) viii. § 11. The authority cited is Anticlides, in his work concerning Alexander the Great. Anticlides was an Athenian writer, who lived soon after the age of Alexander. Bunsen treats the invention of geometry by king Moeris as an historical fact, and declares that, 'like all the other fundamental institutions of Egyptian life, it belonged to the bloom of the old empire,' Egypt, vol. ii. p. 311, Eng. tr.

(23) Diod. i. 15, 16.

(24) i. 94. Compare Bunsen, vol. ii. p. 94, who says that Sasychis 'clearly belongs to the old empire.' Bunsen, ib. p. 65, states that the invention of astronomy was likewise attributed to king Maneros; but I am unable to find any authority for this statement. It is not supported by the passages which Bunsen cites. In Plut. de Is. et Os. 17, he is called the inventor of music, on account of the song Maneros, mentioned in Herod. ii. 79; Athen. xiv. p. 620 A. Compare Müller, Hist of Gr. Lit. c. 3, § 3. According to Pollux, iv. § 54, the Egyptians considered Maneros as the inventor of agriculture, and a disciple of the Muses.

(25) ii. 109. He supposes this art to have been invented in the reign of Sesostris.

(26) xvi. 2, 24; xvii. 1, 3. The same origin is repeated by Servius, ad Ecl. iii. 41. Socrates considered geometry as good only for land-measuring, Xen. Mem. iv. 7, 3; above, p. 112.

(27) i. 81.

of geometry in a different manner. He says that mathematical science originated in Egypt, on account of the leisure which the priests enjoyed for contemplation.(28)

According to Diodorus, the Egyptians claimed the honour of having taught astronomy to the Babylonians. Their account was that, in times of remote antiquity, they had sent out colonies over the whole known world; that Belus, the son of Neptune and Libya, led an Egyptian colony to Babylon; that the priests called Chaldæans were an imitation of the Egyptian priests, maintained at the public charge, and relieved from all labour, and that they, likewise after the model of the Egyptian priests, had become observers of the stars.(29) The author of the treatise on Astronomy (περὶ τῆς Ἀστρολογίης), written in the Ionic dialect, attributed to Lucian, holds that the Æthiopians were the true inventors of this science; that they taught it to the Egyptians; and that the Egyptians taught it to the Babylonians.(30) On the other hand, Josephus conceives that the Egyptians were indebted for their astronomical knowledge to the Chaldæans; for that the patriarch Abraham, who was a Chaldæan, taught arithmetic and astronomy to the Egyptians, and that from Egypt these sciences passed to the Greeks.(31) Proclus considered the astronomical observations of the Egyp-

---

(28) Metaph. i. 1: διὸ περὶ Αἴγυπτον αἱ μαθηματικαὶ πρῶτον τέχναι συνέστησαν· ἐκεῖ γὰρ ἀφείθη σχολάζειν τὸ τῶν ἱερέων γένος. The word συνέστησαν here means, 'assumed consistency and form,' 'were consolidated.'

(29) i. 28, 81.

(30) Lucian, de Astrolog. c. 3—9. The Scholiast on this passage denounces it as contrary to the received opinion of Greece, which assigned the invention of astronomy to the Chaldæans: παρὰ τὰ πᾶσιν ἀνθρώποις σχεδὸν δοκοῦντα ταῦτα· Χαλδαίους γάρ φασι πρώτους ἀστρονομίας ἄρξαι, vol. iv. p. 346, ed. Lehmann.

(31) Ant. Jud. i. 8, § 2. Josephus quotes a passage from Berosus (i. 7, § 2), where Abraham (without being named) is designated as a Chaldæan well-versed in astronomy. See Fragm. Hist. Gr. vol. ii. p. 502. In another place, Ant. Jud. i. 2, § 3, Josephus states that the descendants of Seth, son of Adam, invented Astronomy, and that they engraved their discoveries upon a pillar of brick and a pillar of stone, in order to preserve them from the general destruction which, according to the predictions of Adam, was to arise once from fire and once from water. These pillars remained in the historian's time in the land of Siris.

tians as anterior to those of the Greeks, and the astronomical observations of the Chaldæans as anterior to those of the Egyptians.(³²) Cedrenus, a Greek monk of the eleventh century, says that the Egyptians invented geometry, and the Chaldæans astrology; that the Babylonians were instructed in astronomy by Zoroaster, and that from them it passed to the Egyptians.(³³)

A mythological story told in Diodorus, represents Actis, son of Helius, king of Rhodes, as having taught astronomy to the Egyptians; and it accounts for the apparent superiority of the Egyptians over the Greeks, with respect to astronomical knowledge in later times, by the circumstance that Greece was submerged by a great deluge, which destroyed its ancient civilization, and which did not extend to Egypt.(³⁴)

Another version of the origins of arithmetic and astronomy, which we shall illustrate in a future chapter, assigned them to the Phœnicians.(³⁵)

§ 2 The Greeks were so much in the habit of finding a fabulous origin for every useful art and invention, that the stories which connected the original cultivation of astronomy with the Babylonians, the Egyptians, and the Phœnicians would not, of themselves, deserve more credit than the legends which made Amphion the inventor of music, and Triptolemus the inventor of agriculture.(³⁶) But these popular and uncertified

---

(³²) In Plat. Tim. p. 277 D, p. 671, ed. Schneider.

(³³) vol. i. p. 73, ed. Bonn. Cedrenus, an ignorant compiler, here confounds the Magi with the Chaldæans. Zoroaster is a Persian name. The confusion occurs in other late writers.

(³⁴) v. 57. See above, p. 72.

(³⁵) Below, ch. viii. § 1.

(³⁶) For a collection of such origins, see Plin. N. H. vii. 56; Clem. Alex. Strom. i. 16, § 74. Many Greek writers περὶ εὑρημάτων are enumerated in Fragm. Hist. Gr. vol. iv. p. 692. Compare Mure, Hist. of Gr. Lit. vol. iv. p. 55. 'It was a standard doctrine of the popular Greek antiquaries, that every art or custom, even the most elementary, and such as could hardly fail to spring up simultaneously with the first efforts of a nation to emerge from barbarism, must have had some inventor, or, what is nearly equivalent, some importer from abroad. When the custom was

traditions are corroborated by other statements of a more definite character.

Several of the ancient writers inform us that the Babylonians and Egyptians were in the possession of astronomical observations ascending to a period of remote and unknown antiquity.

Simplicius, an Aristotelian commentator of the sixth century, had, as he informs us, heard that the Egyptians had been in possession of written astronomical observations extending over a period of not less than 630,000 years; and the Babylonians, in like manner, of observations extending over a period of 1,440,000 years.([37]) Consistently with this account, the same writer speaks of astronomical observation having existed for many myriads of years.([38]) We learn, upon the authority of Porphyry, who wrote in the third century, that Callisthenes sent from Babylon to Aristotle a series of astronomical observations reaching back from the time of Alexander the Great over a space of 31,000 years.([39]) Pliny, in order to prove the remote antiquity of writing in Assyria, cites Epigenes, a weighty authority, as stating that the Babylonians were in possession of astronomical observations, for 720,000 years, inscribed upon baked bricks; he adds that Berosus and Critodemus, whose statement was the lowest, made the observations ascend 490,000 years.([40])

Cicero speaks of the written memorials of the Babylonians as including a period of 470,000 years; his context shows that he refers to astronomical observations; ([41]) and in another pas-

---

one of recognised remote antiquity, the title to priority was usually awarded to some mythical hero.' See likewise, p. 101.

(37) Schol. ad Aristot. de Cœl. p. 475 b, ed. Brandis.
(38) Ib. p. 494 b.
(39) Ap. Simplic. ib. p. 503 a.
(40) In the sentence of this passage, ' Litteras semper arbitror *Assyrias* fuisse,' we ought to read *Assyriis*. See Ideler, Chron. vol. i. p. 216. Concerning Epigenes and Critodemus, see Fragm. H. Gr. vol. ii. p. 510. Epigenes is cited only by late writers, and is probably not anterior to Polybius.
(41) De Div. i. 19, cited by Lactant. Div. Inst. vii. 14.

sage of the same treatise he speaks of the Babylonians asserting that their astrological system of divination, upon the births of children, was confirmed by an experience of 470,000 years.([42]) This statement is substantially identical with that of Diodorus, who says that the astronomical observations of the Babylonians reached back for 473,000 years from the landing of Alexander in Asia; ([43]) and it only differs by 20,000 years from that cited by Pliny from Berosus and Critodemus; which, when we are dealing with such high numbers, must be considered a trifling discrepancy.

We are informed that Hipparchus described the Assyrians as having continued their observations for 270,000 years. Iamblichus was not satisfied with this limited period. He asserted that they had observed entire cosmical cycles of the seven planets,([44]) the duration of which was still greater. Diodorus states generally, in reference to an event of 315 B.C., that the Chaldæans of Babylon claimed an antiquity of many myriads of years for their astronomical observations.([45])

The Egyptians, according to Diodorus, were more moderate in their assertions of antiquity than the Babylonians. They affirmed that their regal dynasties had lasted for 4700 years, and that the arts and sciences, including astronomy, had been cultivated in Egypt during that period.([46]) He states, however, that the Egyptians had preserved observations of the stars made during an 'incredible length of time.'([47]) Martianus Capella adopts a statement that astronomy had been practised in secret by the Egyptians for 40,000 years before it was divulged by them to the rest of the world.([48])

Diogenes Laertius informs us that, according to the report

([42]) De Div. ii. 46.   ([43]) ii. 31.

([44]) Ἀσσύριοι δέ, φησὶν Ἰάμβλιχος, οὐχ ἕπτα καὶ εἴκοσι μυριάδας ἐτῶν μόνας ἐτήρησαν, ὥς φησιν Ἵππαρχος, ἀλλὰ καὶ ὅλας ἀποκαταστάσεις καὶ περιόδους τῶν ἑπτὰ κοσμοκρατόρων μνήμῃ παρέδοσαν, Proclus in Tim. p. 31 c, p. 71, Schneider.

([45]) xix. 55.   ([46]) i. 69.
([47]) i. 81.   ([48]) viii. § 812, ed. Kopp.

of the Egyptians, 48,863 years elapsed from the time of Vulcan to that of Alexander the Great; and that during this period there had been 373 eclipses of the sun and 832 eclipses of the moon.([49]) The statement as to the eclipses is as fabulous as the rest; it has no claim to be considered as possessing any astronomical value, or as being the result of actual observation and of contemporary registration.

§ 3 It is stated, moreover, that whereas in other countries the persons who discharged sacerdotal duties were confounded with the rest of the community, and were subject to the duties and burdens of ordinary life; in Egypt, and also in Babylon, the priests formed a distinct caste, which enjoyed legal privileges and a high social position; that they were the companions of the kings; that they possessed an immunity from taxation; that they were relieved from toil, and had leisure for scientific study and meditation; and that from a remote period they habitually observed the stars, recorded their observations, and cultivated scientific astronomy and geometry.([50]) The Egyptian priests are moreover related to have kept registers, in which they entered notices of remarkable natural phenomena, such as the rains in Upper Egypt, which were supposed to cause the inundation of the Nile.([51]) The account of them given by Porphyry embodies the idea of monastic asceticism and purity, and a complete abstraction from secular pursuits.([52])

([49]) Procem. 2. The statement assumes that an eclipse of the sun took place once in every 131 years, and an eclipse of the moon once in every 58 years.

([50]) Respecting the Egyptian priests, see Strab. xvii. 1, § 3, 29, 46; Diod. i. 81; Herod. ii. 37; Porphyr. de Abst. iv. 6—8. Respecting the Chaldæan priests, see Diod. ii. 29—31. Manilius, i. 46, describes the Babylonian and Egyptian priests as having first practised astronomy.

([51]) Strab. xvii. 1, § 5. Compare the passage of Herodotus on the registration of prodigies by the Egyptians, above, p. 70, n. 275.

([52]) Antiphon, a late writer, cited by Porphyr. Vit. Pyth. 8, speaks of the severe regulations, unlike the habits of Grecian life, to which Pythagoras was subjected, in conforming to the life of the Egyptian priests. According to Eudoxus, such was their horror of blood, that they would not associate with butchers and hunters, Porph. Vit. Pyth. 7.

§ 4 Furthermore, it is affirmed by Herodotus that the year, with its division into twelve months, was the invention of the Egyptians. Each month, according to his report, consisted of thirty days; and five complementary days were added, so as to make the Egyptian year consist of 365 days.([53]) A notice in Syncellus states that the addition of the five complementary days was the work of King Aseth, who belonged to the fifteenth dynasty.([54]) According to Manetho, this dynasty reigned from 2607 to 2324 B.C.([55]) Manetho, however, attributed the reform of the Egyptian year to Saites, another king of the same dynasty; his reform is said to have consisted in adding twelve hours to the month, so as to make it of thirty days, and five days to the year, so as to make it of 365 days.([56]) According to Censorinus, a certain King Arminos fixed the year at twelve months and five days; ([57]) that is, at 365 days.

At the town of Acanthæ, near Memphis, was a perforated vessel, which was filled with water by 360 priests on each day in the year.([58]) This custom seems to allude to a year consisting only of 360 days.

Three hundred and sixty pitchers for making funeral libations were placed round the tomb of Osiris in the island of Philæ, which were filled every day by the priests with milk.([59]) This has been referred by Newton to a year of 360 days.([60]) The conjecture is probable; but the time when this religious usage was introduced is uncertain.

If we are to believe the account of the Latin Scholiast on the Aratea of Germanicus, the ancient Egyptian kings took an

---

(53) ii. 4. An ætiological mythus respecting the five complementary days of the Egyptian year is given by Plutarch, De Is. et Osir. 12.

(54) p. 123 D. Compare Lepsius, Chronol. der Ægypter, p. 177.

(55) See Boeckh, Manetho, p. 222, 390.

(56) Fragm. Hist. Gr. vol. ii. p. 570. Compare Boeckh, Manetho, p. 233.

(57) c. 19, where *duodecim* should be read for *tredecim*.

(58) Diod. i. 97.

(59) Diod. i. 22.

(60) See Bailly, Hist. d'Astr. Anc. p. 399.

oath, in the temple of Isis, that they would not add by intercalation to the ancient year of 365 days.(61)

The priests of Thebes declared that the tomb of King Osymandyas had once been decorated with a circle of gold, 365 cubits in circumference, and one cubit in thickness, on which each day of the year was represented. They stated it to have been removed by Cambyses when he invaded Egypt.(62) If this marvellous ornament ever had a real existence,(63) it supposed a year of 365 days.

From these statements it is apparent, that the views of the Greeks respecting the duration of the ancient Egyptian year cannot be reconciled with one another.

The duration ascribed by the Greeks to the year of the Babylonians, and of other oriental nations which are likely to have adopted the Babylonian reckoning, is also various. Thus the circuit of the walls of Babylon was 360 stadia, according to Ctesias, and 365 stadia according to Clitarchus and other historians of Alexander; which numbers were said to be derived from the days of the year.(64) In Persia, 365 youths formed a part of the ceremony of worshipping the sun.(65) According to Plutarch, the number of the concubines of Artaxerxes was 360.(66) The same number is stated by Curtius for the royal concubines at the time of Alexander's expedition;(67) and Diodorus remarks that it was not less than that of the days of the year.(68) It is stated by Herodotus that the annual tribute of the Cili-

---

(61) Deducitur autem a sacerdote Isidis in locum qui nominatur adytos, et jurejurando adigitur neque mensem neque diem intercalandum, quem in festum diem immutarent, sed ccclxv. dies peracturos, sicut institutum est ab antiquis, Arat. vol. ii. p. 71, ed. Buhle.

(62) Diod. i. 49.

(63) Letronne, Sur l'Objet des représent. Zodiac. p. 76, treats the golden circle of Osymandyas as an invention of the Egyptian priests, posterior to Alexander.

(64) Diod. ii. 7. For conjectures concerning the Babylonian year, see Ideler, Chron. vol. i. p. 202—213.

(65) Curt. iii. 7.   (66) Artax. 27.
(67) iii. 8.   (68) xvii. 77.

cians to the Persian king was 360 horses, being one for each day of the year.(69)

According to the account of Polyænus,(70) Hercules had a daughter in India named Pandæa. He gave her a district in the south of India, and distributed the inhabitants into 365 villages, with an obligation to pay the royal tribute in succession on each day of the year. Pliny(71) states that the Indian nation of Pandæ is governed by women, whose race ascends to a daughter of Hercules. They rule over 300 towns, 150,000 infantry, and 500 elephants. The mythology respecting Hercules and his daughter Pandæa is given at great length by Arrian.(72) He says nothing of the 365 villages, but states that Pandæa received from her father 500 elephants, 4000 cavalry, and 130,000 infantry.

§ 5 The astronomical science, which the Egyptians are alleged to have acquired, at a remote period, from the observations and meditations of their priests, and the regulation of the calendar, which this science enabled them to make, are stated to have been communicated to the Greeks. It is indeed intimated that the Egyptian priests regarded their astronomical science as an esoteric and mysterious doctrine, and that they disclosed it to curious strangers with reluctance;(73) but it is affirmed that, for the foundation certainly, and probably for many parts of the superstructure, of their astronomical science, the Greeks were indebted to Egypt. Similar statements are made with respect to Assyrian astronomy.(74) This derivation does not rest merely upon general declarations, but is fortified by detailed accounts of visits of Greek philosophers to Egypt, to Assyria, and to other oriental countries, made for the purpose of profiting by the lessons of the native priests and sages.

Thus Thales is reported to have visited Egypt, and to have

---

(69) iii. 90. See above, p. 18. (70) i. 3, § 4.
(71) vi. 23. (72) Ind. 8, 9.
(73) Strab. xvii. 1, § 29; Mart. Cap. viii. § 812.
(74) See Plat. Epinom. § 7, p. 987. After describing the origin of astronomy in Egypt and Assyria, the writer adds: ὅθεν καὶ πανταχόσε καὶ δεῦρ' ἐξήκει, βεβασανισμένα χρόνῳ μυριετεῖ τε καὶ ἀπείρῳ.

received instruction in astronomy from the priests.[75] Pherecydes of Syros is included among the early Greek philosophers who were the disciples of the Egyptians;[76] but, though he is said to have devoted his thoughts to physics, he was not a proficient in astronomy.

With respect to the lessons derived by Pythagoras from Egyptian and Asiatic teachers, the remark applies, which has been made of mediæval chroniclers, that each successive writer appears to know more of the transaction than his predecessor. The accounts given by the early Greek writers are few and meagre: those of the later writers are many and copious.

Herodotus states that the Orphic and Bacchic rites forbad the interment of a dead body in a woollen covering; and in this respect, he adds, they accord with the Egyptian and Pythagorean ceremonies, which contain a similar prohibition, supported by a sacred legend.[77] This passage proves that Herodotus conceived the superstitious ritual of the Pythagoreans of his time to be connected with Egypt. It is mentioned by Isocrates, that Pythagoras visited Egypt, and that, having become the disciple of the Egyptians, he introduced their philosophy into Greece.[78]

This is the entire information respecting the relations of Pythagoras with Egypt which we receive from writers anterior to the age of Alexander. The poet Callimachus, who lived in the reign of Ptolemy Philadelphus, described Pythagoras as

---

(75) Above, p. 80, 84; Röth, Gesch. der Abendländ. Philos. vol. ii. p. 95, and notes p. 4; Mullach, Frag. Phil. Gr. p. 203.

(76) Joseph. cont. Apion. 1, § 2. Cedrenus, vol. i. p. 165, ed. Bonn, says that he visited Egypt in order to learn theology and physics. Compare C. Müller, Fragm. Hist. Gr. vol. i. p. xxxiv., who discredits the account.

(77) ii. 81. Compare Apul. Apol. p. 495. Quippe lana, segnissimi corporis excrementum, pecori detracta, jam inde Orphei et Pythagoræ scitis, profanus vestitus est. There is a statement as to the doctrine of Pythagoras concerning the purity of linen, in Philostr. vit. Apollon. i. 32. According to Iamblich. vit. Pyth. 100, Pythagoras wore a linen dress.

Concerning a fabulous connexion of Zamolxis and Pythagoras, see Herod. iv. 95—6. Zamolxis was said to have learnt some astronomy from Pythagoras. Some of the Egyptian wisdom was likewise supposed to have penetrated into Thrace, Strab. vii. 3, 5.

(78) Busir. § 30, p. 227, ed. Bekker.

having invented some geometrical problems, and as having introduced others from Egypt into Greece.(79) Strabo informs us that Pythagoras visited Egypt and Babylon for scientific purposes;(80) and Cicero, that he visited Egypt and the Persian magi.(81) According to Diodorus, the Egyptian priests asserted that Pythagoras, Democritus, Œnopides of Chios, and Eudoxus had visited Egypt; and they produced proofs of the presence of these philosophers in that country. They likewise declared that the geometrical and mathematical science of Pythagoras had been derived from Egypt.(82) Justin repeats the account that Pythagoras visited first Egypt, and afterwards Babylon, in order to obtain instruction in astronomy.(83) Ammianus Marcellinus informs us that Pythagoras, Solon, Anaxagoras, and Plato derived instruction from Egypt.(84) The report of Valerius Maximus is similar. According to his version of the story, Pythagoras visited Egypt, and learnt the events of past ages from the registers of the priests. He then went to Persia, and learnt the motions of the stars, and their respective powers and properties, from the magi.(85) Philostratus, in his life of Apollonius, likewise speaks of Pythagoras as having held intercourse with the magi.(86)

Some late writers add another circumstance, that Pythagoras in his visit to Egypt learnt the language of the country, together with the three modes of writing.(87) According to Apuleius, the commonly received account of Pythagoras was, that he visited Egypt, and there learned magic ceremonies, arithmetic, and geometry, from the priests; that he afterwards visited the Chaldæans, and then the Brachmanes of India. The same writer also states that another account represented him to have

---

(79) Ap. Diod. x. 11, ed. Bekker.
(80) xiv. i. § 16.   (81) De Fin. v. 29.
(82) i. 96, 98.   (83) xx. 4.
(84) xxii. 16, § 21—22.
(85) viii. 7, ext. 2.   (86) i. 2.
(87) Antiphon, ap. Diog. Laert. viii. 3; Diogenes on Thule, ap. Porph. Pyth. 12. Concerning this Diogenes, see below, ch. viii. § 5.

been among the prisoners taken by Cambyses in Egypt, and that having been transferred to Persia, he became the disciple of the magi, especially of Zoroaster, the fountain-head of all sacred mysteries.([88]) This latter account makes the visits of Pythagoras to Egypt, and to Persia or Babylon, involuntary, and due only to the accident of his being a prisoner of war. Iamblichus describes his Egyptian visit as voluntary, but says that he was taken from Egypt to Babylon as a prisoner by Cambyses,([89]) and that he there enjoyed the instruction of the magi. The account of Iamblichus is, that Pythagoras left Samos at the age of eighteen, in the reign of Polycrates; that, after visiting some Greek philosophers, he went to Sidon, where he associated with the Syrian priests; that he next passed twenty-two years in Egypt; that he was transferred from Egypt to Babylon, where he passed twelve years; and that he returned to Samos at the age of fifty-six.

According to Porphyry, the received account of Pythagoras was, that he learnt geometry from the Egyptians, arithmetic from the Phœnicians, astronomy from the Chaldæans, and ascetic observances from the Magi.([90]) Polybius remarks, that the priests of the Egyptians, the Chaldæans, and the Magi were highly honoured by the early Greeks:([91]) this remark alludes to the stories of the travels of the ancient Greek philosophers.

Alexander Polyhistor, who lived in the last century before Christ, stated that Pythagoras was the disciple of Nazaratus the Assyrian, and that he also received lessons from the Brachmanes and the Gauls.([92]) By the Gauls he doubtless meant the Druids, who were conceived as sacerdotal philosophers.([93])

([88]) Flor. ii. 15. Suidas, in Πυθαγόρας Σάμιος, states that Pythagoras was the disciple of Abaris the Hyperborean, and of Zares the Magus, and that he likewise received instruction among the Egyptians and Chaldæans.

([89]) Iamblich. vit. Pyth. § 19.

([90]) Porph. vit. Pyth. 6. For the visit of Pythagoras to Sidon, and for his intercourse with the successors of Mochus, see Iambl. Pyth. § 14.

([91]) Ap. Strab. i. 2, 15 (p. 1120, ed. Bekker).

([92]) Clem. Alex. Strom. i. 15, § 70.

([93]) See Diog. Laert. procem. i.; Strab. iv. 4, 4.

According to Plutarch, Œnuphis of Heliopolis was the instructor of Pythagoras. According to Clemens it was Sonchis, the high priest.[94]

Dionysius states that Polycrates gave Pythagoras a letter of recommendation to Amasis;[95] which implies that Amasis was king of Egypt at the time of his visit. Pliny speaks of an obelisk set up at Rome by Augustus, which had been sculptured in the reign of Semneserteus; and he adds the remark, that during this king's reign Pythagoras was in Egypt.[96] No other mention is made of this king: the Egyptologists suggest that his name is an epithet of Amasis;[97] but Pliny evidently supposed it to be that of a king.

Œnopides of Chios, a Pythagorean philosopher, is likewise stated to have derived instruction respecting the sun's course, from Egypt.[98]

Empedocles is mentioned by Pliny and Philostratus as having held intercourse with the Persian magi.[99]

The residence of Democritus in Egypt, and his travels in Asia, have been previously mentioned;[100] he spoke of the Egyptian geometers, but considered them as his inferiors in scientific knowledge.

It is affirmed by Cicero, that Plato visited Egypt for the purpose of learning the science of numbers and astronomy from the priests:[101] this statement recurs in other Latin writers.[102]

---

(94) Plut. de Is. et Osir. 10; Clem. Alex. Strom. i. p. 131.

(95) viii. 3. This story is inconsistent with the account in Iamblichus vit. Pyth. 11, that Pythagoras escaped secretly from Samos, in order to avoid the despotism of Polycrates. Ovid, Met. xv. 60, makes Pythagoras fly from the tyranny of Polycrates.

(96) Plin. N. H. xxxvi. 14.

(97) Röth, Gesch. der Abendl. Philos. vol. ii. p. 69, notes.

(98) Above, p. 132.

(99) Plin. N. H. xxx. i.; Philostrat. vit. Apollon. i. 2.

(100) Above, p. 137.

(101) De Fin. v. 29. He went to Egypt 'discendi causâ,' de Rep. i. 10. His distant travels are alluded to in Tusc. Disp. iv. 19.

(102) See Val. Max. viii. 7, ext. 3; Lucan, x. 181; Quintilian, i. 12, § 15. Amm. Marc. above, n. 84.

According to Apuleius, Plato went to Theodorus at Cyrene, in order to learn geometry, and to Egypt in order to learn astronomy from the priests. He was prevented by wars in Asia from visiting the magi and the Indians.[103] This statement recurs in Diogenes, so far as concerns the visits to Cyrene and Egypt, and the intended visit to the magi.[104] His Egyptian visit, and his obligations to the priests, are also referred to by Philostratus.[105] Clemens describes him as having learnt geometry from the Egyptians, and astronomy from the Babylonians.[106] His Egyptian journey was likewise recognised by Strabo and Plutarch.

The residence of Eudoxus in Egypt appears to be sufficiently attested, but its duration is uncertain; nor is it probable that he accompanied Plato to that country.[107] It is not pretended that Aristotle, or any of the later philosophers, received instruction from the Egyptians, or from any oriental priests.

With respect to the alleged visits of the early Greek philosophers to Egypt, for the sake of profiting by the lessons of the native priests, we may remark, that they have no inherent improbability, so far as distance and difficulty of communication are concerned. A faint knowledge of Egypt is perceptible even in the poems of Homer; the country was opened to the Greeks in the reign of Psammetichus (671—17 B.C.), and a Greek factory was established at Naucratis in that of Amasis (560—26 B.C.). Solon is proved, by the extant remains of his poetry,

---

(103) De Dogm. Plat. i. p. 1040, Valpy.

(104) iii. § 6, 7.

(105) Vit. Apollon. i. 2.

(106) Protrept. c. 6, § 70. Compare Ast, Platons Leben, (Leipzig, 1816), p. 22—25. Above, p. 146; below, p. 279.

(107) See above, p. 145. A statement respecting a mystical doctrine of the Egyptian priests is cited from the Γῆς περίοδος of Eudoxus, by Plut. de Is. et Osir. 6. Concerning this work, see Ideler on Eudoxus, i. p. 200. See likewise above, n. 52.

to have sailed as far as Egypt.([108]) According to Plutarch, he there communed with two learned priests, Psenophis of Heliopolis, and Sonchis of Sais.([109]) The visit of Solon to Egypt is alluded to by Plato in the Timæus.([110]) The physical obstacles to communication with Egypt, even in the sixth century before Christ, were not very formidable.([111]) We could believe the fact, if it were sufficiently attested; but there is a scarcity of contemporary, or even of ancient evidence. The whole life of Pythagoras, as it is delivered to us, is a series of recent fables; and even Egyptologists admit that the visit of Plato to Egypt is doubtful.([112])

There is likewise a similarity in the names of the learned Egyptian priests, which betokens repetition and poverty of invention. Sonchis, the informant of Solon, is also a teacher of Pythagoras; another teacher of Pythagoras is Œnuphis; the teacher of Eudoxus is Chonuphis; that of Plato is Sechnuphis.([113])

The later Greeks appear to have been wanting in that national spirit which leads modern historians of science to contend for the claims of their own countrymen to inventions and discoveries. They rather sought to enhance the glory of their philosophers by claiming for them the merit of having introduced unknown sciences and useful inventions from foreign countries. Long journeys by land were at that time impeded

---

(108) He described himself as having made some stay

Νείλου ἐπὶ προχοῇσι, Κανωβίδος ἐγγύθεν ἀκτῆς.
—Plut. Sol. 26.

(109) Plut. ib. In the treatise, de Is. et Osir. 10, he likewise mentions Sonchis.

(110) § 5, p. 21. Compare Crit. § 7, p. 113.

(111) Alcæus mentioned in his poems that he had visited Egypt, Strab. i. 2, § 30.

(112) See Bunsen, Egypt, vol. i. p. 60, Eng. Tr.
The travels of the early Greek philosophers, and the profound science of the Egyptians and Chaldæans, are doubted by Ukert, Geogr. der Gr. und Röm. vol. i. 1, p. 51, i. 2, p. 90.

(113) See Lepsius, Chron. der Æg. p. 550.

by serious obstacles. The difficulty of the accomplishment, the ardour which it showed in the cause of science, and the love of marvel which exaggerated the excellences of unknown distance, probably furnished the motives for this tendency. Thus Lycurgus, the ancient lawgiver, is reported not only to have visited Crete, Asia Minor, and Egypt, in search of political models for his Spartan reforms, but even to have journeyed to Libya, Iberia, and India; some of which regions could scarcely have been known by name to his countrymen in the eighth or ninth century before Christ.([114])

We perceive, particularly in Diodorus, a disposition to trace everything Greek to Egypt.([115]) Thus he supposes Dædalus to have learnt his architectural skill in that country, and to have built the Cretan after the model of the Egyptian labyrinth.([116]) He likewise describes Orpheus as having derived his religious doctrines from Egypt.([117]) Even Homer, the great national poet, whom Diodorus was contented with taking to Egypt as a visitor,([118]) was, by the general consent of late writers, converted into an Egyptian.([119])

Much patriotic feeling existed in those days as to the comparative antiquity of nations. The Egyptians asserted their claims to be the most ancient of mankind; and their claims were partially admitted.([120]) This circumstance afforded an-

---

([114]) See Plut. Lyc. 4. Plutarch says that Aristocrates, the son of Hipparchus, is the only writer who carried him to the three latter countries. This Aristocrates was a Lacedæmonian, who lived in the first or second century B.C., and wrote a work entitled Λακωνικά, of which several fragments are extant. His authority respecting Lycurgus is at least as good as that of Iamblichus or Clemens respecting Pythagoras. See Frag. Hist. Gr. vol. iv. p. 332. Strabo likewise states that Lycurgus visited Egypt in order to study its laws, x. 4, § 19.

([115]) See Lepsius, Chron. der Æg. p. 41.

([116]) i. 61, 97. The same statement is made by Pliny, xxxvi. 19. Compare Höck's Kreta, vol. i. p. 56—68.

([117]) i. 69, iv. 25.   ([118]) i. 69.

([119]) Ὅμηρον οἱ πλεῖστοι Αἰγύπτιον φαίνουσιν, Clem. Alex. Strom. i. 15, § 6. Compare Heliodor. Æth. iii. 14.

([120]) See Herod. ii. 2; Plat. Tim. § 5, p. 22; Apollon. Rhod. iv. 265—270; Joseph. cont. Apion. i. § 2. Aristotle speaks of the great antiquity of the Egyptians: οὗτοι γὰρ ἀρχαιότατοι μὲν δοκοῦσιν εἶναι, Pol. vii. 10.

other ground for finding the origins of Greek philosophy in Egypt.

After the diffusion of Christianity, a further powerful motive was added for attributing the origins of Greek sciences to the Egyptians, the Phœnicians, the Persians, and the Babylonians. Like these nations, the Jews were *barbarians*; they belonged to a non-Hellenic race; and the Christian apologists sought to remove the prejudice of the Greeks and Romans against a foreign religion, by showing that the Greeks owed their philosophy to foreigners. The merely Jewish writers, such as Josephus, likewise shared in this feeling.

Lactantius expresses his wonder that Pythagoras and Plato, who visited the Egyptians and the magi of Persia in order to learn their sacred rites, should not have visited the Jews for this purpose.[121] This wonder was not shared by other writers, who supposed those philosophers to have drunk at the sacred Jewish fountain. Thus Eusebius states that Pythagoras visited Babylon, Egypt, and Persia, and conversed with the priests and magi: and that his visits were said to have fallen at the time when some of the Hebrews were residing in Egypt and Babylon.[122]

Josephus and Origen concur in stating that Pythagoras was indebted for much of his philosophy to the Jews;[123] and Theodoret, an eminent ecclesiastical writer of the fifth century, speaks of Pythagoras, Anaxagoras, and Plato, as having derived some sparks of the truth from Moses.[124] It is declared by Clemens that the dislike of anthropomorphism attributed to Numa was a belief which descended to him from Moses through the medium of the Pythagorean philosophy;[125] the connexion of

---

[121] Div. Inst. iv. 2.

[122] Præp. Evang. x. 4, p. 470.

[123] Joseph. contr. Apion. i. 22, and Origenes contr. Cels. i. 13. See Fragm. Hist. Gr. vol. iii. p. 36, 41. Compare Porph. vit. Pyth. 11, with Kiessling's note, vol. ii. p. 22.

[124] Græc. Affect. ii. 51, p. 79, ed. Gaisford. The metaphor is borrowed from Clem. Alex. Pæd. ii. 1, § 18.

[125] Strom. i. 15, § 71.

Pythagoras with the doctrines of Moses being not less fabulous than the connexion of Numa with the doctrines of Pythagoras.([126]) The same writer states that Plato learnt legislation and religion from the Hebrews;([127]) and that he was not ignorant of David;([128]) he even calls Plato 'the Hebrew-instructed philosopher.'([129]) Numenius the Pythagorean, who lived in the age of the Antonines, said that Plato was nothing but Moses in an Attic dress.([130]) Eusebius fully adopts the same view, and suggests that Plato may have acquired a knowledge of the Mosaic doctrines from the Jews who resided in Egypt under the Persian dominion.([131]) The same origin for the teaching both of Pythagoras and Plato was likewise claimed in the spurious treatise published under the name of the Jew Aristobulus.([132]) Augustine inclines to the opinion that Plato was acquainted with the Hebrew Scriptures; but he shows that the story of the philosopher having met Jeremiah in Egypt, and having there heard his prophecies, is inconsistent with chronology.([133])

§ 6 The true character both of the Babylonian and the Egyptian priests, as astronomers, seems to have been, that from an early period they had, induced by the clearness of their sky, and by their seclusion and leisure—perhaps likewise stimulated by some religious motive—been astronomical observers. Their observations were rude, and unassisted by

---

([126]) See the Author's Inquiry into the Cred. of the Early Rom. Hist. vol. i. p. 449; Schwegler, Röm. Gesch. vol. i. p. 560.

([127]) Protrept. c. 6, § 70.

([128]) Pæd. ii. 1, § 18.

([129]) ὁ ἐξ Ἑβραίων φιλόσοφος, Strom. i. 1, § 10.

([130]) τί γάρ ἐστι Πλάτων ἢ Μωυσῆς ἀττικίζων; the words of Numenius, cited in Clem. Strom. i. 32, § 150. Suidas in Νουμένιος, Euseb. Præp. Evang. xi. 10. Compare the art. *Numenius* in Dr. Smith's Dict. of Anc. Biogr.

([131]) Præp. Ev. xi. 8.

([132]) Clem. Strom. i. 22, § 150. Compare Valckenaer, Diatribe de Aristobulo, reprinted in the 4th vol. of Gaisford's edition of Eusebius Præp. Evang.

([133]) Civ. Dei, viii. 11, xi. 21.

instruments; and were, doubtless, but irregularly and imperfectly recorded; it may be reasonably suspected that they were directed particularly to phenomena, such as eclipses, to which a superstitious interest attached.([134]) We cannot, consistently with the capacity and tendencies of the Oriental mind, suppose that either of these nations ever rose to the conception of astronomy as a science; that they treated it with geometrical methods; or that they attempted to form a system of the universe founded upon an inductive, or even upon a speculative basis. The knowledge of geometry ascribed to the Egyptians seems merely to have grown out of their skill in land-measuring. All the extant evidence goes to prove that the scientific geometry of the Greeks was exclusively their own invention. It may be doubted whether any Chaldæan or Egyptian priest had a mind sufficiently trained in abstract reasoning to be able to follow the demonstrations of the properties of the conic sections invented by Apollonius.([135])

§ 7 One of the earliest subjects to which astronomical observation was directed must necessarily have been the determination of the solar year. It is highly probable that by observing the equinoxes and solstices,([136]) and the heliacal risings of particular stars, the Chaldæan and Egyptian priests may, at an early period, have approximated closely to its true

---

(134) Herodotus points out the peculiar addiction of the Egyptians to observances connected with the worship of their gods: θεοσεβέες δὲ περισσῶς ἐόντες μάλιστα πάντων ἀνθρώπων, ii. 37.

(135) Brugsch, Hist. d'Egypte, part i. p. 39, admits that the astronomy of the ancient Egyptian priests was not scientific. 'L'astronomie (he says) n'était pas chez eux cette science mathématique qui calcule les mouvements des astres, en construisant les grands systèmes qui composent la sphère céleste. C'était plutôt un recueil d'observations des phénomènes périodiques du ciel et du pays égyptien, dont le rapport réciproque ne pouvait échapper longtemps aux yeux des prêtres, qui observaient dans ces nuits claires de l'Egypte, les astérismes brillants du ciel. Leurs connaissances astronomiques étaient fondées sur la base de l'empirisme, et non sur celle d'une observation mathématique.'

(136) Geminus, c. 6, p. 19, states that the Egyptians particularly observed the solstices.

length, the most important fact with respect to the notation of time for chronological purposes.

Herodotus informs us that, in his time, the Egyptian year consisted of 360 days, to which five complementary days were added, making a year of 365 days.[137] He says nothing of any further correction. If, therefore, Herodotus is accurate in his statement, the Egyptian year in his time was less exact than the year of the Greek octaëteric cycle, which implied a knowledge of the odd quarter of a day.

Strabo states that Plato and Eudoxus journeyed together to Egypt, and passed thirteen years at Heliopolis, for the purpose of extracting the scientific knowledge of the priests; who communicated some of their discoveries, but concealed the chief part. Among the scientific truths which they imparted, was the excess of time beyond 365 days in the year. 'Up to this period (Strabo remarks) the true length of the year was unknown by the Greeks, together with many other truths; until the later astronomers obtained them from Greek translations of the writings of the Egyptian priests. They still continue to derive information from the writings of the Egyptian priests, as well as of the Chaldæans.'[138] This statement is inconsistent with all that we know respecting the progress of astronomy and the history of the calendar in Greece. The Metonic cycle, which was a reform of the previous octaëteric cycle, preceded the visit of Eudoxus to Egypt; and the octaëteric cycle assumed a year of $365\frac{1}{4}$ days. Hipparchus, moreover, who preceded Strabo by a century and a half, determined the length of the year with an exactitude unknown to the native Egyptian priests, and by methods which surpassed their comprehension. No mention is made elsewhere of scientific astronomical treatises by Egyptian priests, or of Greek translations of these treatises. Strabo must conceive them to be of later origin than the visit of Plato and Eudoxus; for he says that these philosophers consumed thirteen years at Heliopolis in Egypt, in a

---

(137) ii. 4. See above, p. 266.   (138) xvii. i. § 29.

partially successful attempt to induce the native priests to divulge their scientific knowledge: whereas, if this astronomical knowledge had been already consigned to written and published treatises, it would have been easy to find Greeks resident in Egypt who could interpret them into Greek.

It does not consist with this account, that the priests of Thebes are represented, both by Diodorus and Strabo, as being singular in using a year of 365¼ days; and in providing for the additional quarter of a day by an intercalary arrangement. It is particularly mentioned by Diodorus, that they determined their year by the sun, not by the moon.(139) In the time of Horapollo, the Egyptian year consisted of 365 days, and an additional day was intercalated every fourth year, as in the Julian calendar.(140)

Strabo admits that the Egyptian priests of his own day were destitute of all scientific and astronomical knowledge; and that they had degenerated into mere guides and *cicerones*, who showed the temples and their curiosities to foreigners. He adds that a certain priest named Chæremon, who accompanied the prefect Ælius Gallus from Alexandria in his ascent of the Nile, pretended to the possession of science, but that his pretensions were generally treated with ridicule and contempt.(141) Ælius Gallus was the intimate friend of Strabo: he was prefect of Egypt in the years twenty-four and twenty-five B.C.

Macrobius indeed declares that the year of the Egyptians had always been correct;(142) but the accounts of its duration are fluctuating,(143) and it is certain that, at the time of the Julian reform, the Egyptian calendar contained a year of 365 days; which, not being rectified by any intercalation, fell one day

---

(139) Diod. i. 50; Strab. xvii. 1, § 46.

(140) διὰ τετραετηρίδος περισσὴν ἡμέραν ἀριθμοῦσιν Αἰγύπτιοι· τὰ γὰρ τέσσαρα τέταρτα ἡμέραν ἀπαρτίζει, i. 5.

(141) xvii. 1, § 29. An attempt to defend the learning of Chæremon is made by Bunsen, Egypt, vol. i. p. 92—4, Engl. tr.

(142) Anni certus modus apud solos semper Ægyptios fuit, aliarum gentium dispari numero pari errore nutabat, Saturn. i. 12, § 2.

(143) See above, § 4.

behind the true time every four years, and was therefore a *vague* or movable year. As soon as the length of the true or fixed solar year was known, it must have been manifest, even to persons unversed in astronomical or mathematical science, that the movable year of 365 days and the fixed year of 365¼ days could not, after starting from the same point, again coincide, until a cycle of 4 × 365, that is of 1460 years, had been completed. This was a matter of simple computation, and was quite independent of astronomical observation. In this light it is viewed by Geminus in his astronomical treatise, who remarks that every Egyptian festival will circulate round the year in a cycle of 1460 years;[144] and Dio Cassius, in his account of the Julian reform of the calendar, appears to refer to the same fact.[145]

§ 8 As the movable Egyptian year was held to have originally begun at the heliacal rising of the Dog-star, which was contemporaneous with the ordinary commencement of the inundation of the Nile—this period was, by late writers, entitled the Canicular or Sothiac period—Sothis being the Egyptian name for the Dog-star, and the month Thoth being the first month of the Egyptian year.[146] Censorinus, writing in the year 238 A.D., remarks that the first of Thoth

---

(144) c. 6, p. 19.

(145) τοῦτο δ' ἐκ τῆς ἐν Ἀλεξανδρείᾳ διατριβῆς ἔλαβε, πλὴν καθ' ὅσον ἐκεῖνοι μὲν τριακονθημέρους τοὺς μῆνας λογίζονται, ἔπειτα ἐπὶ παντὶ τῷ ἔτει τὰς πέντε ἡμέρας ἐπάγουσιν, ὁ δὲ δὴ Καῖσαρ ἐς μῆνας ἑπτὰ ταύτας τε καὶ τὰς ἑτέρας δύο, ἃς ἑνὸς μηνὸς ἀφεῖλεν, ἐνήρμοσεν. τὴν μέντοι μίαν τὴν ἐκ τῶν τεταρτημορίων συμπληρουμένην διὰ πέντε καὶ αὐτὸς ἐτῶν ἐσήγαγεν, ὥστε μηδὲν ἔτι τὰς ὥρας αὐτῶν πλὴν ἐλαχίστου παραλλάττειν· ἐν γοῦν χιλίοις καὶ τετρακοσίοις καὶ ἑξήκοντα καὶ ἑνὶ ἔτει [ἔτεσι ?] μιᾶς ἄλλης ἡμέρας ἐμβολίμου δέονται, Dio Cass. xliii. 26. As this passage stands, Dio states that the period of 365¼ days falls short of the true year, and that it loses a day in 1461 years. In fact, however, this period is slightly in excess of the true year; and it was this excess which necessitated the Gregorian reform of the calendar. The probability is that Dio misunderstood some statement of the difference between the Egyptian year of 365 days and the Julian year of 365¼ days.

(146) See Censorin. c. 18, who considers the Canicular period as founded upon computation. For an account of the Canicular period, see Ideler, Chron. vol. i. p. 126—139; Boeckh, Manetho, p. 18; Bunsen, Egypt, vol. iii. p. 43, 73, Engl. tr.; Lepsius, Chron. der Æg. p. 165.

fell on the 25th of June; and he infers that this current year was the hundredth year of the Canicular cycle, because one hundred years earlier the first of Thoth fell on the 20th of July, which he assumes to be the time of the rising of the Dog-star.(147) His statement is founded on mere computation; it does not justify us in supposing that the Canicular period was used in practice, or that it differed from those astronomical cycles, of which so many varieties were devised by the ancient astronomers. Firmicus applies this cycle to a different purpose: he says that in 1461 years, the sun and moon, and the five planets, return to their original stations, and complete the great year of the universe.(148)

Syncellus cites a passage of an ancient chronicle, in which mention is made of a great cosmical period of 36,525 years; and this period is further divided by 25, so as to reduce it to periods of 1461 years.(149) The entire cycle, which is attributed to the Greeks as well as to the Egyptians, has evidently been formed by the multiplication of 1461 by 25. The 'ancient chronicle' cited by Syncellus, is referred by Boeckh to the interval between the third and fifth centuries;(150) the statement is made on the authority of astrological productions of late origin.(151) It is to be observed that neither this cycle, nor its subdivisions, are in any way connected with the rising of Sirius.

Others identified the great cosmical year with the life of the phœnix;(152) and hence, as we learn from Tacitus, some late

---

(147) Censorin. c. 21. Compare Ideler, ib. p. 127; Clinton, Fast. Rom. ad ann. 238, who rightly reads *Pii* for *Ulpii* in Censorinus.

(148) Quantis etiam conversionibus major ille, quem ferunt, perficeretur annus, qui quinque has stellas, lunam etiam ac solem, locis suis originibus que restituit, qui mille quadringentorum et sexaginta unius annorum circuitu terminatur, Præf. ad Astron.

(149) vol. i. p. 95—7, ed. Bonn, cf. p. 30, 64. The passage is rightly interpreted by Boeckh, Manetho, p. 46.

(150) See Boeckh, ib. p. 52—7.

(151) Concerning these productions, see Bredow, in Syncellus, ed. Bonn, vol. ii. p. 42. For the γενικὰ of Hermes, see Fab. Bibl. Gr. vol. i. p. 87, Harles.

(152) Plin. x. 2; Solin. c. 33, § 11—13; Horapoll. ii. 57.

writers (departing from the authority of Herodotus, who assigned a period of 500 years to the life of the phœnix), made that period consist of 1461 years.([153]) The idea of a great cosmical year, at the close of which the heavenly bodies return to their original stations, occurs in Plato, and is repeated by many subsequent authors.([154]) This fanciful notion seems to have been shared by the Chaldæan astronomers; for Berosus is reported to have declared, that when all the planets met in the sign of Cancer, the world would be submerged by a great deluge; and when they all met in the sign of Capricorn, it would be visited by a great conflagration.([155])

Theon of Alexandria, who lived in the fourth century, in an extant fragment upon the Rising of the Dog-star, reckons an era of Menophres as beginning in the year 1322 B.C.([156]) This year corresponds with the first year of the Canicular

---

([153]) Ann. vi. 28. Compare Herod. ii. 73, who is followed by numerous writers. Other authors, as Claudian and Lactantius, in their poems on the phœnix, double this period, and extend the life of the marvellous bird to 1000 years. See Lepsius, ib. p. 181. Pliny, x. 2, on the authority of Mamilius, a senator, reports that the life of the phœnix lasted for 540 years; and this number is confirmed by Solinus, c. 33, § 12. Lepsius, p. 170, proposes to read MCDLXI for DXL; but his conjecture is arbitrary and improbable. Suidas, in φοῖνιξ, makes the period 654 years. Chæremon, ap. Tzetz. Chil. v. 6, fixes it at 7006 years. Others, according to Solinus, made the life of the phœnix last for 12954 years, § 13, which is the long year of Cicero. Concerning the phœnix, see Horapollo, ed. Leemans, p. 242. A phœnix was exhibited at Rome, in the censorship of the Emperor Claudius, 800 U.C., and the fact was entered in the *Acta Urbis*, the official journal of Rome, Plin. x. 2; Solin. c. 33, § 14.

([154]) See Martin, Timée, tom. ii. p. 78—80; and the authors cited by him; Arago, Pop. Astr. b. 33, c. 43. Cicero, ap. Tac. de Caus. Corr. El. 16, assigns a duration of 12954 years to the great year. Macrob. Comm. Somn. Scip. ii. 11, § 11, fixes it at 1500 years. Sextus Empiricus, adv. Mathem. i. 105, p. 747, Bekker, at 9977 years. Achilles Tatius, c. 18, p. 81, defines the long years of the superior planets as follows: that of Saturn, at 350,635 years; that of Jupiter at 170,620 years; that of Mars at 120,000 years. Concerning the ἀποκατάστασις, or long year, see Horapollo, ii. 57. Ptolemy, Tetrabibl. i. 2, considers the hypothesis of a great year, at the end of which the heavenly bodies are restored to the places where they stood with regard to the earth at its beginning, as groundless and fanciful.

([155]) Sen. Nat. Quæst. iii. 29.

([156]) See Biot, Recherches sur plusieurs Points de l'Astronomie Egyptienne (Paris, 1823), p. 306.

period, whose termination is mentioned by Censorinus (1322 + 139=1461); and therefore it may be inferred that the era of Menophres was a period of 1461 years, calculated backwards from the year 139 A.D., in which the first of Thoth in the Egyptian movable year occupied its proper place at the rising of the Dog-star.

Columella transfers this principle of calculation to the sun's course as measured in days; he declares that a quadriennial period, consisting of 1461 days, brings the sun back to the place of the zodiac from which it commenced its course, and constitutes the great solar year.[157] This fanciful idea is founded on the intercalary period of the Julian calendar.[158]

The Canicular period is not mentioned by Herodotus, or Aristotle, or Aratus; or indeed by any writer anterior to the Christian era.[159] It is not even alluded to by Diodorus, who is so copious on Egyptian antiquities. It is an imaginary cycle, apparently of late origin, though founded on a simple computation. This computation rests exclusively upon a comparison of the year of $365\frac{1}{4}$ days with the year of 365 days; and has no necessary connexion with the rising of the Dog-star. The importance which has been attached to the Sothiac cycle

---

(157) De Re Rust. iii. 6.

(158) Columella quotes Virgil; he appears to have lived in the first half of the first century after Christ. His time is therefore subsequent to the Julian reform.

(159) See Lepsius, ib. p. 167. The treatise περὶ Σώθεως, attributed by Syncellus, vol. i. p. 72, to Manetho, is regarded as spurious by the modern critics. See Boeckh, Manetho, p. 14; Bunsen, Egypt, vol. i. p. 660, E. T.; Lepsius, Chron. der Æg. p. 413; Müller, Fragm. Hist. Gr. vol. ii. p. 512. Boeckh thinks that the dedication to Ptolemy Philadelphus is not earlier than the third century after Christ. There is nothing to show that it referred to the Canicular cycle. Syncellus speaks of the fifth year of Concharis, the twenty-fifth king of the sixteenth dynasty of the Canicular cycle, in Manetho, vol. i. p. 193, ed. Bonn. It is agreed by modern critics that these words are not to be taken as proving the mention of the Canicular cycle by Manetho. See Letronne in Biot, Recherches sur l'Année vague des Égyptiens, p. 570; Mém. de l'Acad. des. Inscr. 1836, vol. xii. part ii. p. 111; Boeckh, Manetho, p. 74—6, 229; Lepsius, Chron. der Æg. p. 419; Letronne, Mém. de l'Acad. xii. 2, p. 111 (1836), thinks that the Sothiac period was unknown to the ancient Egyptians; and that it was

by modern writers, and the computations based upon its supposed antiquity, seem to be equally groundless. What this cycle really attests is, the retention of a civil year of 365 days, after it was known that the true year was longer by a quarter of a day.

Syncellus speaks of Ptolemy having exhibited certain astronomical calculations for a period of 1476 years from the death of Alexander; in other words, by the restoration of one Egyptian year.([160]) This passage shows that Ptolemy regarded the period of 1476 years as a period obtained only by computation; and that he described it as formed by the readjustment of the Egyptian year of 365 days and the true solar year. The number, as reported by Syncellus, is not, however, the exact equation for the two years.

The alleged existence of a year of four months, of three months, of two months, of one month, and even of a single day, in Egypt,([161]) would probably be referred, by the champions of the hypothesis of profound astronomical science among the Egyptian priests, to a primitive period of ignorance, and would by this arbitrary supposition be reconciled with their theory. It is therefore unnecessary to insist on these discordant years, though they rest upon as good evidence as many of the received proofs of the profound science of the Egyptian priests.

§ 9 Diodorus dwells upon the advantages which the hereditary character of the Chaldæan priesthood afforded for the sober cultivation of science, in comparison with the rivalries and disputatious habits of the Greek philosophic sects; and he affirms that, for antiquity and carefulness of astronomical obser-

---

probably even unknown to Manetho, who never employs it: he remarks that it is only cited by Censorinus, Clemens of Alexandria, Chalcidius, and Syncellus; and he supposes that it may not improbably be an invention of the Alexandrine astrologers.

(160) Vol. i. p. 389, ed. Bonn.

(161) See above, p. 32. Boeckh, Manetho, p. 63—64. On the Egyptian year, see La Nauze, Acad. des B. L. xiv. p. 346; and the Memoir of Fourier in the French work on Egypt, tom. i. 803.

vation, the Chaldæans stood pre-eminent among the ancient nations. We are assured by Herodotus that the sundial, and the division of the day into twelve parts (afterwards called hours), for which the sundial afforded a measure, came to the Greeks from the Babylonians.[162] This rude and obvious measure of diurnal time cannot be considered as indicating any great proficiency in astronomy.

The Scholiast to Aratus states that the Greeks derived their tables of stars from the Egyptians and Chaldæans;[163] but what authority there was for this statement, or what was the value or extent of the tables alleged to have been received, we cannot now discover.

The story of the astronomical observations, extending over 31,000 years, sent from Babylon to Aristotle, would be a conclusive proof of the antiquity of the Chaldæan astronomy, if it were true.[164] But, as Delambre has remarked, it has all the appearance of a fable.[165] This remarkable fact was certainly not mentioned by Aristotle himself in any of his published works. If it had been, the mention of it would not have escaped the research of his diligent commentator Simplicius, who quotes it from Porphyry, a writer of the third century. We are required, moreover, to believe not only that Aristotle was silent as to this enormous mass of observations, but that they perished without leaving a trace. Not even Eudemus, or any other of the disciples and immediate successors of Aristotle, seems to have noticed them. Their existence was equally unknown to the astronomers of the Alexandrine School. The authentic reading of the passage exhibits the incredible number of 31,000 years; and this objection to the truth of the story cannot be removed by adopting an unauthorized reduction derived from a Latin translation. Other writers, indeed, who give the portentous numbers of 720,000, or 470,000, or 270,000

---

(162) Diod. ii. 29—31; above, p. 176.
(163) Ad Arat. 752. (164) Above, p. 263.
(165) Hist. Astr. Anc. vol. i. p. 308.

years, for the duration of the series of Babylonian observations, show that Porphyry was moderate in his statement.

That the Babylonian and Egyptian priests, enjoying the advantages of a clear and cloudless sky during a large portion of the year, had made astronomical observations, particularly upon the planets, before the Greeks had begun to observe the heavens systematically, need not be doubted. It is proved by the distinct testimony of Aristotle; who, however, seems to imply that the results of their observations reached the Greeks by means of oral communications.([166])

Eudoxus is said to have derived his knowledge of the motions of the planets from Egypt;([167]) and Conon to have made a collection of eclipses observed by the Egyptians.([168]) But whatever the efforts of the Egyptian priests may have been, they were useless to the scientific School of Alexandria, who must have had access to whatever existed, through the patronage of their Greek kings. Ptolemy never mentions an observation made by a native Egyptian.

Hipparchus may have assigned a duration of 270,000 years to the Babylonian observations, but those which he cited and used for purposes of calculation could not boast of a remote antiquity. The five earliest Babylonian eclipses mentioned by Hipparchus are of the years 721, 720, 621, and 523 B.C.; they are all eclipses of the moon.([169]) The earliest recorded Babylonian observation is fifty-five years subsequent to the commencement of the Olympiads, and thirty-two years subsequent to the foundation of Rome. These were the earliest observations of eclipses known to Hipparchus.([170]) Ptolemy states that a continuous series of observations was in existence in his time up to the commencement of the era of Nabonassar; that

---

(166) See above, p. 157, 162.
(167) Above, p. 156.
(168) Above, p. 196.
(169) See Ideler, Sternkunde der Chaldäer, Berl. Trans. 1814—5.
(170) See Ptolemy, Magn. Synt. iv. 5, 8, p. 243, 244, 269; v. 14, p. 340, 341, ed. Halma. Compare iv. 2, p. 216.

is, up to 747 B.C.(¹⁷¹) How far this series was complete or accurate, he does not inform us; but his words imply that it did not ascend to a higher date. The period of 223 lunations, or eighteen years, in which the eclipses of the moon recur, appears to be attributed by Geminus to the Chaldæans;(¹⁷²) but at what period they were first acquainted with it, is unknown. Ptolemy ascribes the period of eighteen years to the 'ancient mathematicians' generally, without specifying the Chaldæans.(¹⁷³)

The science of Egypt, like the wealth and power of Persia, was found by the Greeks to be a nullity, when it became the subject of certain knowledge and observation, and they were admitted behind the scenes. 'Minuit præsentia famam.' The Alexandrine astronomers discovered the barrenness of the scientific land of Egypt, as the Ten Thousand Greeks discovered the feebleness of the great king.(¹⁷⁴)

If any of the Chaldæan and Egyptian priests had really possessed the profound and exact knowledge of astronomy which is attributed to them in a body, we should probably have heard his name. Some one would have gained an individual reputation: his writings or his discourses would have become celebrated; and he would have been distinguished from the mass.(¹⁷⁵) Neither Plato nor Aristotle, nor any of the earlier writers, however, name any Babylonian or Egyptian, in refe-

---

(171) εἰς τὴν ἀρχὴν τοῦ Ναβονασσάρου βασιλείας, ... ἀφ' οὗ χρόνου καὶ τὰς παλαιὰς τηρήσεις ἔχομεν ὡς ἐπίπαν μέχρι δεῦρο διασωζομένας, Mag. Synt. iii. 6, p. 202, Halma. Compare Ideler, Chron. vol. i. p. 98; Delambre, Hist. Astr. Anc. tom. ii. p. 180.

(172) Gem. c. 15.

(173) Synt. iv. 2. Compare Ideler, Chron. vol. i. p. 206, and Goguet, Sur les Périodes astronomiques des Chaldéens, at the end of his Origines des Lois: also Delambre, Hist. Astr. Anc. tom. ii. p. 143.

(174) See Plut. Artax. 20. Archdeacon Hare remarks, Phil. Mus. vol. i. p. 56: 'It is very remarkable how little of our information about ancient Egypt is derived from the men of letters who lived at the Court of the Ptolemies.' The probable explanation of this fact is that they had no authentic information to give.

(175) Compare the remark of Mr. Grote, Hist. of Gr. vol. iii. p. 390.

rence to astronomical observations: the latter writers mention some names of Egyptian priests, as having imparted instruction to Pythagoras, Plato, and Eudoxus;([176]) but they are mere shadowy forms, without flesh and blood. The largest number of names is connected with Solon,([177]) who was not an astronomer or mathematician. In like manner we hear but few Babylonian names: Alexander Polyhistor assures us that Pythagoras received instruction at Babylon from Nazaratus the Assyrian.([178]) Cidenas, Naburianus, and Sudinas, Chaldæans, whom Strabo states to have been considered by the Greeks as eminent in science,([179]) are not mentioned elsewhere. Seleucus, who lived in the second century before Christ, and who wrote on physical subjects, was a Babylonian; but his training appears to have been Greek.([180])

Much mystical erudition has been bestowed upon the origin of the signs of the zodiac; but by the researches of Letronne and Ideler, the subject has been withdrawn from the transcendental region, and has been reduced within the bounds of intelligible knowledge. A huge frostwork edifice of fanciful conjectures has been melted by Letronne's determination of the date of the zodiac of Tentyra to the reign of Nero.([181]) It is

---

([176]) Sonchis and Œnuphis are called the teachers of Pythagoras; Chonuphis of Eudoxus; and Sechnuphis of Plato. The authorities are Plutarch, Diogenes Laertius, and Clemens. See Lepsius, Chron. der Æg. p. 550.

([177]) Besides Sonchis and Psenophis mentioned by Plutarch, Proclus (in Tim. p. 31 D), names Pateneit at Sais, Ochlaps at Heliopolis, and Ethemon at Sebennytus, referring to 'the histories of the Egyptians.'

([178]) See above, p. 271.

([179]) Strab. xvi. 1, § 6.

([180]) Above, p. 192.

([181]) See Letronne, Sur l'Origine grecque des Zodiaques prétendus Egyptiens. Paris, 1837.
Observations sur l'Objet des Représentations zodiacales qui nous restent de l'Antiquité. Paris, 1824. 8vo.
Recherches pour servir à l'Histoire d'Egypte. Paris, 1823. Introd. p. xiii.
Sur l'Origine du Zodiaque grec. Paris, 1840.
Ideler, Ueber den Ursprung des Thierkreises. Berl. Tr. 1838.
A collection of the separate dissertations and critiques of Letronne is

now admitted, even by Egyptologists, that this and the other zodiacal monuments of Egypt afford no proof of the early cultivation of astronomical science in that country.([182])

Achilles Tatius states that the Babylonian and Egyptian names of the constellations were different from the Greek. The Greeks, he remarks, borrowed many of these names from their own heroic mythology.([183]) He considers, however, the name of the balance, for the claw of the scorpion, to have been derived by the Greeks and Romans from the Egyptians.([184]) This substitution took place about the time of Cicero and Cæsar.([185])

The planets had, doubtless, been named by the Babylonians and the Egyptians, before they received names in Greece. The name of the *sun*, which was sometimes given to Saturn, was of Chaldæan origin.([186]) But the Greek designations, originally mere descriptive epithets, and afterwards borrowed from the native gods, appear to have been of indigenous invention.

The observation of the heavenly bodies, and particularly of eclipses, lay within the range of the obtuse, uninventive, and

---

much to be desired, and would be a valuable present to the student of Egyptian antiquity. See a list of them in Mém. de l'Inst. Acad. des Inscript. tom. xviii. p. 414 (1855). It seems that memoirs by this excellent writer on the Egyptian calendar and on the Julian year, still remain unpublished, ib. p. 417. A similar remark applies to the *Kleine Schriften* of Ideler; which are far superior in value to many collections of miscellaneous writings which have found an editor and a publisher in Germany. Compare Lepsius, Chron. der Æg. p. 4.

(182) See Lepsius, Chron. der Æg. p. 65—84; Brugsch, Hist. d'Egypte, part i. p. 40. The latter states that the zodiac is quite foreign to Egypt, and that its twelve divisions were introduced there by the Greeks in the Alexandrine age. The division of the zodiac is ascribed to the Egyptians in the spurious treatise of Lucian, de Astrol. 6.

(183) c. 39, p. 94.

(184) Fragm. p. 96.

(185) See Ideler, Unters. über die Astr. Beob. der Alten, p. 373; Sternnamen, p. 174; Schaubach, Gesch. der Astr. p. 296; Letronne, Sur l'Origine du Zodiaque grec, p. 20.

(186) Diod. ii. 30 (where the text ought not to be altered); Simplic. ad Aristot. p. 499, ed. Brandis; Hygin. Poet. Astr. iv. 18; Theo Smyrnæus, c. 6, with Martin's comment, p. 87. On the origin of the names of the planets, see Letronne, Orig. du Zod. grec, p. 30.

immovable intellect of an Oriental.([187]) So far as the astronomical observations of the Chaldæans and Egyptians were preserved in an authentic form, and came to the knowledge of the Greeks, the latter, doubtless, turned them to a profitable use. With the addition of their own observations, the Greek astronomers employed them as the foundation for the scientific structure which they erected by means of their own arithmetic and geometry. Hipparchus and Ptolemy may have made fanciful hypotheses for the explanation of the phenomena; but their reasoning was conducted in as scientific a spirit as that of Kepler, Newton, or Laplace. The Chaldæan and Egyptian observers had the merit of furnishing the raw material of observation, within certain limits, to the astronomers of Greece; but the latter alone possessed the scientific genius by which these undigested facts were converted into a symmetrical system.

In estimating the true character of the Egyptian astronomy, it is necessary to avoid all confusion between the Alexandrine School of Astronomers and the native priests. The former was purely Greek; and it was from Sosigenes, a member of this School, that Cæsar derived assistance for the reformation of the Roman calendar. When, therefore, Macrobius says that Cæsar received aid from 'the Egyptian School,'([188]) we must bear in mind that this was the aid of a Greek astronomer of Alexandria.

§ 10 But if the East could not give science to the West, it could give superstition; if it could not give astronomy, it could give astrology. Though it could not guide, it could pervert the human intellect; its soil, though incapable of producing plants fit for the food of man, could generate poisons.

([187]) Delambre, Hist. Astr. Anc. tom. ii. p. 149, considers the Chaldæan astronomers as observers, and not as calculators.

([188]) Saturn. i. 16, § 39.
Lucan, following the traditional idea of a profound Egyptian science, represents Achoreus, a native priest, as addressing a discourse to Cæsar upon cosmology and astronomy, but chiefly upon the cause of the annual inundation of the Nile, x. 194—331.

The authors to whom we are indebted for our accounts of the early Chaldæans are all of late date, and doubtless drew their ideas, to a great extent, from the astrologers under the Empire, or the last century of the Republic.([189]) But our informants are agreed in connecting the early astronomy of the Babylonians with divination from the stars, or astrology. Cicero, in a passage already adverted to, makes their astrological experience ascend to an antiquity of 270,000 years.([190]) The copious account of the Chaldæans given by Diodorus represents them as having from the beginning combined the two characters of diviners and astronomers. They are described as averting evil by rites of lustration, by sacrifices, and by incantations; and as predicting the future by the flight of birds, by the entrails of victims, and by the interpretation of dreams and prodigies. Belesys, described by Ctesias as a distinguished Chaldæan priest who assisted Arbaces the Mede in overthrowing Sardanapalus, about 876 B.C., is represented by him to have been versed in astronomy and in the art of divination, and to have predicted the fate of Sardanapalus by his own observations of the stars.([191]) The Chaldæans appear as soothsayers in the Book of Daniel. They are here coupled with magicians, exorcists, and interpreters of signs.([192]) So we learn from Arrian that the Chaldæans warned Alexander not to enter Babylon,

---

([189]) The word ἀστρολόγος signified an astronomer in the Greek writers anterior to the Christian era. The word *astrologus* has the same sense in the earlier Latin writers. In later times, the distinction which now obtains between the words astrology and astronomy was introduced. See Sextus Emp. ad Math. v. 2, p. 728, Bekker, Simplicius, Schol. Aristot. p. 348 b. Sextus was a writer of the third century. On the origin of the Chaldæan astrology, see Volney, Rech. sur l'Hist. Anc. p. 354 (Œuvres, ed. 1837).

([190]) Nam quod aiunt, quadringenta septuaginta millia annorum in periclitandis experiendisque pueris, quicunque essent nati, Babylonios posuisse, fallunt. De Div. ii. 46.

([191]) Diod. ii. 24, 25. The account of the Assyrian Empire is borrowed by Diodorus from Ctesias. See below, ch. vii. § 2.

([192]) See Dan. i. 4, 17, 20; ii. 2, 27; iii. 6; v. 7, 8, 11, 15. The 'astrologers' of the authorized version does not appear in De Wette's critical translation. See Gesenius, art. *Chaldæa*, in Ersch and Gruber, and Hengstenberg, Authentie des Daniel, p. 338—51.

upon the authority of an oracle which they had received from their god Belus;(193) their prediction, as reported by him, did not profess to be founded upon astronomical or astrological data.

Diodorus, however, proceeds to say that their astronomy had throughout an astrological cast. According to his account, they distinguished the five planets by the appellation of 'Interpreters,' because they foreshowed the destinies of men and nations, and presided over the birth of each individual. They conceived that besides the planets there were thirty stars, which were called 'the Consulting Gods;' twelve of which presided over the signs of the zodiac and the months of the year. Below these are twenty-four stars, of which half are to the north, and visible; half to the south, and invisible: the visible are assigned to the living, the invisible to the dead; and they are denominated the Judges of the World.(194)

Conformably with this view of the Chaldæans, the prophet Isaiah, who lived about 759—710 B.C., addressing Babylon, the 'daughter of the Chaldæans,' says: 'Thou art wearied in the multitude of thy counsels. Let now the astrologers, the stargazers, the monthly prognosticators, stand up and save thee from these things that shall come upon thee.'(195)

---

(193) Arrian, Anab. vii. 16, § 5. Diodorus, however, states that the prediction which the Chaldæan deputation communicated to Alexander on this occasion was derived from their astrological science, xvii. 112. Compare xix. 55. Plutarch, Alex. 73, merely mentions the prediction, without specifying its foundation. An Assyrian magician is mentioned by Theocrit. Id. ii. 162.

(194) Diod. ii. 29—31. In c. 30, Letronne conjectures ὑπὲρ δὲ τὴν τούτων φοράν: De l'Orig. du Zodiac grec, p. 34, but the received reading seems right. The Chaldæans are represented as placing the thirty stars next under the five planets, after them the twenty-four judicial stars, and lowest of all the moon. Letronne likewise alters 30 into 36; a proceeding which seems inadmissible. The passage respecting the twelve gods who preside over the zodiac, presents a difficulty. The words are: τῶν θεῶν δὲ τούτων [viz., the βουλαῖοι θεοί] κυρίους εἶναι φασι δώδεκα τὸν ἀριθμὸν, ὧν ἑκάστῳ μῆνα καὶ τῶν δώδεκα λεγομένων ζῳδίων ἓν προσνέμουσι. Letronne translates 'Ces trente-six dieux conseillers ont pour maîtres douze dieux supérieurs, à chacun desquels est departi un signe du zodiaque et un mois.' The true meaning seems to be: 'Of these gods, twelve are supreme, to each of whom one month and one of the twelve zodiacal signs is assigned.'

(195) xlvii. 13. This chapter belongs to that portion of Isaiah

Clitarchus, cited by Diogenes Laertius, states that the Chaldæans occupied themselves about astronomy and prediction.([196]) The account of Strabo is peculiar: he affirms that the chief body of the Chaldæan philosophers were addicted to astronomy; but that a section of them, whom the others repudiated, practised the genethliac art.([197]) Chaldæa is regarded by Ammianus Marcellinus as the cradle of the true vatication, by which he appears to mean astrology.([198])

In the period anterior to Alexander, the fame of the Chaldæans as soothsayers does not seem to have penetrated into Western Asia; when the Lydians, before the capture of Sardes in 546 B.C., were desirous of obtaining the interpretation of a prodigy, they sent for the Telmessians in Lycia.([199]) The Telmessian diviners are connected with the story of the Gordian Knot: they are stated by Arrian to have been a hereditary caste of expounders of prodigies.([200]) A lost comedy of Aristophanes, entitled the *Telmessians*, appears to have alluded to their character of diviners and augurs.([201]) The city of Telmessus was said to have received from Apollo the gift of teratoscopy.([202]) According to Clemens, the Telmessians invented divination from dreams: according to Tatian, they invented augury by the flight of birds; both these writers agree in declaring that the Carians were the inventors of prediction by the stars.([203])

(xl.—xlvi.) which is considered, by the modern German biblical critics, as the work of a different writer. See De Wette, Einleitung in das A. T., § 208, ed. 5. He is believed to have been coeval with Cyrus, 559 B.C., that is about two centuries after Isaiah. Cyrus is mentioned by name in chapters xliv. and xlv.

(196) Procem. 6.
(197) Strab. xvi. 1, § 6.
(198) Hic prope Chaldæorum est regio altrix philosophiæ veteris, ut memorant ipsi, apud quos veridica vaticinandi fides eluxit, xxiii. 6, § 25.
(199) Herod. i. 78; cf. 84.
(200) Arrian, Anab. ii. 3, 4.
(201) See Dindorf. Aristoph. Frag. p. 186. Cicero de Div. i. 41, speaks of Telmessus as a city, 'in quâ excellit haruspicum disciplina.'
(202) Suidas in Τελμισσεῖς; Eustathius ad Dionys. Perieg. 859.
(203) Clem. Alex. Strom. i. 16, § 74; Tatian, adv. Græc. c. 1.

The earliest allusion in Greek literature to divination from the stars occurs in the Timæus of Plato.[204] Eudoxus became acquainted with the genethliacal astrology of the Chaldæans, and warned his countrymen not to place faith in it;[205] and Theophrastus (though in his character of the Superstitious Man[206] he makes no mention of astrology among the superstitions prevalent at Athens in his time) was, like Eudoxus, aware of its existence as a foreign art practised by the Chaldæars.[207] This mode of divination had not, however, hitherto penetrated into Greece. The anecdote, reported by Gellius, of a Chaldæan prophecy respecting the successes of Euripides, founded on his birth,[208] is the fiction of a time

---

[204] Speaking of the movements of the planets, Plato says : φόβους καὶ σημεῖα τῶν μετὰ ταῦτα γενησομένων τοῖς δυναμένοις λογίζεσθαι πέμπουσι, p. 40 D. These words cannot be held to refer to mere meteorological prediction.

[205] Cic. de Div. ii. 42. See above, p. 158.

[206] Char. xvi.

[207] θαυμασιωτάτην δὲ εἶναί φησιν ὁ Θεόφραστος ἐν τοῖς κατ' αὐτὸν χρόνοις τὴν τῶν Χαλδαίων περὶ ταῦτα θεωρίαν, τά τε ἄλλα προλέγουσαν καὶ τοὺς βίους ἑκάστων καὶ τοὺς θανάτους, καὶ οὐ τὰ κοινὰ μόνον, οἷον χειμῶνας καὶ εὐδίας, ὥσπερ καὶ τὸν ἀστέρα τοῦ Ἑρμοῦ χειμῶνος μὲν ἐμφανῆ γενόμενον ψύχη σημαίνειν, καύματα δὲ θέρους, εἰς ἐκείνους ἀναπέμπει· πάντα δ' οὖν αὐτοὺς καὶ τὰ ἴδια καὶ τὰ κοινὰ προγινώσκειν ἀπὸ τῶν οὐρανίων ἐν τῇ περὶ Σημείων βίβλῳ φησὶν ἐκεῖνος, Proclus ad Tim. p. 285 F.
Concerning the work περὶ σημείων, see Theophrast. vol. i. p. 782, ed. Schneider.

[208] Gell. xv. 20, 1.
Gellius, xiv. 1, reports a discourse against the Chaldæan astrologers, which he heard delivered at Rome by the philosopher Favorinus. In this discourse, Favorinus said : 'Disciplinam istam Chaldæorum tantæ vetustatis non esse, quantæ videri volunt, neque eos principes ejus auctoresque esse, quos ipsi ferant; sed id præstigiarum atque offuciarum genus commentos esse homines æruscatores, et cibum quæstumque ex mendaciis captantes.' Simplicius, ad Aristot. Phys. p. 348 b, ed. Brandis, says, that anciently astronomy was called by the Greeks ἀστρολογία, before the introduction of the apotelesmatic art.

In the following passage of Propertius, the poet seems to trace astrology to an astronomical source; but not to imply that the Pythagorean Archytas and the Alexandrine Conon were astrologers :—

> Certa feram certis auctoribus; aut ego vates
> Nescius ærata signa movere pilâ,
> Me creat Archytæ soboles Babylonius Horos,
> Horos, et a proavo ducta Conone domus.
>   iv. 1, 75—8.

The *ærata pila* is a celestial globe. On the reputation of Archytas

when all genethliac predictions were astrological. The Greek tragedians, who in several places mention magic, (209) never allude to astrology; and Eudoxus, who is reported to have held that the magi were the most illustrious and useful of the philosophic sects, (210) saw the vanity of the astrological craft. An astrological treatise, which was attributed to Helicon of Cyzicus, the contemporary of Plato, was probably a spurious production. (211)

The Chaldæans are stated by Diodorus to have foretold the fate of Antigonus, in 315 B.C. (212) The prediction made a deep impression upon his mind, on account of the astronomical science of the Chaldæans, and of the antiquity of their celestial observations. It was not founded on his nativity, and it had nothing in common with the genethliac predictions of the later Chaldæans.

§ 11 The introduction of astrology into Greece is stated by Vitruvius to have been due to Berosus, who settled in Cos, and opened a school in that island. Vitruvius adds, that Berosus was succeeded by Antipater and Achinapolus, the latter of whom derived prognostics from the time of conception instead of the time of birth. (213)

Berosus the Chaldæan was contemporary with Alexander the Great, but survived him, and lived to dedicate his Babylonian

---

in Italy as a geometer and astronomer, see Horat. Carm. i. 28. The name Horos is borrowed from Egypt.

(209) Thus Œdipus, in Sophocles, calls Tiresias a magician, Œd. T. 387. Iphigenia is described by Euripides as using magic rites, Iph. T. 1335. Compare Orest. 1476; Suppl. 1110.

(210) Plin. xxx. 1. (211) Above, p. 147, n. 27.

(212) xix. 55.

The cheat attempted to be practised on Seleucus Nicator by the magi, with respect to the hour for laying the foundation of the town of Seleucia on the Tigris (Appian, Syr. 58), does not appear to be connected with astrology; though it is so regarded by Letronne, Obs. sur l'Objet des Représ. zodiac. p. 80. A genethliac astrologer did not fix the time for the foundation of a city: though treating the foundation of a city as equivalent to its birth, he might argue from the time of its foundation, and calculate its fortunes. The exact date of the foundation of Seleucia on the Tigris is not known. The reign of Seleucus extends from 306 to 280 B.C.

(213) ix. 7.

history to Antiochus Soter, who began to reign in 280 B.C.(²¹⁴) A medical school was attached to the temple of Æsculapius at Cos, the birthplace of Hippocrates; and it is not improbable that this seminary of physical science may have attracted Berosus to the island. He appears to have subsequently removed to Athens; for Pliny states the fame of his predictions to have been such that the Athenians erected to him a statue with a gilt tongue in their Gymnasium.(²¹⁵) Of Antipater and Achinapol is nothing is known beyond what is stated by Vitruvius. Their names are of Greek forms, and do not lead to the supposition that they were Chaldæans.

We are informed by Pliny, that Berosus fixed the limit of human life at 116 years, in reference to astrological science.(²¹⁶)

Berosus was an astronomer, as well as an astrologer; he is said by Josephus to have written in Greek concerning the astronomy and philosophy of the Chaldæans;(²¹⁷) and it is apparently in reference to his writings on astronomy that Seneca designates him as the translator of Belus.(²¹⁸) Several astronomical doctrines are ascribed to him. Thus he taught that the moon is a sphere, half of which is igneous; that its light is not derived from the sun; that it has three movements, one in longitude, one in latitude, but irregular, like the movements of the planets; and a third in an orbit; that its phases are produced by the alternate conversion of the luminous and opaque sides of the moon to the earth; and that these conversions always coincide with the moon's conjunction with the sun; and that an eclipse of the moon likewise takes place when the non-ignited part is turned to the earth.(²¹⁹)

---

(214) C. Müller, Frag. Hist. Gr. vol. ii. p. 495; Clinton, F. H. vol. iii. ad ann. 279; Ideler, Chron. vol. ii. p. 599. The lifetime of Berosus may be conjecturally fixed at 340—270 B.C. See Letronne, Orig. du Zod. grec, p. 54.
(215) vii. 37.
(216) vii. 50.
(217) Contr. Apion. i. 19.
(218) Nat. Quæst. iii. 29. Berosus, qui Belum interpretatus est. See above, p. 258, n. 14.
(219) Vitruvius, ix. 2, who is very copious on the lunar doctrine of Berosus, Cleomed. ii. 4. Stob. Ecl. Phys. i. 26; Euseb. Præp. Ev. xv. 51;

§ 12 From the time of Berosus, astrology gradually gained ground among the Greeks, and passed into general acceptance as one of the modes of divination. Panætius, who lived in the second century B.C., mentioned that Archelaus and Cassander, two contemporary astronomers, were singular in rejecting this pseudo-science.[220] It was particularly cultivated by Posidonius, and the Stoic philosophers of that age.[221] From Greece the art of astrology passed to Italy and Rome. Cato the Elder, in his treatise on agriculture, lays it down that the *villicus*, or overseer of a farm, ought not to consult any augur, soothsayer, or Chaldæan.[222] Astrology was regarded by the Roman State as a foreign and unauthorized superstition; and as early as 139 B.C., before the time of the Gracchi, Cneius Cornelius Hispallus, the prætor peregrinus, issued an edict expelling the Chaldæans from Rome and Italy. The same functionary expelled the Jews from Rome, on the ground that they had endeavoured to introduce the worship of Jupiter Sabazius.[223] This prohibition, however, did not long remain in force. Cneius Octavius, who was slain in the time of Marius, had a

---

Plut. Plac. Phil. ii. 29; Galen, Phil. Hist. c. 15. This doctrine concerning the eclipses or phases of the moon is called a Babylonian and Chaldæan doctrine by Lucretius:—

Ut Babylonica Chaldæûm doctrina, refutans
Astrologorum artem, contra convincere tendit.
v. 719—727.

Apuleius, De Deo Socrat. ad init., says of the moon : Sive illa proprio seu perpeti candore, ut Chaldæi arbitrantur, parte luminis compos, parte alterâ cassa fulgoris.

Diodorus, on the other hand, states the Chaldæan doctrine to be like that of the Greeks; namely, that the moon shines with borrowed light, and that lunar eclipses are caused by the shadow of the earth, ii. 31.

(220) Cic. de Div. ii. 42.
(221) See Augustin. C. Dei, v. 2.
(222) Haruspicem, augurem, hariolum, Chaldæum, ne quem consuluisse velit, De R. R. c. 5, § 4. A passage from Ennius, in which diviners are treated with contempt, is cited by Cic. de Div. i. 58; but the words 'de circo astrologos' appear to belong to Cicero; see Ribbeck, Trag. Lat. Rel. p. 44.
(223) Val. Max. i. 3, 2. The worship of Sabazius was Phrygian, and it was celebrated with nocturnal orgies, like the Bacchanalia at Rome. This Hispallus is mentioned by Appian, Lib. 80.

Chaldaic prediction upon his person at the time of his death;(224) and Sylla received several prophecies from the Chaldæans.(225) L. Tarutius Firmanus, the friend of Varro, was versed in astrology, and calculated the nativity of Rome;(226) and Cicero speaks of many predictions given by the Chaldæans to Pompey, Crassus, and Cæsar, which were never accomplished.(227)

Nigidius Figulus was distinguished as an astrologer in the later part of the Republic:(228) he is said to have foretold the future greatness of Augustus.(229) The latter, when he was at Apollonia, likewise consulted an astrologer named Theogenes, and obtained from him a highly favourable prediction.(230)

The astrologers were generally known to the Romans by the national name of Chaldæans.(231) Thus, by degrees, the name Chaldæan lost its national signification, and came to

---

(224) Plutarch, Mar. 42.

(225) Plutarch, Sylla, 5, 37.

(226) See the Author's Inquiry into the Cred. of the Early Rom. Hist. vol. i. p. 393. Cicero, De Div. ii. 47, describes him as 'in primis Chaldaicis rationibus eruditus.'

(227) De Div. ii. 47. The predictions of the Chaldæans respecting length of life are alluded to by Cic. Tusc. i. 40; Horat. Carm. i. 11.

(228) Dio Cass. xlv. 1.; Augustin. C. D. v. 3. Lucan introduces him as predicting the civil war:—

> At Figulus, cui cura deos, secretaque cœli
> Nosse fuit, quem non stellarum Ægyptia Memphis
> Æquaret visu, numerisque moventibus astra.—i. 639—41.

A Greek translation of a tonitrual Diary of Figulus—in which the prophetic meaning of lightnings is explained for each day of the year, beginning with June, and ending with May—is given in Lydus de Ostentis, c. 27, p. 306—331, ed. Bonn. The contents resemble those of the astrological almanacs of this and other countries.

(229) Dio, ib. et lvi. 25; Suet. Aug. 94. Augustus likewise consulted two diviners, probably astrologers, named Areius and Athenodorus, Dio Cass. lii. 36.

(230) Suet. ib.

(231) Compare Juvenal, vi. 553—6:—

> Chaldæis sed major erit fiducia: quicquid
> Dixerit astrologus, credent a fonte relatum
> Hammonis; quoniam Delphis oracula cessant,
> Et genus humanum damnat caligo futuri.

Censorinus, c. 8, speaks of astrology as an exclusively Chaldæan art:

denote an astrologer, even though he might not ; of Babylonian extraction.([232]) The majority of them, indeed, appear to have been Greeks, or at least to have borne Greek names.([233]) They rose to the zenith of their influence in the early part of the Empire. They received the countenance of the powerful,([234]) including even the emperors; nevertheless their craft was always considered illicit, and edicts banishing them from Italy were from time to time issued.([235]) In later times also, astrology, though condemned by the Church, and regarded as impious,([236]) was upheld by the European princes, who culti-

---

Bardesanes, ap. Euseb. Præp. Evang. vi. 10, vol. ii. p. 85, calls the astrologers οἱ Χαλδαΐζοντες.

> Te Persæ cecinere magi, te sensit Etruscus
> Augur, et inspectis Babylonius horruit astris,
> Chaldæi stupuere senes.
> Claudian de Quart. Cons. Honor. v. 145—7.

(232) Hence Cicero, in speaking of the ancient Chaldæans of Babylon, thinks it necessary to say: 'Non ex artis, sed ex gentis vocabulo nominati,' De Div. i. 1.

(233) Theogenes, an astrologer consulted by Augustus, was a citizen of Apollonia. Thrasyllus, the astrologer of Tiberius, was a Phliasian; see Martin, ad Theon. p. 69. His son was consulted by Nero, Tac. Ann. vi. 22. Ptolemæus, consulted by Otho (Tac. Hist. i. 22; Plut. Galb. 23), and Seleucus, consulted by Vespasian (Tac. Hist. ii. 78) appear from their names to be Greeks. Sueton. Oth. 4, confounds Ptolemæus with Seleucus. Heliodorus, an astrologer, employed by the emperor Valens as an instrument of his cruelties (Ammian. Marcellin. xxix. 2, § 6) is another Greek name. On the other hand, the name of Barbillus, much trusted by Vespasian (Dio Cass. lxvi. 9) is Roman. See above, p. 297.

(234) Juvenal, iii. 42, enumerating the dishonest arts, an ignorance of which disqualifies a person from making a livelihood at Rome, says: Motus astrorum ignoro.

(235) Tacitus speaks of the astrologers as genus hominum potentibus infidum, sperantibus fallax, quod in civitate nostrâ et vetabitur semper et retinebitur, Hist. i. 22. According to Juvenal, vi. 560—4, the frequent punishments of the astrologers, and the dangers which they incurred, increased the confidence of their dupes. Concerning the legal proscription of the astrologers under the Empire, see Becker, Handbuch der Röm. Alt. vol. iv. p. 100—102. Tiberius was the first emperor who expelled the mathematici, Suet. Tib. 36. See the advice of Mæcenas to Augustus, Dio Cass. lii. 36.

(236) Astrology was condemned by the Christian fathers and by the Church, Gothofred. ad Cod. Theodos. vol. iii. p. 146; Bingham's Antiq. of the Christ. Church, xvi. 5, 1. Astrologers are declared unworthy of baptism in the Apostolic Constitutions, viii. 32; Bingham, ib. ii. 5, 8. Com-

vated its professors, and consulted them with respect to future events.

§ 13 When the Chaldæan astrology travelled westward, it was introduced into Egypt; but it seems to have been an exotic in that country, not less than in Greece. There is indeed a passage of Herodotus, which has been considered as proving the early prevalence of astrology among the Egyptians. This historian remarks it as a peculiar practice of the Egyptians to regard each month and day as sacred to some deity; and from the day of a man's nativity to determine his fortunes, character, and mode of death. He adds that this idea had been borrowed by some Greek writers of poetry.(237) There is no doubt that prediction from births is characteristic of the Chaldæan astrology, and so far there is a resemblance between astrology and the Egyptian mode of divination described by Herodotus. But Herodotus makes no allusion to any connexion with the stars in the Egyptian genethlialogy of his time; nor is it likely that the Greek writers of poetry referred to (whom Lobeck conjectures to have been poets of the Orphic and Pythagorean School)(238) attempted to initiate their countrymen in any apotelesmatic mysteries.

After a time, however, the Egyptians appropriated this superstitious craft;(239) and among the numerous forgeries of

---

pare Tertullian, Apol. 35. Suidas, in Διόδωρος, cites an account of a bishop of Tarsus who wrote against astronomers and astrologers in the reigns of Julian and Valens. His treatise was entitled, Κατὰ Ἀστρονόμων καὶ Ἀστρολόγων καὶ Εἱμαρμένης. Astrology is treated as unchristian by Augustine, Conf. iv. 3. Isidorus proscribes it on account of its fatalist tendency, Orig. iii. 70, § 39, 40.

(237) ii. 82. The expression, for poets, is singular: οἱ ἐν ποιήσει γενόμενοι.

(238) Aglaopham. p. 427.

(239) Flavius Vopiscus, in his life of Saturninus, c. 7, gives the following character of the Egyptians: 'Sunt enim Ægyptii viri ventosi, furibundi, jactantes, injuriosi, atque adeo vani, liberi, novarum rerum, usque ad cantilenas publicas, cupientes, versificatores, epigrammatarii, *mathematici, haruspices*, medici.' This description doubtless applies particularly to that part of the population which spoke Greek. Vopiscus wrote between 290 and 300 A.D. Gibbon remarks that 'the people of Alexandria, a various mixture of nations, united the vanity and inconstancy of the

ancient books which the first five centuries of the Roman Empire produced, were astrological treatises attributed to names of Egyptian antiquity.

Necepso, a king, and Petosiris, a priest, of Egypt, were fabled to have written on astrological doctrine at some remote period. They are stated by Pliny to have fixed the limit of human life at 124 years, on grounds derived from the number of degrees in the zodiac.(240) Julius Firmicus speaks of Petosiris and Necepso as divine men and priests, the latter being likewise king, who were ancient masters and teachers of astrology.(241) He classes them with Æsculapius, Abraham, and Orpheus;(242) and he ascribes to Necepso a particular cultivation of medico-astrology. A treatise of this character under the name of Necepso is still preserved.(243) Petosiris had become proverbial for the name of an astrologer as early as the times of Nero and Domitian.(244) The spurious Manetho declares that his astrological poem is a versification of the prose writings of Petosiris, whose wisdom he extols.(245)

Greeks with the superstition and obstinacy of the Egyptians,' Decl. and Fall, c. 10.

(240) vii. 50.   (241) viii. 5, p. 216.

(242) Præf. ad lib. iv. p. 84.

(243) A preface by a certain Harpocration 'in regis Necepsi librum de xiv. tum lapidum tum herbarum remediis, secundum zodiaci signorum ordinem,' is extant in manuscript; see Fabr. Bib. Gr. vol. i. p. 70, ed. Harles. The science of medical astrology is attributed by Ptolemy to the Egyptians, Tetrabibl. i. 2. A medico-astrological treatise ascribed to Hermes Trismegistus is extant, Ideler, Med. Minor. vol. i. p. 430. There is a similar treatise by Galen, entitled περὶ κατακλίσεως προγνωστικὰ ἐκ τῆς μαθηματικῆς ἐπιστήμης, vol. xix. p. 529, ed. Kühn. A collection of the passages relating to Necepso and Petosiris is given in Marsham, Can. Chron. ed. Lips. 1676, p. 474—481; Fabr. Bibl. Gr. vol. iv. p. 160, ed. Harles. Compare Salmas. de Ann. Clim. p. 353.

(244) Lucillius, Anth. Pal. xi. 164, who was contemporary with Nero, Juven. vi. 581, and Ruperti's note.

(245) See Manetho, Apotelesmatica, i. (v. ed. Köchly) 11, v. (vi.) 10. The poem of Manetho has been lately twice edited by Köchly; once in the volume of Poetæ Bucolici et Didactici, in Didot's Collection of Greek Authors (Paris, 1851), and again, separately, in the Teubner Collection, Lips. 1858. Köchly enters into an elaborate investigation of the date of the composition of the poem, in the preface to the former edition, and fixes it to the reign of Alexander Severus, 222—235 A.D. (p. xvii.)

Lydus, a writer of the sixth century, likewise speaks of Petosiris as eminent among the ancient Egyptian sages in the art of divination.(246) The titles of the works attributed to him are stated by Suidas. They are, 1. Extracts from the Sacred Books; 2. On Astrology; 3. On the Egyptian Mysteries.(247) Necepso is included in the list of Manetho, and is stated to have reigned six years, from 672 to 666 B.C. He is mentioned by Ausonius as the instructor of the magi in their mysterious doctrine.(248)

Letronne justly remarks, that no trace of astrology occurs in the remains of the scientific Alexandrine School down to the Roman conquest of Egypt; and that there is no mention of a natal theme calculated for any of the Ptolemies. This fact, he adds, is the more important, inasmuch as the writings of the Egyptian astrologers must have been deposited in the Alexandrine library from the reign of Ptolemy Philadelphus, and have been accessible to the astronomers of Alexandria in Greek translations.(249) It may, however, be likewise observed, that there is no trace of astrology having been practised by native Egyptians before the Christian era, and that the astrological treatises of Egyptian priests, whether in their original language or in Greek translations, which Letronne supposes to have been deposited in the Alexandrine library, are probably quite imaginary.

---

(246) De Ostentis, c. 2, p. 271, ed. Bonn.

(247) Suid. in Πετόσιρις. The word Petosiris occurs in a verse of the Danaïdes of Aristophanes, ap. Athen. iii. p. 114 C:—

καὶ τὸν κύλλαστιν φθέγγου καὶ τὸν πετόσιριν.

Κύλλαστις was an Egyptian name for a species of bread. Meineke (Fragm. Com. Gr. vol. ii. 1047) supposes that Danaus addresses this verse to his daughters, instructing them to use peculiar Egyptian words, and to pass themselves off as Egyptians. There is nothing in the verse of Aristophanes to indicate that an astrologer is meant. The word appears to be formed from *Osiris*.

(248)   Quique Magos docuit mysteria vana Necepsus.
                                                     Auson. Epist. xix.

(249) Observations sur l'Objet des Représentations zodiacales, p. 78.

The only Egyptian astrologer who occurs in history is one whom Antony had with him in Italy; but his name is not mentioned; and whether he was an Alexandrine Greek, or a native Egyptian, does not appear.(250)

The miracle of the thundering legion, 174 A.D., was by some attributed to the magic arts of the Chaldæans;(251) Dio ascribed it to Arnuphis, an Egyptian magician;(252) but this is an act of thaumaturgy, and has no connexion with divination.

Diodorus makes a power of predicting future events from the planetary movements, as connected with the births of living beings, an essential part of the science of the Egyptian priests in the most ancient times.(253) Cicero likewise describes astrology as being of remote antiquity in Egypt;(254) but these statements are not confirmed by any historical fact which has reached our time.

Dio Cassius states that the Egyptians were the authors of the planetary names for the days of the week; that these names were unknown to the ancient Greeks, but that the use of them, though of recent introduction, was universal in his time, especially among the Romans.(255) The use of these names can be traced in the Roman authors from the early part of the Empire.(256) Tibullus designates the Sabbath, or the last day of the week, by the name of Saturn's day.(257) Tacitus mentions the same appellation, and accounts for the plane-

---

(250) Plut. Anton. 33.

(251) Lamprid. vit. Heliogabal. 9; Claudian, vi. Cons. Honor. 348.

(252) lxxxi. 8.

(253) i. 81. See likewise Lucan, above, n. 228.

(254) Eandem artem etiam Ægyptii longinquitate temporum innumerabilibus pæne sæculis consecuti putantur, De Div. i. 1.

(255) xxxvii. 18. Compare Joseph. Apion. ii. 39.

(256) See Hare on the Names of the Days of the Week, in the Philological Museum, vol. i. p. 1. Compare Grimm, D. M. p. 87.

(257) i. 3, 18. Compare Ideler, Chron. vol. i. p. 178, ii. p. 177.

tary names of the days of the week by the septenary number.(258)

Justin Martyr calls the first day of the week the day of the sun,(259) and so does Tertullian.(260) The same name is used by Constantine, the two Valentinians, and the two Theodosii, in laws relating to the observance of the day.(261) It is noted, however, that the earlier and proper appellation among Christians was the Lord's day.(262) The Eastern Church, from the earliest time, observed the Sabbath, or the seventh day of the week, as an additional festival, and with nearly the same solemnities as the Lord's day.(263) The fanciful analogy between the seven days of the week and the seven planets was founded, like the analogy with the tones in the musical scale, on numerical coincidence, not upon an astronomical basis.(264) The principle of the order in which the five planets are assigned to the days of the week (namely, Mars, Mercury, Jupiter, Venus, and Saturn) is obscure,(265) notwithstanding the two explanations

---

(258) Seu quod e septem sideribus, queis mortales reguntur, altissimo orbe et præcipuâ potentiâ stella Saturni feratur; ac pleraque cœlestium vim suam et cursum septimos per numeros conficiunt, Hist. v. 4. Lydus, de Mens, ii. 3, says that the Chaldæans and Egyptians derived the names of the days of the week from the number of the planets.

(259) Apol. ii. p. 98, 99.

(260) Apol. c. 16, ad Nation. i. 13. Compare Sozomen. i. 8, τὴν κυριακὴν καλουμένην ἡμέραν, ἣν Ἑβραῖοι πρώτην τῆς ἑβδομάδος ὀνομάζουσιν, Ἕλληνες δὲ ἡλίῳ ἀνατιθέασι.

(261) See Bingham's Antiq. of the Christ. Ch. xx. 2, 1.

(262) Solis die quem Dominicum rite dixere majores, Cod. Theod. viii. 8, 3, xi. 7, 13.

(263) Bingham, xx. 3; Lardner, Cred. of the Gosp. Hist. c. 85, § 12.

(264) Above, p. 131. An analogy was discovered between the planets and the faculties of the human mind, Macrob. in Somn. Scip. i. 12, § 14, but this was independent of numerical coincidence.

(265) Plutarch, in the 7th chapter of the 4th book of his Symposiac Problems (not now extant), discussed the question why the names of the planets, when transferred to the days of the week, were not taken in their natural order: διὰ τί τὰς ὁμωνύμους τοῖς πλάνησιν ἡμέρας, οὐ κατὰ τὴν ἐκείνων τάξιν, ἀλλ' ἐνηλλαγμένως ἀριθμοῦσιν. The five planets are placed in this order: 3, 1, 4, 2, 5. A scheme by which the succession of the days of the week is explained on astrological principles, is given by Arago, Pop. Astron. b. 33, c. 4.

reported by Dio; but neither of these explanations connects it with astrological ideas.

Some of the extant Roman calendars of the Imperial period mark certain days as 'dies Ægyptiaci.' These were unlucky days;(266) but there is nothing to show that they had an astrological origin.

§ 14 The Chaldæan astrology was founded upon the principle that a star or constellation presided over the birth of each individual, and either portended his fate, or shed a benign or malign influence upon his future life. This heavenly inspector was one of the five planets, or one of the signs of the zodiac. The fixed stars played no part in the astrological scheme, except so far as they composed the zodiacal signs.(267)

The star which looked upon the child at the hour of his birth

---

(266) See the subject explained in Salmasius, de Ann. Climacter. p. 815—6. For the Roman calendars, see Græv. Thes. Ant. Rom. vol. viii. An allusion to prophetic Egyptian dreams occurs in Claudian, Eutrop. i. 312, ii. præf. 39, but they are unconnected with astrology.

(267) A copious exposition of the astrological system may be found in the following works:—

1. The poem of Manilius; a Roman, who wrote under Tiberius. He declares himself to be the first poet who had treated astrology, i. 6, 113, ii. 53—7, iii. ad init. The extant poem consists of five books of hexameter verse; the sixth book is lost. See Fabr. Bib. Lat. vol. i. p. 499; Bernhardy, Röm. Litt. p. 452.

2. The Tetrabiblos of Ptolemy; of which there are numerous translations. The original Greek was published at Basil. 1553, 1 vol. 12mo. Concerning this treatise, see Delambre, Hist. Astr. Anc. tom. ii. p. 543.

3. The treatise of Julius Firmicus, in eight books. Julius Firmicus lived in the time of Constantine. Notwithstanding the poem of Manilius, he states that he was the earliest writer on astrology in the Latin language, Præf. lib. iv. p. 84; Præf. lib. v. p. 115, viii. 33, p. 244. Basil. 1551, fol.

4. The Apotelesmatica of Manetho, a Greek hexameter poem, in six books. See above, p. 302, n. 245.

5. Paulus of Alexandria, a writer of the fourth century, whose work, entitled Εἰσαγωγὴ εἰς τὴν Ἀποτελεσματικήν, was published at Wittenberg, 1586, 4to. Apollinarius, an unknown astronomer mentioned by Paulus in his preface, is also named by Achilles Tatius, p. 82.

A full account of the astrological system is given incidentally by Sextus Empiricus, adv. Mathem. v. 12, p. 730, ed. Bekker. Geminus, c. i. p. 4 and 5, and Censorinus, c. 8, likewise expound the method of predicting from nativities. A passage concerning the astrological influence of the constellations occurs in Petronius, Satyr. c. 39; and see the epigram of Philodemus, Anth. Pal. xi. 318, who was contemporary with Cicero.

was called the horoscopus;(268) and the solution of the natal problem by an astrologer bore the appellation of *thema natalicium*. As the divination was invariably founded upon a birth, the Chaldæan astrologers were called *genethliaci*, and their craft *genethlialogy*.(269) As their methods were geometrical, they likewise went by the name of *mathematici*. Astrology was also denominated the *apotelesmatic* art, in allusion to the ἀποτελέσματα, or effects, of the stars.

---

(268) ὡροσκόπος μὲν οὖν ἐστὶν ὅπερ ἔτυχεν ἀνίσχειν καθ' ὃν χρόνον ἡ γένεσις συνετελεῖτο, Sextus Empiricus, adv. Math. v. 13, p. 730, Bekker.

> Tertius æquali pollens in parte, nitentem
> Qui tenet exortum, qua primum sidera surgunt;
> Unde dies redit, et tempus describit in horas.
> Hinc inter Graias horoscopos editur urbes,
> Nec capit externum, proprio quia nomine gaudet.
>   Manilius ii. 826—30.

The same poet lays it down, as an astrological canon, that the influence of a star depends upon its place in the heaven:—

> Omne quidem signum sub qualicunque figurâ
> Partibus inficitur mundi: locus imperat astris,
> Et dotes noxamve facit.—ii. 856—8.

The method of observation is thus described by Sextus Empiricus, ib. v. 27, p. 733. νύκτωρ μὲν ὁ Χαλδαῖος ἐφ' ὑψηλῆς ἀκρωρείας ἐκαθέζετο ἀστεροσκοπῶν, ἕτερος δὲ παρῂδρευε τῇ ὠδινούσῃ μέχρις ἀποτέξοιτο, ἀποτεκούσης δὲ εὐθὺς δίσκῳ διεσήμαινε τῷ ἐπὶ τῆς ἀκρωρείας. ὁ δὲ ἀκούσας καὶ αὐτὸς παρεσημειοῦτο τὸ ἀνίσχον ζῳδίον ὡς ὡροσκοποῦν, μεθ' ἡμέραν δὲ τοῖς ὡροσκοπίοις προσεῖχε καὶ ταῖς τοῦ ἡλίου κινήσεσιν. The metallic δίσκος was used for making a noise as a signal. Sextus afterwards adds, v. 50, ἀρχὴ τοίνυν καὶ ὥσπερ θεμέλιος τῆς Χαλδαϊκῆς ἐστὶ τὸ στῆσαι τὸν ὡροσκόπον. The horoscopus is alluded to by Horace:

> Seu Libra seu me Scorpios *aspicit*
> Formidolosus.—Carm. ii. 17.

(269) That the astrologers were called *genethliaci* is stated by Gell. xiv. 1, ad init. and Augustin. de Doctr. Christ. ii. 21. The verb γενεθλιαλογῶ is applied to the Chaldæans by Strab. xvi. 1, 6. They are called γενεθλιαλόγοι by Hieronymus in Daniel, c. 2, vol. v. p. 484, ed. Francfort. The word γενεθλιαλογία is used as synonymous with astrology by Joseph. Ant. Jud. xviii. 6, 9. *Genethlialogia* has the same sense in Vitruv. ix. 7. Compare Amm. Marc. xxix. 2, § 27. Hence the word γένεσις acquired the sense of *fate*. Salmasius, de Ann. Climact. p. 801; comp. p. 87. The words γενέθλη and γενέθλια bore in later times the sense of *birthday* (Etym. Magn. in v.); Appian, B.C. iv. 34, uses γενεθλιάζω for the celebration of a birthday.

The connexion of astrology with births is alluded to by Juvenal, xiv. 248, Nota mathematicis genesis tua, and by Tacit. Ann. vi. 22; Spartian. vit. Sever. 2.

That the time of birth was a necessary foundation of the Chaldæan astrology, is implied in the arguments of those who seek to prove its futility. Cicero inquires how the astrologers explain the diversity of character and fortune in persons born at the same moment;(270) and Augustine particularly insists on the case of twins, whose fates ought to be identical, if the genethliac theory were true:(271) another writer points out that, where a national character prevails, all persons cannot have been born under the same starry influence.(272) The idea of divination from birth was even transferred to an entire community; and Tarutius Firmanus, at the request of Varro, calculated the fortunes of Rome, from its natal day, on the 21st of April, when the festival of Palilia was celebrated.(273) The nativity of the world was likewise brought within the grasp of astrological computation; and the sign Aries was supposed to be upon the meridian at this great natal hour.(274)

The proper influence of each planet was determined by the professors of the genethliac art; hence Horace and Persius oppose the benign influence of Jupiter to the malign influence of Saturn;(275) to which Propertius, in making a similar contrast, adds the pernicious influence of Mars.(276) Ovid speaks

---

(270) De Div. ii. 45.
(271) Civ. Dei, v. 2—5. In like manner Persius:—

Geminos, horoscope, varo
Producis genio.—vi. 18.

(272) Bardesanes ap. Euseb. Præp. Evang. vi. 10.
(273) See above, p. 299.
(274) Aiunt in hac ipsâ geniturâ mundi ariete medium cœlum tenente horam fuisse mundi nascentis, Cancro gestante tunc lunam. Post hunc sol cum Leone oriebatur, cum Mercurio Virgo, Libra cum Venere, Mars erat in Scorpio, Sagittarium Jupiter obtinebat, in Capricorno Saturnus meabat. Sic factum est, ut singuli eorum signorum Domini esse dicantur, in quibus, cum nasceretur, fuisse creduntur. Macrobius, Comm. in Somn. Scip. i. 21, § 24. The Mundi thema is given in Jul. Firm. iii. 1. The Egyptians made the first of Thoth, or the rising of Sirius, the birthday of the world; Boeckh, Manetho, p. 20. Virgil conceives the world to have commenced with the spring, Georg. ii. 336.
(275) Horat. Carm. ii. 17; Pers. v. 50.
(276) iv. i. 83.

of a child being born under a sinister planet, not under the sun or the moon, under Jupiter, Venus, or Mercury, but under Saturn or Mars;[277] and Juvenal opposes the favourable aspect of Venus to the disastrous aspect of Saturn.[278]

Augustus was born under the influence of the constellation of Capricornus; and hence this constellation appears on some of his coins.[279]

§ 15 The diffusion of the Chaldæan astrology in Greece and Italy in the last two centuries before the Christian era was favoured by several causes, of which the most operative appear to have been, 1. Its resemblance to the meteorological astrology of the Greeks; 2. The belief in the conversion of the souls of men into stars; 3. The cessation of the oracles; 4. The belief in a tutelary genius.

1. The Greeks, from their habit of using the risings and settings of stars for the notation of annual time, had recorded the coincidence of certain states of the weather with the appearance or disappearance of certain stars. This notation, which, if properly understood, was strictly scientific, and signified only coincidence in time, degenerated into the superstitious idea of

---

[277] Ibis, 211—8. In Amor. i. 8, 29, he opposes the beneficent influence of Venus to the sinister influence of Mars.

[278] vi. 569. Ptolemy lays it down that Jupiter and Venus are benefic, Saturn and Mars malefic, the Sun and Mercury neutral, Tetrabib. i. 5.

Seneca, Consol. ad Marc. 18, describes the five planets as determining the fortunes of men. 'Videbis quinque sidera diversas agentia vias, et in contrarium præcipiti mundo nitentia: ex horum levissimis motibus fortunæ populorum dependent, et maxima ac minima perinde formantur, prout æquum iniquumve sidus incessit.' The distinction between the benefic, malefic, and neutral planets is pointed out by Plut. de Is. et Osir. 48.

[279] Suet. Aug. 94; Manil. ii. 509. Germanicus Cæsar, in his Aratea, has the following verses respecting Capricorn.

> Hic, Auguste, tuum genitali corpore numen
> Attonitas inter gentes, patriamque paventem
> In cœlum tulit, et maternis reddidit astris.
> v. 552—4.

Augustus was born in 63 B.C., at the time of the Catilinarian conspiracy.

Achilles Tatius, c. 23, says that the genealogi look to the signs of the zodiac, and to the planets moving in the zodiac.

an occult connexion between the star and the weather. The sign was converted into a cause.(280)

The signs of weather thus noted by the Greeks in their calendars were called ἐπισημασίαι. In the calendars of Ptolemy and Geminus,(281) which are extant, the prognostics are given with their respective authorities. Thus Geminus mentions that, according to Eudoxus, on the ninth day of Cancer, the wind is southerly; that on the tenth day of Virgo there is stormy weather with lightning; that from the fourth of Pisces, the north wind blows for thirty days; that, according to Callippus, on the third of Aries, there is rain or snow. Similar notices occur in the calendar of Ptolemy; resembling the predictions of weather in a modern almanac.(282)

Predictions of this sort were not superstitious; but they were founded on an insufficient induction, and were fallacious. So Theophrastus informs us that when Mercury appears in winter, it portends cold; when in summer, heat.(283) Similar meteorological fancies occur in Aristotle, which the great philosopher doubtless believed to rest on scientific proof. Thus he lays it down that the moon produces a winter and a summer in the month, by the increase and diminution of its light, as the sun produces winter and summer in the year;(284) and he holds

---

(280) Pliny considers it as certain, that the stars exercise a meteorological influence: 'Quis æstates et hiemes, quæque in temporibus annuâ vice intelliguntur, siderum motu fieri dubitet? Ut solis ergo natura temperando intelligitur anno, sic reliquorum quoque siderum propria est quibusque vis, et ad suam cuique naturam fertilis. Alia sunt in liquorem soluti humoris fecunda, alia concreti in pruinas, aut coacti in nives, aut glaciati in grandines; alia flatus, alia teporis, alia vaporis, alia roris, alia rigoris, ii. 39.

(281) P. 36, 42, ed. Petav. Compare the calendar of Aëtius, a medical writer of the fifth century, ib. p. 216. The calendar of Ptolemy is edited and illustrated by Ideler, Berl. trans. 1816—7. The scientific character of meteorological prediction from the moon and stars is asserted by Ptolemy, Tetrabib. i. 2.

(282) Concerning this class of predictions, see Vitruv. ix. 6.

(283) De Sign. Pluv. c. 3, § 9, vol. i. p. 796, ed. Schneider.

(284) De Gen. An. ii. 4, and iv. 2, ad init.
The moon was considered a nocturnal sun. γίνεται γὰρ ὥσπερ ἄλλος ἥλιος ἐλάττων [the moon], Aristot. Gen. An. iv. 10. ἡ γὰρ σελήνη νυκτὸς οἷον ἥλιός ἐστι, Theophrast. de Sign. Pluv. c. i. § 5, vol. i. p. 783,

that the temperature and the winds are governed by the periodical courses of the sun and moon.([285]) A further step was made by the Ceans, who, according to Heraclides Ponticus, used to observe the rising of the Dog-star, and to predict, from its comparative splendour or dimness, whether the year would be healthy or pestilential.([286])

It should be observed that the Chaldæans themselves predicted the weather for fixed days of the year.([287]) They had likewise an astro-meteorological period, consisting of twelve solar years, in which they conceived the phenomena of weather to recur.([288])

Geminus points out at length, and with great perspicuity, that the predictions of weather in the Greek calendars are founded on observations, but that they only represent its average and ordinary state, and do not always come true. Hence, he remarks, an astronomer is not to be blamed if his weather prophecies are not fulfilled: whereas if his prediction of an eclipse or of the rising of a star proves erroneous, he is justly censured; for it is founded on a method which leads to certainty and precision. It is generally believed, he adds, that Sirius produces the heat of the dogdays; but this is an error, for the star merely marks a season of the year when the sun's heat is the greatest. A star, he says, is a sign of certain weather, as a beacon is a signal of war, without being its cause.([289]) Sextus Empiricus likewise contrasts the scientific predictions of weather by Eudoxus and Hipparchus with the superstitious prophecies of human affairs by the Chaldæan astrologers.([290])

Schneider. Hence Ennius, ap. Cic. de Div. i. 58, uses *sol albus* for the moon.

(285) Ib. iv. 10.   (286) Cic. de Div. i. 57.

(287) Columella, xi. 1, § 31.

(288) Proxima est hanc magnitudinem, quæ vocatur δωδεκαετηρίs, ex annis vertentibus duodecim. Huic anno Chaldaico nomen est, quem genethliaci non ad solis lunæque cursus, sed ad observationes alias habent accommodatum; quod in eo dicunt tempestates, frugumque proventus, sterilitates item morbosque circumire, Censorin. c. 18.

(289) c. 14, p. 32.

(290) Adv. Math. v. 1—3, p. 728, Bekker. Isidorus, in his work on

It is manifest that astro-meteorology would serve to facilitate the entrance of the Chaldæan astrology; but that the former rested on an apparently scientific basis is shown by the fact that it still lingers among us. The belief in the moon's influence upon the weather belongs to natural astrology;[291] a belief which was already developed in antiquity.[292]

2. Another article in the religious faith of a Greek or Roman, which might induce him to give a willing reception to the Chaldæan astrology, was the persuasion that an affinity existed between the stars and the souls of men; that the ethereal essence is divine; that the souls of men are taken from this reservoir, and return to it at death; and that the souls of the more eminent of mankind are converted into stars.[293]

Shortly after the death of Julius Cæsar, during the celebration of games in his honour by Augustus, a brilliant comet appeared in the north, and was visible for seven days. This comet was generally believed by the Roman people to be the soul of Cæsar translated into the heaven.[294]

Lucan supposes the soul of Pompey, after his death, to mount to the region where the souls of men endowed with

---

origins, distinguishes natural from superstitious astronomy: 'Inter astronomiam autem et astrologiam aliquid differt. Nam astronomia cœli conversionem, ortus, obitus, motusque siderum continet, vel quâ ex causâ ita vocentur. Astrologia vero partim naturalis, partim superstitiosa est. Naturalis, dum exsequitur solis et lunæ cursum vel stellarum, certasque temporum stationes. Superstitiosa vero illa est, quam mathematici sequuntur, qui in stellis augurantur, quique etiam duodecim signa cœli per singula animæ vel corporis membra disponunt, siderumque cursu nativitates hominum et mores prædicare conantur, iii. 26. Isidorus lived in the seventh century.

(291) See on this subject Lardner's Handbook of Astronomy, by Dunkin, § 215.

(292) See Arat. 778—818; Virg. Georg. i. 424—37; Plin. xviii. 79.

(293) See Aristoph. Pac. 832—3; Virg. Georg. iv. 219—227. Varro held 'ab summo circuitu cœli ad circulum lunæ æthereas animas esse astra ac stellas, eos cœlestes deos non modo intelligi esse, sed etiam videri: inter lunæ vero gyrum et nimborum ac ventorum cacumina aëreas esse animas, sed eas animo non oculis videri; et vocari heroas et lares et genios.' Augustin. Civ. Dei, vii. 6. Several of the ancient philosophers supposed the soul to be of an igneous nature, Stob. Ecl. Phys. i. 41.

(294) Suet. Cæs. 88; Plin. N. H. ii. 23 (where the words of Augustus himself are cited); Ovid, Met. xv. 746, 845—51; Dio Cass. xlv. 7; Seneca Nat. Quæst. vii. 17. Compare Virg. Ecl. ix. 47: Ecce Dionæi processit Cæsaris astrum.

superhuman virtues have been converted into stars.(295) So Claudian describes Theodosius as ascending through the spheres of the planets until he reaches the portion of heaven where the fixed stars welcome him as a new comer, to be added to their band.(296)

One of the opinions respecting the Milky Way, versified by Manilius, is that it is formed of the souls of illustrious men, who after their death have been received into the heaven.(297)

That the stars are of a divine nature was the general tenet of the ancient philosophers: even Aristotle believed that they are divine beings, endowed with independent volition. (298) Heraclitus is reported to have held that the soul is a spark taken from the stellar essence.(299)

3. The cessation of the native oracles, and the consequent demand for a substitute, likewise contributed to the diffusion of astrology.(300)

4. Another circumstance which tended to promote the spread of astrology in Greece was the belief in the existence of a peculiar genius which took charge of every person from his birth, and whose life ceased with that of his human client.(301)

---

(295) ix. 1—9.   (296) De Tert. Cons. Honor. 162—174.

(297) An fortes animæ dignataque nomina cœlo,
Corporibus resoluta suis, terrâque remissa,
Huc migrant ex orbe; suumque habitantia cœlum
Æthereos vivunt annos, mundoque fruuntur.—i. 756—59.

(298) Above, p. 163.

(299) Scintilla stellaris essentiæ, Macrob. in Somn. Scip. i. 14, § 19. According to Stob. Ecl. Phys. i. 41, p. 322, Gaisf.: Ἡρακλείδης φωτοειδῆ τὴν ψυχὴν ὡρίσατο.

(300) See Plut. de Def. Or. The cessation of the oracles is expressly mentioned by Juvenal as a cause of the predilection for astrology, in the verses cited above, p. 299. Lucan represents the priestess of Apollo as declaring the silence of the Delphi oracle:—

Muto Parnassus hiatu
Conticuit, pressitque deum.

And see the rest of the passage, v. 131—140.

(301) ἅπαντι δαίμων ἀνδρὶ συμπαρίσταται
εὐθὺς γενομένῳ, μυσταγωγὸς τοῦ βίου
ἀγαθός.

Menander, ap. Meineke, Frag. Com. Gr. vol. iv. p. 238. Origenes, contr. Cels. 8: Ἑλλήνων μὲν οὖν οἱ σοφοὶ λεγέτωσαν δαίμονας εἰλη-

This genius was naturally placed in relation with the star which presided over the child's birth.(³⁰²)

§ 16 Astrology, as practised in the first centuries of the Roman Empire, was an intricate and abstruse system. Its professors were popularly called *mathematicians*. It involved more reasoning, and demanded more constructive ingenuity, than the modern pseudo-sciences of phrenology and homœopathy. Even for his own objects of superstitious imposture, a Chaldæan or an Egyptian would have been unable, without the assistance of a higher intelligence, to produce so refined an instrument as the genethliac astrology. It required the intellectual gifts of the Greeks to give it the precision and finish, and the false lustre of a scientific method, which it obtained in their hands.(³⁰³) At a later time, the Arabians received it from the Greeks, but simply handed it down to posterity without either exposing the fraud, or adding to its ingenuity.(³⁰⁴)

χέναι τὴν ἀνθρωπίνην ψυχὴν ἀπὸ γενέσεως. Compare Seneca, Epist. 110, § 1; Amm. Marc. xxi. 14. 'Genius est Deus, cujus in tutelâ, ut quisque natus est, vivit. Hic, sive quod, ut genamur, curat, sive quod una genitur nobiscum, sive quod nos genitos suscipit ac tuetur; certe a genendo genius appellatur.'—'Genius autem ita nobis assiduus observator appositus est, ut ne puncto quidem temporis longius abscedat, sed ab utero matris exceptos ad extremum vitæ diem comitetur,' Censorinus, c. 3.

(302)   Scit genius, *natale comes qui temperat astrum*,
         Naturæ deus humanæ, mortalis in unum-
         Quodque caput, vultu mutabilis, albus et ater.
                                        Horat. Epis. ii. 2, 187.

The words, 'mortalis in unumquodque caput,' mean that the guardian genius does not survive his ward.

(303) On the connexion of Chaldæan doctrine with Pythagorism and Neo-Platonism, see Lobeck, Aglaopham. p. 922, 929.

(304) See Gibbon, c. 52, vol. vii. p. 38. On the stationary character of the Arabian astronomy, see Delambre, Hist. de l'Astron. du Moyen Age, p. 95, 455.

Chapter VI.

EARLY HISTORY AND CHRONOLOGY OF THE EGYPTIANS.

§ 1   IN support of the alleged antiquity of the astronomical observations of Assyria and Egypt, and of the astronomical science attributed to their priests, an appeal is made to the primitive civilization of those two empires, and to their lists of kings reaching through long lines of consecutive dynasties. It becomes necessary, therefore, to advert to the early history and chronology of these countries, for the purpose of ascertaining how far they serve to prop up the tottering edifice of Assyrian and Egyptian science.

The ancient chronology of Egypt has been rendered the subject of elaborate investigations by recent critics, and it will therefore be convenient to begin with an inquiry into its authenticity.

Reckoning back from the conquest of Egypt by Cambyses, and its annexation to the Persian empire, we find a period of 145 years, extending from 670 to 525 B.C., during which the succession of its independent sovereigns appears to rest on historical grounds.(¹) Their names, the years of their respective

---

(1) Mr. Kenrick, in his work on ancient Egypt, makes the following observations upon the history subsequent to the time of Psammetichus: 'Herodotus, when he resumes the history of Egypt after the reign of Sethos, remarks that from this time forward he shall relate that in which the Egyptians and other nations agree. Previously to this time there was no other testimony to control the accounts which the "Egyptians and the priests" gave of their own history; no Greek had advanced beyond Naucratis, and no record was left even of the imperfect knowledge of Egypt which they might thus have gained. The effect is immediately visible, and we have henceforth a definite chronology, an authentic succession of kings conformable to the monuments, and a history composed of credible facts,' vol. ii. p. 381. Niebuhr, in like manner, remarks upon the Egyptian history of Herodotus, that 'the whole narrative of the period before

reigns, and some real events connected with each, have been preserved to us by the Greek writers.

The succession is as follows, according to Herodotus:—

	B.C.
Psammetichus	670—616
Neco	616—600
Psammis	600—595
Apries	595—570
Amasis	570—526
Psammenitus	526—525(²)

The commencement of this period is 186 years before the birth of Herodotus. The average duration of the six reigns is about twenty-four years, including one king dethroned in the first year of his reign.

Of the six kings in question, Amasis was within the limits of the traditions which were tolerably fresh at the commencement of Greek prose literature. He had relations with Crœsus, and Polycrates of Samos: he sent presents to several of the Greek cities, and subscribed to the rebuilding of the temple of Delphi.(³) He was fond of the Greeks, and gave them the factory of Naucratis:(⁴) the circumstance of his mummified body being taken from its tomb by Cambyses, and being subjected to insults, was likely to be remembered in Egypt.(⁵) Of

---

Psammetichus is without value; but from that time it is historical and excellent,' Lect. on Anc. Hist. vol. i. p. 44, ed. Schmitz.

(2) See Clinton, F. H. at the years 670, 616, and 525; Boeckh, Manetho, p. 332.

(3) Herod. ii. 180, 182, iii. 47.

(4) Ib. ii. 178. Strabo, however, appears to state that Naucratis was occupied by the Milesians in the reign of Psammetichus, xvii. 1, § 18.

(5) iii. 16. Polyæn. viii. 29, states that Cyrus having demanded the daughter of Amasis in marriage, Amasis sent him Nitetis, the daughter of Apries, whom he had dethroned. After the death of Amasis, Nitetis informed her husband Cyrus of the deceit, and urged him to take vengeance of Psammetichus, the son of Amasis. Cyrus died before he could accomplish this purpose; but it was executed by Cambyses, who restored the family of Apries to the throne. This story appears to have been borrowed from Dinon and Lyceas: the account of Ctesias was that Cambyses, not Cyrus, demanded the daughter of Amasis; Athen. xiii. p. 560; Fragm. Hist. Gr. vol. ii. p. 91, vol. iv. p. 441.

Apries less is known; but he is stated to have made war upon Syria, and to have been dethroned by Amasis.(⁶) He is the Pharaoh Hophra of the contemporary prophet Jeremiah.(⁷) The preceding king, Psammis, is stated by Herodotus to have received an embassy of Eleans, who boasted of the Olympic Games.(⁸) This synchronism is consistent with the date assigned for the institution of the games in question.

The reign of Neco is marked by his attempt to dig a canal from the Nile to the Red Sea; by his alleged circumnavigation of Africa; and by his successful invasion of Syria. He is stated to have dedicated a reminiscence of a victory in Syria, in the Temple of Apollo at Branchidæ, near Miletus.(⁹) His expedition against Assyria is mentioned by the prophet Jeremiah, and in the Books of Kings and Chronicles.(¹⁰) The lifetime of Jeremiah, according to the Biblical chronology, coincides with the reigns of Neco and Apries, as determined by the chronology of Herodotus and Manetho. The death of Josiah, king of Judah, who is represented in the Books of the Kings and Chronicles as slain in the battle with Pharaoh Necho, likewise coincides, according to the chronology of those books, with the reign of Neco, according to Herodotus. The coincidence with Manetho is not entire, but the discrepancy is inconsiderable.(¹¹)

---

(6) Herod. ii. 161—171.

(7) Jerem. xliv. 30. The prophecies of Jeremiah appear to have been composed during the years 628—586 B.C.

(8) ii. 160. (9) Herod. ii. 158—9.

(10) Jerem. xlvi. 2; 2 Kings xxiii. 29—35; 2 Chron. xxxv. 20—27. Megiddo in the Books of Kings and Chronicles evidently agrees with Magdolum in Herodotus; but the Jewish town of Migdol near the lake of Galilee, not the town of Lower Egypt near Pelusium, must be intended, though Herodotus may have confounded the two. See Rawlinson's Herod. vol. ii. p. 246; Dr. Smith's Dict. of Anc. Geogr. arts. *Magdolum* and *Jerusalem*, vol. ii. p. 17. Josephus, Ant. x. 5, 1, makes Mendes in Egypt the place of the battle, which town was, he states, in the dominions of Josiah.

(11) The death of Josiah is placed by Clinton, Fast. Hell. vol. i. p. 328, and Winer, Bibl. Real-Wört. art. *Juda*, at 609 B.C. The reign of Neco, according to Herodotus, extends from 616 to 600 B.C.; according to Manetho, from 604 to 598. See a comment on the discrepancies between the chronologies of Herodotus and Manetho for this period in Boeckh,

Psammetichus first opened Egypt to the Greeks;(¹²) and he permitted a colony of Ionians and Carians to establish itself near the city of Bubastis. Herodotus states that the Greeks were well acquainted with Egyptian affairs since this epoch.(¹³) It is likewise related by Herodotus, that after the Scythians had defeated Cyaxares and the Medes, they advanced as far as Palestine; but that Psammetichus, king of Egypt, induced them by presents and entreaties to retire, and to desist from their intention of invading his kingdom.(¹⁴) The date of this event is fixed by Larcher at 628 B.C.; the same critic places the reign of Cyaxares (which, according to Herodotus,(¹⁵) lasted forty years,) from 634 to 594 B.C. If these determinations are correct, the lifetimes of Psammetichus and Cyaxares are coincident; and the date of the Scythian invasion of Palestine is consistent both with the Egyptian and the Median chronology. The invasion of Judæa by a northern equestrian people is likewise alluded to in two remarkable passages of the contemporary prophet Jeremiah.(¹⁶)

The Psammetichus who succeeded Periander in the dynasty of Corinthian despots, and who ruled about 585 B.C., probably took his name from the king of Egypt.(¹⁷) The Corinthian ruler might have been born during the reign of the Egyptian

---

Manetho, p. 331—351. Compare Des Vignoles, Chronologie de l'Histoire Sainte, vol. ii. p. 146—150. According to Manetho, Fragm. Hist. Gr. vol. ii. p. 593, Neco took Jerusalem, and carried Joachaz the king a prisoner to Egypt. This agrees with 2 Kings xxiii. 33—4; 2 Chron. xxxvi. 4.

(12) On the exclusiveness of the Egyptians, and their repulsiveness towards the Greeks, see Herod. ii. 41, 91.

(13) ii. 154, cf. 147. Concerning the Egyptian kings from Psammetichus to Psammenitus, see Grote's Hist. of Gr. vol. iii. p. 429—447; Rawlinson's Herodotus, vol. ii. p. 381—390.

(14) i. 105; Strabo, xvii. 1, § 18, states that Psammetichus was contemporary with Cyaxares the Mede.

(15) i. 106. See Larcher, Hérodote, tom. vii. p. 150.

(16) iv. 6, vi. 22—4.

(17) Aristot. Pol. v. 12; Nicol. Damasc. 60; Fragm. Hist. Gr. vol. iii. p. 393.

king, and friendly commercial relations have subsisted between Corinth and Egypt during the reign of the latter.

The account of the series of kings from Psammetichus to Amasis, given by Diodorus, agrees generally with that in Herodotus. Diodorus represents the kings before Psammetichus as exercising a Chinese repulsion towards foreigners, and says that Egypt was opened to strangers, and especially to Greeks, by this king. He describes Apries as separated from Psammetichus by four generations,[18] apparently counting both extremes inclusively; he states that Apries reigned twenty-two years, and was dethroned by Amasis, whose reign lasted fifty-five years; and that Amasis died in Olymp. 63.3=526 B.C. at the time of the invasion of Egypt by Cambyses.[19]

According to Manetho, the names and duration of reigns of the six kings, beginning with Psammetichus, are as follows:—

Psammetichus	54
Nechao	6
Psammuthis	6
Vaphris	19
Amosis	44
Psammecherites	$\frac{1}{2}$
	$129\frac{1}{2}$ [20]

(18) Neco is included in this interval. Elsewhere Diodorus mentions him as the son of Psammetichus, and as having begun the canal from the Nile to the Red Sea, i. 33.

(19) Diod. i. 67—8. Herodotus states that Apries reigned twenty-five years, ii. 161; and that Amasis reigned forty-four years, and died a short time before Cambyses invaded Egypt, iii. 10. See Clinton, F. H. ad ann. 616.

(20) Fragm. Hist. Gr. vol. ii. p. 593. These numbers are according to Africanus: according to Eusebius the Manethonian years of the reigns stand thus:—

Psammetichus	44 years
Necao	6
Psammuthis	17
Vaphris	25
Amosis	42
	134

Compare Boeckh, Manetho, p. 331—351.

Reckoning this period at 130 years, it differs by only fifteen years from the chronology of Herodotus.

§ 2 Up to this point, we may consider the Egyptian chronology as determined, within moderate limits of error, upon trustworthy evidence. We have next to inquire how our different authorities represent that chronology for the period anterior to the reign of Psammetichus, reckoning back from the epoch 670 B.C., 106 years after the commencement of the Olympiads.

Herodotus visited Egypt, and had personal communication with the priests, from whom his information respecting the ancient history and chronology of the country was, as he himself declares, derived.[21] The date of his visit is about 450 B.C.

According to his report of their information, Egypt was at a remote period subjected to the rule of eight gods of the first order, of whom Pan was one. These deities were succeeded by twelve gods of the second order, who were produced 17,000 years before the reign of Amasis. Hercules was in the second order; but Bacchus belonged to a later period: he preceded the reign of Amasis by only 15,000 years. Herodotus is perplexed by this statement, because, according to the chronology received among the Greeks, Bacchus lived only 1600, Hercules only 900, and Pan only 800 years before his own time. However, he tells us that the Egyptians assured him that error was impossible, as their predecessors had preserved a contemporaneous notation of the years during this entire period.[22]

The priests further informed Herodotus that Men was the first king of Egypt; that he turned the course of the Nile, and built the city of Memphis: and that he was succeeded by 330 kings, whose names they read to him from a book, or roll of papyrus. With the exception of Queen Nitocris, who reigned

---

(21) See ii. 99, 142. The communications of Herodotus with the Egyptian priests were doubtless carried on through a Greek interpreter. On the class of ἑρμηνεῖς in Egypt, see Lepsius, Chron. der Æg. p. 247—9.

(22) Herod. ii. 43, 145.

during this period, these monarchs were undistinguished, and left no memorials of themselves.(23) Men and his 330 successors, as we shall see presently, are supposed to have reigned from 11,400 to 1080 B.C., a period of 10,320 years.

With regard to the twelve kings who succeeded, and who nearly filled up the interval before Psammetichus, the priests had much more to tell Herodotus than they were able to communicate respecting their 330 predecessors. These twelve kings (whose names are subjoined) reigned during a period of 400 years: the dates placed respectively opposite their names are calculated on the assumption that each represents a generation.(24)

After Men.		B.C.
331	Mœris	1080
332	Sesostris	1046
333	Pheros	1013
334	Proteus	980
335	Rhampsinitus	946
336	Cheops	913
337	Chephren	880
338	Mycerinus	846
339	Asychis	813
340	Anysis	780 (25)
341	Sabacos	746
342	Sethon	713
	Last year of his reign	680

---

(23) Herod. ii. 4, 101.

(24) This is the calculation which, as we shall show presently, Herodotus himself makes. He states, however, that Cheops reigned fifty and Chephren fifty-six years (ii. 127—8), and apparently that Mycerinus reigned only seven years (ii. 133). The meaning of the period of 150 years mentioned in c. 133, is not explained. In calculating the dates, I have assigned thirty-four years to the first king of the century, and thirty-three to each of the two others, in order to avoid fractions.

(25) It is difficult to understand the passage of Herodotus, ii. 140, which seems to make an interval of more than 700 years between the concealment of Amyrtæus in 455 B.C. (Clinton ad ann.), and the reign of Anysis, and would therefore raise Anysis to 1155 B.C. It has been proposed to substitute 300 or 500 for 700 in the text of Herodotus. See

Mœris, the first of these twelve kings, was excavator of the lake which bore his name; likewise the builder of the pyramids in this lake, and of the gateway of the Temple of Vulcan.([26]) His successor Sesostris was a great conqueror. He sailed along the shores of the Red Sea until he reached a sea innavigable from shoals. He likewise marched with a vast army across Asia Minor, and subdued Scythia and Thrace. He erected some columns which attested the extent of his march, and were partly extant in the time of Herodotus: and he left behind him an Egyptian colony at Colchi. He dug the canals with which Egypt was intersected, and he made an equal division of the land of Egypt among the inhabitants. The belief in the exploits of Sesostris was deeply rooted in Egypt; for the priest of Vulcan is said to have refused to permit the erection of a statue of Darius, in front of the statues of Sesostris and his queen, on the ground that Sesostris had conquered the Scythians, whereas Darius had failed in his attempt upon this people.([27])

Sesostris was succeeded by his son Pheros, who was punished with blindness for his impiety, and was restored to sight in consequence of a marvellous cure directed by an oracle. Pheros gave two obelisks, of great height, to the Temple of the Sun.([28])

The next king was Proteus, who reigned at the time of the Trojan War. The priests informed Herodotus, partly on the authority of Menelaus himself,([29]) that Paris and Helen, on

---

Schweighäuser ad loc.; Niebuhr, Lect. Anc. Hist. vol. i. p. 68, and Volney, Recherches nouvelles sur l'Histoire ancienne, Chron. des Egyptiens, c. i, Œuvres, ed. 1837, p. 515, approve of 300.

([26]) Herod. ii. 101, 149, 150. Herodotus states that at the time of his visit to Egypt, 900 years had not elapsed since the death of Mœris, ii. 13. Assuming that Herodotus visited Egypt about 450 B.C., we obtain about 1350 B.C. for the date of the death of Mœris, which does not accord with the result of the calculation founded upon the succession of kings.

([27]) ii. 102—110. The same story respecting the statue of Darius is told by Diodorus, i. 58, who calls Sesostris Sesoösis.

([28]) ii. 111.

([29]) ἔφασαν (οἱ ἱερεῖς) πρὸς ταῦτα τάδε, ἱστορίῃσι φάμενοι εἰδέναι παρ' αὐτοῦ Μενέλεω, ii. 118.

their voyage from Greece to Troy, were driven by contrary winds to the coast of Egypt; that Proteus dismissed Paris, but detained Helen; and that Helen, who never went to Troy, was afterwards surrendered by Proteus to Menelaus. The priests assured Herodotus that they knew the truth of this narrative, partly from inquiries, and partly from the events having occurred in their own country; and Herodotus gives his reasons for considering it to be historical. The work of Proteus was the temenos of the Temple of Vulcan at Memphis.[30]

Proteus was succeeded by Rhampsinitus, whose riches were immense. The robbery of his treasury is the subject of a story which resembles in its character one of the Arabian Nights, and which has for its moral the surpassing wisdom of the Egyptians. His descent to Hades, where he plays at dice with Ceres, furnished the material for a festival-legend. The monuments attributed to him were the gateway of a temple, and two statues.[31]

The next two kings were Cheops and Chephren, brothers; the former of whom reigned for fifty, and the latter for fifty-six years. These kings built two great pyramids: they compelled all the people to labour at these works; and they closed the temples. Egypt, which up to this time had been mildly governed, endured grievous oppression for 106 years.[32]

Chephren was succeeded by Mycerinus, son of Cheops. He was a mild king; he re-opened the temples, and relieved the people from their sufferings. He left a pyramid, but smaller than that of his father. Legendary stories were told respecting the figure of a cow, in which the body of his daughter was entombed, and twenty wooden statues of his concubines. His reign lasted only six years.[33]

Asychis came next, who built the most richly-decorated of the four gateways of the Temple of Vulcan, and a pyramid of

---

(30) Herod. ii. 112—120.   (31) ii. 121—124.
(32) ii. 124—128.   (33) ii. 129—135.

brick, which was thought superior to the pyramids of stone.[34] His successor was Anysis, a blind king, who was driven from his throne by Sabacos, king of the Æthiopians. Anysis lay hid in an island in the marshes for fifty years; at the end of which time Sabacos voluntarily evacuated the country in consequence of a vision. Sabacos built the Temple of Bubastis.[35]

The last of the 342 kings who reigned after Men was Sethon, a priest of Vulcan, who treated the warrior class with indignity, and deprived them of their lands. During his reign, Sanacharib, king of the Arabians and Assyrians, invaded Egypt; but his army was defeated, in consequence of an irruption of field mice, which gnawed the weapons of his soldiers, and made them unserviceable. A statue of Sethon, with a mouse in its hand, was preserved in an Egyptian temple in the time of Herodotus.[36]

Between the reign of Sethon and that of Psammetichus, there intervenes the period of the Dodecarchy, respecting which Herodotus relies on Greek as well as on Egyptian informants. During this period, Egypt was divided into twelve districts, and each district was governed by a king. Psammetichus was one of these twelve kings; and with the assistance of Carian and Ionian mercenaries, he ultimately succeeded in deposing the other

---

(34) Herod. ii. 136.

(35) ii. 137—140. Hosea, King of Israel, near the beginning of his reign, became tributary to Shalmaneser, King of Assyria; but he shortly afterwards withheld his tribute, and entered into treasonable correspondence with So, King of Egypt, 2 Kings xvii. 3, 4. According to Clinton, Hosea began to reign in 730 B.C. This date agrees with the date of Sabaco, which results from the Egyptian account in Herodotus, viz. 746—713 B.C. The interval between the alliance of Hosea with So, and the expedition of Sennacherib, according to the Biblical chronology, is about sixteen years.

(36) ii. 141. The invasion of Judæa by Sennacherib, and his return to Nineveh after the miraculous destruction of his army, took place in the fourteenth year of King Hezekiah, according to Isaiah xxxvi., xxxvii.; 2 Kings xviii., xix.; Joseph. Ant. x. 1; that is, in 712 B.C. See Clinton, vol. i. p. 327, and Winer. This accords with the date obtained above, p. 321, from Herodotus for the reign of Sethon; but the Biblical account does not represent Sennacherib as having invaded Egypt. Josephus supposes Sennacherib to have invaded both Judæa and Egypt, and the same account was given by Berosus, cited in Josephus.

rulers, and subjecting all Egypt to his dominion. Under the Dodecarchy the Labyrinth was built, which Herodotus considers the greatest work in Egypt—greater even than the pyramids.[37]

The priests informed Herodotus that, from the first king, Men, to king Sethon, there was a series of 341 generations; the detailed statements previously given by him make this number 342, if both extremities are included. The Theban priests likewise showed to Herodotus (as they had previously shown to Hecatæus, the historian)[38] a series of 345 wooden statues of high priests, reaching down to the priest last deceased; these high priests all bore the same name, and each was the son of his predecessor. Herodotus considers each of these royal successions as a generation, and reckons three generations to a century. If we adopt his principle of calculation, we arrive at the following chronological scheme, as representing the result of the account given to him by the Egyptian priests:—

	B.C.
Dynasty of eight gods of the first order	not stated
Twelve gods of the second order	17,570
Reign of Bacchus	15,570
Men	11,400
After him	
329 kings, until	
Mœris	1080
After him	
Eleven kings, until	
The Dodecarchy	680
Psammetichus	670 [39]

---

(37) ii. 147—152.

(38) The birth of Hecatæus fell about 550 B.C. His visit to Egypt may be supposed to have been about sixty years before that of Herodotus.

(39) The calculation of Herodotus assumes 341, not 342 generations; although the latter number clearly results from his own statements. See Lepsius, Chron. der Ægypter, p. 259. Moreover, after stating that three generations make a century, he reckons $341 \times 33\frac{1}{3} = 11,340$, whereas the product of these numbers is $11,366\frac{2}{3}$. Either, therefore, Herodotus made

The time occupied by the Dodecarchy is not stated by Herodotus; in the above computation, it has been taken at ten years.([40])

§ 3 Plato states it to be a literal truth, that some of the works of art in Egypt are 10,000 years old:([41]) he likewise describes Solon as having heard from an Egyptian priest that the city of Sais was proved by existing records to have lasted for 8000 years; and that the historical registration of the Egyptians ascended to an epoch long anterior to that of the Greeks.([42]) The Epinomis likewise speaks of the astronomical observations of the Egyptians and Assyrians as ascending to an infinite antiquity.([43])

§ 4 The next authority on Egyptian chronology is Manetho. The accounts of Manetho's life are not very clear or satisfactory; but he appears to have been a native of Sebennytus, in Egypt; to have held a high position in the native priesthood; and to have lived in the reigns of the first two Ptolemies (306—247 B.C.) ([44]) Syncellus says that Manetho wrote his work in imitation of that of Berosus; having been either his contemporary, or a little subsequent to him; and that it was dedicated to Ptolemy Philadelphus.([45])

---

an error of computation, or the numbers in his text have been corrupted. See ii. 142—3.

The statement of Mela, i. 9, is derived inaccurately from Herodotus; see Boeckh, Manetho, p. 94, n. 3.

(40) The interval between the resignation of Sabaco and the accession of Psammetichus is made by Diod. i. 66 to consist of seventeen years; namely, anarchy two years, and Dodecarchy fifteen years.

(41) Leg. ii. § 3, p. 656.

(42) Tim. p. 23, A. E. Compare Macrob. Comm. in Somn. Scip. ii. 10, § 14.

(43) χρόνος μυριετὴς καὶ ἄπειρος, c. 9, p. 987.

(44) See Fragm. Hist. Gr. vol. ii. p. 511; Boeckh, Manetho, p. 11; Bunsen's Egypt, vol. i. p. 60, Engl. tr.; Lepsius, Chron. der Æg. p. 405.

(45) Vol. i. p. 29, ed. Bonn. Bunsen, Egypt, vol. i. p. 97, Engl. tr., says that 'Manetho under the first Ptolemies opened up to the Greeks the treasures of Egyptian antiquity,' and that 'none of the later *native* historians can be compared with him.' It does not, however, appear that there were any native historians of Egypt after Manetho. Concerning the date of Berosus, see above, ch. v. § 11.

Manetho published a work on Egyptian history, composed in the Greek language, and derived, as he himself stated, from sacred writings in the native tongue. It was divided into three parts or volumes.(46) This work is lost, but a summary of it has been preserved by the chronographers; their accounts of its contents are often inconsistent, and it is probable that they may, in many cases, have relied upon secondhand authorities, without consulting the original work. The different versions have been investigated with great care, and much critical acumen, by Prof. Boeckh, whose restoration of the entire chronological scheme will be assumed as the basis of the following remarks.

Manetho begins his Egyptian chronology at the year 30,627 B.C., but at the outset of this period he places three dynasties of gods, four dynasties of demigods, and one dynasty of manes, which together occupy 24,925 years.(47) The first mortal king, Menes, commences his reign in the year 5702 B.C. Menes and his successors are distributed into thirty-one dynasties, ending with 333 B.C., the last year of Darius Codomannus. Our attention will now be confined to the first twenty-five dynasties, ending with the year 680 B.C.

According to the chronological scheme of Manetho, Egypt was governed during 5022 years, between 5702 and 680 B.C., by twenty-five dynasties, containing 439 kings. Of these 439 kings, 346 are unnamed; of the remaining 93, the names are expressed.(48) For each dynasty, even where the names are not given, it is stated to what city of Egypt (as Memphis, Thinis, Elephantina) the kings belong; where they are foreigners (as Phœnicians or Æthiopians), this circumstance is mentioned. To the

---

(46) Fragm. Hist. Gr. ib. p. 511—2.

(47) Boeckh alters this number into 24,837, in order to produce a multiple of the Canicular period (1461×17), Manetho, p. 84, 93, 385.

(48) In this enumeration, no account is taken of the seventh dynasty, consisting of 70 unnamed Memphite kings, who reigned seventy days (*i.e.*, as it appears, one day each). See Boeckh, Manetho, p. 153; Fragm. Hist. Gr. vol. ii. p. 555. The alteration of 70 into 5, in Eusebius, is probably merely a reduction made on grounds of probability.

names of twenty-seven kings, a short notice of an event connected with their reign is attached.

With the exception of these notices, the remains of Manetho furnish us only with a chronological series of kings, stating the number of years during which each king reigned, or each dynasty lasted, but in more than three cases out of four, not even mentioning the king's name. Its general character is that of a chronology of anonymous kings arranged in dynasties.

The following extracts contain all the historical notices in this list, from which their number, character, and value may be appreciated:—

'Menes the Thinite, first king of dynasty 1 (5702—5450 B.C.). He was torn in pieces by a hippopotamus, and perished.

'Athothis, son of Menes, second king of dynasty 1. He built the palace at Memphis. He was skilled in medicine, and wrote a work on anatomy, still extant.

'Uenephes, son of Kenkenes, fourth king of dynasty 1. In his time a great famine prevailed in Egypt. He built the pyramids near Cochome.([49])

'Semempses, son of Miebis, seventh king of dynasty 1. In his time Egypt was afflicted with a great calamity ($\phi\theta o \rho \acute{a}$).

'Boethus, first king of dynasty 2 (5449—5148 B.C.) In his time a great opening of the earth took place near Bubastus, and many persons perished.

'Cæechos, second king of dynasty 2. In his time the bull Apis, in Memphis, the bull Mnevis, in Heliopolis, and the goat at Mende, first received divine honours.

'Binothris, third king of dynasty 2. In his time it was ruled that women might succeed to the throne.

'Nephercheres, seventh king of dynasty 2. It is fabled that in his time the waters of the Nile were for eleven days mixed with honey.([50])

---

(49) Nothing is known of these pyramids. Compare the guesses of Bunsen, Egypt, vol. ii. p. 60.

(50) Bunsen, vol. ii. p. 106, Eng. tr., thinks that this story probably originated in some natural phenomenon recorded in the annals. The king under whom this phenomenon occurred is called Binoris by

'Sesochris, eighth king of dynasty 2. This king was five cubits in height, and three palms in width.

'Necherophes, first king of dynasty 3 (5147—4934 B.C.). In his time the Libyans revolted against the Egyptians, but surrendered themselves as prisoners through fear at a marvellous increase in the size of the moon.

'Tosorthrus, second king of dynasty 3. This king was considered an Æsculapius by the Egyptians, on account of his medical science; he also invented the use of polished (or worked) stones in building; and he paid attention to painting.

'Suphis, second king of dynasty 4 (4933—4651 B.C.). He built the great pyramid, which Herodotus declares to have been built by Cheops. He was guilty of arrogance towards the gods; he wrote the sacred book, which I, Africanus, acquired as a precious remnant of antiquity, when I was in Egypt.([51])

'Othoes, first king of dynasty 6 (4402 to 4200 B.C.). This king was slain by his body-guards.

'Nitocris, sixth of dynasty 6. This queen was the most spirited and most beautiful woman of her time; her complexion was fair; she erected the third pyramid.

'Achthoes was a more tyrannical king than any of his predecessors, and oppressed all the Egyptians; at last he was seized with insanity, and was killed by a crocodile.

'Sesostris, third king of dynasty 12 (3404—3245). He subdued all Asia in nine years, and Europe as far as Thrace. He left everywhere monuments of his conquests, by engraving on pillars the male parts of generation for the warlike, and the female for the unwarlike nations. His exploits were such that he was ranked by the Egyptians next after Osiris.([52])

Joannes Antiochenus, Fragm. Hist. Gr. vol. iv. p. 539, whose words stand thus: Ἐπὶ Βινώριος βασιλέως Αἰγύπτου φασὶ τὸν Νεῖλον μέλιτι ἁ ῥυῆναι. Read μέλιτι ιά ἡμέρας ῥυῆναι.

(51) See Boeckh, ib. p. 177.

(52) The substance of this passage is repeated in Joannes Antiochenus, Fragm. Hist. Gr. vol. iv. p. 539.

Lachares, fourth king of dynasty 12. He built the labyrinth in the Arsinoite nome, as a place of sepulture for himself.

'The fifteenth dynasty consisted of shepherd kings. They were Phœnicians; the first of them was named Saites, and gave his name to the Saite nome. They built a city in the Sethroite nome,[53] from which they subdued Egypt, and took Memphis. (2607—2324 B.C.)

'Amos, first king of dynasty 18 (1655—1327 B.C.). A remark of Africanus is appended to this king, that the exodus of Moses from Egypt fell under his reign.[54]

'Misphragmuthosis, sixth king. The deluge of Deucalion took place in his time.

'Amenophis, eighth king. He was reputed to be Memnon, of the speaking statue.

'Thuoris, sixth king of dynasty 19 (1326—1184 B.C.). This king is called Polybus, the husband of Alcandra, mentioned in Homer.[55] In his time Troy was taken.

'Petubates, first king of dynasty 23 (814—726 B.C.). In his time the first Olympiad was celebrated.

'Osorcho, second king of dynasty 23. This king is called Hercules by the Egyptians.

'Bocchoris, dynasty 24 (725—720 B.C.). In his time a lamb spoke.

'Sabacon, first king of the Æthiopian dynasty 25 (719—680 B.C.). He took Bocchoris prisoner, and caused him to be burnt to death.'

The notices relating to Jewish history, appended to Nechao II. and Vaphris, in the 26th dynasty, have been already mentioned.

So far as Manetho is concerned, these passages contain the

---

(53) This nome was in or near the Delta: see Strab. xvii. 1, § 24.
(54) See Boeckh, Manetho, p. 177, 195.
(55) See Odyssey, iv. 126. Phylo, one of the handmaidens of Helen, carries a silver basket, which had been given to her by Alcandra, the wife of Polybus. The latter was a native of the Egyptian Thebes, where the inhabitants are very wealthy.

entire history of Egypt for more than 5000 years; that is to say, for nearly double the period from the first Olympiad to the present day.

§ 5 After Manetho, among the authorities on Egyptian chronology, comes Eratosthenes, who was called to Alexandria by Evergetes, the third of the Ptolemies, and died about 196 B.C. He may be considered to have belonged to the generation next succeeding Manetho.([56])

The Egyptian chronology of Eratosthenes consists of a list of thirty-eight successive kings of Thebes, beginning with Menes, A.M. 2900, and ending with Amuthartæas, 3975 A.M.; it extends, therefore, over a period of 1076 years, from 2600 to 1524 B.C.,([57]) which gives about twenty-eight years to each king. Apollodorus, to whom we owe this list, states that Eratosthenes derived it from historical memorials written in the Egyptian language, and preserved at Diospolis, and that he translated it into Greek at the command of King Ptolemy.([58])

§ 6 After Eratosthenes, the next leading authority on Egyptian antiquity, now extant, is Diodorus; who visited Egypt in 60 B.C.,([59]) and composed his history about twenty years afterwards. He was separated from Eratosthenes by an interval of about a century and a half.

With respect to his account of the divine dynasties of Egypt, our attention may be confined to those points which are connected with chronology. He states that the interval of time

---

(56) Above, p. 198.

(57) See Fragm. Hist. Gr. vol. ii. p. 540, 545, 549, 554, 558, 565, 612. These extracts of Eratosthenes are preserved in Syncellus, and Syncellus makes the year of the world 5500 coincide with the Christian era; see vol. ii. p. 279, ed. Bonn.

(58) Syncellus, vol. i. p. 171, 279, ed. Bonn. The number $\gamma\mu'\epsilon$ in the first of these passages must be corrected from the second, as is remarked by Lepsius, Chron. der Æg. p. 511. The correction had been made in the Latin translation.

(59) Diodorus visited Egypt in Olymp. 180, in the reign of Ptolemy XIII., i. 44, 46, 83. See Clinton, F. H. ad ann. 60, 43. Concerning the Egyptian chronology of Diodorus, see Bunsen's Egypt, vol. i. p. 156, Eng. tr.; Lepsius, Chron. der Æg. p. 245.

between Osiris and Isis and the reign of Alexander the Great, was reckoned by some at more than 10,000, by others at somewhat under 23,000 years.[60] The latter number is likewise stated, on the authority of the Egyptian priests, as the interval between the reign of Helius and the invasion of Asia by Alexander.[61] In another passage he informs us that Egypt was computed to have been governed by gods and heroes for nearly 18,000 years; and that the human reigns, beginning with that of Mœris, had lasted nearly 5000 years up to Olymp. 180 (60 B.C.); most of these kings having been native Egyptians.[62] Elsewhere he says that Egypt had been governed by native kings for more than 4700 years.[63] The number 18,000 appears to be the difference between 23,000 and 5000: we may therefore assume that the more prevalent chronology for the divine and human dynasties, received by the informants of Diodorus, was 18,000 years for the former, and 5000 years for the latter: but that some placed the duration of the divine dynasties at only 5000 years; so that the divine and human dynasties together made up only 10,000 years.

Diodorus states that the native rulers of Egypt were 475 in number, of whom 470 were kings, and five were queens. He adds that concerning all of these, the priests were possessed of registers in the sacred books, handed down by each priest to his successor from ancient times; and that these registers contained a description of the physical appearance and moral character of each king, together with an account of the acts of his reign.[64]

The historian proceeds to specify the most remarkable in this series of 475 sovereigns. The first is Menas, who teaches religion to the people, and is also the author of luxurious living. He is succeeded by fifty-two kings, whose joint reigns amount to more than 1400 years, being an average of about

---

(60) Diod. i. 23.   (61) ii. 26.   (62) i. 44.
(63) i. 69.   (64) i. 44.

twenty-seven years each; nothing worthy of record occurred during their reigns. This long line of obscure princes was followed by Busiris, and eight descendants, of whom the last bore the same name as the original progenitor. They were the founders of the great and wealthy city of Thebes.(65) After an interval of seven kings, Uchoreus succeeds, the founder of Memphis.(66)

Ægyptus, the next king, was fabled to be the offspring of the daughter of Uchoreus, and of the river Nile, in the form of a bull. He was a good and just king, and gave his name to the country.(67)

After an interval of twelve generations, Mœris succeeds. He builds the great propylæa at Memphis, and excavates lake Mœris.

Another interval of seven generations elapses, and Sesoösis becomes king, the Sesostris of Herodotus, Manetho, and the other Greek writers. Diodorus informs us that great discrepancy existed, both among the Greek historians and the Egyptian priests and hymnologists, concerning the exploits of Sesoösis; he adds that his own narrative is founded on the most probable accounts, and those which agree best with the extant monuments of this king. According to the narrative of Diodorus, the greatness of Sesoösis was predicted by signs at his birth: he was brought up by a severe discipline, and trained to martial exercises :(68) and even during his father's lifetime he

---

(65) i. 45. Diodorus says that the stories respecting the cruelty of Busiris are fabulous, and that they had their origin in the inhospitality of the Egyptians to foreigners before the reign of Psammetichus, i. 67. According to another and inconsistent story reported by Diodorus, Busiris was a satrap of the N. E. part of Egypt, under King Osiris, i. 17.

(66) The words τῶν δὲ τούτου τοῦ βασιλέως ἀπογόνων in c. 50, would naturally refer to Osymandyas; but Osymandyas is introduced incidentally, and no fixed place in the series is assigned to him. I have therefore followed Lepsius in supposing that the seven unnamed kings intervene between Busiris II. and Uchoreus.

(67) It seems doubtful whether Diodorus means to fix the place of Ægyptus in the series, c. 51.

(68) Diodorus tells us that Sesoösis, and the youths who participated in his discipline, were not allowed any food until they had run 180 stadia = 22¼ miles (i. 53).

subdued Arabia, and the chief part of Libya. Having succeeded to the throne, he made extensive and long-sighted preparations for foreign conquest. He divided Egypt into thirty-six nomes, and appointed collectors of the royal revenues in each. At the same time he sought, by various beneficent measures, to render himself beloved by the people. He collected an army of 600,000 infantry, 24,000 cavalry, and 27,000 war chariots. With this vast host he conquered the Æthiopians; afterwards, with the aid of a fleet of 400 ships, he subdued India, as far as the country beyond the Ganges, the whole of Western Asia, most of the Cyclades, and Europe as far as Thrace. Stories are told similar to those in Herodotus, respecting the memorials of his expedition and the Colchian colony; the engraving on the monuments is described as in Manetho. From this triumphant expedition he returned at the end of nine years, and by it, according to Diodorus, established the reputation of the greatest military conqueror who ever lived. He was considered the author of all the regulations relating to the military system of Egypt.[69]

Diodorus informs us that a town in Egypt, named Babylon, was so called from some Babylonian captives, brought to Egypt by Sesoösis, who revolted on account of the severity of their taskwork, but were pardoned; and that another Egyptian town named Troy, was so called from Trojan prisoners, who were brought to Egypt by Menelaus, and obtained their liberty.[70] He admits, however, that, according to Ctesias, these two towns were founded by Babylonians and Trojans who accompanied Semiramis to Egypt.

Sesoösis likewise executed great works; he covered the

---

[69] Diod. i. 94.

[70] A similar account of the origins of the Egyptian Babylon and Troy is given by Strabo, xvii. 1, § 30, 34. The Egyptian Babylon and Troy are mentioned by Steph. Byz. in Βαβυλὼν and Τροία.

Villehardouin calls Egypt the 'land of Babylon.' Sismondi says that this appellation was derived from the Egyptian town of Babylon, referring to William of Tyre, l. xix. c. 13. See his Histoire des Rép. Ital. c. 14, vol. ii. p. 371, ed. 1818.

country with a complete system of canals; he built a wall 1500 stadia (= 187 miles) in length, on the eastern frontier, from Pelusium to Heliopolis; he constructed a ship of cedar wood, 280 cubits long, its external surface ornamented with gilding, and its internal with silver, which he consecrated to a god at Thebes: he also erected two stone obelisks 120 cubits in height, and two monolithic statues of himself and his queen, 30 cubits in height.([71])

Sesoösis the Great is succeeded by his son, of the same name, concerning whom Diodorus tells the same legendary stories which are related by Herodotus of Pheron, the son of Sesostris.([72])

At this point there is an interval of numerous kings, who did nothing worthy of being recorded. The next king who is thought deserving of mention is named Amasis; he is a tyrannical prince, and is dethroned by Actisanes, the Æthiopian king. The latter governs mildly; and his leniency towards certain robbers, to whom he assigns a residence in the desert, and on whom he inflicts no other punishment than the loss of their noses, furnishes an explanatory legend for the town Rhinocolura.([73])

After the death of Actisanes, the sceptre returns to a native king, Mendes, by some called Marrus. He constructed the Great Labyrinth, which served as the model for the Labyrinth of Crete, built by Dædalus: it remained entire at the time of Diodorus.([74]) An anarchy, of five generations, ensues; after which a man of low birth, named Ceten, obtained the

---

([71]) Diod. i. 53—58.

([72]) Diod. i. 59. Compare Herod. ii. 111. The verse of Ausonius, Epist. xix., 'Et qui regnavit sine nomine mox Sesoöstris,' seems to mean that the celebrated Sesostris was succeeded by a king of the same name, who did not support the great reputation of his predecessor.

([73]) Diod. i. 60. This legend recurs in Strabo, xvi. 2, § 31. On the other hand, Seneca represents the mutilation of the noses of a whole population as the cruel act of a Persian king, De Irâ, iii. 20. Concerning Rhinocolura, see Dr. Smith's Dict. of Anc. Geogr. in v.

([74]) i. 61. See above, p. 275, n. 116.

throne. Diodorus determines that he is identical with the king called Proteus by the Greeks, who was contemporary with the Trojan War.(75)

The next king, Remphis, was a lover of wealth, and accumulated 400,000 talents of gold: but owing to his miserly spirit he expended nothing in offerings to the gods, or in works useful to men.(76)

For the next seven generations the kings, with a single exception, were devoted to inglorious ease, and did nothing which deserved mention in the sacred registers. This exception was Nileus, who improved the channel of the river by various useful works, and gave it his own name, its previous appellation having been Ægyptus.(77)

Chemmis was the successor of Nileus: he reigned fifty years, and constructed the great pyramid.(78) He was succeeded by his brother Cephren, or by his son Chabryes, who reigned fifty-six years, and built the second pyramid.(79)

After them comes Mycerinus, the son of Chemmis: he began the third pyramid, but left it unfinished. This king mitigated the cruelty of his predecessors, and treated the people with mildness.(80) These three kings clearly correspond with the Cheops, Chephren, and Mycerinus of Herodotus.

Mycerinus was succeeded by Bocchoris, a king of mean appearance, but of great mental capacity, and distinguished as a lawgiver and judge.(81) A long interval elapses; after which Sabaco, the Æthiopian, a mild and humane sovereign, ascends the throne: he abolished capital punishment, and employed the convicts in public works. His voluntary resignation of the throne, and the motive for it, are described as in Herodotus.(82)

An anarchy of two years ensues; and is followed by the

---

(75) i. 62. Concerning Proteus as an Egyptian king, see Herod. ii. 112; Eurip. Helena, Apollod. ii. 5, 9; Tzetz. ad Lyc. 112, 124, 820.
(76) Diod. i. 62.   (77) i. 63.   (78) i. 63.
(79) i. 64.   (80) i. 64.   (81) i. 65, 94.
(82) i. 65. The story about the oracle being fulfilled by a libation from

Dodecarchy, to which Diodorus assigns fifteen years; Psammetichus, one of the Dodecarchs, makes himself king, as in the account of Herodotus.[83]

Diodorus states that the human reigns of Egypt had, in 60 B.C., lasted 5000 years: and in another place that the native kings reigned more than 4700 years; the difference between which numbers allows for the interval between Cambyses and his own time. We may, therefore, place the first year of Menas, according to the chronology of the Egyptian priests, as reported by Diodorus, at about 5000 B.C. He informs us further, that the native sovereigns who reigned during the period from Menas to Cambyses were 475 in number, but of these he only names twenty-one; the rest are too obscure for notice, and nothing memorable was recorded of them in the sacred registers.

The statements of Diodorus render it necessary that we should fix the first year of Menas at 5000 B.C. But it is difficult to reconcile this date with the details of his chronological scheme. He assigns 1400 years to the fifty-two unnamed kings after Menas, which gives about twenty-six years to each reign; and 106 years to the reigns of Chemmis and Cephren. If we deduct 1506 from 5000 years, and 54 from 475 kings,[84] we obtain 3494 years, to be divided among 421 kings; which gives an average of only about $8\frac{1}{4}$ years apiece.

The chronological canon of Diodorus, restored upon the assumption that each king, the duration of whose reign is not stated, reigned $8\frac{1}{4}$ years, will be as follows:—

---

the brazen helmet of Psammetichus, told in Herodotus, is contemptuously repeated by Diodorus, as 'a fable of some of the ancient historians.'
Bunsen, Egypt, vol. i. p. 147, Eng. tr., thinks that Mneues and Sasychis, mentioned by Diod. i. 94 as lawgivers, are regarded by him as kings. But Diodorus clearly distinguishes them from the next three lawgivers, Sesoösis, Bocchoris, and Amasis, all of whom he specially designates as kings.

[83] Diod. i. 66.
[84] I include the five queens among the kings.

KINGS.	YEARS.	B.C.
1. Menas	8¼	5000
53. Fifty-two unnamed kings . .	1400	
54. Busiris I.	8¼	
61. Seven unnamed kings . . .	57¾	
62. Busiris II.	8¼	
70. Eight unnamed kings . . .	66	
71. Uchoreus	8¼	
72. Ægyptus	8¼	
84. Twelve unnamed kings . . .	99	
85. Mœris	8¼	
92. Seven unnamed kings . . .	57¾	
93. Sesoösis I.	8¼	
94. Sesoösis II.	8¼	3254
Total . . .	1746½	

With respect to the remaining kings, 301 in number, of whom only fourteen are named, it is impossible to restore his chronological scheme, on account of the long intervals of uncertain duration, between Sesoösis II. and Amasis, and between Bocchoris and Sabaco.([85])

§ 7 Lastly, there is the scheme of the 'Ancient Chronicle,' cited by Syncellus, which makes the entire Egyptian chronology consist of 36,525 years, with the statement that this number is a multiple of the years of the Canicular cycle (1461 × 25). The chief part of this colossal number is assigned to the præ-human period. Nileus, the son of Vulcan, reigns 30,000 years; Saturn and the other twelve gods reign 3984 years; and eight demigods reign 217 years, making altogether, for the superhuman period, 34,201 years. There then follows a period of fifteen

---

([85]) The restoration of the canon of the Egyptian kings, according to Diodorus, made by Larcher, in his Chronologie d'Hérodote (Trad. d'Hérodote, tom. vii. p. 73, ed. 1802), seems to me to proceed on wholly erroneous principles. According to Larcher, Diodorus places the first year of Menas at 14,940 B.C. The general principle of his restoration is to assign thirty years to each king (475 × 30 = 14,250). He entirely disregards the statements of Diodorus as to the total duration of the Egyptian monarchy.

generations of the Canicular cycle, occupying 443 years, to which neither divine nor human government is assigned. Next in order are fifteen dynasties of mortal kings, reaching to the reign of Nectanebus, in 341 B.C. These fifteen mortal dynasties ought to occupy a period of 1881 years, if the numbers for the divine and heroic dynasties are correct. The figures in the extant text, however, give only 1697. About the sum total of 36,525 no doubt can exist, as the factors of the product are stated. The chronological result of the scheme may therefore be thus exhibited:—

	YEARS.	B.C.
Superhuman reigns	34,201	
Intermediate period	443	2665—2223
Fifteen human dynasties	1,881	2222— 341
	36,525 [86]	

The antiquity of this chronicle was first suspected by Des Vignoles:[87] Letronne declared it to be the production of a Jew or a Christian, subsequent to Ptolemy the astronomer, who wished to reduce the Egyptian dynasties into harmony with the Biblical chronology.[88] Boeckh has lately investigated the subject at length, and has proved, by convincing arguments, that this 'Ancient Chronicle' was composed in the interval between Eusebius and the two chronographers Anianus and Panodorus;[89] that is, during the fourth and fifth centuries.

§ 8 It now remains for us to compare these different ac-

---

[86] See Fragm. Hist. Gr. vol. ii. p. 534; Boeckh, Manetho, p. 40—57. In Boeckh's table, p. 41, the number for the twenty-sixth dynasty is erroneously stated as 117, instead of 177 (as C. Müller has remarked); and this error has affected the subsequent calculations.

Concerning the scheme of this 'Ancient Chronicle,' see Des Vignoles, Chron. de l'Hist. Sainte, vol. ii. p. 659.

[87] Vol. ii. p. 659, 663.

[88] See Biot, Recherches sur l'Année vague des Egyptiens, p. 569.

[89] Manetho, p. 52—57. His conclusion is accepted by C. Müller, Fragm. Hist. Gr. vol. ii. p. 536.

Concerning Anianus and Panodorus, see Bunsen, Egypt, vol. i. p. 208, Eng. tr.

counts of Egyptian antiquity, and to attempt to determine their claims to credibility.

The native Egyptian race seems to have resembled the black African races, but nevertheless to have been distinct in its character. It belonged to a more intelligent type, and at an early period attained to a higher civilization than they have ever reached. Its affinity, however, was with the Asiatics, not with the Greeks; it approximated to the neighbouring Arabians, Syrians, and Assyrians.([90]) The Egyptians participated in the Oriental type: they had writing, but no literature or history. The lively and ingenuous, but simple mind of Herodotus was imposed upon by their readiness in fabricating explanatory stories about their buildings, and statues of men, and was led to believe in their historical knowledge.([91]) Their pretensions to a remote antiquity, ascending to thousands of years, were admitted by the Greeks,([92]) and induced ancient writers to speak of their

---

([90]) Some of the ancients made the Nile the boundary of Asia and Africa, Strab. i. 2, § 25, 30, i. 4, § 7. Others included the whole of Egypt in Asia, Plin. N. H. v. 9. According to Juba, ap. Plin. vi. 29, there was an affinity between the Egyptians and Arabians. Mr. Kenrick remarks, in reference to the ethnological character of the ancient Egyptians: 'The Egyptians may be said to be intermediate between the Syro-Arabian and the Ethiopic type; but a long gradation separates them from the negro. The evidence derived from the examination of the skulls of the mummies approximates the Egyptians rather to the Asiatic than the African type,' Ancient Egypt, vol. i. p. 98. Sir Gardner Wilkinson says: 'In manners, language, and many other respects, Egypt was certainly more Asiatic than African,' Ancient Egyptians, vol. i. p. 3. Brugsch, Histoire d'Egypte, part i. p. 1, thinks that the Egyptians were of the Caucasian race, and came originally from Asia. Æschylus, Supp. 719, 745, describes the crew of the Egyptian ship which arrived at Argos, as being of black colour; but he appears to refer to the rowers, who probably were Æthiopian slaves. Manilius represents the Egyptians as less dark in colour than the Æthiopians:—

Jam proprio tellus gaudens Ægyptia Nilo
Lenius irriguis infuscat corpora campis.—iv. 726.

Herod. ii. 104, says that the Colchians resemble the Egyptians in having black skins and woolly hair (μελάγχροες καὶ οὐλότριχες). Aristotle classes the Æthiopians and Egyptians together, and says that both are splay-footed, and woolly-haired, Problem. xiv. 4. These are both negro characteristics. Ammianus Marcellinus assigns a dark colour to the Egyptians. Homines Ægyptii plerique subfusculi sunt et atrati, xxii. 16, § 23.

([91]) Herod. ii. 77.

([92]) See above, p. 275.

historical records.(93) But for the existence of an authentic contemporary registration in Egypt, made by an intelligent historian, we have no sufficient warranty.(94)

The historical information respecting Egyptian antiquity is represented to us as derived exclusively from the priests, and from their sacred books, preserved in the archives of temples. Now the Egyptian priests had the character of an Oriental sacerdotal caste, like the Chaldæans, the Magi, or the Brahmins. Their knowledge was suited to a country in which there was neither freedom of thought nor activity of mind; which produced nothing useful, and which contributed nothing to the progress of mankind; whose despotic masters employed all the surplus labour of the people in constructing pyramids and labyrinths, and other colossal works, destitute of any rational destination, and only intended to perpetuate their own memory.(95)

If the priests and their sacred books are not admitted to be trustworthy authorities upon Egyptian antiquity, the whole basis of our supposed knowledge fails. From them the information of Herodotus, Manetho, Eratosthenes, and Diodorus was equally derived. We learn from the express testimony of Herodotus himself, that the priests not only gave him an oral ac-

---

(93) Theophrast. ap. Porph. de Abst. ii. 5; Cic. de Rep. iii. 9. The registration of prodigies mentioned in Herod. ii. 82, has nothing to do with political history. See above, p. 70, n. 275.

(94) This fact is admitted by Bunsen: 'If then (he says) the sacred books of the Egyptians contained no single section of pure history, we cannot wonder that we hear of no historical work of that people before Manetho; that is, before they came in contact with the genius of Hellas. These books contained all that the Egyptians possessed of science or historical lore,' Egypt, vol. i. p. 23, Engl. tr. Compare p. 254.

(95) Aristotle enumerates the Egyptian pyramids among the great works executed by despotic rulers, for the purpose of keeping the people employed, and of making them poor, Polit. v. 11. Herod. ii. 135, says of the pyramid built by the daughter of Cheops, ἰδίῃ καὶ αὐτὴν διανοηθῆναι μνημήιον καταλιπέσθαι. Diodorus accounts for the splendid tombs of the Egyptians, as compared with their dwelling-houses, by saying that the former were considered as intended for perpetual, the latter for temporary occupation, i. 51. Bunsen, however, states that the pyramids do not in general contain any sepulchral chamber, and he thinks that their destination is unknown, vol. ii. p. 389. Diodorus, ii. 7, states that a vast tomb of Ninus, the primitive Assyrian king, was erected by his wife Semiramis, nine stadia high, and ten stadia wide.

count of their history,(⁹⁶) and answered questions which he put to them concerning the visit of Menelaus and Helen to Egypt,(⁹⁷) but also that they read to him the names of 330 kings from a book,(⁹⁸), and that they showed him the statues of a series of 345 high priests, in proof of the truth of their chronological statements.(⁹⁹) Manetho, as we learn from Josephus, himself announced that his Egyptian history was translated from sacred books.(¹⁰⁰) Eratosthenes derived his Egyptian chronology from the same source.(¹⁰¹) Diodorus likewise professes to have obtained his information respecting ancient Egypt from authentic registers of the native priests;(¹⁰²) and by the assistance of these memorials to give more credible and trustworthy information than that to be found in Herodotus and other previous histories.(¹⁰³)

---

(96) In the whole narrative, ii. 99—146, Herodotus repeatedly refers to the oral information of the priests:—Αἰγυπτίους ἔρχομαι λόγους ἐρέων—κατὰ τὰ ἤκουον—οἱ ἱρέες ἔλεγον—ὡς ἔλεγον οἱ ἐν Σάϊ πόλι ἱρέες—κατὰ τῶν ἱρέων τὴν φάτιν.

(97) ἔλεγον δέ μοι οἱ ἱρέες ἱστορέοντι τὰ περὶ Ἑλένην, γενέσθαι ὧδε, ii. 113. εἰρομένου δέ μευ τοὺς ἱρέας, εἰ μάταιον λόγον λέγουσι οἱ Ἕλληνες τὰ περὶ Ἴλιον γενέσθαι, ἢ οὔ· ἔφασαν πρὸς ταῦτα τάδε, ἱστορίῃσι φάμενοι εἰδέναι παρ' αὐτοῦ Μενέλεω, ii. 118.

(98) κατέλεγον οἱ ἱρέες ἐκ βύβλου, ii. 100.

(99) ἀριθμέοντες ὦν καὶ δείκνυντες οἱ ἱρέες ἐμοὶ, ἀπεδείκνυσαν παῖδα πατρὸς ἑωυτῶν ἕκαστον ἐόντα, ἐκ τοῦ ἄγχιστα ἀποθανόντος τῆς εἰκόνος διεξιόντες διὰ πασέων, ἕως οὗ ἀπέδεξαν ἁπάσας αὐτάς, ii. 143.

(100) Μανεθὼν δὲ ἦν τὸ γένος ἀνὴρ Αἰγύπτιος, τῆς Ἑλληνικῆς μετεσχηκὼς παιδείας, ὡς δῆλός ἐστι· γέγραφε γὰρ Ἑλλάδι φωνῇ τὴν πάτριον ἱστορίαν, ἔκ τε τῶν ἱερῶν, ὥς φησιν αὐτός, μεταφράσας, καὶ πολλὰ τὸν Ἡρόδοτον ἐλέγχει τῶν Αἰγυπτιακῶν ὑπ' ἀγνοίας ἐψευσμένον, Contr. Apion. 1. § 12. After ἱερῶν, the word γραμμάτων, or γραφῶν, or βίβλων, seems to have fallen from the text (see Lepsius, ib. p. 534). Μανεθὼς, ὁ τὴν Αἰγυπτιακὴν ἱστορίαν ἐκ τῶν ἱερῶν γραμμάτων μεθερμηνεύειν ὑπεσχημένος, ib. § 26.

(101) ὧν τὴν γνῶσιν, φησὶν [Ἀπολλόδωρος χρονικός], ὁ Ἐρατοσθένης λαβὼν Αἰγυπτιακοῖς ὑπομνήμασι καὶ ὀνόμασι κατὰ πρόσταξιν βασιλικὴν τῇ Ἑλλάδι φωνῇ παρέφρασεν οὕτως, Syncell. vol. i. p. 171, Bonn. ἡ τῶν λή βασιλέων τῶν κατ' Αἴγυπτον λεγομένων Θηβαίων, ὧν τὰ ὀνόματα Ἐρατοσθένης λαβὼν ἐκ τῶν ἐν Διοσπόλει ἱερογραμματέων παρέφρασεν ἐξ Αἰγυπτίας εἰς Ἑλλάδα φωνήν, ib. p. 279. For ἱερογραμματέων in this passage, the sense seems to require ἱερῶν γραμμάτων. Compare the passage of Josephus, § 26, cited in the last note.

(102) See above, p. 322.

(103) ὅσα μὲν οὖν Ἡρόδοτος καί τινες τῶν τὰς Αἰγυπτίων πράξεις συνταξαμένων ἐσχεδιάκασιν, ἑκουσίως προκρίναντες τῆς ἀληθείας τὸ παραδοξολογεῖν καὶ μύθους πλάττειν ψυχαγωγίας ἕνεκα, παρήσομεν, αὐτὰ δὲ τὰ παρὰ τοῖς ἱερεῦσι τοῖς κατ' Αἴγυπτον ἐν ταῖς ἀναγραφαῖς γεγραμμένα φιλοτίμως ἐξητακότες ἐκθησόμεθα, i. 69. This passage refers to the general description of Egypt.

Now, if the Egyptian priests had, in the time of Herodotus, been in possession of complete and authentic written records of their ancient history, it must be supposed that these records would be carefully preserved in the temple archives, and that they would be identical with the records translated by Manetho and Eratosthenes, about two centuries afterwards, and consulted by Diodorus at the interval of another century and a half. But on comparing the accounts of the history and chronology given by these four writers, we find the utmost discrepancy between them.

According to the report of the priests to Herodotus, the divine dynasties governed Egypt for at least 5600 years; and these reigns were stated to have been the subject of contemporaneous registration, not less than the subsequent human reigns.([104]) Men, the first king, and 330 unnamed successors, reigned 10,320 years, from 11,400 to 1080 B.C.; and twelve named kings, from Mœris to Sethon, reigned 400 years, from 1080 to 680 B.C. In this scheme, the total number of kings, from Men to Sethon, is 343.

Manetho gives to the divine and semi-divine dynasties a duration of 24,925 years; the first king, Menes, begins to reign in 5702 B.C. He, and 438 successors, occupying twenty-five dynasties, reign till 680 B.C. The total number of kings during this period is 439.

Eratosthenes enumerates thirty-eight Egyptian kings of Thebes, beginning with Menes. They reign 1076 years, from 2600 to 1524 B.C.

According to the scheme preserved by Diodorus, the divine and semi-divine dynasties reign for either 18,000 or 5000 years. Menas, the first king, begins his reign in 5000 B.C. Until the time of Cambyses, 525 B.C., the total number of kings is 475, or 470 to Psammitichus.

These four schemes harmonize in making Menes the first

([104]) ii. 145.

human king of Egypt; but they differ widely in the dates which they assign to him, which are respectively as follows:—

Date of Menes according to	B.C.
Herodotus	11,400
Manetho	5702
Eratosthenes	2600
Diodorus	5000

When the discordance is on so vast a scale, the dates of Manetho and Diodorus, differing from each other by only 700 years—that is to say, by about the period from the era of the Olympiads, or of the Foundation of Rome, to the Birth of Christ, or from the Norman Conquest of England to the reign of George the Third—seem to be nearly identical.

The average duration of the reigns, moreover, and the number of kings, differ widely in these several schemes.

Average duration of Reigns according to	Years.
Herodotus	$33\frac{1}{3}$
Manetho	$11\frac{1}{2}$
Eratosthenes	28
Diodorus	$9\frac{1}{4}$

Number of Reigns from Menes to Psammitichus, according to	
Herodotus	343
Manetho	439
Diodorus	470

Here, again, the divergence of Diodorus from Manetho, being only to the amount of thirty-one reigns—that is to say, as many reigns as from Hugues Capet to Louis XIV., or from William the Conqueror to George the First—appears quite trivial, and, in the midst of such utter uncertainty, to be a close approximation to agreement.

If we attempt to compare the names of the kings, and to determine their mutual correspondence in the several lists, the discrepancy is still greater, and the confusion still more hopeless.

The names of the immediate successors of Menes are not

stated in Herodotus and Diodorus; but they appear in the lists both of Manetho and Eratosthenes, and these differ entirely, except as to Menes himself and his successor. The first ten names, with the years of each reign, stand thus in the two lists:—

Manetho.		Eratosthenes.	
Kings.	Years of reign.	Kings.	Years of reign.
1. Menes	62	1. Menes	62
2. Athothis	57	2. Athothes I.	59
3. Kenkenes	31	3. Athothes II.	52
4. Uenephes	23	4. Diabaes	19
5. Usaphaedus	20	5. Pemphos	18
6. Miebidus	26	6. Momcheiri	79
7. Semempses	18	7. Stoechus	6
8. Bieneches	26	8. Gosormies	26
9. Boethus	38	9. Mares	26
10. Kaiechos	39	10. Anoyphis	20[105]

But it is in the later period, in the centuries immediately preceding Psammitichus, subsequent to the reigns of David and Solomon, and in part even later than the poems of Homer and the era of the Olympiads, that this discrepancy of the lists is most remarkable. The nearest approach to agreement is between Herodotus and Diodorus; but how near that approach is, will appear from the following comparison:—

Herodotus.		Diodorus.
1. Mœris	1080 B.C.	1. Mœris.
		2. An interval of 7 generations.
2. Sesostris		3. Sesoösis I.
3. Pheros		4. Sesoösis II.
		5. An interval of numerous kings.

---

(105) Bunsen alters Διαβαῆς and Πεμφῶς, in Eratosthenes, upon conjecture, into Μαεβαῆς and Σεμψῶς, in order to make them harmonize with Μιεβιδὸς and Σεμέμψης in Manetho. In like manner he metamorphoses Μόμχειρι into Σεσορχέρης, Γοσορμίης into Σεσόρτασις, and 'Ανωϋφίς into *Αν ἢ Σωϋφίς.

HERODOTUS.	DIODORUS.
	6. Amasis.
	7. Actisanes.
	8. Mendes.
	9. An anarchy of 5 generations.
4. Proteus	10. Ceten, called Proteus by the Greeks.
5. Rhampsinitus	11. Remphis.
	12. An interval of 7 generations, during which Nileus is king.
6. Cheops	13. Chemmis.
7. Chephren	14. Cephren.
8. Mycerinus	15. Mycerinus.
9. Asychis	16. Bocchoris.
10. Anysis	17. A long interval.
11. Sabacos	18. Sabaco.
12. Sethon	
13. Dodecarchy	19. Anarchy and Dodecarchy.
14. Psammitichus	20. Psammitichus.

With regard to Manetho, the discrepancy is so complete, that it is difficult to institute any comparison. The following is the series of the last 27 kings in his list immediately preceding Psammitichus. He mentions neither the Dodecarchy, nor Cheops, nor Mœris; and he places Sesostris at a much earlier period:—

  1. Smendes . . . . . 1048 B.C.
  2. Psusennes.
  3. Nephercheres.
  4. Amenophthes.
  5. Osochor.
  6. Psinaches.
  7. Psusennes.
  8. Sesonchis.
  9. Osorthon.
 10—12. Three unnamed kings.

13. Takelothis.
14—16. Three unnamed kings.
17. Petubates . . . . . 776 B.C.
18. Osorcho.
19. Psammus.
20. Zet.
21. Bocchoris.
22. Sabacon.
23. Sebichos.
24. Tarcus.
25. Stephinates.
26. Nechepsos.
27. Nechao I.
28. Psammitichus.

In this list, the four kings from 24 to 27, Tarcus, Stephinates, Nechepsos, and Nechao, correspond with the period of the Dodecarchy in Herodotus and Diodorus. Psammitichus is stated by Herodotus to be the son of Neco.[106] Sebichos corresponds with the Sethon of Herodotus. Sabacos occurs both in Herodotus and Diodorus. Bocchoris is wanting in Herodotus, but is named by Diodorus. At this point all correspondence with either historian ceases. The names from Smendes to Zet, all posterior to 1048 B.C., the period of the reign of David in the history of the Jews, are peculiar to Manetho.

Now it is to be observed, that these discordant schemes all profess to be derived from the same authentic source. They cannot be reconciled by any legitimate methods of criticism, and yet there is no satisfactory ground for preferring one to another.[107] We are not entitled to assume that any one of our authorities was intentionally deceived by the priests,[108] or that

---

[106] ii. 152.

[107] Bunsen prefers Eratosthenes to Manetho, as a guide to Egyptian chronology, Egypt, vol. i. p. 134, vol. ii. p. 19, Eng. Tr. Lepsius, on the other hand, prefers Manetho to Eratosthenes, Chron. der Æg. p. 407. Both preferences are equally arbitrary.

[108] Lepsius lays it down that no intentional deceit was practised by the Egyptian priests upon Herodotus, Chron. der Æg. p. 248.

he reported or transcribed his information incorrectly. Having therefore no sufficient reason for selecting any one of these systems, we are compelled, by the laws of historical evidence, to reject them all.

It remains to be seen whether the events attributed by our authorities to the history of Egypt before Psammitichus supply the defect of credibility in the chronological schemes.

The facts reported by Herodotus for this period have in general the character either of fabulous stories, involving incredible marvels, or of monumental legends, accounting for the origin of some building or constructive work, extant in the historical time. Many of these narratives are repeated, with slight variations, by Diodorus. The short notices appended to the names of some of the kings in Manetho are still more puerile and worthless than the fuller accounts in Herodotus and Diodorus, and seem to be the clumsy productions of recent fiction.

As examples of stories incredible from their marvellous character, we may instance the account of Pheros, in Herodotus, transferred by Diodorus to Sesoösis II., exhibiting the interference of the gods for his punishment and cure; likewise the adventures of the thief in the reign of Rhampsinitus, resembling a story in the Arabian Nights. Several of the notices in Manetho partake of the same character. Thus Menes is said to have been torn in pieces by a hippopotamus, which is a herbivorous, not a carnivorous animal; under Nephercheres, the waters of the Nile were for eleven days mixed with honey; Sesochris was five cubits high; under Necherophes there was a preternatural enlargement of the moon; and under Bocchoris a lamb spoke. The latter incident is likewise mentioned by Ælian, who informs us that the Egyptians were particularly proud of this prodigy; he adds, moreover, that the lamb of Bocchoris had eight feet and two tails.([109]) The medical science

---

(109) Nat. An. xii. 5. Joannes Antiochenus likewise mentions the speaking lamb of Bocchoris: he adds that Sabacon, king of the Æthiopians, took Bocchoris prisoner, and either burned him or flayed him alive. Fragm. Hist. Gr. vol. iv. p. 539.

of Athothis and Tosorthrus, the medical treatise of the former, and the book of Suphis, may likewise be classed among the incredible stories.

The extensive conquests of Sesostris, and his vast armies, likewise belong to the marvellous figments of Egyptian antiquity, and are unworthy of any credit. According to Herodotus, Sesostris was the next king after Mœris, which, according to his computation by generations, would place Sesostris about 1046 B.C. Another statement of Herodotus respecting the time of Mœris, would indeed lead to the inference that Sesostris reigned as early as 1300 B.C.;[110] but the former date agrees best with the scheme of Egyptian chronology exhibited by him. The date assigned to Sesostris by Diodorus is much higher, viz., 3625 B.C. Aristotle lays it down that the reign of Sesostris was long anterior to that of Minos;[111] which latter reign he doubtless conceived as prior to the Trojan War. Strabo, who says that the canal from the Nile to the Red Sea was by some attributed to Sesostris, places him 'before the Trojan War,' without any more precise specification.[112] In the list of Manetho, Sesostris stands in the 12th dynasty, 3320 B.C.

A great Egyptian conqueror, described by Apollonius Rhodius as having overrun Europe and Asia, and having planted a colony at Colchi,[113] is stated by the Scholiast to be Sesostris, or, as he was called by Dicæarchus, Sesonchosis. Dicæarchus represented this Sesonchosis to have succeeded Orus, the son of Isis and Osiris, and to have reigned 2500 years before Nilus, who again reigned 436 years before the First Olympiad; so that the date of Sesonchosis, according to Dicæarchus, is 3712 B.C. He further attributed to Sesonchosis the general character of a civilizer, and described him as having first enacted the Egyp-

---

(110) See above, p. 332, n. 26.

(111) πολὺ γὰρ ὑπερτείνει τοῖς χρόνοις τὴν Μίνω βασιλείαν ἡ Σεσώστριος, Polit. vii. 10.

(112) xvii. 1, § 25. This report also occurs in Pliny, N. H. vi. 29.

(113) iv. 272.

tian law which made trades hereditary, and as having taught mankind to ride on horseback.(114)

The notion of a great Egyptian conqueror, who overran all Asia as far as the Ganges, and subdued the western part of Europe,(115) is not only quite inconsistent with the Jewish annals, which reach back, in an uninterrupted stream, at least as far as Solomon and David;(116) but also with the earliest traditions of authentic Greek history. Central Asia had not, to the knowledge of the Greeks, been opened by any Western invader, before the time of Alexander. The monuments cited by Herodotus in proof of the conquests of Sesostris, are not more conclusive evidence than the reliques of Jason, Ulysses, and Æneas, preserved in various towns of Italy, which were held to demonstrate the former presence of those heroes on the Italian coast. Herodotus had seen one of these monuments in Palestine; he mentions two others as extant in his time—one between Ephesus and Phocæa, the other between Sardis and Smyrna; but he declares that the rest had disappeared.(117) In the time of Strabo, other monuments had been connected with

---

(114) Schol. Apollon. Rhod. iv. 272, 276, 277. Compare Fragm. Hist. Gr. vol. ii. p. 235; Bunsen, vol. i. p. 682. The words, φησὶ δὲ Δικαίαρχος ἐν δευτέρῳ καὶ Ἑλληνικοῦ βίου καὶ Σεσογχώσιδι μεμεληκέναι, ought to be written thus, after Bunsen and C. Müller: φησὶ δὲ Δικαίαρχος ἐν βίῳ Ἑλλάδος καὶ πολιτικοῦ βίου Σεσογχώσιδι μεμεληκέναι.

According to Virgil, Georg. iii. 115—7, the Lapithæ taught the art of riding. Pliny states that Bellerophon invented the art of riding, that Pelethronius invented the bridle and the saddle, and that the Centaurs first taught the art of fighting on horseback, vii. 56. Bellerophon is introduced in this fabulous origin on account of his connexion with Pegasus; see Pindar, Olymp. xiii. 65. Compare Hygin. fab. 274. The art of riding was attributed to Sesostris as being a conqueror. The *bellator equus* was connected with military pursuits; see Virg. Æn. i. 444, iii. 539.

(115) Megasthenes stated that Sesostris invaded Europe, but denied the truth of his expedition to India, Strab. xv. 1, § 6. His conquest of the Getæ is mentioned by Val. Flac. Argonaut. v. 419. He subdued most of the Greek States, according to Athenodorus, Fragm. Hist. Gr. vol. 3, p. 487. Eustathius ad Dionys. Perieg. p. 80, ed. Bernhardy, regards Sesostris rather in the light of a traveller than of a conqueror, and says that he constructed geographical maps, which he gave, not only to the Egyptians, but even to the Scythians.

(116) Josephus, Bell. Jud. iv. 3, 10, represents Ananus, the high priest, as declaring that the Jews had never been subjugated by the Egyptians.

(117) ii. 106.

the name of Sesostris: one near Dira, on the African side of the Straits of Babelmandel, commemorated his passage across the sea from Æthiopia to Arabia, whence he overran all Asia; many ditches, or fortifications, bearing the name of this fabulous Egyptian conqueror, and other records of his march through Æthiopia, were known to Strabo.[118] These monuments resemble the Memnonea in Asia, which were considered as works, or tombs, of the fabulous prince Memnon, the son of Aurora;[119] the Jasonia in Media, said to have been built by Jason;[120] and the great works attributed to queen Semiramis.[121]

The affinity of the Colchians to the Egyptians, which Herodotus deduces from the expedition of Sesostris,[122] is, like the affinity of the Etruscans to the Lydians, repeated by later authors,[123] but is probably no better authenticated. Herodotus infers the affinity of the Colchians and Egyptians, from both nations being dark skinned and woolly haired, and from their having the customs of circumcision and wearing linen in common; but he speaks with doubt as to the manner in which the colony was planted by Sesostris.

The Egyptians do not appear to have been a warlike and conquering nation: with a few exceptions, they extended their arms neither to the West nor to the East; they did not permanently subjugate any people beyond the Valley of the Nile. Their kings employed the surplus labour of the country in exe-

---

[118] xvi. 4, § 4, xvii. 1, § 5.

[119] Concerning the Memnonea, see Diod. ii. 22; and compare Thirlwall, in Phil. Mus. vol. ii. p. 146. The monuments of Sesostris were confounded with those of Memnon, Herod. ii. 106.

[120] Strab. xi. 13, § 10; ib. 14, § 12; Grote, Hist of Gr. vol. i. p. 329.

[121] See below, ch. vii. § 7.

[122] ii. 104.

[123] The Egyptian colony of Colchi is mentioned in Apollon. Rhod. iv. 272—81, and Scymnus Chius, ap. Schol. ad loc. Strabo refers to the affinity of the Egyptians and Colchians, xi. 2, § 17. Also Dionys. Perieg. 689, Ammian. Marcellin. xxii. 8. Diod. i. 28, first states that the Colchians were a colony of Egyptians who migrated separately, without reference to Sesostris; afterwards, i. 55, he repeats the account of Herodotus. Ritter, Vorhalle Europ. Völkergesch. p. 40, doubts the story of the Sesostrian foundation of Colchi.

cuting gigantic but useless works of construction.([124]) They had a military caste,([125]) but it was employed in coercing the king's subjects, not in extending his dominions. Nevertheless, the Greeks conceived Sesostris as a mighty conqueror, who, as Diodorus tells us, harnessed kings to his chariot:([126]) even Aristotle represents him as the author of the hereditary separation of the military race from the civil population in Egypt.([127])

Such was unquestionably the predominant Greek conception of Sesostris. But though predominant, it was not universal; for Nymphodorus of Syracuse, in a work on the Institutions of Barbarous Nations, which he appears to have composed in the time of Ptolemy Philadelphus (283—247 B.C.), states that Sesostris introduced customs in Egypt for the purpose of breaking the spirit of the men, and of rendering them effeminate,([128]) similar to those which Herodotus supposes to have been established by Cyrus for insuring the obedience of the Lydians.([129])

When Germanicus visited Egypt, for the sake of seeing its vast remains of antiquity, one of the priests interpreted to him an inscription in native Egyptian—doubtless hieroglyphic—characters upon the walls of a building at Thebes. The inscription, as interpreted to Germanicus, declared that king Rhamses, with an army of 700,000 men, conquered Libya, Æthiopia, Media, Persia, Bactriana, and Scythia; and extended his rule over Syria, Armenia, Cappadocia, and the rest of Asia Minor: it then proceeded to enumerate the tributes of the subject

---

(124) Strabo says of the Egyptian temples: πλὴν γὰρ τοῦ μεγάλων εἶναι καὶ πολλῶν καὶ πολυστίχων τῶν στύλων, οὐδὲν ἔχει χαρίεν οὐδὲ γραφικὸν, ἀλλὰ ματαιοπονίαν ἐμφαίνει μᾶλλον, xvii. 1, § 28.

(125) Concerning the military caste, see Herod. i. 164—8.

(126) i. 58. Pliny tells the same story, xxxiii. 3. Menander Protector, the continuator of Agathias (about 590 A.D.), repeats the story, and mentions the saying of one of the harnessed kings, who directed the attention of Sesostris to the revolution of the wheels of his chariot, Fragm. Hist. Gr. vol. iv. p. 210.

(127) Pol. vii. 10.

(128) Fragm. Hist. Gr. vol. ii. p. 380.

(129) i. 135.

nations.([130]) Pliny states that Rhamses was contemporary with the taking of Troy.([131]) The Rhamses of this supposed inscription is only another version of Sesostris; the whole account is manifestly a vainglorious fiction of the priest, who could indulge his imagination, and excite the wonder of Germanicus, without the fear of detection.

Josephus identifies Sesostris with Shishak, king of Egypt,([132]) who, according to the Biblical history, took Jerusalem, and plundered the Temple, in 980 B.C.([133]) This identification is wholly inconsistent with the dates obtained from the Greek writers for Sesostris.

The reports respecting king Bocchoris are likewise discordant. Diodorus places Bocchoris low in his series of kings; Manetho places him next before Sabaco, only a few reigns before Psammitichus. On the other hand, Lysimachus of Alexandria,([134]) followed by Tacitus,([135]) states that it was Bocchoris who expelled the Jews from Egypt. This statement, taken in connexion with the Biblical history, would refer his reign to a much earlier period. Lysimachus himself dated the reign of Bocchoris at about 1700 years before the Christian era.([136]) Bocchoris is described by Diodorus as a great lawgiver and judge;([137]) and he is reported to have imitated his father, who was distinguished by the simplicity of his diet.([138])

---

(130) Tac. Ann. ii. 59, 60. The visit took place in 19 A.D. Compare Merivale's Hist. of Rome under the Empire, vol. v. p. 78.

(131) xxxvi. 8. The place of Ramses, or Ramesses, in the list of Manetho is uncertain. Boeckh places Ramses the Great at 1411 B.C. See his Manetho, p. 294—9.

(132) Ant. viii. 10, § 2. Compare Marsham, Can. Chron. p. 376.

(133) See 1 Kings xiv. 25; 2 Chron. xii.; and compare Winer in Sisak. The date is the fifth year of Rehoboam.

(134) Ap. Joseph. contr. Ap. i. 34; Fragm. Hist. Gr. vol. iii. p. 334.

(135) Hist. v. 3.

(136) Joseph. contr. Ap. ii. 2. The reduction of this number by Böckh, Manetho, p. 325, is inadmissible.

(137) Diod. i. 65, 79, 94.

(138) Alexis, ap. Athen. x. p. 418 D; Diod. i. 45; Plut. de Is. et Os. 8. Concerning Alexis, see Fragm. Hist. Gr. vol. iv. p. 299. The father of Bocchoris is named Neochabis, according to Alexis; Tnephachthus, according to Diodorus; and Technactis, according to Plutarch.

Ælian, on the other hand, declares that the fame of Bocchoris for justice and piety was undeserved, and that he was in reality a wicked and sacrilegious king.([139]) Plutarch, moreover, states him to have been so severe a judge, that Isis placed an asp on his head, in order to warn him against injustice.([140])

A large part of the history of the Egyptian kings, in Herodotus and Diodorus, is made up of architectural legends.([141]) Each king is the author of some constructive work, to which his name is attached. Menes is the founder of Memphis, of which town he appears indeed to be the eponymous hero. Mœris excavated the lake which bore his name.([142]) The pyramids and the labyrinth are assigned to various kings. According to Herodotus and Diodorus, the three great pyramids were built by Cheops (or Chemmis) and Chephren, two brothers,([143]) and Mycerinus, the son of Cheops. Manetho states that Uenephes built the pyramids near Cochome; and that Suphis, and not Cheops, was the builder of the great pyramid.

The legendary character of the stories respecting the founders of the pyramids appears from the following passage of Diodorus:—'With regard to the pyramids, there is no agreement either among the native authorities or the Greek historians. Some say that they were built by Chemmis, Cephren, and Mycerinus; some assign them to other names—as the great pyramid to Armæus, the second to Amosis, and the third to Inaros. Some, again, say that the third pyramid is the tomb of Rhodopis, the courtezan, which was built by a contribution of several of the monarchs, her former lovers.'([144]) Pliny reports the same

---

([139]) De Nat. An. xi. 11.

([140]) De vitioso pudore, c. 3.

([141]) Lepsius remarks that the Egyptian history of Herodotus is chiefly composed of narratives concerning the authors of remarkable monuments, Chron. der Æg. p. 249.

([142]) Concerning the lake of Mœris, see Bunsen, vol. ii. p. 338–368, Eng. tr.

([143]) Herodotus states that two brothers reigned 106 years, which is an impossibility. The variation mentioned in Diodorus that the first brother was succeeded by his son (i. 64), was devised in order to obviate this difficulty.

([144]) i. 64.

conflict of testimony, and declares that the true names of the builders of these gigantic masses have been obliterated by time.([145])

Rhodopis, the courtezan, to whom the building of the third pyramid was attributed, was a contemporary of Sappho, or even of Amasis.([146]) A fabulous story, told by Strabo, represents her as becoming the king's wife.([147]) Manetho assigns the third pyramid to Nitocris, a queen of the sixth dynasty, who reigned in 4211—4200 B.C. Herodotus makes her one of the predecessors of Mœris.([148]) She is likewise included by Eratosthenes in his list: he makes her equivalent to Minerva Victrix.([149])

Various founders are assigned for the Labyrinth. According to Diodorus, it was erected by Mendes, who reigned before the builders of the pyramids.([150]) Herodotus attributes its construction to the twelve kings of the Dodecarchy; an origin which may have grown out of the circumstance that it consisted of twelve halls.([151]) Manetho states that Lachares, of the twelfth dynasty, whose reign falls in 3272 B.C., built the Labyrinth, as a sepulchre for himself. The report of Pliny is, that it was constructed by king Petesuchis, or king Tithoes, 3600 years before his time: he adds, that Demoteles stated it to be the palace of Moteris, and Lyceas to be the sepulchre of

---

([145]) Qui de iis scripserunt sunt Herodotus, Euhemerus, Duris Samius, Aristagoras, Dionysius, Artemidorus, Alexander Polyhistor, Butoridas, Antisthenes, Demetrius, Demoteles, Apion. Inter omnes eos non constat a quibus factæ sint, justissimo casu oblitteratis tantæ vanitatis auctoribus, xxxvi. 12. Concerning the writers cited in this passage (all of whom, with the exception of Herodotus, are subsequent to Plato), see Fragm. Hist. Gr. vol. ii. p. 99.

([146]) Herod. ii. 134—5, repeated in Plin. xxxvi. 12.

([147]) xvii. 1, § 33.

([148]) Herod. ii. 101. The severe taskwork imposed by Nitocris upon her Egyptian subjects is alluded to in Dio Cass. lxii. 6.

([149]) Frag. Hist. Gr. vol. ii. p. 554.

([150]) i. 61.

([151]) ii. 148. He is followed by Pliny and Mela, i. 9, who call it the work of Psammitichus.

Mœris.(¹⁵²) The names of Petesuchis, Tithoes, and Moteris, do not occur in any list of Egyptian kings; the two last are altered on conjecture by Bunsen.(¹⁵³)

Josephus mentions that at the time of the Roman war with Judæa, the inhabitants of the city of Chebron considered it more ancient than Memphis, inasmuch as its age then amounted to 2300 years.(¹⁵⁴) If Menes was the founder of Memphis, the date of the foundation of Memphis supposed in this passage is wholly inconsistent with the date of Menes, as given by Herodotus, Manetho, and Diodorus.(¹⁵⁵)

Much of what is called Egyptian history has evidently been borrowed from the Greek mythology. Thus even the native priests spoke of a king Proteus, who lived at the time of the Trojan war, and they gave a detailed account of his relations with Menelaus and Helen. Their story, however, is only a rationalized form of the marvellous sea-god of the Odyssey, the servant of Neptune, endowed with prophetic powers.(¹⁵⁶) Thuoris is stated in Manetho to have been the Polybus of the Odyssey.(¹⁵⁷) Thon, the husband of Polydamna, likewise mentioned in the Odyssey, was converted into a king of the Canopic mouth of the Nile, who offered violence to Helen, and was in consequence killed by Menelaus.(¹⁵⁸) Busiris, infamous for his human sacrifices, was celebrated in Greek legend. Apollo-

---

(152) xxxvi. 13. Concerning Demoteles, see Fragm. Hist. Gr. vol. iv. p. 386. Concerning Lyceas, see ib. p. 441.
For an account of the labyrinth, see Bunsen, Eg. vol. ii. p. 313—327, Eng. tr.

(153) See Egypt, vol. i. p. 698, vol. ii. p. 308, Eng. tr.

(154) Bell. Jud. iv. 9, § 7.

(155) See above, p. 344.

(156) See Od. iv. 385, 456, 468. Euripides, in his tragedy of *Helena*, follows Herodotus in making Proteus king of Egypt, and represents him as succeeded by a king named Theoclymenus.

(157) Above, p. 330. Compare Lepsius, Chronol. der Æg. p. 298.

(158) Od. iv. 228, and Hellanicus, cited in the Scholia. King Thon, or Thonus, was the eponymus of the town Thonis, Strab. xvii. i. § 16; Diod. i. 19. Other passages from the Αἰγυπτιακὰ of Hellanicus, are collected in Frag. Hist. Gr. vol. i. p. 66; but the citation in the Scholia to the Odyssey respecting Thonus is not mentioned.

dorus describes him as having, in consequence of his cruelty, been slain by Hercules.([159]) Hesiod, however, made Busiris eleven generations earlier than Hercules,([160]) and Isocrates places him more than 200 years before Perseus.([161]) The etymological fable which explained the name Rhinocolura must have been of Greek origin.([162]) Queen Nitocris is assigned by the Greek writers to Babylon as well as to Egypt.([163]) The fable of Horus having instituted the year of three months, from the duration of the seasons, to which we shall advert presently, is likewise manifestly of Greek origin.([164]) The names Nilus or Nileus,([165]) and Ægyptus, for Egyptian kings, are probably Greek, and not native.

The general character of the Manethonian lists of kings is that, assuming them to be authentic, they exhibit only chronology without history. This is likewise to a great extent the character of the accounts in Herodotus and Diodorus: Herodotus passes over the 329 successors of Menes without so much as mentioning their names; Diodorus names only twenty-one out of 475 monarchs, from Menas to the invasion of Cambyses. Herodotus and Diodorus indeed present us with something which resembles a narrative of the acts of a few of the more noted kings; but the Egyptian dynasties of Manetho are a mere bead-roll, or string of names, accompanied, at rare

---

([159]) ii. 5, 11. Compare Lepsius, ib. p. 271.

([160]) Ap. Theon. Progymn. c. 6, fragm. 31, Marckscheffel.

([161]) Busir. p. 228. Seleucus, an Alexandrine grammarian, wrote a treatise, περὶ τῆς παρ' Αἰγυπτίοις ἀνθρωποθυσίας, Athen. iv. p. 172 D. Compare Fragm. Hist. Gr. vol. iii. p. 500. Busiris was the eponymus of the Egyptian town of the same name, which was probably connected with *Osiris* and *Petosiris*.

([162]) Lepsius remarks that this name was modified by the Greeks from its original barbarous form, and that the fable about the mutilation of noses was subsequently invented by them for its explanation, ib. 295.

([163]) Herod. i. 185. Compare Clinton, F. H. vol. ii. p. 278, who treats the Babylonian Nitocris as a historical personage.

([164]) Below, ch. vi. § 9.

([165]) The Nile is mentioned as a river by Hesiod, Theog. 338. Homer is ignorant of the name.

intervals, with a notice of some fabulous event. Such naked lists, even if they were founded upon contemporary registration, would be valueless for historical purposes. Assuming the names of the kings, and the lengths of their reigns, to be authentic, they are a mere chronological measure, or canon. We should gain nothing from a list of victors at the Olympic Games if nothing else was preserved to us of Greek antiquity. To be told that Saites, Bnon, Pachnan, Staan, Archles, and Aphobis were the six kings of the fifteenth dynasty, and reigned over Egypt from 2607 to 2324 B.C., conveys no available information.[166] We should learn as much from an authentic account of the succession of a breed of crocodiles or hippopotami in the Nile, or of a series of sacred apes in a temple, for the same period. Some modern philosophers have erroneously thought that history can be written without names, and be reduced to a series of moral forces, of a certain duration and intensity. But if it is impossible to write history without names, it is equally impossible to reduce history to mere names. The Manethonian lists, at the best, exhibit a royal phantasmagoria, without the bone and muscle of history. The kings who reign for a period of 5000 years 'come like shadows, and so depart.' They are nothing more than a long procession of regal spectres.

But these lists of Egyptian kings are not entitled to credit. They are presented to us without any sufficient voucher of authenticity. If the priests had possessed any one list founded on contemporary registration, and preserved by an uninterrupted tradition, the reports of successive Greek writers would not have been conflicting; nor would it have been reserved for Manetho to publish it to the world. There is no example of history founded on contemporary registration being reduced to mere chronology. The lists of primitive kings which appear in the ancient chronologists, as the Athenian, Sicyonian, and

---

[166] Bunsen acknowledges that 'series of dynasties and kings do not give us a history,' vol. ii. p. 244, Eng. tr.

Alban kings, are the products of late fiction. The list of Manetho must, in like manner, be regarded as the result of his own invention; aided, doubtless, by some traditionary names and stories received from his predecessors.

The fabrication of imaginary lists of kings was not confined to Egyptian priests or Greek antiquarians. It was renewed by the mediæval chroniclers at a time when the spirit of historical curiosity had been revived, but when historical criticism was in its second infancy. A series of fabulous British kings, beginning with Brutus, the son of Silvius and grandson of Æneas, who migrated to Britain, and ending with the invasion of Julius Cæsar, was promulgated by Geoffrey of Monmouth in the twelfth century. Shakspeare probably considered king Lear and his three daughters as equally historical personages with Henry the Fourth and Richard the Third; Spenser reviews the entire line of British kings, from Brute to Uther, in his 'Fairy Queen;'[167] and Milton, in the first book of his 'History of England,' inserts the account of these kings, whom he declares to be 'attested by ancient writers from books more ancient,' and to be 'the due and proper subject of story.' He says that their existence is 'defended by many, denied utterly by few:' and that it would be an unreasonable excess of incredulity to reject them altogether.[168] Nevertheless, this series of

---

(167) F. Q. b. 2, canto 10.

(168) Milton, Hist. of Engl. b. 1, p. 5, Prose Works, vol. iv.: 'But now of Brutus and his line, with the whole progeny of kings, to the entrance of Julius Cæsar, we cannot so easily be discharged; descents of ancestry long continued, laws and exploits not plainly seeming to be borrowed, or devised, which on the common belief have wrought no small impression; defended by many, denied utterly by few. For what though Brutus and the whole of the Trojan pretence were yielded up, ... yet those old and inborn names of successive kings never any to have been real persons, or done in their lives at least some part of what so long hath been remembered, cannot be thought without too strict an incredulity. For these, and those causes above mentioned, that which hath received approbation from so many, I have chosen not to omit. Certain or uncertain, be that upon the credit of those whom I must follow, so far as keeps aloof from impossible and absurd, attested by ancient writers from books more ancient, I refuse not as the due and proper subject of story.'

supposed British kings has been so completely exploded, since the time of Milton, that a modern historian passes them over in silence, as undeserving even of mention.

Fordun, likewise, who composed the primitive history of Scotland, at the end of the fourteenth century, gave a series of fabulous kings, forty-five in number, beginning with Fergus I. in 320 B.C., and ending with Fergus II. in 404 A.D. These forty-five kings, however, whose reigns extend over 700 years, are, like the Manethonian kings, mere names; no event is associated by Fordun with their reigns. This list of early Scottish kings was altered by Hector Boethius, who wrote in the early part of the sixteenth century, by reducing the number of kings, and modifying their names. He also garnished their reigns with fictitious events. Boethius was followed by Buchanan as to the number and names of the kings;([169]) and their portraits adorn the walls of the palace of Holyrood.

Julius Africanus, a Christian writer of the early part of the third century, treats the Manethonian chronology of Egypt as a figment, which had been manufactured in imitation of the Babylonian chronology of Berosus.([170]) Syncellus follows Africanus: he speaks of Manetho as 'the author of false chronology, and a glorifier of the Egyptian nation:' he states that 'Manetho's historical writings were full of fiction, and were fabricated by him in imitation of Berosus, with whom he was nearly contemporary.' ([171]) We shall see, in the following chapter, what was the character of the Babylonian history of Berosus.([172])

([169]) Concerning the fabulous kings of Scotland, see Innes, Critical Essay on the Northern Inhabitants of Scotland (Lond., 1729, 2 vols. 8vo), vol. i. p. 211—214. This work contains a detailed and instructive account of the process by which this imaginary series of kings was fabricated. The list of kings of Scotland from Fergus I. in 330 B.C. to James VI. in 1567, with their respective dates, is prefixed to Buchanan's History, in the collected edition of his works, 2 vols. 4to. Lugd. Bat. 1725.

([170]) Ap. Syncell. vol. i. p. 32, ed. Bonn.

([171]) Ib. p. 27, 229.

([172]) Ch. vii. § 3, 7.

It is clear that the work of Manetho was not highly prized by his contemporaries or immediate successors; for he is never mentioned by Diodorus or Strabo, both of whom travelled in Egypt. The earliest authors who cite him as a historical writer are Josephus and Plutarch, who lived in the first century after Christ.

The tendency of the Oriental mind to enormous numerical exaggeration is perceptible both in Babylonian and Egyptian antiquity. In addition to the authors already cited, Heraïscus wrote the Primitive History of Egypt, for a period exceeding 30,000 years:([173]) a spurious epistle from Alexander the Great to Olympias, professing to be founded on the information of Leo, an Egyptian priest, assigned a long period of time to the Egyptian kings, from sacred books. The same epistle stated the duration of the Assyrian kingdom at more than 5000 years.([174])

When once the guidance of positive historical evidence is abandoned, it is as easy to imagine a large number of years as a small one. It is as easy to put down fictitious dynasties for 30,000, or 10,000 years, as for 1000 or 500. Similar statements of vast periods of time occur in the Hindu writers. They divide the present age of the world into four periods, denominated yugs, which together amount to no less than 3,892,911 years, down to the year 1817 of the Christian era.([175]) The Tyrians reckoned 30,000 years from the beginning of the world.([176]) Zoroaster was reported to have lived 5000 years before the Trojan War, and to have written

---

([173]) καὶ συγγραφὴν δὲ ἔγραψεν Αἰγυπτίων ὠγυγίων πράγματα περιέχουσαν οὐκ ἐλαττόνων ἐτῶν ἢ τριῶν μυριάδων, ἀλλὰ πλειόνων ὀλίγῳ, Suidas in Ἡραΐσκος.

([174]) Augustin. C. D. viii. 5, 27, xii. 10; Westermann, de Epistol. Script. ii. p. 9; Minucius, c. 21.

([175]) This subject is fully illustrated by Mr. Mill, Hist. of Brit. India, b. 2, c. 1.

([176]) Africanus, ap. Syncell. vol. i. p. 31, Bonn. The long periods of Phœnician antiquity are alluded to in general terms by Joseph. Ant. i. 3, 9.

two million verses.(177) Hermes was fabled to have composed twenty thousand books.(178)

The accumulation of large numbers in order to produce an imposing effect is the mark of a barren and inactive, rather than of a lively and inventive imagination. It may be compared with the Oriental tendency to produce architectural effect by immense piles, destitute both of utility and beauty;(179) or to suggest the idea of power by representing the statues of the gods with a hundred hands or feet. In this respect the elegant and chastened fancy of the Greek mythology is remarkably contrasted with the clumsy and lumbering fictions of the East.(180)

§ 9 Such being the improbabilities and inconsistencies apparent upon the face of the accounts handed down as schemes of ancient Egyptian chronology, it is natural that attempts should have been made to give credibility to these accounts by some method of reduction or alteration.

Among both ancient and modern writers the main purpose

---

(177) Plin. xxx. 1; Plut. de Is. et Osir. 46.

(178) Lepsius, Chron. der Æg. p. 40, 45; Fab. Bib. Gr. vol. i. p. 86, Harless.

(179) Mr. Fergusson says of the Egyptian pyramids: 'As examples of technic art, they are unrivalled among the works of men, but they rank among the lowest if judged by the æsthetic rules of architectural art. The same character belongs to the tombs and buildings around them: they are low and solid, and possess neither beauty of form nor any architectural feature at all worthy of attention or admiration, but they have lasted nearly uninjured from the remotest antiquity, and thus have attained the object their builders had principally in view when they designed them,' Handbook of Architecture, p. 223.

(180) 'I asked if they knew anything about the cave on the other side of the hill; on which the old Gossain [Hindoo hermit], with an air of much importance, said, that nobody had ever seen its end; that 2000 years ago a certain Raja had desired to explore it, and set out with 10,000 men, 100,000 torches, and 100,000 measures of oil, but that he could not succeed; and if I understood him rightly, neither he nor his army ever found their way back again! These interminable caves are of frequent occurrence among the common people of every country. But the centenary and millesimal way in which the Hindoos express themselves, puts all European exaggeration to the blush. Judging from the appearance of the cave, and size of the hill which contains it, I have no doubt that a single candle, well managed, would more than light a man to its end and back again,' Heber's Journey through India, vol. i. p. 267.

has been to diminish the vast periods of time assigned to Egyptian antiquity, either on account of their character of wild exaggeration, or with the more special object of making the chronology of Egypt harmonize with that of the Old Testament.

The favourite contrivance for this reduction among the ancients was to affix an arbitrary meaning on the word 'year;' and to assume that the year of ancient Egypt was not the solar year, but some shorter period. This period was taken at four months, at three months, at two months, at one month, and even at one day.[181] Two of these hypotheses are expounded in the following passage of Diodorus:—

'The Egyptian priests (he says) reckon the time from the reign of Helius to the expedition of Alexander into Asia at about 23,000 years. They report that each of the most ancient gods reigned more than 1200 years, and of the later gods not less than 300. So great a length of reigns being incredible, some of the priests resort to the supposition that, owing to the ignorance of the motion of the sun, the year was anciently measured by a single circuit of the moon. Hence they argue that, if the year consisted only of thirty days, it was possible for a man to live 1200 of these years; being equivalent to 100 years according to the ordinary mode of reckoning: a term of life which many persons exceeded. With regard to the reigns of later date, they suppose the year to have consisted of four months, according to a division into the three seasons of spring, summer, and winter: which gives a similar result for a reign of 300 years.'[182]

The statement that the Egyptian year consisted originally

---

(181) Concerning this method of reduction, see Uhlemann, Handbuch der ägyptischen Alterthumskunde, Part iii. p. 47.

(182) i. 26. This explanation is repeated by Eusebius, ap. Syncell. vol. i. p. 73, ed. Bonn, with this exception,—that he supposes the second class of years to consist of 3 months. The statement respecting the division of the Egyptian year according to the seasons recurs in Censorinus, c. 19; only he supposes 4 seasons and a year of 3 months. Solinus, i. § 34, and Augustin. C. D. xii. 10, xv. 2, mention an Egyptian year of 4 months. Compare Boeckh, Manetho, p. 63—89; above, p. 32.

of one month, and afterwards of four months, is repeated by Plutarch, who remarks that the great antiquity of the Egyptians, and the length of time which they assign to a generation, are to be explained by the use of years containing only one month.[183] Censorinus, if his present text be correct, states that the original Egyptian year was of two months, but that it was afterwards increased by king Iso to four months.[184] He adds that Horus the Egyptian was by some considered as the author of the year of three months, defined by the seasons ($ὧραι$).[185]

Varro supposed the ancient Egyptian year to have been one circuit of the moon, not of the sun; and he explained by this hypothesis the statements that Egyptian kings had lived 1000 years—a life of 1000 months being within the limits of nature.[186] This explanation is repeated and adopted by Pliny.[187]

Julius Africanus mentions a similar contrivance of historians of his own time, who reduced a period of 9000 years, assigned by Manetho to the reign of Vulcan, to 727¾ years, by assuming that a year was equal to a lunar month.[188]

The report that the Egyptian year consisted of only one month is as early as Eudoxus, who may perhaps have used it for reducing the long periods of Egyptian chronology in Plato.[189]

---

(183) Num. 18.

(184) In Ægypto quidem antiquissimum ferunt annum bimestrem fuisse, post deinde ab Isone rege quadrimestrem factum, novissime Arminon ad tredecim menses et dies quinque perduxisse, c. 19. The best manuscript has *menstrem* for *bimestrem*; whence Jahn, with probability, restores *menstruum*. In the latter part of the passage, *duodecim* ought to be read for *tredecim*, with Scaliger and Salmasius. See above, p. 266.

(185) In Horapollo, ii. 89: τὸ δὲ ἔτος κατ' Αἰγυπτίους τεττάρων ἐνιαυτῶν, the sense seems to require ὡρῶν for ἐνιαυτῶν.

(186) Ap. Lactant. Div. Inst. ii. 12 (1000 months=83⅓ years). Macrobius in Somn. Scip. ii. 11, § 6, remarks that 'mensis lunæ annus est.'

(187) N. H. vii. 48. He refers in this passage to years of 6 months, to the Arcadian year of 3 months, and to the Egyptian year of 1 month. See above, p. 32.

(188) Ap. Syncell. vol. i. p. 32, ed. Bonn.

(189) Above, p. 33.

The hypothesis that the ancient Egyptian year consisted only of a day, is the invention of later and more unscrupulous writers.(190)

A similar method was likewise applied, by some early Christian writers, to the reduction of the lives of the patriarchs, as stated in the Book of Genesis. These writers, as we are informed by Augustine,(191) supposed the year to have been formed by the square of six, which number represented the days of the creation. They held that the original year consisted of thirty-six days, which, multiplied by ten, afterwards formed the solar year of 360 days. Hence the original year would be one-tenth of the common year; so that when it is stated in Genesis that Adam was 230 years old at the birth of his son Seth, and that Seth was 205 years old at the birth of his son Enos,(192) the meaning is, that their respective ages were 23 years and 20 years 6 months. This method of reduction is controverted and refuted, by Augustine, who shows that it is inconsistent with the meaning of the author of Genesis. He maintains that the day, month, and year of Genesis, correspond exactly with the day, month, and year, as received in his own time.(193)

Augustine likewise rejects the long periods assigned to Egyptian and Babylonian antiquity, and lays it down that the Egyptian chronology is incredible, even if the ancient Egyptian year be computed at four months.(194) The contrivance of Varro for the same purpose of reduction is, moreover, repudiated by

---

(190) See above, p. 33, n. 126.

(191) Civ. Dei, xv. 12, 14, where the numbers differ from those in the received text of Genesis.

(192) See Gen. v. 4, 5, 6.

(193) Prorsus tantus etiam tunc dies fuit, quantus et nunc est, quem viginti et quatuor horæ diurno curriculo nocturnoque determinant; tantus mensis, quantus et nunc est, quem luna coepta et finita concludit; tantus annus, quantus et nunc est, quem duodecim menses lunares additis propter cursum solarem quinque diebus et quadrante, consummant, C. D. xv. 14.

(194) Civ. Dei, xii. 10.

Lactantius.(195) Africanus, another Christian writer, discredits the enormous periods of Egyptian chronology, which he conceives to be derived from some astronomical cycle.(196) The views expressed in the latter passages are, to a considerable extent, founded on the inconsistency of the heathen chronologies of Assyria and Egypt with the age assigned to the Creation and the Deluge by the books of the Old Testament.

According to the uncanonical book of Enoch, the year was originally equivalent to a week, and this mode of numeration lasted until 1286 A.M.(197)

The fifteen human dynasties, of the 'Ancient Chronicle,' lasting only 1881 years, and the ten dynasties of Syncellus, from Menes to Amosis (or Amasis), lasting only 2160 years,(198) have, in like manner, been reduced to these amounts from the larger numbers of earlier chronographers, in order to bring them into conformity with the chronology of the Old Testament.

§ 10 The modern critics who have operated upon the ancient Egyptian chronology have employed less simple and more multifarious methods of reduction.(199)

One method, first devised by Sir John Marsham, has been to arrange some of the dynasties, which Manetho meant to be successive, in parallel lines, and to cancel a portion of the time by supposing the reigns to be contemporary. Now this method proceeds on the hypothesis that we have to do only with naked

---

(195) Div. Inst. ii. 12.

(196) An extract from Africanus, in Syncellus, vol. i. p. 31, headed περὶ τῆς τῶν Αἰγυπτίων καὶ Χαλδαίων μυθώδους χρονολογίας, commences thus: Αἰγύπτιοι μὲν οὖν ἐπὶ τὸ κομπωδέστερον χρόνων περιττὰς περιόδους καὶ μυριάδας ἐτῶν κατὰ θέσιν τινὰ τῶν παρ' αὐτοῖς ἀστρολογουμένων ἐξέθεντο.

(197) Syncell. vol. i. p. 60, ed. Bonn. See Salmas. de Ann. Clim. p. 654.

(198) Namely, 2776 to 4936 A.M. See Syncellus, vol. i. p. 397; Fragm. Hist. Gr. vol. ii. p. 507. Lepsius lays it down that the 'Ancient Chronicle' and the spurious book of Sothis were intended to harmonize the Egyptian chronology with that of the Old Testament, Chron. der Æg. p. 522, 546. The short Egyptian year is considered as fictitious by Ideler, Chron. vol. i. p. 93.

(199) Compare Uhlemann, Handbuch, p. 49.

chronology, and that the historical element is altogether wanting. If we had the history of two neighbouring States, we should know not only the names of the sovereigns, and the length of their reigns, but we should be informed of their relations with one another. We should hear of their mutual differences and wars; or of their treaties, friendly intercourse, and royal intermarriages. From the history of England and Scotland, which has been preserved by contemporary writers, it is certain that Robert Bruce was contemporary with Edward I., and that Queen Mary was contemporary with Queen Elizabeth. Authentic German history shows in like manner that the Elector Frederic of Saxony was contemporary with the Emperor Charles V., and that Frederic the Great was contemporary with Maria Theresa. No historian, however paradoxical, could put forth a scheme of chronology in which Robert Bruce, Edward I., Queen Mary, and Queen Elizabeth, the Elector Frederic, the Emperor Charles V., Frederic the Great, and Maria Theresa, should be placed in a consecutive series, all divided from each other by wide chronological intervals.

§ 11 The principal manipulator of the ancient Egyptian chronology is Baron Bunsen, who, in his recent work on Egypt, has avowedly applied the method of Niebuhr to Egyptian antiquity.[200] Now the method with which Niebuhr treated the early history of Rome, was to reject the historical narrative handed down by ancient, and generally received by modern writers; and to substitute for it a new narrative reconstructed on an arbitrary hypothetical basis of his own.[201] Everything

---

[200] See the critique on Bunsen's Egypt, in The Quarterly Review, vol. cv. p. 382.

[201] See Introd. vol. i. p. xl. Engl. tr., and the introductory verses, addressed to Niebuhr:

Grosses hast du zerstört, doch Grösseres wieder gebauet;
Als spätklügelnden Trugs täuschendes Bild du zerschlugst.

Thus translated by Mr. Lockhart:

Great was what thou didst abolish; but greater what thou hast erected
High on the ruins of fraud, shattered for aye by thy blow.

that is original and peculiar in Niebuhr's historical method, and in its results, is indeed unsound. But it possessed advantages, when employed in the transmutation of Roman antiquity, which are wanting to it when applied to Egyptian antiquity. The early Roman history, whatever may be its authenticity, presents at least a full and continuous narrative, most parts of which are related in discordant versions by different classical writers. As none of these versions rests on an ascertained foundation, or can be traced to coeval attestation, great facility is afforded for ingenious conjecture, for bold and startling combinations, for hypothetical reconstruction by means of specious analogies, and for the display of imposing paradox and dazzling erudition. But the so-called history of ancient Egypt consists of little more than chronology. It is, for the most part, merely a string of royal names. Now this is a most unattractive field for the hypothetical historian: he is condemned to make bricks without straw. Instead of demolishing and rebuilding constitutions, instead of creating new states of society out of obscure fragments of lost writers, he is reduced to a mere arithmetical process. Accordingly, the operations of Bunsen and other modern critics upon the ancient history of Egypt, rather resemble the manipulation of the balance-sheet of an insolvent company by a dexterous accountant (who, by transfers of capital to income, by the suppression or transposition of items, and by the alteration of bad into good debts, can convert a deficiency into a surplus), than the conjectures of a speculative historian, who undertakes to transmute legend into history.

Egyptology has a historical method of its own. It recognises none of the ordinary rules of evidence; the extent of its demands upon our credulity is almost unbounded. Even the writers on ancient Italian ethnology are modest and tame in their hypotheses, compared with the Egyptologists. Under their potent logic, all identity disappears; everything is subject to become anything but itself. Successive dynasties become contemporary dynasties; one king becomes another king, or several other kings, or a fraction of another king; one name becomes another name; one number becomes another number; one place becomes another place.

In order to support and illustrate these remarks, it would be necessary to analyse Bunsen's reconstruction of the scheme of Egyptian chronology. Such an analysis would be inconsistent with the main object of the present work; but a few examples will serve to characterize his method.

Sesostris is the great name of Egyptian antiquity. Even the builders of the pyramids and of the labyrinth shrink into insignificance by the side of this mighty conqueror. Nevertheless, his historical identity is not proof against the dissolving and recompounding processes of the Egyptological method. Bunsen distributes him into portions, and identifies each portion with a different king. Sesostris, as we have already stated, stands in Manetho's list as third king of the twelfth dynasty, at 3320 B.C., and a notice is appended to his name, clearly identifying him with the Sesostris of Herodotus. Bunsen first takes a portion of him, and identifies it with Tosorthrus (written Sesorthus by Eusebius), the second king of the third dynasty, whose date is 5119 B.C., being a difference in the dates of 1799 years—about the same interval as between Augustus Cæsar and Napoleon. He then takes another portion, and identifies it with Sesonchosis, a king of the twelfth dynasty; a third portion of Sesostris is finally assigned to himself. It seems that these three fragments make up the entire Sesostris; who, in this plural unity belongs to the Ancient Empire; but it is added that the Greeks confounded him with Ramesses, or Ramses, of the New Empire, a king of the nineteenth dynasty, whose date is 1255 B.C.; who, again, was confounded with his father, Sethos; which name again was transmuted into Sethosis and Sesosis.[202]

Lepsius agrees with Bunsen, that Sesostris in the Manethonian list, who stands in the twelfth dynasty, at 3320 B.C., is not

---

[202] Bunsen, Egypt, vol. ii. p. 89—93, 292—304, 554—5; vol. iii. p. 170.

Sesostris; but, instead of elevating him to the third dynasty, brings him down to the nineteenth dynasty, and identifies him with Sethos, 1326 B.C.; chiefly on account of a statement of Manetho, preserved by Josephus, that Sethos first subjugated Cyprus and Phœnicia, and afterwards Assyria and Media, with other countries further to the east.([203]) Lepsius, moreover, holds that Ramses, the son of Sethos, was, like his father, a great conqueror, but that the Greeks confounded both father and son under the name of Sesostris.([204])

We therefore see that the two leading Egyptologists, Bunsen and Lepsius, differing in other respects, agree in thinking that Sesostris is not Sesostris. The notice appended to his name in Manetho, which identifies him with the Sesostris of Herodotus, Diodorus, and other Greek writers, is regarded by Lepsius as spurious.([205]) But here their agreement stops. One assigns Sesostris to what is called the Old, the other to what is called the New Empire, separating his respective dates by an interval of 3793 years. What should we think, if a new school of writers on the history of France, entitling themselves Francologists, were to arise, in which one of the leading critics were to deny that Louis XIV. lived in the seventeenth century, and were to identify him with Hercules, or Romulus, or Cyrus, or Alexander the Great, or Cæsar, or Charlemagne; while another leading critic of the same school, agreeing in the rejection of the received hypothesis as to his being the successor of Louis XIII., were to identify him with Napoleon I. and Louis Napoleon?

Herodotus informs us that one of the accounts respecting the third pyramid was, that its founder was a Greek courtezan,

---

(203) Contr. Apion. i. § 15. Compare Boeckh, Man. p. 300.

(204) Chron. der Æg. p. 278—288. Lepsius uses the story told by Manetho in Josephus, of Armais, brother of Sethosis, as an argument that Sethosis is Sesostris, on account of its agreement with the story told in Herod. ii. 107, and Diod. i. 57, of the brother of Sesostris. It is true that the treachery of the brother is common to both stories: but all the details differ. Mr. Kenrick identifies the Sesostris of Herodotus with Rameses III. of the 18th dynasty, Ancient Egypt, vol. ii. p. 271.

(205) Ib. p. 286.

named Rhodopis, who was once the fellow-slave of Æsop, and who afterwards became the wife of Charaxus, brother of the poetess Sappho. This story would suppose the third pyramid to have been built about 600 B.C. Manetho, on the other hand, attributes the construction of the third pyramid to queen Nitocris, whose reign began in 4211 B.C. This entire discrepancy of person, nation, and time, might seem to set all reconcilement at defiance. The feat is however accomplished with facility by the Egyptologists. It happens that Manetho states Nitocris to have been a beautiful woman, with a fair complexion.[206] Now Rhodopis was beautiful, and her name means 'rosy-cheeked;' therefore Nitocris and Rhodopis are one and the same person.[207] Whatever ingenuity this mode of argument may possess, it is wanting in novelty; for it is clearly anticipated by Fluellen's argument in Shakspeare, proving that Alexander the Great was born at Monmouth: 'There is a river in Macedon, and a river in Monmouth; and there is salmons in both.'

Lepsius has a different solution of this difficulty. He refuses to identify the ancient queen Nitocris of the sixth dynasty with Rhodopis of the sixth century B.C.; but he discovers that a king Mencheres had a queen Nitocris; he identifies Mencheres with Psammitichus II., who reigned between Neco and Apries; he thinks that Psammitichus-Mencheres had two wives, Nitocris and Rhodopis; he conjectures that Nitocris, who did not build the pyramid, was first confounded with her co-wife Rhodopis, the real builder of the pyramid; and was afterwards confounded by the ignorant Greek dragomans at Sais with Nitocris, the ancient queen. By this series of hypotheses he explains how the ancient queen was assumed to be the builder of the third pyramid.[208] It may be remarked, upon this edifice of conjec-

---

[206] ξανθὴ τὴν χροίαν, Fragm. Hist. Gr. vol. ii. p. 554.

[207] Bunsen, vol. ii. p. 211, who is much pleased at finding that in this discovery he had been anticipated by Zoega. The view of Zoega and Bunsen is approved by Mr. Kenrick, Ancient Egypt, vol. ii. p. 152.

[208] Chron. der Æg. p. 303—308

tures, that it implies the notice in Manetho to be spurious; for the translator of ancient Egyptian records could not have been deceived by ignorant dragomans at Sais.

Bunsen's work on Egypt is a book of metamorphoses.[209] By his method, Agamemnon or Achilles might be identified with Alexander the Great, Pompey might be identified with Cæsar, and Hannibal with Scipio. Such identifications as that of William the Conqueror with William of Orange, or of St. Louis with Louis XVI., would be so obvious and natural as not to require formal proof, and would be disposed of in a parenthesis, if this mode of dealing with evidence were transferred to modern history.

It may, however, be said that injustice is done to Bunsen's restorations by comparing the dates as they stand in Manetho, because these dates are themselves reformed by Bunsen; and that different results would come out if the comparison were made according to Bunsen's own reconstructed scheme.

According to the canon of Manetho, as restored by Boeckh, the first dynasty after the gods, demigods, and manes, begins with the year 5702 B.C. This is the first year of the reign of Menes.[210] The series of royal dynasties is carried down uninterruptedly till the first year of the reign of Alexander the Great, 332 B.C., and therefore the entire time is 5370 years.

---

(209) Bunsen identifies Athothis the Second, the third king in the list of Eratosthenes, with Kenkenes the third king in the first dynasty of Manetho, remarking that 'the difference in the names is no argument against their agreement,' vol. ii. p. 43, Eng. tr.

In the third dynasty of Manetho, the third king is Mesochris. Bunsen says that Mesochris is the same as Sesochris, and that Sesochris is the same as Sesorcheres. He then identifies this transmuted king with the Mares of Eratosthenes, and thus forms a new king whom he calls Sesorcheres-Mares. He next identifies Sesorcheres-Mares with the lawgiver Sasychis in Diod. i. 94 (who evidently is not a king, see above, p. 260); and lastly, he identifies Sesorcheres-Mares-Asychis with the Sasychis of Herodotus, who is placed near the end of his series. (Egypt, vol. ii. p. 76, 94—6, Eng. tr.)

(210) Bunsen lays it down that the historical age of Egypt begins with Menes, vol. ii. p. 63, Eng. tr. Niebuhr, on the other hand, considers everything before the 18th dynasty (1655 B.C.) unhistorical. Lectures on Ancient History, vol. i. p. 42—3, ed. Schmitz.

Upon the authority of an obscure and controverted passage in Syncellus, which appears to state that the period of 113 generations, described in the thirty dynasties of Manetho, amounts altogether to 3555 years,[211] Bunsen assumes that the sum 5370 cannot express consecutive reigns, and that the difference between the two numbers, viz., 1815 years, is due to the fact that a portion of them are concurrent.[212] After a long and intricate series of arbitrary arithmetical operations—which, differently applied, would enable the chronologist to bring out any result which might be desired—he produces the following scheme from Menes to Alexander the Great:—

Duration of the Old Empire	1076 years.
Middle	922[213]
New	1296
	3294

This computation supposes the New Empire to begin at 1632, the Middle Empire to begin at 2554, and the Old Empire to begin at 3630 B.C. Assuming the number of 3555 years stated in the passage of Syncellus to be the total duration of the Egyptian kingdom until 340 B.C., the first year of Menes would fall in 3895 B.C.

The method of reduction employed by Bunsen affords a remarkable proof of the insecure foundation upon which the schemes of Egyptian chronology rest. We have detailed reports of the Manethonian dynasties, handed down to us by the ancient chronographers, which, according to Boeckh's critical restoration, made the Egyptian monarchy, from Menes to Alexander, last 5375 years. In a single passage in Syncellus, of dubious interpretation, this sum is incidentally mentioned as

---

(211) Vol. i. p. 97—8, ed. Bonn.
(212) Egypt, vol. i. p. 86, 98, 130, 134, vol. ii. p. 182.
(213) Bunsen hesitates between two hypotheses, one of which gives 922 years to the middle empire, and another which gives 512 years. He mentions a third hypothesis of Rougé, which lengthens the period to 2017 years, vol. ii. p. 450, Engl. tr.

amounting to 3555 years. The difference is no less than 1820 years—about the period which has elapsed since the Christian era. Which of these two numbers is the authentic number of Manetho?

> Grammatici certant, et adhuc sub judice lis est.

Bunsen decides for 3555, and he is supported by Lepsius. Boeckh considers the passage of Syncellus as corrupt, and proposes a conjectural emendation, which entirely alters its meaning. He holds that the number 3555 does not refer to Manetho. Lepsius concurs with Boeckh in considering the passage as corrupt, but he proposes a different emendation, and holds that the number 3555 does refer to Manetho.[214] Such are the insoluble difficulties which arise respecting chronology dissociated from history, handed down by conflicting authorities, and reduced to an arithmetical puzzle.

There is something attractive to a writer in this discretionary power of dealing with the history of bygone ages. His imagination is captivated with the faculty of creating or annihilating dynasties by a stroke of his magic pen; he becomes, in the language of the ancient astrologers, a 'chronocrator.' He likewise appears to possess a sort of reflex second sight, by which he is able to look back into the unknown past, and to discern images invisible to ordinary eyes. He can evoke a great mediæval period of antiquity, which has hitherto been wrapt in oblivion. If his pretensions to these gifts are admitted, and if he succeeds in imposing on the credulity of his readers by his familiar handling of subjects remote from ordinary studies, he is regarded as a historical seer, elevated far above those obscure chroniclers

---

(214) The following are the material words of Syncellus: τῶν γὰρ ἐν τοῖς τρισὶ τόμοις ριγ´ γενεῶν ἐν δυναστείαις λ´ ἀναγεγραμμένων, αὐτῶν ὁ χρόνος τὰ πάντα συνῆξεν ἔτη ,γφνέ. Boeckh reads ἀναγεγραμμένων αὐτῷ, and for ὁ χρόνος he proposes ὁ Ἀνιανός, Man. p. 137. Lepsius expunges the words αὐτῶν ὁ χρόνος. Mr. Palmer, Egyptian Chronicles, vol. i. p. 121, concurs in reading αὐτῷ, and for ὁ χρόνος corrects ὁ χρονογράφος, understanding Eratosthenes to be meant. C. Müller, Fragm. Hist. Gr. vol. ii. p. 537, entirely rejects the hypothesis of Bunsen respecting the 3555 years.

who occupy themselves with digesting the occurrences of well-attested history.

§ 12 In the absence of all decisive evidence with respect to Egyptian antiquity, handed down by the classical authors, it has been attempted to supply the defect by supposing that the Egyptians possessed an ancient indigenous literature, which has perished, but which was accessible to the writers from whom we derive our information.

Thus it has been asserted that the books stated to have been carried in the sacred processions of Egypt, were records of ancient events and mysteries.([215]) But these books, which are described only by Clemens of Alexandria, a Christian writer of the second and third centuries, must be considered as of late concoction, and as chiefly of Greek manufacture. The two books of Hermes, borne by the Chanter; the four astronomical books of Hermes, borne by the Horoscopus, who likewise carried a sundial and a palm;([216]) another book, with ink and a reed, borne by the Hierogrammateus; the ten books relating to ceremonies of worship, borne by the Stolist, and the ten books concerning the gods and the education of the priests, which the Prophet was bound to learn, cannot be regarded as genuine remains of remote Egyptian antiquity. These books are stated by Clemens to be forty-two in number, thirty-six of which are learnt by the ministers of religion: the remaining six, containing treatises on medicine, are learned by the Pastophori, or shrine-bearers.([217])

Hermes Trismegistus was a fiction of the Neo-Platonic

([215]) See Bunsen's Egypt, vol. i. p. 9, Eng. tr.

([216]) See Leemans on Horapollo, p. 134. Horapollo, i. 3, states that in the hieroglyphic writing, the palm denoted a year.

([217]) Clem. Alex. Strom. vi. 4, § 35. The Hermetic books mentioned in this passage are considered as of recent fabrication, by Fabricius, Bibl. Gr. vol. i. p. 48, ed. Harles. The Hierostolist, the Hierogrammateus, the Prophet, and the Horologus, are mentioned as different orders of the Egyptian priesthood, by Porphyr. de Abstin. iv. 8. He likewise speaks of the Pastophori. The Horologus of Porphyry corresponds with the Horoscopus of Clemens.

philosophy. He was the supposed mystical author of all wisdom and knowledge; the source of all intellectual light. He was the offspring of the amalgamation of the Oriental and Hellenic philosophies; and much of the Oriental spirit of mysticism and of exaggeration is visible in the circle of ideas in which he moved. Iamblichus([218]) attributes to him 20,000 books—upon the authority of Seleucus; but adds, that Manetho assigned to him 36,525 books:([219]) as this last number coincides with the total number of years of Egyptian chronology in the 'Ancient Chronicle,' where it is avowedly obtained by multiplying the 1461 years of the Canicular cycle by 25, we perceive that this mystical mode of computing the number of the books of Hermes was of late date.

It may be observed that methods for reducing the large number of the books of Hermes, similar to those used for reducing the long periods of time in Egyptian chronology, have been employed by modern critics. Thus Bochart understands lines to be meant, and Hornius pages of papyrus.([220])

Several works on astrology were likewise attributed to Hermes:([221]) an astrological treatise on medicine, still extant, bears his name;([222]) and the 'Kyranian books' (βίβλοι Κυρανίδες) cited in the book of Sothis, attributed to Manetho, were a Hermetic treatise of Materia Medica, of which a Latin translation has been published, and the Greek original of which is preserved.([223])

Pamphilus introduced, in his treatise on Botany, some

---

(218) De Myst. viii. 1.
(219) See Boeckh, Manetho, p. 17.
(220) Fabric. ib. p. 86.
(221) The following are titles of two Hermetic treatises on astrology: Hermetis Libri duo de Revolutionibus Nativitatum. Hermetis Aphorismi sive Centum Sententiæ Astrologicæ, Fabric. ib. p. 67—8. The latter is printed at the end of Julius Firmicus De Astrolog. ed. Basil. fol. 1551, p. 85.
(222) See Ideler's Medici Græci Minores, vol. i. p. 387, 430.
(223) Fabric. ib. p. 69. See Dr. Schmitz's art. *Hermes Trismegistus*, in Dr. Smith's Dict. of Anc. Biogr. and Mythol.; and compare Boeckh, Manetho, p. 55.

charms and incantations of Egyptian origin; also some Egyptian and Babylonian names of plants. He likewise referred to a treatise ascribed to Hermes the Egyptian, including thirty-six sacred plants of the horoscopus (assuming three horoscopes to each sign of the zodiac), which, says Galen, are evidently a mere invention of the fabricator of the book.(224) It appears, therefore, that Galen regarded the Hermetical books of astrological botany as the work of an impostor.

The Hermetic books were, in the opinion of Fabricius, the forgeries of a Jew, or of a semi-Platonic semi-Christian writer, who lived about the beginning of the second century after Christ.(225)

Theophrastus, in his treatise on Stones, says that the registers respecting the Egyptian kings make mention of some emeralds of immense size, which were sent as a present by the king of the Babylonians.(226) The size mentioned is inconsistent with the stones in question being emeralds, or even malachite: they may have been some green marble. The notice does not prove the existence of contemporary chronicles of political events in Egypt.

§ 13 But a more potent support to the theory of authentic records of Egyptian antiquity is invoked by the Egyptologist from the hieroglyphical inscriptions preserved upon the ancient buildings of Egypt. It is certain that if these inscriptions were made by public authority; if they contain a record of contemporary events; and if the language in which

---

(224) De Simpl. Medicum. vi. prooem. vol. xi. p. 792, 793, 798, ed. Kühn. Concerning Pamphilus, see Dr. Smith's Dict. in v.

(225) Ib. p. 65. Baumgarten Crusius, in a Programm de Librorum Hermeticorum origine atque indole, Jena, 1827, remarks: 'In Hermeticis libris, certum est nonnisi Platonismum illum recentiorem repeti, et Ægyptiacarum rerum præter Deorum nomina nihil deprehendi,' p. 6.

(226) De Lapid. § 24, ed. Schneider. Repeated by Pliny, xxxvii. 5. For . . . νους γάρ φασι, Schneider adopts the conjecture, ἔνιοι γάρ φασι. See his note, vol. iv. p. 557. Bunsen, Egypt, vol. i. p. 3, proposes Νεκάῳ γάρ φασι. If this conjecture were correct, the register would not be of great antiquity; but there is no reason for inserting a proper name; and if a proper name is inserted, any other royal name, such as Μηνῇ or Μυκερίνῳ, would be equally suitable.

they are written can be read and interpreted correctly, they furnish a solid basis of trustworthy history and chronology.

The first postulate for the historical use of the hieroglyphical inscriptions manifestly is, that they should be correctly read and interpreted.

Now it seems to be a necessary condition for the intelligibility of a language, that its tradition should have been preserved unbroken, either in writing or orally. A language cannot in general be restored, if a period has intervened during which it was entirely forgotten and unknown. For example, if the Basque were to become extinct as a living language, and if its dictionaries and grammars were lost, no means would exist for interpreting the remains of its literature.

A confusion of thought on this subject is sometimes created by the use of the term 'dead languages,' as applied to the Greek and Latin. But these languages, although they are *dead*, in the sense that they have not for some centuries been spoken, are nevertheless connected with the ages of Plato and Cicero by a living tradition. From the time when Greek and Latin were spoken languages, each generation has handed down a knowledge of them to its successors. During the period which intervened between the use of Latin as a living language and the invention of printing, there were no bilingual grammars and dictionaries for facilitating its acquisition; and its knowledge must have been imparted exclusively by oral teaching. But immediately after the introduction of printing, grammars and dictionaries for the interpretation of Latin into the modern languages of Europe were composed,[227] and thus the authentic tradition of the language has been perpetuated in a written form. The Greek language was in like manner continued by an unbroken tradition.[228]

---

(227) See Nouv. Biogr. Univ. art. *Calepino*.

(228) The tradition of the Greek language was never completely interrupted, even in Italy; see Sismondi, Hist. des Rép. Ital. c. 41, tom. vi. p. 150, ed. 1818.

The question as to the possibility of interpreting a language whose tradition has been lost, is further confused by a deceptive analogy derived from the process of deciphering.([229]) A cipher is a contrivance for disguising the alphabetical writing of a known language by a conventional change of characters. The explanation of this conventional change is called the *Key*. If a document written in cipher falls into the possession of a stranger ignorant of the Key, and if he can conjecture with tolerable certainty the language in which it is written, he can proceed to apply to it the rules for deciphering, which are founded upon the comparative frequency of certain letters and certain words in the given language.([230]) This process, if the document be tolerably long, is almost infallible. It is difficult to devise a cypher, sufficiently simple for frequent use, which cannot be deciphered by a skilful and experienced decipherer. But this operation supposes the language to be understood; it is a merely alphabetical process; it does not determine the meaning of a single word; it merely strips the disguise off a word, and reproduces it in its ordinary orthography. No process similar to deciphering can afford the smallest assistance towards discovering the signification of an unknown word, written in known alphabetical characters. The united ingenuity of the most skilful decipherers in Europe could not throw any light upon an Etruscan or Lycian inscription, or interpret a single sentence of the Eugubine Tables. In like manner, assuming an Egyptian hieroglyphical text to be correctly read into alphabetical characters, no process of deciphering could detect the meaning of the several words.

The first question for our consideration therefore is, whether the tradition of the hieroglyphical characters and language is altogether lost.

([229]) 'Une langue à l'égard d'une autre est un chiffre où les mots sont changés en mots, et non les lettres en lettres: ainsi une langue inconnue est déchiffrable,' Pascal, Pensées, vol. i. p. 232.

([230]) The principal rules for deciphering a document in the English language are given in the Encycl. Britannica, art. *Cipher*.

The earliest account of the method and signification of the hieroglyphical writing which has descended to us, is by Chæremon, the sacred scribe, who has been already mentioned as having lived in the first half of the first century after Christ.(231) His statement, reported to us by Tzetzes, is attributed by him to the Æthiopians, but evidently refers to the Egyptians. We learn from Suidas that Chæremon wrote a treatise on hieroglyphics.(232)

'The Æthiopians have no letters, but instead of them use the figures of various animals and parts of animals. The origin of this mode of writing was that the sacred scribes of early times, wishing to conceal the doctrines of Natural Theology, transmitted them to their children by means of allegorical representations and symbols of this kind, as Chæremon, the sacred scribe, informs us. Thus a woman playing on the tambourine denoted joy; a man holding his chin with his hand, and bending his head downwards, denoted grief; calamity was denoted by an eye weeping; destitution, by two empty hands stretched out; the east, by a serpent creeping out of a hole; the west, by a serpent entering a hole; resurrection, by a frog; life, also sun and god, by a hawk; mother, time, and heaven, by a vulture; king, by a bee; birth, by a beetle; earth, by an ox; government and superintendence, by a lion's head; necessity, by a lion's tail; year, by a stag, or by a palm tree; increase, by a child; decay, by an old man; force, by a bow; and thousands of others of like nature.'(233)

---

(231) Concerning the journey of Chæremon with Ælius Gallus, see above, p. 280. Chæremon was the instructor of the Emperor Nero, with Alexander of Ægæ, Suidas in Ἀλέξανδρος Αἰγαῖος. Nero was born in 37 A.D. Chæremon was also the teacher of Dionysius of Alexandria, and preceded him in the headship of the Alexandrine Library. Dionysius lived from Nero to Trajan, Suidas in Διονύσιος Ἀλεξανδρεύς.

(232) Suid. in Χαιρήμων and ἱερογλυφικά. Concerning Chæremon, see Birch, in the Transactions of the Roy. Soc. of Lit. vol. iii. p. 385, who says that Chæremon was 'undoubtedly well versed in the knowledge of the sacred characters of Egypt,' and Fragm. Hist. Gr. vol. iii. p. 495.

(233) Ap. Tzetz. Exeges. in Iliad. in the appendix to Draco de Metris, ed. Hermann, p. 123.

[SECT. 13.] OF THE EGYPTIANS.

There is likewise extant a work which contains a detailed exposition of 119 hieroglyphic symbols, and which professes to have been composed by Horapollo the Niloan, in the Egyptian language, and to have been translated into the Greek language by a writer named Philip. In this treatise the hieroglyphics are described as having a symbolical value, but as being the metaphorical, not the simple representatives of ideas or objects. Thus a serpent devouring its own tail is stated to represent the world; Isis, the Dog-star, to represent the year; two men holding each other by the right hand, to represent concord; a hippopotamus to represent time, a ladder to represent a siege. Leemans, the recent editor of Horapollo, conjectures that the author of the work was Horapollo, a Byzantine grammarian, who lived at the close of the fourth century. Concerning the translator Philip nothing is known.

A similar account of the Egyptian hieroglyphics is given by Ammianus Marcellinus, who lived in the fourth and fifth centuries. In the 17th book of his history he introduces a description of the Egyptian obelisks, together with a Greek translation of the hieroglyphic inscription upon an obelisk in the Roman circus, borrowed from the work of a certain Hermapion. Ammianus informs us that the ancient Egyptians did not use alphabetical letters, but that a single character represented a noun or a verb, sometimes even an entire sentence. In illustration of this statement he says that a vulture signified *nature*, and a bee making honey signified a *king*.[234]

---

(234) 'Formarum innumeras notas, hieroglyphicas appellatas, quas ei [on the obelisk] undique videmus incisas, initialis sapientiæ vetus insignivit auctoritas. Volucrum enim ferarumque, etiam alieni mundi, genera multa sculpentes, ad ævi quoque sequentis ætates ut patratorum vulgatius perveniret memoria, promissa vel soluta regum vota monstrabant. Non enim, ut nunc litterarum numerus præstitutus et facilis exprimit quicquid humana mens concipere potest, ita prisci quoque scriptitarunt Ægyptii; sed singulæ literæ singulis nominibus serviebant et verbis, nonnunquam significabant integros sensus. Cujus rei scientiæ in his interim sit duobus exemplum. Per vulturem naturæ vocabulum pandunt; quia mares nullos posse inter has alites inveniri rationes memorant physicæ: perque speciem apis mella conficientis indicant regem; moderatori cum jucunditate aculeos quoque innasci debere his signis ostendentes, et similia plurima,' xvii. 4.

Tacitus says that the Egyptians first expressed ideas by the figures of

The traditionary accounts of the Egyptian hieroglyphics which have reached us, therefore represent the system as ideographic, and not as alphabetical. As the system of Champollion, which has been generally adopted by the Egyptologists, is founded on the supposition that the hieroglyphic writing is in great part phonetic or alphabetic, it becomes necessary for them to reject the only positive evidence which has descended to us from antiquity. Accordingly the subject is thus treated by Bunsen. After expounding the hypothesis of Champollion, of a secret character in the exclusive possession of the priests, mixed with the phonetic characters, he proceeds thus:

'The work of Horapollo, dating from a comparatively recent age, also clearly proves the existence and nature of this secret character. While few of the explanations it offers are confirmed by the monuments, the greater part are contradicted both by them and by the Book of the Dead. The explanations themselves are little better than arbitrary subtleties, or false, cabalistic mysticism, the simple and historical meaning being palpable and obvious, while the very hieroglyphical representations which he describes are chiefly borrowed from that secret character, and consequently do not apply to the monuments and books.'(235) A similar judgment on the treatise of Horapollo is passed by Mr. Sharpe, in his recent work on Egyptian Hieroglyphics. 'As the greater part of the characters which he describes are not found in any of the numerous inscriptions known to us, and as most of the meanings are such that it is scarcely possible they could have existed on the monuments at all, the work must be, both on external and internal evidence, rejected as of little worth.'(236)

---

animals, and that they laid claim to the invention of writing: 'Primi per figuras animalium Ægyptii sensus mentis effingebant (ea antiquissima monimenta memoriæ humanæ impressa saxis cernuntur), et litterarum semet inventores perhibent,' Ann. xi. 14. This passage does not show whether Tacitus understood the hieroglyphics to be ideographic or phonetic.

(235) Egypt, vol. i. p. 341, Engl. tr. In p. 554, however, he appears to reject Champollion's hypothesis of a secret character.

(236) p. 24 (London, 1861).

The system of Champollion assumes that the hieroglyphics are mainly phonetic; that for the most part they are alphabetical characters expressing sounds. It proceeds from the interpretation of the name *Ptolemy* on the Rosetta stone, of the name *Cleopatra* on the obelisk of Philæ, and of the names and titles of some of the Roman emperors. The early publications of Champollion were almost exclusively confined to a theory of reading the hieroglyphics: he did not at first attempt to interpret the words which he read. The second edition of his 'Précis du Système hiéroglyphique des anciens Egyptiens,' published in 1828, did not exceed the limits which have been just described. What is called the 'great discovery' of Champollion was, substantially, his doctrine that the hieroglyphics can be read into words, and, for the most part, are not symbols of ideas.[237]

But Champollion, having laid this foundation, and having shown that the Egyptian hieroglyphs are, in general, symbols, not of ideas but of alphabetical sounds, proceeded to complete his system by restoring the grammar and dictionary of the forgotten language. 'The development of the important fact in the decipherment (says Mr. Birch) gradually revealed the whole grammar and system of the language, and placed the future inquirer on the route for the decipherment and interpretation of the whole.'[238] 'The French Government (Mr. Birch continues), ever alive to the interests of learning, sent him, at the head of a scientific mission to Egypt, to rescue the rapidly

---

(237) Bunsen, Egypt, vol. i. p. 320, Engl. tr., lays it down that Champollion's discovery of the old Egyptian language and character, is 'the greatest discovery of the century.' In p. 325 he speaks of Champollion's 'immortal letter to Dacier.' The language of Brugsch is similar: 'Hoc lapide [the Rosetta stone] detecto postquam omnium animi ad spem enucleandi tandem illud monstruosum et perplexum per tot sæcula quasi involucris involutorum genus signorum arrecti sunt, unus vir Champollio Franco-Gallus exstitit qui mirâ sagacitate incredibilique studio adjutus totam hieroglyphorum rationem nullâ fere parte relictâ luce clarius explanavit et exposuit,' Inscript. Rosettan. p. 2. Concerning the phonetic system of Champollion, see Schwartze, das alte Ægypten, vol. i. part 1, p. 199, 253.

(238) Introduction to the Study of the Egyptian Hieroglyphs (Lond. 1857), p. 201.

vanishing monuments of the country from oblivion, by copying them, and to illumine the world by explaining them. Provided with a simple and efficacious system, and the experience of an examination of a great number of texts, Champollion translated with a marvellous facility the inscriptions submitted to him. He at once saw the purport of the hieratic monument of M. Sallier at Aix, containing the campaign of Rameses against the Sheta. He read with fluency the different inscriptions on the monuments, and gave life to their mute forms. His most remarkable reading was the name of Judah Malcha (the kingdom of Judah) on the wall of Karnak, amidst the prisoners of Sheshak. His 'Lettres Ecrites'(239) are full of new translations, illustrative of the mythology, history, ethnography, manners, and customs of the Egyptians as they were, and declare themselves.'(240)

The researches of Champollion in the field of hieroglyphics were unhappily interrupted by a premature death. He died in 1832, at the age of 42. His grammar and dictionary of the ancient Egyptian language were posthumous publications, and appeared under the editorship of his brother, M. Champollion-Figéac, in 1836—1841. His system has however been adopted, and his researches have been continued, with zeal and diligence, by Egyptologists, not only in his own country, but also in Germany and England. The comprehensive work of Baron Bunsen upon Egypt exhibits the latest and most improved version of the Egyptian alphabet and vocabulary as now received by the Egyptologists of the highest authority.(241)

---

(239) These letters were published at Paris in 1833.
(240) Ib. p. 202.
(241) 'This Egyptian dictionary, which, thanks to Mr. Birch's erudition, is the fullest in existence, has been arranged in the usual alphabetical order, and a reference given to the monument or text where each word occurs. The hieroglyphical scholar, therefore, will now find in this work, not only the most complete Egyptian grammar extant, but also the most complete dictionary; as well as the only existing collection of all the hieroglyphic signs, arranged in their natural and historical order, and explained according to the monuments quoted or referred to,' Egypt, vol. i. p. 477.

The Egyptian vocabulary of Bunsen contains 685 words, read alphabetically from the hieroglyphic writing, and interpreted into English. The following will serve as specimens:—

aa. 1. To knit. 2. To wring out. 3. A kind of gazelle. 4. To be born of. 5. To wash. 6. Water for washing. 7. An arm. 8. Place, abode. 9. A noble.

ab. 1. Flesh, viand. 2. A nosegay (?) 3. A horn. 4. To butt. 5. Pure, to purify. 6. Ivory. 7. A rhinoceros. 8. A calf. 9. Thirst.

The Egyptian alphabet, as exhibited by Bunsen, consists of signs of four classes, viz., 1. Ideographics. 2. Determinatives. 3. Phonetics. 4. Mixed.

The Ideographics are 620 in number. Some of these are simple pictures or outlines of the object signified, similar to those in a child's primer: thus the figure of a cow signifies *cow,* the figure of an elephant signifies *elephant,* a figure of the lips signifies *lips,* a figure of the ears signifies *ears.* In other cases, the idea represented bears only a certain analogy to the sign: thus a man bending down means 'to beseech, to beg;' a man clad with a panther-skin means 'a high priest, or judge;' a man overthrowing an Asiatic foreigner means 'to subdue;' a divinity with horns, plumes, staff, and whip, means 'to terrify;' an ass-headed god, holding clubs, has the same meaning.

The Determinatives are 164 in number. A determinative is explained to be an ideographic sign, which is annexed to a word expressed by phonetic signs, for the purpose of fixing and defining its signification. Thus 'the sun's disk (*ra*) was affixed to several words or signs which express the divisions of time regulated by the sun, as *hr,* day, *hunnu,* hour; or those expressive of light, as *ht,* to illumine. The disk, so employed, does not express the word of which it is the symbol; it only determines the meaning of the preceding phonetic sign, the sense of which would otherwise remain doubtful to the reader, owing to the various significations of the same Egyptian roots.'[242] Thus a

---

[242] Bunsen, ib. vol. i. p. 535.

figure of a duck is a determinative of, 1. waterfowl; 2. birds generally; 3. flying animals, a scarab; 4. to doctor; 5. to sleep. A figure of a tree is a determinative of names of trees; a figure of a sword is a determinative of cutting actions; a figure of a seal is a determinative of 'to shut' and 'to enclose;' a figure of a club is a determinative of names of foreigners, 'to create,' and 'wicked.'

The class of Phonetics consists of 27 alphabetic, and 103 syllabic signs. To these Bunsen adds 100 signs, constituting the Later Alphabet, which he conceives to have been begun after the twentieth dynasty, but to have been greatly augmented under the Roman dominion.

The Mixed class is composed of hieroglyphic groups, in which the principal element is ideographic, but a phonetic complement is annexed to it. Of these Bunsen enumerates 55.

The system of reading the hieroglyphic characters, as expounded by the Egyptologists, is flexible and arbitrary. It involves the hypothesis of homophones; that is to say, of a plurality of signs for the same sound.([243]) It likewise involves a mixture of ideographic and phonetic symbols.

But if we concede that Champollion's method of reading the hieroglyphic characters, as developed and improved by his successors, is sound, we are yet far removed from the interpretation of the written language thus recovered. Even assuming that we can read the hieroglyphic characters as they were read by the Egyptians themselves, it does not follow that the words so read can be explained.([244]) We may have the remains of a language which are legible, but which we are unable to understand. This is the state of the remains of the Etruscan language, which being written in Greek letters are as legible as a modern news-

---

([243]) See Champollion, Précis, ed. 2, p. 364. In p. 448 he lays it down that 'chaque voix et chaque articulation pouvaient . . . être représentées par plusieurs signes phonétiques différens, mais étant des signes homophones.' Compare Schwartze, i. 1, p. 254—61, and concerning the variations in the writing of proper names, p. 450—9.

([244]) The distinction between these two problems is recognised by Schwartze, i. 2, p. 953.

paper, but which we are nevertheless wholly unable to interpret.(245) The remains of the Lycian language are in a similar position.(246) It is likewise the state, to a great extent, of the Oscan and other inscriptions of South Italian dialects, and of the Eugubine Tables. The attempts even of the most accomplished modern linguists to explain these inscriptions must be regarded, by an impartial judge, as utter failures.

If the explanations of Horapollo are rejected as altogether fanciful and erroneous, and as founded upon a false method, the direct tradition of the language must be admitted to be altogether broken. If the true meaning of the words is to be restored, this restoration can only be effected by indirect means. It is admitted that the Rosetta stone, though it contains a Greek translation of a hieroglyphic inscription, partially preserved, affords little assistance for the general interpretation of the lost language of ancient Egypt.(247) In this apparent destitution of means for recovering the key to the language, Champollion conceived that the tradition might be revived by a comparison of the Coptic language. He went so far as to lay it down that the Coptic language is the ancient Egyptian written in Greek letters.(248) Schwartze, moreover, who has investigated the subject with great diligence and learning, and who was a com-

---

(245) Concerning the Etruscan alphabet, see Müller's Etrusker, vol. ii. p. 291; and concerning the Etruscan language, vol. i. p. 58—70.

(246) Concerning the Lycian language and inscriptions, see Fellows, Account of Discoveries in Lycia (1841), p. 427, 468.

(247) The hieroglyphic portion of the Rosetta inscription is that which has suffered the most. It consists of 14 lines, only 4 of which are entire. The remaining 10 are more or less mutilated. An edition of the hieroglyphic portion of the inscription was published by Brugsch at Berlin in 1851 (Inscriptio Rosettana Hieroglyphica, 4to.). He gives the reading of each word in Roman letters, with its Latin version. He likewise appends a table containing the hieroglyphic alphabet and syllabic signs: but he gives no explanation of the process by which he determines the meaning of the several words. Without this essential part of the proof, the interpretation must be considered as arbitrary.

(248) 'La langue copte...est l'ancien égyptien écrit en lettres grecques,' Précis, p. 361. He repeats this assertion in p. 368. Compare Schwartze, p. 268, 272, 277, 285, 953. The view of Champollion on this subject is shared by Peyron; see his Lexicon Linguæ Copticæ (Taurin. 1835), Præf. p. xi.

plete master of the Coptic language, so far as its published remains extend, has collected proofs of the identity of the Coptic and ancient Egyptian.([249]) He considers the Coptic as one of the most ancient languages in the world;([250]) and as having an affinity both with the Semitic and Indo-Germanic families of languages;([251]) but he places the origin of this affinity at an ante-historical period.([252]) He conceives the Coptic to have been the language of the ancient Egyptian priests.([253])

On the other hand, it is to be observed that the remains of the Coptic language which have reached us do not ascend beyond the third century of the Christian era; that they are exclusively of an ecclesiastical or liturgical character, and that they belong to a sphere of ideas from which the ancient religion and polity of Egypt are altogether excluded.([254]) Schwartze, himself a firm believer in the close affinity of the Coptic with the language of ancient Egypt, thus candidly enumerates the difficulties which surround this hypothesis :—

'The circumstances (he says) which characterize the literature of the Coptic language must make the prudent inquirer mistrustful as to inferences founded upon its relation to other languages. The Coptic first appears after the Christian era. Although the language can be dated back from the early centuries of our era, yet the chief part of the extant literature belongs to a much later period. The Coptic is destitute of the advantage, possessed by other languages, of ancient inscriptions engraved on brass and stone. Numerous deviations from its original form are therefore possible. We have likewise to bear

---

(249) Ib. p. 969—972.   (250) P. 2036.
(251) P. 979—991, 991—7, 1033.
(252) P. 2019.   (253) P. 2036.
(254) 'The dialect of the Christian Egyptians, or Copts, is but the younger branch of the Egyptian language, the latest form of the popular dialect, although, from the age of the Ptolemies downwards, mixed with Greek words and forms, and, since the third or fourth century, written with an alphabet, containing only five old Egyptian, in addition to the twenty-four principal Greek letters,' Bunsen, ib. p. 259. The Coptic versions of the New Testament and of the Pentateuch were first published at Oxford, by Dr. Wilkins, the editor of the Concilia.

in mind the political, social, religious, and scientific state of Egypt before and after the origin of the Coptic literature—namely, the overthrow of the national throne for a period of nearly 1000 years, the hostile and friendly influences of the successive foreign governments, the settlement in Egypt of numerous foreigners belonging to different nations, the destruction of the ancient religion and priesthood, manifest traces of an ancient peculiar language of the priesthood, different from the language of the people, the translation of the sacred writings of a new foreign religion into the popular dialect, the probability that the disciples of the new religion belonged to the lower classes of the community, increasing ignorance among the Copts, the reception of numerous Greek words in the Coptic;—if all these circumstances are weighed, they cannot fail to create an unfavourable impression with respect to the soundness of methods of interpretation for the ancient Egyptian language founded upon the Coptic.'[255]

Schwartze, indeed, proceeds to say that a long and profound study of Coptic had removed this unfavourable impression, and had convinced him of the substantial identity of the Coptic with the ancient Egyptian.

More recent interpreters of the hieroglyphics, however, lay less stress upon the Coptic as a key to the mysteries of the lost language. Thus, Mr. Birch, in his Introduction to the Study of Egyptian Hieroglyphs, published in 1857, gives the following account of the method by which the signification of words written in the hieroglyphic character is discovered:—

'In the great hieroglyphical inscriptions of the temples, in which the determinatives are complete pictures of the idea intended to be conveyed, even if the word was not to be found in the Coptic, there was no difficulty in the interpretation of the idea conveyed; but there was a much more serious difficulty in the discovery of the meaning of words of which no equivalent

---

[255] P. 2015. Schwartze also complains of the imperfect publication of the remains of Coptic literature, p. 2039—40.

Coptic form remained, and of which there was only a generic determinative, a difficulty which some have conceived no genius could surmount. The Coptic, after all, was a very feeble aid, and was probably used by Champollion rather as a justification to the world of the truth of his statements than as the means of his interpretations. Yet another mode of analysis, the inductive, still remained for words no longer to be found in the Coptic dictionaries. 'After examining the numerous passages in which certain uninterpreted words appear, it is evident that the meaning which suits them in all instances is the true one. In proportion as the knowledge of the hieroglyphs increases, the conditions of the unknown groups are more restricted, and their meaning more easily discovered.'[256]

Mr. Goodwin, in an article on Hieratic Papyri, published in the Cambridge Essays for 1858, thus explains the method of arriving at the same result:—

'The radical letters of a large number of Egyptian words are now known, and of them a considerable portion may be traced in the Coptic, some without alteration, others under various disguises, of which the laws have been well ascertained. A good many words have their congeners in Hebrew and Syriac. Practically, the meaning of a word is first approached by help of the determinative and the context in which it occurs; if a similar root can be discovered in Coptic or a cognate language, the value thus obtained may be confirmed or modified. In many cases, however, these languages leave us without assistance. It is obvious that every word of which the meaning is once assured, must lead the way to the discovery of others; and thus the completion of the old Egyptian vocabulary is by no means a hopeless task; and it is one in which daily advances are being made.'—(p. 228.)

To whatever extent the Coptic may be applied in restoring the signification of words read from the hieroglyphic writing, the application can only be made by analogy, and by assuming

---

(256) Introduction, p. 246.

that the Coptic word and the ancient Egyptian word have a common root.

Now, where the tradition of a language is lost, but its affinity with a known language is ascertained or presumed, the attempts to restore the significations of words proceed upon the hypothesis that the etymology of the word can be determined by its resemblance, more or less close, to a word in the known language, and that the etymology of the word is a certain guide to its meaning. But although there is a close affinity between etymology and meaning, yet etymology alone cannot be taken as a sure index to meaning. When the signification of a word is ascertained, it is often difficult to determine the etymology. The Lexilogus of Buttmann, the Romance Dictionary of Diez—in fact, any good etymological vocabulary—will furnish ample evidence of this truth. But when the process is inverted, and it is proposed to determine the signification of the words of an entire language from etymological guesses, unassisted by any other knowledge, the process is necessarily uncertain and inconclusive, and can be satisfactory only to a person who has already made up his mind to accept *some* system of interpretation.([257])

Thus in Italian the word *troja* signifies a sow. Diez refers the origin of this word to the old Latin expression, *porcus Trojanus*, which meant a pig, stuffed with other animals, and served for the table; the name being an allusion to the Trojan horse. He conceives that this phrase first became *porco di troja*, and afterwards *troja* simply, with the signification of a pregnant sow.([258]) Assuming this etymology to be true, what possible ingenuity could have enabled anybody to invert the process, and to discover the meaning by the etymology, if the meaning were unknown?

([257]) On the uncertainty of this method, see the judicious remarks of Schwartze, ib. p. 1003, and the passage of Quatremère, quoted by him, p. 976.

([258]) See Diez, Roman. Wörterb. p. 356; and compare Macrob. Sat. ii. 9.

If the method of restoring the meaning of a lost language by etymological guesses from words of similar sound in a known language is uncertain and deceptive, the method of guessing the signification of a word from the context is equally unsatisfactory. This method is applicable to the determination of obscure and forgotten words in a language of which the general tradition has been preserved. For example, the meaning of the terms ἤλεκτρον and νυκτὸς ἀμόλγῳ, in Homer, may be determined within narrow limits by a comparison with the context of the passages in which they respectively occur. But where the entire language has been lost, the context itself is uncertain, and the uncertainty of uncertain conjectures built upon other uncertain conjectures increases at a compound ratio.

If the true significations of the words of a lost language can by any process be restored—if the dictionary of such a language can ever be reconstructed by hypotheses—the method must be slow and laborious, and can only be founded on a wide and patient induction. Hence the sudden illumination as to the meanings of the hieroglyphic words, which Champollion appears, from the description above cited, to have received on reaching the soil of Egypt, wears a suspicious appearance. It suggests a doubt whether having, as he believed, lifted the veil from the alphabetic value of the hieroglyphs, his imagination did not outrun his judgment as to the possibility of discovering the signification of the unknown sounds which he had disclosed. The wide discrepancy between the views of Champollion and those of Mr. Birch and Mr. Goodwin, respecting the extent to which the Coptic is available as an instrument for the interpretation of the hieroglyphs, is also apparent. If, however, Champollion could read hieroglyphic texts with so much facility and certainty, how comes it that his method has not been found sufficient by his successors?

The enormous duration of the hieroglyphic system, without any material change, is another part of the modern theory which it is difficult to reconcile with probability or experience. Champollion declares that the use of the hieroglyphic writing was not confined to the priests, that it was not a mysterious secret reserved to the sacred caste, but that it was common to all the educated classes of the community.(259) But he tells us that this system lasted for twenty-two centuries without undergoing the smallest modification.(260) That a system of so much intricacy—consisting of ideographic, syllabic, phonetic, and determinative symbols, with a large class of homophones, or alternative signs for the same sound—should have remained in common use by a whole nation for twenty-two centuries without alteration, is utterly incredible.

Bunsen, indeed, supposes that changes in the hieroglyphic writing took place at successive periods, and he undertakes to point out the dynasties under which these changes occurred. Thus he speaks of the hieroglyphic alphabet as constituted in the eighteenth dynasty, and of an obvious and remarkable change which commenced in the twentieth dynasty.(261) If, however, the views of the Manethonian dynasties propounded above be sound, these supposed periods are fictitious and unreal. A theory of hieroglyphical interpretation which implies the historical truth of the Manethonian dynasties rests on a foundation of sand.

The late period to which the supposed discoveries of valuable records upon the walls of Egyptian edifices have been reserved, likewise raises a presumption against their reality. If the Greeks and Romans had been informed that important historical records existed upon Egyptian buildings, or in the archives of Egyptian temples, is it likely that, during their

---

(259) Précis, p. 425.

(260) 'Parmi les monumens égyptiens connus jusqu' à ce jour, ceux qui remontent à l'époque la plus reculée, ont été exécutés vers le 19ᵉ siècle avant l'ère vulgaire, sous la 18ᵉ dynastie, et ils nous montrent déjà l'écriture comme un art essentiellement distinct de la peinture et de la sculpture, avec lesquelles il reste confondu chez les peuples à peine échappés à l'état sauvage. L'écriture égyptienne de ces temps éloignés étant la même que celle des derniers Egyptiens, il faut croire que ce système graphique était déjà arrivé à un certain degré de perfection absolue, puisque, pendant un espace de vingt-deux siècles à partir de cette epoque, il ne paraît point avoir subi la moindre modification,' ib. p. 329.

(261) Egypt, vol. i. p. 554, Eng. tr.

supremacy in Egypt, they should not have taken steps for procuring translations of them?([262]) The traditions of the language and of the hieroglyphical writing were still alive. The Greek court of the Ptolemies, in particular, distinguished itself for its patronage of literature, and might be considered as having a sort of national interest in the ancient glory and power of Egypt. The priest who interpreted the Egyptian inscription at Thebes to Germanicus was indeed an impostor; but it is clear that Germanicus and his attendants believed that the art of interpreting the hieroglyphic characters was still possessed by the priests.([263])

The Emperor Severus, when visiting Egypt in 202 A.D., collected all the books which contained anything mysterious, from the different temples, and shut them up in the tomb of Alexander the Great, in order that they might not be read.([264]) As this measure was intended to prevent popular excitement,([265]) and was similar in policy to the destruction of the prophetic books at Rome by Augustus,([266]) it may be inferred that the books seized by Severus were composed in the native language, and therefore that the hieroglyphic or hieratic writing was at that time generally understood in Egypt.

Ammianus Marcellinus, a historian of the 4th century, inserts in his text a Greek translation from the hieroglyphical inscriptions upon an obelisk, which had been removed from Egypt, and was in his time standing in the Roman circus. He

---

(262) Pythagoras is stated to have learnt the native language, and the three modes of writing, in his visit to Egypt, above, p. 170, n. 87. Democritus likewise composed a treatise on the sacred writings at Meroë, see Mullach's Democritus, p. 126.

(263) Tac. Ann. ii. 60. See above, p. 352. An attempt to explain why the Romans should not have occupied themselves about the historical records of ancient Egypt is made by Bunsen, Egypt, vol. i. p. 152—8, Engl. tr.

(264) Dio Cass. lxxv. 13. During this visit, Severus inspected Memphis, the statue of Memnon, the pyramids, and the labyrinth, with much care. Spartian. Vit. Sever. 17.

(265) Concerning the prohibition of prophecies and magic by the Imperial government, see Bingham, Ant. of the Church, xvi. 5.

(266) Suet. Oct. 31.

cites it from the work of Hermapion, probably a contemporary writer, of Egyptian origin.([267]) The work of Horapollo upon the hieroglyphics likewise professes to be a Greek translation by a certain Philip from the original Egyptian.([268])

Now if these traces of ancient civilization were really preserved, in intelligible writing, upon the ancient monuments of Egypt, is it likely that they should not have been brought to light? We may be sure that Manetho and others, who inquired into Egyptian antiquities, were as ignorant of the existence of copious historical native records on the walls of ancient buildings, as Hipparchus and Ptolemy were of the existence of ancient astronomical observations by the native priests. Whoever calmly considers the long possession of Egypt by the two most civilized nations of antiquity, while the sacred language and writing of the ancient Egyptians were still perpetuated by an unbroken tradition, will be slow to believe that these supposed treasures, if they really existed, could have remained untouched, or that they would have been left to be opened by the laborious investigation of modern archæologists, more than 1500 years after the key to this secret had been lost.

The Egyptologists do not indeed pretend that any great amount of historical knowledge has been hitherto derived from hieroglyphic inscriptions. They profess to have read certain names of kings, which they identify variously with names in Manetho's lists: but they do not assert that the inscriptions furnish either a coherent chronology, or events in the reigns of the kings. Brugsch, in his work on the primeval history of Egypt, lays it down that the ancient Egyptians had no era, that

---

([267]) xvii. 4. In § 8, Ammianus speaks of the hieroglyphic writing upon the obelisks. The inscription is restored by K. O. Müller, Arch. der Kunst, § 224. Compare Uhlemann, Handb. part i. p. 18.

([268]) Leemans, Prolegom. ad Horapoll. p. xvii. has the following remark upon the preservation of the knowledge of the hieroglyphic language: 'Temporibus serioribus, sacris Christianorum invalescentibus, cognitionem saltem aliquam hieroglyphicorum apud occultos veterum rituum Ægyptiacorum cultores servatam fuisse, probabile est; præcipue apud eos, qui ad scholas et sectas Gnosticorum Ægyptiacas pertinerent.'

they denoted events only by the year of the king's reign, and that this mode of reckoning affords no materials for a chronological system.([269]) The meagreness of the historical information which Bunsen([270]) and Brugsch profess to have extracted from the hieroglyphical inscriptions, must be apparent to every reader. Bunsen, indeed, speaks of ancient Egypt as the 'monumental' nation; but its monuments are colossal buildings, not intelligible inscriptions containing historical records.([271]) If the hieroglyphical writings which have been interpreted have been interpreted correctly, and if they may be taken as a sample of the rest, we may be satisfied that there is nothing worth knowing. The work of Sir Gardner Wilkinson, upon ancient Egypt, which speaks to the eye, is far more instructive than the efforts to address the mind through the restored language of the Egyptians. It may be feared that the future discoveries of the Egyptologists will be attended with results as worthless and as uncertain as those which have hitherto attended their ill-requited and barren labours. The publication of an inedited Greek scholiast or grammarian might be expected to yield more fruit to literature than the vague phrases of Oriental adulation or mystical devotion which are propounded to us as versions of hieroglyphic inscriptions.

---

(269) Hist. d'Egypte, part i. p. 241.

(270) The utmost that Bunsen can bring himself to say is, that 'he individually persists in believing that the Egyptian monuments contain chronological notations;' but he admits 'that we do not yet understand them sufficiently to build any system upon representations of so problematical a character,' Egypt, vol. ii. pref. p. xii. Eng. trans.

(271) 'We have no traces (says Niebuhr) of the Egyptians having ever had a history of their own. They had indeed a chronology (?), but true history they had not; and this observation is confirmed by what has been found in the newly explained inscriptions since the discovery of the art of deciphering the hieroglyphics. We might have expected to find in the inscriptions on the obelisks records of the exploits of the kings; but we nowhere meet with historical accounts. There are indeed historical representations; but they are not accompanied by historical inscriptions, and in most cases the representations have nothing at all to do with history.' Lectures on Ancient History, translated by Dr. Schmitz, vol. i. p. 63.

Chapter VII.

EARLY HISTORY AND CHRONOLOGY OF THE ASSYRIANS.

§ 1 IT now remains for us to examine the accounts of Assyrian antiquity, which have been handed down by the ancient authors, and to consider how far they serve to support the hypothesis of the existence of scientific astronomy at Babylon in remote times.

Herodotus, who is copious upon the primitive kings of the Egyptians, Medes, and Lydians, is brief upon the Assyrian history. This brevity seems to have been due to his intention of composing a separate work on Assyrian affairs;(1) an intention which he never executed. He assigns a period of 520 years to the dominion of the Assyrians in Upper Asia; and he states that the Medes were the successors of the Assyrians.(2)

Herodotus mentions a long series of Babylonian kings, but reserves a detailed account of them for his separate Assyrian history.(3) The only two sovereigns whom he names in this context are Semiramis and Nitocris. He states that the latter was the wife of Labynetus, whose son of the same name was king when Cyrus took Babylon(4) (538 B.C); and that the

---

(1) i. 106, 184. See Mure's Hist. of Gr. Lit. vol. iv. p. 332. Larcher doubts whether the Assyrian history of Herodotus was ever written, Hérodote, tom. vii. p. 147. Dahlmann and Bähr are of the same opinion. Müller, Hist. of Gr. Lit. vol. i. p. 354, thinks that it was composed and published. Mr. Rawlinson holds the same opinion, Herod. vol. i. p. 29, 249.

(2) i. 95.

(3) τῆς δὲ Βαβυλῶνος ταύτης πολλοὶ μέν κου καὶ ἄλλοι ἐγένοντο βασιλέες, τῶν ἐν τοῖσι Ἀσσυρίοισι λόγοισι μνήμην ποιήσομαι . . . ἐν δὲ δὴ καὶ γυναῖκες δύο, i. 184.

(4) i. 188. Labynetus, the king of the Babylonians who mediated between Cyaxares and Alyattes after the eclipse of Thales (585 B.C.) is identified by C. O. Müller, Rhein. Mus. 1827, p. 295, with Nebucadnezzar, see Herod. i. 74, 77. Above, p. 86.

former reigned five generations before Nitocris :(⁵) so that her lifetime would fall about a century before the capture of Babylon, 638 B.C. He makes Belus and Ninus successive kings of the Heraclide dynasty of Lydia ;(⁶) and does not connect them with Assyria; though he mentions the town of Ninus as the ancient Assyrian capital ;(⁷) and states that one of the gates of Babylon was called after the name of Ninus, another after the name of Belus.(⁸) Elsewhere he speaks incidentally of Sardanapalus, King of Ninus, as possessing large treasures, accumulated in a subterranean treasury.(⁹) The place of Sardanapalus in the Assyrian series of Herodotus is unfixed.

§ 2 A copious and connected account of the early Assyrian kings, and of their chronology, was first presented to the Greeks by Ctesias, a Greek physician, who resided at the Court of Artaxerxes, King of Persia, about the year 415 B.C.

He wrote a history of Persia, in 23 books, the first six of which contained a history of Assyria. In this work he made use of the records in the royal archives ;(¹⁰) and it can be reasonably assumed, that, though he may have been credulous and deficient in judgment, he followed the best accredited account of early Assyrian history which was then current in Central Asia. He considered the Persian as the successors of the Assyrian kings, and he inserted at the end of his work a list of the entire series from Ninus and Semiramis to Artaxerxes.(¹¹) Photius, to whom we owe an abridgment of the latter books of the Persica of Ctesias, omits the introductory part relating to Assyria. Diodorus, however, in his account of Assyrian history, has taken Ctesias for his guide, and we are thus

---

(5) i. 184. The story of the tomb over a gate of Babylon, opened by Darius, which Herodotus tells of Nitocris, c. 187, is transferred by Plutarch to Semiramis, Reg. et Imp. Apophth. p. 173. Repeated by Stob. Flor. x. 53.

(6) i. 7.   (7) i. 178.
(8) iii. 155.   (9) ii. 150.
(10) Diod. ii. 22. Compare Clinton, F. H. vol. ii. p. 307.
(11) Ctesias, ap. Phot. Bibl. Cod. 72, ad fin : κατάλογος βασιλέων ἀπὸ Νίνου καὶ Σεμιράμεως, μέχρι Ἀρτοξέρξου, ἐν οἷς καὶ τὸ τέλος.

enabled to ascertain the principal features of the Assyrian history of that writer.

According to Ctesias, the Assyrian empire was established by Ninus. Before him, there were kings in Asia; but they had performed no memorable act, and their names were not preserved. Ninus was a great conqueror; he subdued all Asia from the Tanais and the Nile in the West, to India and Bactria in the East. He likewise founded the city of Ninus (or Nineveh), and called it after his own name. Not being satisfied with the empire of all Western Asia, he invaded Bactria with an army of 1,700,000 infantry, 210,000 cavalry, and nearly 10,600 war chariots. In this campaign Ninus received important assistance from a woman distinguished for her beauty and courage, the wife of one of his officers, named Semiramis. He made her his own wife, and she bore him a son named Ninyas. Ninus died soon afterwards, and left the kingdom to Semiramis. She executed many great works, which preserved her name in later times; she likewise became the founder of Babylon and other cities.[12] Semiramis was a conqueror, as well as her husband: she overran Egypt, Northern Africa, and Æthiopia.[13] She also invaded India, with an army of 3,000,000 infantry, 500,000 cavalry, and 100,000 chariots; but was forced to return with the loss of two-thirds of her army. Semiramis died at the age of 62, after a reign of 42 years, and was succeeded by her son Ninyas.[14] This prince renounced the energetic and warlike policy of his father and mother; he passed his time in his palace, devoted to luxurious enjoyment. Ninyas was succeeded by 30 kings, in a regular series from

---

[12] Pliny, vi. 3, states that Melita in Cappadocia was built by Semiramis; repeated by Solin. c. 45, § 4. Solinus, c. 54, § 2, says, that she built Arachosia. Two cities of Petra, in Arabia, named Besamnæ and Soractia are stated by Pliny, vi. 28, to have been founded by Semiramis. A mountain of Carmania bore the name of Semiramis, Arrian. Peripl. Mar. Rub. p. 20.

[13] Plutarch, de Alex. Fort. ii. 3, describes Semiramis as a martial queen, who sailed along the Red Sea, and subdued the Æthiopians and Arabians.

[14] Diod. ii. 1—20.

father to son,(15) whose names Diodorus does not transcribe, because they did nothing worthy of being recorded. The only event during their reigns which had been rescued from oblivion was the expedition sent by Teutamus to his vassal, King Priam, under the command of Memnon, when Troy was besieged by the Greeks; and this was said to rest upon the authority of the court registers of the Assyrian kings.(16) The last of these 30 kings was Sardanapalus, celebrated for his luxury, who was besieged in Nineveh by Arbaces the Mede and Belesys the Chaldæan, and who destroyed himself on the pyre, in order to avoid defeat. With him ended the Assyrian empire, after a duration of more than 1300 years.

The duration of the Median empire from Arbaces to Astyages is reckoned by Ctesias at 282 years:(17) so that, assuming the accession of Astyages to be fixed at 594 B.C., the 1300 years of the Assyrian empire, according to the chronology of Ctesias, lasted from 2176 to 876 B.C.

§ 3 Berosus has been already mentioned as a Babylonian astronomer and astrologer, who introduced astrology into Greece, and who lived in the times of Alexander the Great and of Antiochus Soter.(18) Besides his labours on the science of the stars, he composed a history of Babylon, containing a cosmogony and an account of the primitive Assyrian kings; he

---

(15) Diod. ii. 21, 28, states that the Assyrian monarchy lasted for 30 generations, from Ninus to Sardanapalus; and in c. 23, that Sardanapalus was the 30th king from Ninus. Syncellus, however, quotes Diodorus as saying that Sardanapalus was the 35th king from Ninus, and that the monarchy lasted for 45 generations, vol. i. p. 312, 313, Bonn. Both these latter numbers appear to be corrupt. Cephalion, cited by Syncell. vol. i. p. 316 (repeated in Euseb. p. 41), states that Ctesias enumerated only 23 of the *rois fainéants*, who succeeded Ninyas; see Fragm. Hist. Gr. vol. iii. p. 626; but the number 30 in Diodorus was probably derived from Ctesias.

(16) περὶ μὲν οὖν Μέμνονος τοιαῦτ' ἐν ταῖς βασιλικαῖς ἀναγραφαῖς ἱστορεῖσθαί φασιν οἱ βάρβαροι, Diod. ii. 22.

Concerning the synchronism of Teutamus with the Trojan War, see below, § 7.

(17) Ap. Diod. ii. 32—4. Compare Clinton, F. H. vol. i. p. 261.

(18) Above, p. 296; Fragm. Hist. Gr. vol. ii. p. 495; Clinton, F. H. vol. iii. p. 505.

declared his book to be founded upon records carefully preserved in Babylon, which went back for a period somewhat exceeding 150,000 years.([19])

The work of Berosus is lost, and his Assyrian chronology is principally known to us through the extracts preserved in the Greek chronographers, whose reports do not always coincide. He stated that the Babylonians reckoned past time by long periods; and that, in their language, Sarus denoted a period of 3600 years, Nerus a period of 600 years, and Sossus a period of 60 years. The ten first Assyrian kings, in the list of Berosus, reigned before the deluge, and the duration of their reigns was stated by him in Sari. The series is as follows:—

		Sari.	Years.
1.	Alorus	10	36,000
2.	Alaparus	3	10,800
3.	Amelon	13	46,800
4.	Ammenon	12	43,200
5.	Amegalarus	18	64,800
6.	Daonus	10	36,000
7.	Enedorachus	18	64,800
8.	Amempsinus	10	36,000
9.	Otiartes	8	28,800
10.	Xisuthrus	18	64,800
	Total	120	432,000 ([20])

The only events attributed by Berosus to this portentous antediluvian period are the periodical appearances of certain

---

(19) Syncell. vol. i. p. 25; Alexander Polyhistor, ib. p. 50, ed. Bonn. The number in the Armenian Eusebius is, 'ab annorum myriadibus ducentis et quindecim,' *i.e.* 2,150,000; but the text of Syncellus appears preferable. Josephus, contr. Apion. i. 19, speaks of Berosus as ταῖς ἀρχαιοτάταις ἐπακολουθῶν ἀναγραφαῖς. Agathias, ii. 24, mentions Berosus as having composed the primitive history of the Assyrians.

(20) This list is preserved, with close agreement, by Euseb. p. 5, ed. Mai, and Syncellus, vol. i. p. 69—71, ed. Bonn. See Fragm. Hist. Gr. vol. ii. p. 499—500, vol. iv. p. 280, 281. Eusebius cites from Alexander Polyhistor, Syncellus from Abydenus and Apollodorus. Some discrepancies occur in the proper names; but all the three versions agree in the total of 10 kings and 120 sari.

amphibious monsters, compounded of a man and a fish, like the mermen of medieval fiction, who passed the night in the Red Sea, but came upon land during the daytime.([21]) The prevailing name for these monsters was Annedotus; but some of them received other denominations.

From the deluge, Berosus enumerated a series of 86 kings of Assyria, whom he named in their order. The names of the two first are alone preserved, Euechius,([22]) who reigned 4 neri, or 2400 years; and his son Chomasbelus, who reigned 4 neri 5 sossi, or 2700 years. The reigns of these 86 kings are said to have lasted 34,080 years.([23])

According to Berosus, as reported in Eusebius, the reigns of these 86 kings were followed by an invasion of the Medes, who captured Babylon, and established a dynasty of Median kings. This dynasty lasted for eight reigns and 224 years, and was succeeded by 11 kings, the duration of whose reigns cannot be stated, in consequence of a defect in the text of Eusebius. After them came 49 Chaldæan kings, who reigned 458 years; then 9 Arabian kings, who reigned 245 years; and lastly 45 kings, who reigned 526 years. The next kings are Phul, Senecharib and his son Asordanes, and Marudach Baldanes.

The early Assyrian chronology of Berosus is differently reported by Syncellus, though both he and Eusebius profess to follow the authority of Alexander Polyhistor. According to

---

([21]) Concerning the god Dagon, whom Milton describes as

'Sea-monster, upward man,
And downward fish,'—(Par. Lost, b. i.)

see Winer in v. Derceto likewise was a fish-goddess, Diod. ii. 4; Ovid, Met. iv. 45; Plin. v. 32; Lucian, de Deâ Syr. c. 14. Concerning the sanctity of fishes in Syria, see Ovid, Fast. ii. 461—472; Hygin. Poet. Astron. ii. 30; Lucian, ib. c. 45, and the notes in Lehmann's edition, vol. ix. p. 393; Manil. iv. 580—583.

([22]) This name is written Εὐήχιος in Syncellus according to Dindorf's text, and *Evexius* in the Armenian Eusebius, the Greek χ and the Latin x being confounded.

([23]) Syncellus, vol. i. 147, states the period at 9 sari, 2 neri, and 8 sossi $= 32,400 + 1200 + 480 = 34,080$. In Eusebius the sum is stated as 33,091, Fragm. Hist. Gr. ib. p. 503.

Syncellus, among the 86 kings, who reigned 34,080 years, only the two first, Euechius and Chomasbelus, were native Chaldæans; the other 84 were Medes. There follows a dynasty, of 7 Chaldæan kings, at the head of which is Zoroaster: they reign for 190 years.[24]

Omitting, therefore, the ten antediluvian kings from Alorus to Xisuthrus, who reigned 432,000 years, the Assyrian chronology of Berosus, so far as it can be restored from our extant sources, appears to have been as follows:—

	Years of reigns.
Euechius, Chomasbelus, and 84 other kings	34,080
8 Median kings	224
11 kings	(text imperfect)
49 Chaldæan kings	458
9 Arabian kings	245
Semiramis	not stated.
45 kings	526
Phul	not stated.
Senecharib	18
Asordanus	8
Sammughes	21
Sardanapalus	21
Nabopolassar	21[25]
Nabucodrossor	43
Amilmarudoch	2
Neriglissar	4
Laborosoarchod	9 months.
Nabodenus	17
Cyrus	9[26]

(24) Syncell. vol. i. p. 147. Clinton, F. H. vol. i. p. 270, note d, considers the account of Syncellus as confused and mutilated, and prefers that of Eusebius.

(25) C. Müller, Fragm. Hist. Gr. vol. ii. p. 505, supplies in the text of Eusebius, 'et Nabopolassarus annis viginti.' But in the following extract of Josephus, Berosus is cited as giving 21 years to Nabopolassar: hence restore in Eusebius, 'et Nabupolassarus viginti annis et uno.'

(26) See Fragm. Hist. Gr. vol. ii. p. 509; Clinton, F. H. vol. i. p. 272.

§ 4 The astronomical canon, which enumerates the Assyrian kings, with the years of their reigns, from Nabonassar to Cyrus, was regarded by the Alexandrine astronomers as authentic, and may reasonably be assumed to have been founded on the best extant evidence. In this canon the series of kings is as follows:—([27])

	Years of reign.	B.C. First year of reign.
Nabonassarus	14	26 Feb. 747
Nadius	2	733
Chinzer and Porus	5	731
Ilulæus	5	726
Mardocempadus	12	751
Arceanus	5	729
First interregnum	2	724
Bilibus	3	722
Aparanadius	6	699
Rhegebelus	1	693
Mesesimordacus	4	692
Second interregnum	8	688
Asaridinus	13	680
Saosduchinus	20	667
Ciniladanus	22	647
Nabopollassarus	21	625
Nabocolassarus	43	604
Illoarudamus	2	561
Nericasolassarus	4	559
Nabonadius	17	555
Cyrus	9	538

§ 5 The Assyrian chronology adopted by Syncellus is as follows. He begins the series with Euechius and Chomasbelus, identifying Euechius with Nimrod, and he states the reigns thus:—

---

(27) See Halma's Ptolemy, vol. iii.; Des Vignoles, Chron. d'Hist. Sainte, vol. ii. p. 349; Ideler, Chron. vol. i. p. 98, 220.

	Years of reign.	
Euechius	6⅓	2776 A.M.
Chomasbelus	7	
Porus	35	
Nechubes	43	
Nabius	48	
Oniballus	40	
Zinzerus	46	
Total	225⅓	

These kings were succeeded by an Arabian dynasty of six kings, viz.:—

Mardocentes	44 years of reign.
Mardacus	40
Sisimordacus	28
Nabius	37
Parannus	40
Nabunnabus	25
Total	214

The Arabian dynasty was succeeded by forty-one native kings, who reigned 1460 years. The first of these was Belus, and the last was Macoscolerus, otherwise called Sardanapalus. The entire time from Euechius to Macoscolerus is reckoned by Syncellus as 1899 years, from 2776 to 4675 A.M. (2724 to 825 B.C.) ([28])

The forty-one kings, from Belus, Ninus, Semiramis, and Ninyas, to Sardanapalus, inclusive, are enumerated by Syncellus, together with the durations of their respective reigns.([29])

§ 6 According to the list of Eusebius, there are thirty-six kings from Ninus to Sardanapalus, making a period of 1240 years. Syncellus prefixes Belus to his list, and gives him a reign of fifty-five years: his kings from Ninus to Sardanapalus

---

([28]) Sync. vol. i. p. 169, 172.
([29]) Ib. p. 181, 193, 203, 232, 277, 285, 293, 301, 312.

are forty, being four more than those of Eusebius: and their reigns occupy 1462 years. With the exception of the four additional kings in the list of Syncellus, the names and places of the kings in the two lists agree, and the durations of the reigns nearly agree. Several of the names are Greek, as Sphærus, Amyntas, Lamprides, Dercylus, Laosthenes.(30)

§ 7 Ctesias was the earliest Greek writer who gave a full narrative of Assyrian history. Although his chronology differed from that of Herodotus, his account seems to have been generally adopted by his countrymen, and to have formed the basis of the received version of Assyrian history current in later times. The foundation of the Assyrian Empire by Ninus and Semiramis, its extinction under Sardanapalus, and its duration for 1300 years, were accepted by the prevailing belief of antiquity, as fully represented in Diodorus, and partially exhibited by other writers.(31)

---

(30) See Clinton, F. H. vol. i. p. 267. The list of Assyrian kings in the Chronicon of Cassiodorus agrees with that in Eusebius from Ninus to Mithræus.

(31) Strabo, xvi. i. § 2, says that Ninus founded the city of Ninus, and that his wife Semiramis founded Babylon: that they bequeathed the empire to their successors down to Sardanapalus, when it passed to the Medes. (In this passage the words καὶ 'Αρβάκου should be expunged; or we should read, μετέστη δ' εἰς Μήδους ὕστερον ὑπ' 'Αρβάκου.) Nicolaus Damascenus, who lived in the time of Augustus, states that the Assyrian Empire lasted from Ninus and Semiramis to Sardanapalus, and was overthrown by Arbaces and Belesys, Fragm. Hist. Gr. vol. iii. p. 357. Justin exactly follows Ctesias: he makes Ninus the founder of the Assyrian Empire; Semiramis the builder of Babylon; and Sardanapalus the last king. He gives the name of Arbactus to the Satrap of Media, who overthrew Sardanapalus, and founded the Median Empire, i. 1—3. Velleius, i. 6, states that the Assyrian Empire was founded by Ninus and Semiramis, who built Babylon; that they were succeeded by 33 kings, in regular succession from father to son, the last of whom was Sardanapalus; that Sardanapalus was overthrown by Pharnaces the Mede; and that the Assyrian Empire lasted 1070 years. The latter number is considered corrupt by Mr. Clinton. For *Pharnaces* we should probably read *Arbaces*. Orosius, i. 1, states that Ninus, son of Belus, founded the Assyrian Empire; that Ninus lived 3184 years after Adam, and 2015 years before the birth of Christ. The conquering career of Ninus and Semiramis is described by Orosius after Ctesias: he introduces the period of 1300 years, but apparently misapplies it (i. 4). He states that Ninus reigned 52 years, and Semiramis 42 years, and that Semiramis founded Babylon, ii. 3; that Sardanapalus, the last king of Assyria, reigned 64 years before the building of Rome

The scheme of Assyrian kings from Ninus to Sardanapalus was adopted by Syncellus, but with variations to which we shall advert lower down.

Ninus, and his queen Semiramis, appear to be purely fabulous beings. The name of Ninus is derived from the city; he is the eponymous king and founder of Nineveh:([32]) and stands to it in the same relation as Tros to Troy, Medus to Media,([33]) Perseus to Persia,([34]) Ægyptus to Egypt, Lydus to Lydia,([35]) Mæon to Mæonia,([36]) Romulus to Rome. His conquests, and those of Semiramis, are as unreal as those of Sesostris. It is characteristic of these fabulous conquerors, that although they are reported to have overrun and subdued many countries, the history of those countries is silent on the subject. Sesostris is related to have conquered Assyria; and the king of Assyria was doubtless one of those whom he harnessed to his chariot. But the history of Assyria makes no mention of Sesostris. Semi-

---

(817 B.C.); that he was overthrown by his satrap Arbatus the Mede, and burnt himself on the pyre, and that the empire passed to the Medes, after having remained with the Assyrians for 1164 years, i. 19, ii. 3. The period of 1300 years, in i. 4, seems to be the interval between the foundation of the Assyrian Empire and the foundation of Rome; according to the figures subsequently given by Orosius this interval would be 1164 + 64 = 1228 years. Col. Mure remarks that 'the dates of Ctesias [for the Assyrian Empire] have been preferred, with occasional slight variations, by almost all the subsequent native Greek chronologers,' Hist. of Lit. of Gr. vol. iv. p. 334.

(32) This is the statement of Ammian. Marcellin. xxiii. 6, § 22, and of Steph. Byz. in Νίνος.

(33) Medus was the founder of the Persian Empire, according to Æsch. Pers. 765. Medus, the progenitor of the Medes, was also called the son of Medea and Jason, Strab. xi. 13, § 10.

(34) Herod. vii. 61, says that Perseus, son of Jupiter and Danae, married Andromeda, and that their son Perses gave his name to the Persian people. Compare Apollod. ii. 4, § 5. Perseus, the son of Jupiter and Danae, was the reputed founder of Tarsus, Amm. Marc. xiv. 8, 3. Josephus, B. J. iii. 9, § 3, states that the marks of the chains by which Andromeda was fastened to the rocks were still shown on the sea-shore near Joppa, in Syria. Josephus, who did not receive the Greek mythology, observes that these marks attest, not the truth, but the antiquity, of the legend.

(35) According to Herod. i. 171, Lydus, Mysus, and Car, were brothers, and gave their names to the three nations, cf. i. 7. Xanthus, ap. Dion. Hal. Ant. Rom. i. 28, likewise derives the Lydians from Lydus.

(36) 'Dicti post Mæona regem Mæones,' Claudian, Eutrop. ii. 245. Concerning Mæon, see Diod. iii. 58.

ramis is related to have conquered Egypt; but the history of Egypt makes no mention of Semiramis. Osymandyas, king of Egypt, is reported to have made war upon the revolted Bactrians, with an army of 420,000 men; but his expedition is not recorded in the history of the countries which he must have invaded in order to reach this distant region.([37]) To find a parallel for this state of things in the annals of modern nations, we must suppose that the history of France described Napoleon as a great conqueror, who subdued or overran Germany, Russia, Italy, and Spain; but that the histories of Germany, Russia, Italy, and Spain were silent as to such a person, and described each country as undisturbed by any aggressor, at the time when the history of France represented Napoleon to have been occupied in performing these great deeds.

Phœnix of Colophon, a choliambic poet, who lived about 300 B.C.,([38]) conceived Ninus as an indolent and luxurious king, and confounded him with Sardanapalus.([39]) He attributes to Ninus the epitaph on the theme, 'Let us eat and drink, for tomorrow we die,' which Chœrilus and other authors more consistently ascribe to Sardanapalus.

The birth of Semiramis is enveloped with marvels. She is the daughter of a goddess and of a man; the mother is ashamed of her amour with a mortal, and exposes her newborn infant in a lonely place; the babe is fed for a year by doves; and at the end of this time is found, like Romulus and Remus, by shepherds, who carry her to the manager of the royal flocks.([40])

---

(37) Diod. i. 47—8. Osymandyas is said to have had a tame lion, which fought by his side in battle. The use of lions in war is ascribed by Lucretius to the Parthians:—

    Et validos Parthi præ se misere leones,
    Cum doctoribus armatis, sævisque magistris,
    Qui moderarier his possint, vinclisque tenere.—v. 1309—11.

This custom is inconsistent with the nature of the lion, and never occurs in authentic history.

(38) Paus. i. 9, § 7.

(39) Ap. Athen. xii. p. 530 E. Compare Naeke's Chœrilus, p. 196, 226; Anth. Plan. iii. 27, App. 97, ed. Jacobs. Above, p. 259.

(40) Diod. ii. 4. Derceto, the Syrian goddess, is the mother of Semi-

Ctesias represented Ninus as having bequeathed his kingdom to Semiramis. Dinon, however, and other historians, stated that she was a beautiful courtezan, who became the wife of Ninus, and persuaded him to entrust her with supreme power for five days; that on the first day she gave a splendid banquet to the principal grandees and the commanders of the troops; and that, having gained them over, she threw her husband into prison, put him to death, and made herself queen.([41]) A story followed by Conon represented Semiramis as the daughter, not the wife, of Ninus.([42]) Some writers mentioned by Macrobius likewise adopted the same story.([43])

Semiramis was not only the mythical founder of Babylon,([44])

---

ramis, and her birth and exposure are supposed to take place near Ascalon, in Syria. Justin. xxxvi. 2, however, says that Damascus was the birthplace of Semiramis, and the cradle of the Assyrian kings. Lucian, de Deâ Syriâ, c. 14, speaks of Derceto as the mother of Semiramis. He likewise mentions the sanctity of the dove in Syria, and states that the Syrians abstain from eating its flesh. Tibullus alludes to the same fact:—

> Quid referam, ut volitet crebras intacta per urbes
> Alba Palæstino sancta columba Syro?—i. 7, 17.

Compare Hygin. fab. 197.

([41]) Diod. ii. 20; Ælian, V. H. vii. 1; Plutarch, Amator. 9. Compare Frag. Hist. Gr. vol. ii. p. 89. Dinon lived soon after 350 B.C. Like Ctesias, he wrote a work entitled Περσικά. Echion, or Aetion, who flourished in 352 B.C., painted a celebrated picture of Semiramis, in her nuptial attire, raised from the condition of a slave to the rank of queen, Plin. xxxv. 10. According to Hygin. fab. 240, Semiramis killed her husband Ninus.

([42]) Narr. c. 9.   ([43]) In Somn. Scip. ii. 10, § 7.

([44]) This statement, repeated by Diodorus from Ctesias, is followed by Strab. ii. 1, § 31, xvi. 1, § 2; Plutarch, de Alex. Fort. ii. 3, and Solinus, c. 56, § 1. Propertius has the following verses on the subject:—

> Persarum statuit Babylona Semiramis urbem,
>   Ut solidum cocto tolleret aggere opus.
> Et duo in adversum immissi per mœnia currus,
>   Ne possent tacto stringere ab axe latus.
> Duxit et Euphratem medium, qua condidit arces,
>   Jussit et imperio surgere Bactra caput.—iii. 11, v. 21—26.

The brick wall of Semiramis is likewise alluded to by Ovid:—

> Ubi dicitur altam
> Coctilibus muris cinxisse Semiramis urbem.—Met. iv. 57.

Ovid places the tomb of Ninus at Babylon, ib. 88. The brick wall of Babylon, built by Semiramis, is included among the Seven Wonders of the

but her name was attached to numerous buildings, roads, canals of irrigation, and great works in Central Asia.(45)

Berosus contradicted the statement of the Greek historians; he denied that Semiramis founded Babylon, and that she was the author of the marvellous works attributed to her.(46) Stephanus of Byzantium says that Babylon was founded by a wise man of the same name as the city, the son of Belus; and that, according to Herennius, Semiramis lived above a thousand years after the foundation of Babylon.(47) The author here referred to is Herennius Philo, who wrote at the end of the first century after Christ. Moses of Chorene cites a certain Maribas of Catana as having explored Chaldæan histories for the adven-

---

World, by Hygin. fab. 223. It is also mentioned as the work of Semiramis by Schol. Juven. x. 171. Semiramis is alluded to by Claudian as the founder of Babylon:—

> Claras Carthaginis arces
> Creditur, et centum portis Babylona superbam
> Femineus struxisse labor.　　　　　　In Eutrop. i. 334.

Quintus Curtius says that Babylon was founded by Semiramis, and not, as is generally believed, by Belus, v. 1, § 24. According to Ammian. Marcellin. xxiii. 6, § 23, the walls of Babylon were built by Semiramis, and the citadel by Belus. The account of Abydenus was that the walls of Babylon were built by Belus, and afterwards restored by Nabricodrossorus, Fragm. Hist. Gr. vol. iv. p. 283. Dorotheus, in the astrological verses at the end of Köchly's Manetho, has ἀρχαίη Βαβυλὼν Τυρίου Βήλοιο πόλισμα. Orosius considers Babylon to have been founded by Nimrod the giant, and to have been restored by Ninus or Semiramis, ii. 6. This is an attempt to combine the Biblical and classical accounts.

(45) Diod. ii. 13, 14; Lucian, de Deâ Syr. c. 14; Strab. ii. 1, § 26, xi. 14, § 8, xiii. 2, § 7, xii. 3, § 37. In xvi. 1, § 2, Strabo says: τῆς Σεμιράμιδος, χωρὶς τῶν ἐν Βαβυλῶνι ἔργων, πολλὰ καὶ ἄλλα κατὰ πᾶσαν γῆν σχέδον δείκνυται, ὅσα τῆς ἠπείρου ταύτης ἐστί, τά τε χώματα, ἃ δὴ καλοῦσι Σεμιράμιδος, καὶ τείχη καὶ ἐρυμάτων κατασκευαὶ καὶ συρίγγων τῶν ἐν αὐτοῖς καὶ ὑδρείων καὶ κλιμάκων καὶ διωρύγων ἐν ποταμοῖς καὶ λίμναις καὶ ὁδῶν καὶ γεφυρῶν. Herodotus says that she made the dykes of the Euphrates near Babylon, i. 184. A ditch of Semiramis on the Euphrates, is mentioned by Isidorus Characenus, ap. Geogr. Gr. Min. vol. i. p. 247, ed. C. Müller.

(46) Ap. Joseph. contr. Apion. i. 20. According to Euseb. Chron. p. 36, the Chaldæans do not include Ninus and Semiramis among their royal names.

(47) In v. Βαβυλών. Compare Volney, Rech. Nouv. sur l'Hist. Anc. Œuvres, p. 481; Salmasius, Exerc. Plin. p. 866 E; Dr. Smith's Dict. of Anc. Biog. and Myth. art. *Philon*, 7. The text of Stephanus has χιλίοις δύο, which must be rendered 1002. Volney, however, and Movers, Phönizier, vol. ii. p. 253, translate it as if it were δισχιλίοις. The statement recurs in Eustath. ad Dionys. Perieg. 1005, where the number is 1800.

tures of Semiramis;(48) but we know nothing as to the authors or dates of these works.

According to one story, Semiramis took Babylon by diverting the course of the Euphrates.(49) She is also reported to have vigorously repressed a revolt of the inhabitants of this city.(50)

The accounts of the great constructions of Semiramis are not more veracious than the stories told in Herodotus of the great works executed by queen Nitocris, the mother of the king of Babylon defeated by Cyrus.(51) Philostratus has also a fabulous story of an ancient queen of Babylon, named Medea, who made a tunnel under the Euphrates.(52)

The idea of a military queen, who carries her conquering standards from the Indus to Mount Atlas,(53) is foreign to authentic Oriental history, and may be classed with the accounts of the Amazons, or with the story of the Pandæ, an Indian nation governed by a line of female sovereigns who traced their descent to a daughter of Hercules.(54)

---

(48) Fragm. Hist. Gr. vol. iii. p. 627.
(49) Frontin. Strat. iii. 7, 5. (50) Val. Max. ix. 3, ext. 4.
(51) i. 185—188. (52) Vit. Apollon. i. 25.
(53) According to Nearchus, Alexander the Great believed that Semiramis invaded India, and returned with only twenty men, Strab. xv. 1, § 5, ib. 2, § 5. Megasthenes, however, stated that Semiramis died before her intended expedition, ib. § 6. According to Solinus, c. 49, § 3, Panda in Sogdia was the furthest limit of the Indian expeditions of Bacchus, Hercules, Semiramis, and Cyrus. Plutarch, de Is. et Os. i. 24, compares the exploits of Semiramis with those of Sesostris.

(54) See above, p. 268. The numerous children of Hercules were supposed to have been all sons, Apollod. ii. 7, § 8. Claudian, inveighing against an eunuch consul, declares that many Oriental states have been governed by queens, but that no Oriental state was ever governed by an eunuch:—

Sumeret illicitos etenim si femina fasces,
Esset turpe minus. Medis levibusque Sabæis
Imperat hic sexus, Reginarumque sub armis
Barbariæ pars magna jacet. Gens nulla probatur
Eunuchi quæ sceptra ferat.   In Eutrop. i. 319—324.

The supposed queen of the Medians here alluded to, appears to be Medea, see Strab. xi. 13, § 10; Diod. iv. 57. With respect to the Sabæan queens, Claudian alludes not to the queen of Sheba who visited Solomon, 1 Kings x.; but to the queens of Meroe, also called Saba, who are said to have borne the common name of Candace; see Strab. xvii. i. § 54; Dio Cass.

The notions of the Greeks and Romans concerning Semiramis were discordant; for while she was represented as a hardy warrior,([55]) and as a founder of large cities, and the author of great engineering works, she was also conceived as a sort of Messalina, devoted to licentious and even incestuous love, and murdering her lovers after she had dismissed them from her embraces.([56]) She likewise furnished several mythical origins. She was supposed to have introduced the loose and flowing dress of the Orientals;([57]) to have established the Asiatic custom of

---

liv. 5; Acts viii. 29; Fragm. Hist. Gr. vol. iv. p. 351. Some Romans sent to Æthiopia by Nero, reported (among other things), 'Regnare feminam Canagen, quod nomen multis jam annis ad reginas transiit,' Plin. vi. 29. Claudian repeats the same sentiment lower down, v. 427, where he says of eunuch rulers,—
> 'Auroram sane, quæ talia ferre
> Gaudet, et assuetas sceptris muliebribus urbes,
> Possideant.'

The poet's dictum is, however, inconsistent with fact. The lawless habits of the East, and the seclusion of women, have in general been incompatible with the rule of queens.

([55]) Diodorus mentions an equestrian statue of Semiramis spearing a leopard, ii. 8. Ælian, Var. Hist. xii. 39, speaks of Semiramis as a huntress of lions and leopards. According to Ctesias, Semiramis was the inventor of the war-galley, Plin. vii. 58.

([56]) Ctesias stated that the mounds of Semiramis were the tombs of her lovers, whom she buried alive, Joann. Antiochen. ap. Fragm. Hist. Gr. vol. iv. p. 539. According to Justin. i. 2, and Agathias, ii. 24, she cherished an incestuous love for her son Ninyas. Her promiscuous and incestuous loves and the murders of her lovers are mentioned in Oros. i. 4. Ovid places Semiramis in juxtaposition with Lais, Am. i. 5, 11, and Juvenal with Cleopatra, ii. 108. Gabinius is called a Semiramis by Cicero, de Prov. Consul. 4, in reference to his libidinous conduct in his Syrian province. The unnatural love of Semiramis for a horse is mentioned by Juba, ap. Plin. viii. 42; Fragm. Hist. Gr. vol. iii. p. 472. The 'impuri et meretricii mores' of Semiramis are noticed by Moses Choren. i. 16. Dante places Semiramis among the *lussuriose*:—
> La prima di color, di cui novelle
> Tu vuoi saper, mi disse quegli allotta,
> Fu imperatrice di molte favelle.
> A vizio di lussuria fu sì rotta,
> Che libito fe lecito in sua legge,
> Per torre il biasmo in che era condotta.
> Ell' è Semiramis, di cui si legge
> Che succedette a Nino e fu sua sposa:
> Tenne la terra che il Soldan corregge.
>   Inf. canto V. ver. 52—60.

([57]) Diod. ii. 6; Justin. i. 2. The story of her wearing man's clothes is alluded to in Claudian, Eutrop. i. 339, and Oros. i. 4.

marriages between mothers and sons;(58) and to have initiated the use of eunuchs.(59)

Alexander Polyhistor likewise derived the name of Judæa from Judas, one of the sons of Semiramis.(60) This origin of the name is different from that of the Greek writers mentioned by Tacitus, who traced the Jews to Mount Ida, in Crete, making them a Cretan colony; and conceived that the name Ἰουδαῖοι was lengthened from Ἰδαῖοι.(61) The latter derivation may be compared with that of the medieval chroniclers, who traced the Egyptian Pharaohs to the island of Pharos. Other mythologists made Judæus and Hierosolymus the sons of the Egyptian god Typhon.(62)

Hellanicus transferred some of the attributes of Semiramis to an ancient queen Atossa, whom he described as assuming male attire, as introducing the use of eunuchs, as martial in her habits, and as subduing many nations.(63)

The foundation of Babylon was attributed by some writers to Belus, not to Semiramis. Abydenus made him the first king of Assyria, and interposed five kings between him and Ninus.(64) According to Syncellus, Belus was the immediate predecessor

---

(58) Oros. i. 4. The prevalence of marriages between mothers and sons among the Magi is stated in an extract attributed to Xanthus, Fragm. Hist. Gr. vol. i. p. 43. This custom of the Magi is carried still further by Catullus:—

> Nam magus ex matre et nato gignatur oportet,
> Si vera est Persarum impia relligio.—Carm. 90.

The practice of incest is attributed to the barbarians generally by Eurip. Andromm. 173—6. Ptolemy, Tetrabibl. ii. 3, considers incest with mothers as characteristic of the nations of Southern Asia.

(59) Ammian. Marcellin. xiv. 6, § 17; Claudian, Eutrop. i. 339—345, attributes the invention to Semiramis, or to the Parthians. Hellanicus, fragm. 169, ed. C. Müller, assigned it to the Babylonians; Clearchus to the Medes, Athen. xii. p. 514 D. An etymological mythus in Steph. Byz. in Σπάδα gives it to the Persians. Xenophon ascribes the use of eunuchs as a bodyguard of the Persian kings to the institution of Cyrus, Cyrop. vii. 5, § 60—5. Josephus says that Nebuchadnezzar made some of the Jewish youths eunuchs, Ant. x. 10, 1.

(60) Steph. Byz. in Ἰουδαία: Fragm. Hist. Gr. vol. iv. p. 364.
(61) Hist. v. 2.   (62) Plut. de Is. et Osir. 31.
(63) Fragm. 163a et b, ed. Müller.
(64) Fragm. Hist. Gr. vol. iv. p. 284.

of Ninus; and he was generally regarded in later times as a primitive king of Babylon.([65]) His name was connected with the lofty temple in the middle of the city; a gate of Babylon was likewise called after him,([66]) and he was regarded as the primitive teacher of astronomy to the Assyrians.([67]) Belus is the Hellenized form of Baal, who was worshipped in Syria as well as in Assyria.([68]) The early mythology of the Greeks connected Belus with Africa, rather than with Asia. Thus Æschylus, in his tragedy of the *Supplices*, describes Belus, the son of Libya, as the father of Ægyptus and Danaus.([69]) According to Apollodorus,([70]) Agenor and Belus were the sons of Neptune and Libya; Agenor became king of Phœnicia, and Belus king of Egypt. The early logographer, Pherecydes, likewise establishes an affinity between Agenor, Belus, Ægyptus, and Danaus, though by different links.([71]) Pausanias explains the presence of Belus at Babylon, by saying that he derived his name from the Egyptian Belus, son of Libya.([72]) Writers of the historical school transferred him to Asia: thus Herodotus places his name both among the primitive rulers of Persia and in the series of the Heraclide kings of Lydia;([73]) Virgil makes him the father of Dido, and the first of the Tyrian kings;([74]) Alexander of

---

(65) Babrius, part i. prooem. ii., says that the Æsopian fable was invented by the ancient Assyrians, οἳ πρίν ποτ' ἦσαν ἐπὶ Νίνου τε καὶ Βήλου.

(66) Herod. i. 181, iii. 156, 158. Diodorus, ii. 8, 9, states that Jupiter was called Belus by the Babylonians. This statement recurs in Agath. ii. 24. Some precious stones found in Assyria received the name of Belus, from the great god of the country, Plin. xxxvii. 55, 58.

(67) See above, p. 258.

(68) See Dr. W. Smith's Dict. of the Bible, in *Baal*; Winer, in *Baal* and *Bel*.

(69) Suppl. 314—20. Pausanias mentions a monument of Ægyptus, the son of Belus, at Patræ, where he took refuge, in order to avoid Danaus at Argos, vii. 21, § 13.

(70) i. 4.   (71) Fragm. Hist. Gr. vol. i. p. 83, fr. 40.

(72) iv. 23, § 10.

(73) i. 7, vii. 61, 150. He says that the Persians were originally called Cephenes by the Greeks, from their king Cepheus, son of Belus; that Perseus, the son of Jupiter and Danae, married Andromeda, daughter of Cepheus, and that his son Perseus gave his name to the nation.

(74) Æn. i. 622, 729.

Ephesus, an author contemporary with Cicero, spoke of Belus as the founder of towns in the island of Cyprus.([75])

The early Greeks were acquainted with the Phœnician and Egyptian coasts; but they knew nothing of Central Asia. Strabo thinks that Homer was ignorant of the Assyrian and Median empires; for that if he had heard of the wealth of Babylon and Nineveh and Ecbatana, he would have mentioned it, as he mentioned the wealth of Phœnicia and of the Egyptian Thebes.([76]) The same writer discredits the accounts of primitive Oriental history given by Herodotus and the early Greek authors.([77])

Although Nebuchadnezzar reigned over Assyria from 604 to 561 B.C., and was therefore contemporary with Solon; and although he twice took Jerusalem, laid siege to Tyre, and advanced as far as Egypt; yet his name and exploits were unknown to the early Greeks. The only Greek writer stated to have mentioned him is Megasthenes, who lived about 300 B.C.; and he assigns to Nebuchadnezzar the fabulous exploit of having subdued North Africa and Iberia.([78])

A fragment of Alcæus, ingeniously restored by Otfried Müller, describes the poet's brother, Antimenidas, as having fought in the army of the Babylonians. This event is referred by Müller to one of the western expeditions of Nebuchadnezzar; but Alcæus does not appear to have mentioned the name of the Babylonian king under whom his brother fought.([79])

The only Assyrian kings between Ninus and Sardanapalus, of whom we have anything beyond the name, owe this comparative fame exclusively to their being brought into connexion with the Greek mythology. Perseus, the son of Danae, having been defeated by Bacchus, the son of Semele, is reported to have landed on the Assyrian coast with 1000 ships, and to have

---

([75]) Steph. Byz. in Λάπηθος; Meineke, Anal. Alex. p. 375.
([76]) xv. 3, § 23.   ([77]) xi. 6, § 2, 3.
([78]) Ap. Joseph. Ant. x. 11, 1; Contr. Apion. i. 20; Syncell. vol. i. p. 419, ed. Bonn.
([79]) See Rheinisches Museum, 1827, p. 287.

taken refuge with king Belimus, in the 640th year after the foundation of the Assyrian Empire.(80) At a later date, the expedition of the Argonauts occurred during the reigns of the Assyrian kings Panyas and Mithræus,(81) the latter of whom reigned a thousand years after Semiramis.(82) According to Diodorus and Cephalion, the Trojan War took place during the reign of Teutamus, the successor of Mithræus; Priam was a satrap of the Assyrian Empire, and sent to Teutamus for assistance after the death of Hector; Cephalion even gives the text of the letter in which this request was made. Teutamus granted the request, and despatched Memnon, the son of Tithonus, with an army, to his assistance.(83) Syncellus states that Babius,

---

(80) Cephalion, ap. Fragm. Hist. Gr. vol. iii. p. 626.
(81) Ibid.   (82) Ibid.
(83) Diod. ii. 22; Cephalion, ib. Compare Syncellus, vol. i. p. 285, Bonn. Diodorus says that Teutamus was the twentieth king from Ninyas. According to the Greek account, Tithonus and Priam were brothers, sons of Laomedon, Diod. iv. 75. The Homeric hymn to Venus, v. 219—239, describes Tithonus, of the royal family of Troy, as being carried away by Aurora, and becoming her companion; but does not make him the father of Memnon. According to the ancient epic account in the Æthiopis of Arctinus, copied by Quintus Calaber, in his second book (see Thirlwall in Phil. Mus. vol. i. p. 147), Memnon, the Æthiopian, the son of Aurora, comes to Troy from the borders of Ocean, and is there slain by Achilles. He is a purely mythical personage, and has no connexion either with Tithonus or with the Assyrian Empire. The account in the Chrestomathia of Proclus is, Μέμνων δὲ ὁ Ἠοῦς υἱὸς ἔχων ἡφαιστότευκτὸν πανοπλίαν παραγίνεται τοῖς Τρωσὶ βοηθήσων. This view of Memnon is rendered in the verse of Virgil:—

    'Eoasque acies, et nigri Memnonis arma.'—Æn. i. 489.

Compare Heyne, Exc. xix. Hesiod placed the tomb of Memnon in Phrygia, in accordance with the cyclical legend, Fragm. p. 292, ed. Marckscheffel. Ælian, Hist. An. v. 1, likewise speaks of the tomb of Memnon in the Troad. It was on the banks of the Æsepus, Paus. x. 31, § 6; Oppian, de Aucup. i. 6. Subsequently, however, the monuments of Memnon were transferred to other parts of Asia. Aristotle, Pepl. 55, describes Memnon, the son of Tithonus and Aurora, as buried on the banks of the Belus in Syria. According to Josephus, Bell. Jud. ii. 10, § 2, there was a monument of Memnon on the river Beleus, not far from Ptolemais, in Syria (Ace or *Acre*). Oppian, Cyneg. ii. 150, places the temple of Memnon in Assyria. Susa was likewise denominated the Memnonian city, and its acropolis and palace were called after Memnon's name, Welcker, Ep. Cyclus, vol. ii. p. 214. It was likewise said to have been founded by Tithonus, the father of Memnon, Strab. xv. 3, § 2. According to Hygin. fab. 223, the palace of Cyrus at Ecbatana, one of the Seven Wonders of the World, was built by Memnon. Memnon was described as coming from Susa to Troy, Paus. x. 31, 7. This must have been a Greek fiction. The Æthiopes of Homer

otherwise Teutamus, or Tautanes, the Second, called by the Greeks Tithonus, a later king, sent his son Memnon to assist Priam; and he adds that it was to this king that the letter of Priam was addressed.(84)

The account of the connexion between Memnon and Teutamus appears to have been derived from Ctesias; and to the account of Ctesias we may probably trace the theory of Plato respecting the origin of the Trojan War, who, in his treatise of *Laws*, states that the Trojans, relying on the assistance of the Assyrians of Nineveh, provoked the aggression of the Greeks.(85)

The story of Memnon coming to the aid of Priam was engrafted on Assyrian history by the license of Greek mythologists,(86) as the story of Helen and Proteus was engrafted upon Egyptian history; and as the story of Æneas was engrafted upon Roman history. Belus, Ninus, and Semiramis were of Oriental origin, though moulded by Greek fiction; the same remark likewise applies to the last king Sardanapalus, whose death on the pyre has been by some critics derived from the representation of an Assyrian divinity.(87)

Callisthenes, the contemporary of Alexander the Great,

---

and of the early epic poets were a nation which lived at the extremity of the world, on the shores of the circumfluous ocean; see Völcker's Homerische Geographie, p. 87. Hence the black races to the east and south of Egypt were called Æthiopians; Memnon was made a negro; and his name was introduced into the Greek mythology of Egypt; see Welcker, ib. p. 211; Thirlwall, ib. p. 152. Æschylus, Prom. 807, speaks of a black tribe at the extremity of the earth, near the river Æthiops.

(84) Syncell. vol. i. p. 293, Bonn. Compare Des Vignoles, vol. ii. p. 266—270.

(85) De Leg. iii. 6, p. 685. This may be compared with the theory of Thucydides, that the Trojan War was owing to the overweening power of Agamemnon, i. 9. In both cases, the object is to find a more probable cause than the abduction of Helen.

(86) Niebuhr, indeed, holds that the account of the kingdom of Troy being a fief of the Assyrian Empire of Nineveh, and of the king of Troy being a vassal of the Assyrian king, is 'a correct historical idea,' Lectures on Ancient History, vol. i. p. 24, ed. Schmitz; but Welcker justly denies it any historical foundation, Ep. Cyclus, vol. ii. p. 213.

(87) See K. O. Müller, Sandon und Sardanapal, in the Rheinisches Museum, vol. iii. p. 22 (1829); Movers, Phönizier, vol. i. p. 458 (1841), vol. ii. p. 289 (1849).

stated in his history of Persia, that there were two kings named Sardanapalus, one of whom was energetic and brave, while the other was effeminate.(⁸⁸) This contrivance for reconciling the discordances of fiction was often employed by the ancients; but we know nothing from any other quarter of a warlike Sardanapalus.

The other obscure kings who were enumerated by Ctesias, and whose names are probably preserved in the extant lists of Eusebius and Syncellus, resemble the obscure kings in the Egyptian lists of Herodotus and Diodorus. They performed nothing worthy of record, and therefore their reigns are a historical blank. Such is the apology made for the barrenness of the traditionary account. It is, however, an entire mistake to suppose that because the king of a great empire is weak and indolent, his reign will be devoid of events. His weakness and his indolence may prevent him from attacking his neighbours; but they may induce his neighbours to attack him, or may tempt his subjects to rebel against him. Experience has proved that the reigns of weak sovereigns are by no means uneventful. The fall of the Assyrian Empire itself is ascribed to the contempt of Sardanapalus inspired into Arbaces the satrap, by seeing him spinning among the women.(⁸⁹)

---

(88) Photius and Suidas in Σαρδανάπαλος. Compare Geier, Alex. Hist. Script. p. 244. The statement as to the existence of two Sardanapali is cited from the Persica of Hellanicus in Schol. Aristoph. Av. 1022, Fragm. Hist. Gr. vol. i. p. 67; but the Scholiast appears to have mistaken Hellanicus for Callisthenes; the story of the effeminate king Sardanapalus was probably unknown in Greece at the time of Hellanicus. Hesychius, in v., states that there were two Sardanapali. It is possible that the idea of a warlike Sardanapalus may have arisen from a confusion with Ninus. We know from the verses of Phœnix already cited, that Ninus and Sardanapalus were sometimes confounded. Hyginus, fab. 243, has the following passage: Semiramis in Babylonia, *equo amisso*, in pyram se conjecit. For *equo amisso*, we must read either *equo admisso* or *regno amisso*; but in either case we must suppose a confusion between Semiramis and Sardanapalus. A confusion of Nitocris and Semiramis has been pointed out above, p. 398, n. 5. The hypothesis of a double Sardanapalus has been adopted by many modern critics, in order to remove chronological difficulties, Winer, B. R. W. art. *Assyrien*, vol. i. p. 122. The hypothesis of a double Ninus will be mentioned below, p. 420. Larcher supposes that there were several queens named Semiramis, Herod. i. note 437. Mr. Rawlinson supposes two queens of that name, Herod. vol. i. p. 322.

(89) Aristot. Pol. v. 10; Diod. ii. 24.

The discordance between the accounts of the profane writers of antiquity, respecting the history of the Assyrian Empire, is so great, that it is scarcely possible to institute any comparison between them. They seem to relate to different countries; so rare are the points of agreement. They differ in the duration of the empire; the time and mode of its foundation; the time and mode of its overthrow; the names of the kings, their acts, and the duration of their reigns. In this state of confusion and conflict, modern chronologists, starting from the assumption that the traditionary accounts are substantially true, have acted upon the principle adopted by a court of justice in construing written instruments, that it will give effect to as much of the document as it can, 'ut res magis valeat quam pereat.' They have therefore attempted to harmonize these accounts as far as is practicable; and for this purpose they have resorted to two contrivances. They have, as in other cases of discordant testimony, multiplied the subject, and distributed the evidence. They have supposed a double Assyrian Empire in relation to time—an old and a new empire: they have likewise supposed a double Assyrian Empire in relation to space—one empire whose capital is Nineveh, another empire whose capital is Babylon. Grant these hypotheses, and four discordant schemes of Assyrian history are provided each with a separate compartment.

Des Vignoles, in his work on the Chronology of Sacred History, exhibits at length his arrangement of the Assyrian dynasties, from the foundation of the empire to its final extinction. He places first the old Assyrian Empire, of 1459 years, beginning with Belus and Ninus, and ending with Sardanapalus, whom he supposes to have reigned from 915 to 900 B.C.[90] He adopts the account of Ctesias and other Greek writers, that Arbaces with the assistance of the Medes dethroned Sardanapalus; but instead of assuming that Babylon and Ninus became subject towns of the Median capital

---

[90] Chronologie de l'Histoire Sainte, vol. ii. p. 7, 210, 284.

Ecbatana, he supposes Belesys, the confederate of Arbaces, to have been appointed Governor of Babylon, without tribute, and to have been succeeded by a line of virtually independent satraps, down to Nabonassar, who in 747 B.C. declared himself King of Babylon.([91]) He makes no attempt to supply the names of these semi-independent rulers. The kings between Nabonassar and Cyrus in the Astronomical Canon from from 747 B.C. downwards are considered by Des Vignoles as kings of Babylon exclusively.([92]) With regard to Nineveh, he supposes that after the death of Sardanapalus it became the capital of a second Assyrian kingdom founded under a second King Ninus. The name of Ninus, the founder of the second Assyrian Empire, is obtained from a fragment of Castor, the chronographer, who probably lived about the time of Cicero :([93]) but with this single exception no king of the second Assyrian Empire of Nineveh, subsequent to Sardanapalus, is mentioned in any profane writer. The list of Assyrian Ninevite kings in Des Vignoles, after Ninus the Second, first presents a chasm of more than a century, and is then filled up with the names of Assyrian kings, incidentally mentioned in the sacred history of the Jews, beginning with Phul or Pul, and ending with Esarhaddon. He supposes Esarhaddon to have been the last of the Ninevite kings of the second Assyrian Empire; to have transferred the seat of his empire to Babylon in 680 B.C., and to have consolidated the Ninevite and Babylonian kingdoms, which thus again became one Assyrian kingdom from 680 to 538 B.C.([94])

According to the scheme of Assyrian chronology adopted in the *Art de vérifier les Dates*, Eucchous, the successor of Nimrod, is king of Babylon. He and his successors are kings of Babylon from 2575 to 1993 B.C. In this year, Belus, who had

---

(91) Ib. p. 378.   (92) Ib. p. 11, 367.

(93) καταλήγομεν ἐπὶ Νίνον τὸν διαδεξάμενον τὴν βασιλείαν παρὰ Σαρδαναπάλλου, Castor, ap. Syncell. vol. i. p. 387, Bonn.

(94) Ib. p. 323, 389.

ruled over Assyria for 30 years, conquers Babylon, and annexes it to his kingdom. Belus is succeeded by Ninus and Semiramis, and the joint kingdom of Assyria and Babylon lasts from 1993 to 793 B.C., when Empacmes or Eupales, otherwise called Sardanapalus, is dethroned by Arbaces and Belesys; and the kingdom was again divided: the kingdom of Assyria being governed by Phul, the kingdom of Babylon by Belesys. The former kingdom was overthrown by the Medes, in the reign of Chinaladanus, also called Sarak and Sardanapalus, in 625 B.C.: the latter kingdom lasted until the reign of Nabodanius, when it was overthrown by Cyrus (538 B.C.)[95] According to this hypothesis, there is first a separate kingdom of Babylon; then there is a kingdom of Assyria including Babylon; afterwards there are separate kingdoms of Assyria and Babylon in parallel lines until 625 B.C., at which year the kingdom of Assyria becomes extinct, but the kingdom of Babylon lasts 87 years longer, until 538 B.C.

These conjectural plans of Assyrian history may serve to exemplify the manner in which the subject is treated by modern critics. Every successive critic exercises the same latitude of discretion in the construction of hypotheses, but each rejects the views of his predecessors, and propounds a new scheme of his own. Thus Larcher identifies Sardanapalus with Phul, and places him at 765 B.C.[96] He considers Nabonassar as the founder of the kingdom of Babylon in 747 B.C., and makes Semiramis his wife.[97] Volney adopts another system of Assyrian chronology;[98] Clinton a fourth,[99] and Mr. Rawlinson a fifth.[100] Mr. Clinton has quite a peculiar arrangement. He

---

[95] See Art de vérifier les Dates (Paris, 1819, 8vo), tom. ii. p. 338—364.

[96] Trad. d'Hérodote, tom. vii. p. 148, 595. Larcher holds that the era of Nabonassar marks the commencement of Babylonian independence, ib. p. 158, 167.

[97] Ib. p. 167, 171, 182.

[98] Œuvres, p. 409—514 (ed. 1837).

[99] Fast. Hell. vol. i. App. c. 4.

[100] Herodotus, vol. i. p. 432.

supposes a separate kingdom of Babylon, which both begins and ends before the Ninevite kingdom of Assyria. The chronology of this separate Babylonian kingdom is thus stated by him:—

	YEARS.	B.C.
Conquest of Babylon by the Medes:		
8 Median kings	224	2233
11 kings	69	2009
49 Chaldæans	458	1940
9 Arabians	245	1482
Ended		1237
Total duration	996	

The chronological scheme for the parallel Ninevite kingdom of Assyria is stated as follows:—By 'the Empire' is meant the union of Nineveh and Babylon under one king:—

	YEARS.	B.C.
Ninus—Assyrian monarchy		2182
Before the Empire	675	1912
During the Empire, 24 kings	526	1237
Sardanapalus		876
After the Empire, 6 kings	105	711
Capture of Nineveh		606
Total duration	1306	

The reader will observe that the kingdom of Babylon is made to end in the year at which the Assyrian 'Empire' is made to begin.[101] Nevertheless, Mr. Clinton states that Babylon had always kings of her own from the earliest times; and he arranges the kings of the astronomical canon from Nabonassar to Nebuchadnezzar, in a parallel column to the Assyrian kings from Pul, 769 B.C., to the capture of Nineveh in 606 B.C. He identifies Sardanapalus with the Assyrian (not the Babylonian) king Nabuchodonosor, and places him at 630 B.C.[102]

---

(101) See Clinton, ib. p. 282. (102) Ib. p. 278.

§ 8 It is possible that Nineveh and Babylon may at certain periods have been each the seat of an independent kingdom; but there is no evidence of such a fact. Herodotus considered Nineveh to have been the original capital of the Assyrian Empire, and after its downfall to have been succeeded by Babylon.([103]) He calls Sardanapalus king of Nineveh.([104]) He describes Phraortes, king of Media, as making an expedition against the Assyrians of Nineveh, who formerly were masters of all that country, but whose allies had then revolted from them; and as losing his life and army in the enterprise. It is further stated that Cyaxares, the son of Phraortes, made an expedition against Nineveh, in order to avenge his father's death; but that while he was besieging the city he was attacked by an invading army of Scythians; that a battle between the Medes and Scythians ensued, in which the Medes were defeated; and that they lost the Empire of Asia for twenty-eight years. Herodotus proceeds to relate that when the Medes had recovered their empire, and reduced the nations over which they had previously ruled, they took Nineveh, and subjugated the Assyrians with the exception of the province of Babylon.([105]) Herodotus considers queen Nitocris, with her husband Labynetus, and her son of the same name, who reigned over Babylon after the destruction of Nineveh by the Medes, to be the heads of the Assyrian Empire. He likewise appears to consider Semiramis, who was queen of Babylon five generations, or about 150 years, before Nitocris, and many kings of the same city, as Assyrian rulers.([106])

([103]) τῆς δε Ἀσσυρίης ἐστὶ τὰ μέν κου καὶ ἄλλα πολίσματα μεγάλα πολλά· τὸ δὲ ὀνομαστότατον καὶ ἰσχυρότατον, καὶ ἔνθα σφι, Νίνου ἀναστάτου γενομένης, τὰ βασιλήια κατεστήκεε, ἦν Βαβυλών, i. 178. The seat of the king's palace, in an Oriental country, is the seat of government.

([104]) ii. 150.   ([105]) i. 102—106.

([106]) i. 184, 187, 188. In c. 187, he describes Nitocris as making a tomb for herself, with an inscription beginning thus: τῶν τις ἐμεῦ ὕστερον γινομένων Βαβυλῶνος βασιλέων ἦν σπανίσῃ χρημάτων. In c. 188, he says that Cyrus made war on the son of this Nitocris, ἔχων τε τοῦ πατρὸς τοῦ ἑωυτοῦ τοὔνομα Λαβυνήτου καὶ τὴν Ἀσσυρίων ἀρχήν. Labynetus the Babylonian, mentioned in i. 74, appears to be the husband of Nitocris.

The account of Ctesias is, that Ninus was the founder of Nineveh, and his wife Semiramis of Babylon; but, like Herodotus, he supposes Nineveh to be the seat of Assyrian government, for the pyre of Sardanapalus, the last king, is at Nineveh, not at Babylon.[107] He conceives the Empire of Central Asia to have passed from the Assyrians to the Medes at the death of Sardanapalus; and Arbaces to have been the first king of the Median dynasty which governed Assyria as a subject province.[108] Eusebius and Syncellus suppose a single continuous line of Assyrian kings to reach from the earliest times to Sardanapalus: but they indicate no distinction between a Ninevite and a Babylonian kingdom.[109] The only trace of a distinction between the two is to be found in Herodotus, who describes Cyaxares as taking Nineveh, and reducing all the Assyrians, with the exception of Babylon and its district, in 606 B.C. He seems to have supposed that Babylon retained its independence, as head of a fragment of the Assyrian Empire, until 538 B.C., when it was taken by Cyrus. His narrative, however, excludes the idea that Nineveh and Babylon were ever at the same time the seats of independent kingdoms. Pliny declares that Babylon was long the capital of the Chaldæan nations.[110]

There is a statement in the apocryphal book of Tobit, which, assuming it to be historical, would prove that Babylon and Nineveh, if not heads of independent kingdoms, were at least rival cities, and sometimes made war upon each other. The statement is, that Nineveh was captured by Nebuchadnezzar and Ahasuerus, kings of Babylon, during the lifetime of Tobit.[111] This statement differs from that of Herodotus, re-

---

(107) Diod. ii. 26—28. Diodorus, indeed, is not clear; for he speaks of Ninus being upon the Euphrates. Herod. ii. 150, likewise calls Sardanapalus king of Nineveh.

(108) Diod. ii. 32.   (109) Clinton, F. H., vol. i. p. 267.

(110) Babylon Chaldaicarum gentium caput diu summam claritatem obtinuit in toto orbe, vi. 30.

(111) c. 15, ad fin. Ahasuerus in this passage is construed to mean Cyaxares. See Winer, art. *Ahasuerus*. But Cyaxares was not king of Babylon.

specting the capture of Nineveh by Cyaxares and the Medes. The ordinary receipt for reconciling these two statements would be to assume that they refer to two different sieges. Mr. Clinton, however, resorts to another expedient. He compounds them into one, and declares that Nineveh was taken by the united forces of the Medes and Babylonians.([112])

During the Jewish captivity, commencing from the reign of Nebuchadnezzar, in 604 B.C.,([113]) Babylon was the seat of the Assyrian government until it was taken by Cyrus, and incorporated with the Persian Empire. The captive Jews were transported to Babylon or its neighbourhood, and there established themselves; the colony remained in this district, until a few of the tribes availed themselves of the permission of Cyrus to return to Judæa.

Berosus, in the long extract respecting Nebuchadnezzar, cited by Josephus, treats him throughout as king of Babylon, and speaks of his embellishing the city.([114])

Syncellus says that, before the death of Sardanapalus, the Assyrians held both the Chaldæans and Medes in subjection; but that after this event the Assyrians were sometimes subject to the Medes, and sometimes to the Chaldæans.([115]) This statement, again, is inconsistent with the concurrent existence of two Assyrian kingdoms; one ruled from Nineveh, and the other from Babylon.

The account of Strabo is, that Nineveh ceased to exist immediately after the overthrow of the Assyrian Empire, and that it was a much larger city than Babylon.([116])

§ 9 Volney lays it down, that the duration of the Assyrian

([112]) Ib. p. 275. Mr. Grote thinks that Babylon was in some sort of dependence upon Nineveh, but was governed by kings or chiefs of its own, Hist. of Gr. vol. iii. p. 386.

([113]) See Dr. Smith's Dict. of the Bible, art. *Captivity*. Winer, art. *Exil*, places the beginning of the general captivity at 588 B.C.

([114]) Fragm. Hist. Gr. vol. ii. p. 506.

([115]) Vol. i. p. 387, Bonn. For τῶν δύο λοιπὸν, read τῶν δύο λοιπῶν.

([116]) xvi. 1, § 3.

Empire, and the dates of its commencement and termination, form the greatest difficulty of ancient history.(117) He proceeds to investigate this thorny subject, and to form one of the many hypothetical schemes of Assyrian chronology. Like other critics, he assumes that the truth lies concealed in some of the extant accounts, and that the difficulty consists in its discovery. But what if the problem is insoluble, because the truth has been altogether lost?

So far as notices of Assyrian history occur incidentally in the contemporary Jewish chroniclers, we have a firm footing of evidence to rest upon. But the earliest king of Assyria recorded in the Biblical history, is Phul, who lived about 772 B.C. Before this date, the Assyrian chronology rests exclusively on the testimony of the classical historians, and of the later chronographers who chiefly relied upon their authority. These are the schemes of Herodotus, Ctesias, and Berosus—inconsistent with one another, and equally destitute of credible attestation. There is no valid reason for supposing that Herodotus, about 450 B.C., and Ctesias, about 420 B.C., were able to obtain authentic accounts of the ancient Assyrian Empire, which Herodotus supposed to have lasted 520,(118) and Ctesias to have lasted 1300 years.

With respect to the 122 kings in Berosus, immediately preceding Phul, to whom average reigns of a moderate length are assigned, and above all, to the 86 kings, beginning with Euechius, who are stated to have reigned altogether 34,080 years; that is, on an average, 396 years each; it is impossible to discover any legitimate ground for considering them historical. It should likewise be borne in mind, as illustrative of the method of

---

(117) Œuvres, p. 409.

(118) Appian, Hist. Rom. Præf. 9, states that the duration of the Assyrian, Median, and Persian empires put together falls short of 900 years. 900 years, counted back from 331 B.C., the date of the battle of Arbela, would give the year 1231 B.C. for the commencement of the Assyrian Empire. Macrobius, in Somn. Scip. ii. 10, § 11, places Ninus about 2000 years before his own time, which was the 5th century after Christ.

Berosus, that he assigned 432,000 years to ten antediluvian kings.

The view of Berosus expressed by Syncellus is that he was an impostor, who sought to glorify his own country by attributing to it an antiquity greater than that claimed by any other nation, and that his Assyrian chronology was a figment of his own invention. He attributed a similar character to Manetho, whom he considered as the imitator of Berosus.[119] These opinions may be ascribed to the wish of Syncellus to discredit all schemes of chronology which could not be reconciled with the Biblical chronology: they appear nevertheless to rest on solid critical grounds. Berosus seems to have published his astronomical writings in the form of a translation from a work of the primitive king Belus;[120] which fact proves that he founded even his scientific doctrines upon a fabulous basis.

From about the era of the Olympiads we have the incidental notices of Assyrian kings in the historical and prophetic books of the Old Testament, whose dates are determined by the synchronism of the Jewish kings. We have likewise the list of the Chaldæan kings from Nabonassar, in 747 B.C., to Cyrus, in the astronomical canon.

The following is a table of the Assyrian kings mentioned in the Old Testament, together with their respective dates. As it is compiled from incidental notices of these rulers, it cannot be regarded as a continuous series:—

	B.C.
Phul	772
Tiglath Pileser	741
Shalmaneser	722

---

[119] See vol. i. p. 29, 67, 71, Bonn. In p. 56 he designates Berosus as ὁ τὰ Χαλδαϊκὰ ψευδηγορῶν, and speaks of his Χαλδαϊκὴ τερατολογία, ib. He says, p. 68, that the writers on Chaldæan antiquity consider the accounts of Egyptian antiquity to be fabulous, and that the writers on Egyptian antiquity hold a similar opinion with respect to the accounts of Chaldæan antiquity. Neither regards the other, ἀλλ' ἕκαστος τὸ ἴδιον ἔθνος καὶ τὴν πατρίδα δοξάζων ἀράχνας ὑφαίνει.

[120] See above, p. 297, n. 218.

	B.C.
Sargon	
Sennacherib	714
Esarhaddon, son of Sennacherib	
Baladan, king of Babylon, contemporary with Hezekiah, 725—696 B.C.	700
Nebuchadnezzar	605—561
Evil Merodach, son of Nebuchadnezzar	561
Belshazzar, last king of Babylon, deposed in	538

§ 10 The Astronomical Canon, for the period before Cyrus, probably contains authentic materials; but it is a complete historical puzzle. The period of Nabonassar was certainly known to Berosus; for the latter stated that Nabonassar destroyed the records of his predecessors, in order that the reigns of the Chaldæan kings might be reckoned from his time.([121]) This explanation has been deservedly rejected by modern critics as fabulous;([122]) but it proves that the era of Nabonassar was recognised in the time of Berosus.

Ptolemy informs us that the four earliest eclipses observed at Babylon were recorded to have occurred in the first and second years of Mardocempadus, and in the fifth year of Nabopolassar.([123]) This statement implies contemporary registration; and if the date was a part of the record, the reigns and names of Mardocempadus and Nabopolassar must be considered as resting on certain attestation.

The name of Nabonassar, from whom the era is denominated, is unknown to us from any other source. It is not mentioned elsewhere that any Nabonassar was king either of Assyria or of Baby-

---

([121]) ἀπὸ δὲ Ναβονασάρου τοὺς χρόνους τῆς τῶν ἀστέρων κινήσεως Χαλδαῖοι ἠκρίβωσαν, καὶ ἀπὸ Χαλδαίων οἱ παρ' Ἕλλησι μαθηματικοὶ λαβόντες, ἐπειδὴ, ὡς ὁ Ἀλέξανδρος καὶ Βήρωσσός φασιν οἱ τὰς Χαλδαϊκὰς ἀρχαιολογίας περιειληφότες, Ναβονάσαρος συναγαγὼν τὰς πράξεις τῶν πρὸ αὐτοῦ βασιλέων ἠφάνισεν, ὅπως ἀπ' αὐτοῦ ἡ καταρίθμησις γίνεται [leg. γίνηται] τῶν Χαλδαίων βασιλέων, Syncell. vol. i. p. 390.

([122]) Des Vignoles, vol. ii. p. 368.

([123]) Synt. iv. 5, v. 14 (vol. i. p. 243—5, 340, Halma). The word used with reference to the first three eclipses is ἀναγέγραπται.

lon. Of the other eighteen kings between Nabonassar and Cyrus, not more than five or six can be identified with any known name, either in sacred or profane writers. In this state of things, modern critics have been driven to the expedient of assuming that the kings of the Canon were kings merely of Babylon, and of putting them in a parallel column to the kings of Assyria. This expedient is in the highest degree arbitrary; for when a continuous line of kings terminating in Cyrus, and the other kings of the Persian Empire is presented to us, the natural supposition is, that the predecessors of Cyrus were, like him, the sovereigns of a great empire, and not the chieftains of a single city.

It is, moreover, important to compare the last eight names before Cyrus in the list of Berosus and in the astronomical canon. The names of Asordanus and Asaradinus nearly agree in sound, but the lengths of their reigns differ. In the seven next, the agreement of the years of the reigns is so close, that it cannot be fortuitous. The names of the kings likewise correspond closely in Nos. 4, 5, 7, and 8.

Berosus.	Years.	Astr. Canon.	Years.
1. Asordanus	8	1. Asaridinus	13
2. Sammughes	21	2. Saosduchinus	20
3. Sardanapalus	21	3. Ciniladanus	22
4. Nabopolassar	21	4. Nabopollassar	21
5. Nabucodrossor	43	5. Nabocolassar	43
6. Amilmarudoch	2	6. Illoarudamus	2
7. Neriglissar	4	7. Nericasolassar	4
[Laborosoarchod, 9 months]			
8. Nabodenus	17	8. Nabonadius	17
9. Cyrus	9	9. Cyrus	9

Now, Berosus does not profess to give a merely Babylonian dynasty. The kings in his series belong to the Assyrian Empire. It is clear, however, that the seven kings preceding Cyrus in Berosus coincide chronologically with the corresponding kings of the Canon. It follows that the kings of the Canon cannot be considered as exclusively Babylonian kings.([124])

Assuming the capture of Babylon by Cyrus to be fixed at 538 B.C., the chronology of Berosus would place the accession of Sennacherib at 693 B.C., and the reign of Phul immediately before that of Sennacherib. This agrees tolerably well with the Biblical chronology, with respect to Sennacherib, who is described as invading Palestine in 714 B.C. With respect to Phul, the Biblical accounts place him at 770 B.C., and separate him from Sennacherib by Tiglath Pileser and Shalmaneser.([125])

The result of this investigation is, that the whole of the accounts of Assyrian history and chronology, handed down to us by the classical writers for the periods anterior to the capture of Nineveh by Cyaxares, and of Babylon by Cyrus, are destitute of authentic support, and unworthy of credit. With regard to the period subsequent to 772 B.C., we have some authentic notices in the historical books of the Old Testament; but the Chaldæan kings of the Astronomical Canon, from Nabonassar, in 747 B.C., to Cyrus, in 538 B.C., and the Assyrian kings of Berosus for the same period, are of an uncertain historical character.([126])

Like the Egyptian chronology, the Assyrian chronology comes down to us in the shape of divergent lists of kings, dissociated from history, and these are subjected by modern critics to

---

(124) Dodwell and Des Vignoles have conjectured that Berosus was the author of the Astronomical Canon down to Alexander. See Ideler, Chron. vol. i. p. 222. This supposition is adopted by Volney, Œuvres, p. 498. But the close agreement between the numbers of Berosus and those of the canon for the last kings before Cyrus, and the wide divergence in the names of the kings, show that the list of Berosus and the canon could not have been framed by the same person. Müller, Rhein. Mus., ib. p. 293, thinks that Berosus founded all his dates upon the era of Nabonassar.

(125) Niebuhr exaggerates when he speaks of 'the strikingly exact agreement of the statements respecting the later Assyrian Empire, which are derived from the work of Berosus, and the historical books of the Old Testament,' Lect. on Anc. Hist. vol. i. p. 12, ed. Schmitz.

(126) On the chronological difficulties in the arrangement of the later Assyrian kings, see Clinton, vol. ii. p. 301 sq.

a free and discretionary treatment, in which names and numbers go for little. It is admitted that the accounts cannot be received in the form in which they have reached us; and therefore they must undergo transmutation before they can pass for historical. 'In tracing the identity of Eastern kings (says Mr. Clinton), the times and the transactions are better guides than the names; for these, from many well-known causes (as the changes which they undergo in passing through the Greek language, and the substitution of a title or an epithet for the name), are variously reported; so that the same king frequently appears under many different appellations.'[127]

In general, all that is recorded of an Assyrian king is his name and the length of his reign. It is easy to identify him with another king, either better known or equally unknown; and if the reigns differ in length, to alter the text, or to suppose that the father admitted the son to a share of his power during the latter part of his life, and that this concurrent period is omitted in one statement, and included in the other.

Voltaire said that etymology is a science in which consonants go for little, and vowels for nothing. It may in like manner be said that chronology, as treated by the restorers of Assyrian and Egyptian antiquity, is a science in which numbers go for little, and names for nothing.

It would be a vain task to follow all the changes in the phantasmagoria of Assyrian antiquity; but a few examples of the diversity of hypotheses, all equally unsupported by positive testimony, may be given.

Larcher thinks that Nabonassar, whose reign began in 747 B.C., founded the kingdom of Babylon, and that Semiramis was his wife.[128] Volney, on the other hand, thinks that Semiramis was born in 1241 B.C., and entered the harem of Ninus in 1221 B.C.[129] Larcher identifies Sardanapalus with Phul, and

---

[127] F. H., vol. i. p. 277.
[128] Hérodote, tom. vii. p. 167, 171, 182. Larcher, however, supposes that there was more than one Semiramis.
[129] Œuvres, p. 489.

places him at 747 B.C.(130) Numerous critics identify Sardanapalus with Esarhaddon.(131) Clinton appears to identify Sardanapalus with Esarhaddon, with Nabuchodonosor of Assyria, and with Saracus, at the dates of 711, 650, and 630 B.C.(132) Des Vignoles places Sardanapalus at 915 B.C., and identifies him with no other king. Syncellus says that Thonos Concoleros was called Sardanapalus; Esarhaddon is identified with Asaridinus of the Canon by Des Vignoles;(133) but this identification is altogether rejected by Larcher.(134) Nabopolassar is identified with the Labynetus I. of Herodotus by Larcher;(135) but Mr. Clinton identifies Labynetus I. with Nebuchadnezzar, and Labynetus II. with the Nabonadius of Berosus and the Canon, and the Belshazzar of Daniel.(136)

Some kings of Assyria are incidentally mentioned, who find no place even in the copious and discordant lists of names which have been handed down to us. Thus we are informed by Macrobius that the statue of Helius was translated to Assyria from Heliopolis, when Senemures or Senepos was king of Egypt; and that the statue was removed by Opias, the envoy of Deleboris, king of the Assyrians, with a party of Egyptian priests, the chief of whom was named Partemetis.(137) Alexander Polyhistor and Bion related that the dynasty of Ninus and Semiramis became extinct with Beleus the son of Dercetades; that the sceptre was then seized by a certain Beletaras, who had been superintendent of the royal gardens, and that it remained with his descendants down to Sardanapalus.(138) Deleboris, Dercetades, Beleus, and Beletaras are names unrecognised in Assyrian chronology.

It must not be assumed that any authentic memorials of the early Assyrian history were in existence when Herodotus and

---

(130) Ib. p. 148, 595.
(131) Des Vign. vol. ii. p. 322.
(132) Vol. i. p. 278.
(133) Vol. ii. p. 322, 388.
(134) Vol. vii. p. 183.
(135) Ib.
(136) Ib. p. 278.
(137) Macrob. Sat. i. 23, § 10.
(138) Fragm. Hist. Gr. vol. iii. p. 210, iv. p. 351. A river in Syria was named Beleus, above, p. 416, n. 83.

Ctesias collected their information. Oral tradition would not have carried them back with safety for much more than a century;[139] and we have no reason to suppose that any contemporary chronicles or registers, of a historical nature, had been composed and preserved. The imputation of ignorance with respect to early Assyrian history, which Des Vignoles makes on Herodotus,[140] is doubtless well founded; but it is like the similar imputation, with respect to primitive Italian ethnology, which Niebuhr makes upon Polybius. The materials of knowledge did not exist, and all attempts to ascertain the truth would have been fruitless.[141]

In history, as in philosophy, it is important to fix the boundaries within which knowledge can be attained, and not to waste the time of writers and readers in vain endeavours to determine facts, of which no credible testimony exists, and of which the memory has perished. Researches into ancient history, which lead to merely negative results, are important and useful, as well as similar researches which lead to positive results. They distinguish between fiction—which, however diverting, instructive, or elevating, can never be historical—and reality, which is a necessary attribute of a historical narrative.

§ 11 If we examine the records of Greek and Roman history, and the accounts of the history of other nations which have been preserved and handed down to us by the classical writers, we are unable to find any authentic evidence of events ascending higher than about the era of the Olympiads, 772 B.C. It is even difficult to fix any event resting on a certain tradition which can be carried up to so high a date. The poems of Homer are probably anterior to this era. They describe a state

---

(139) Compare the remark of Mr. Rawlinson, Herod. vol. i. p. 68.

(140) Vol. ii. p. 175, 182.

(141) Larcher, ib. vol. vii. p. 145, thinks that the uncertainties of Assyrian chronology will not be removed, until some 'ouvrage précieux' is discovered. There is, however, no reason for supposing that the writers of the later periods of antiquity were in possession of any information respecting Assyrian chronology which has not descended to our time.

of society which has already reached a considerable degree of intelligence and civilization; but they fix no trustworthy dates as to previous time.

The only other evidence as to the duration of mankind which is furnished to us by profane writers, is the existence of great works in Babylon and Egypt, which had been executed before the time of Herodotus, and which may be considered as implying necessarily a long period for their construction.

If we suppose a state of society in which there is a strong government, and in which the arts necessary for the support of a large population—the production of food, the manufacture of clothing, the construction of habitations—are regularly carried on; if we suppose, further, that the people have acquired habits of absolute submission to the despotic ruler, and that there is no class of freemen, but that the entire community are his slaves; we have all the conditions requisite for the execution of great works, provided that the government possesses sufficient capacity and skill for the organization of labour on a large scale, and that it does not employ its command over its subjects for warlike purposes. Now the empire which was established on the banks of the Euphrates and the Tigris, and the kingdom of Egypt, appear, at an early period, to have reached a state which coincides with this description. The Oriental form of rule, from the earliest times of which we have any account, has always been purely despotic: the people have been the absolute slaves of the kings. Although the Oriental civilization has never succeeded in passing certain narrow limits, in respect of government, law, literature, and science, yet in respect of the useful arts it has made considerable progress. Among these arts, there is none in which it has so much excelled as the art of building. Many of the Oriental cities not only evince considerable proficiency in the constructive art, but even contain edifices remarkable for architectural beauty in original and indigenous styles.

The ancient writers have left us no account of Nineveh; but the walls of Babylon, and its great palaces and temples were still standing when the city was visited by Herodotus; and his description is that of an eye-witness,(142) though his numerical statements of magnitudes must be considered as exaggerated. The walls of Babylon were unquestionably of great height and extent;(143) but they must have been mere aggregations of earth or of unbaked bricks, for all trace of them has disappeared. The pyramids and other great buildings of Egypt, being made of stone, have remained to our day in a state not very different from that in which they were seen by Herodotus.

The Egyptians do not seem to have been a warlike nation, the great conquests of Sesostris and Rhamses are (as we have already observed) purely fabulous. Their country was fertilized by the natural irrigation of their beneficent river;(144) and they were thus relieved from the necessity of that system of artificial irrigation which the more energetic and intelligent inhabitants of Mesopotamia were enabled to employ.(145) Hence, after the agricultural work of the year was performed, a large amount of surplus labour remained at the disposal of the government.

Ancient Egypt may be considered as a great *latifundium*, or plantation, cultivated by the entire population as the king's slaves. Whatever part of his slaves could be withdrawn, either

---

(142) He speaks of the temple of Jupiter Belus, as ἐς ἐμὲ ἔτι ἐόν, i. 181, and he describes himself as having conversed with the Chaldæan priests of this temple, c. 183.

(143) Aristotle, Pol. iii. 3, considers Babylon to have been so large as to exceed the dimensions of a city, and rather to have resembled a nation. He mentions a story that a portion of its inhabitants were not aware of its being taken till the third day after the event. The story which he cites is told by Herod. i. 191, and is referred to the capture by Cyrus; but the third day is not specified.

The prophet Jeremiah says, with reference to the capture of Babylon: 'One post shall run to meet another, and one messenger to meet another, to show the King of Babylon that his city is taken at one end,' li. 31. The distance from the city wall to the king's palace is described as so great, that a single foot messenger is not sufficient to carry the news from one point to the other.

(144) Euripides, at the beginning of the Helena, describes the Nile as fulfilling the functions of rain in Egypt.

(145) See Grote, vol. iii. p. 392, 401.

permanently or temporarily, from agriculture and other necessary labours, could be devoted to works of construction. To these they would be assigned, and at these they would work under the lash of the royal drivers.(146)

We may conceive what was the command of labour enjoyed by an Oriental king, in one of the monarchies which we are considering, from the accounts of the western expeditions of Darius and Xerxes, both of which are historical, though our knowledge of the former is comparatively imperfect.

The expedition of Darius against the Scythians took place about 515 B.C.: his army was reckoned at 700,000 men, and his fleet at 600 ships. Notwithstanding his command of this large fleet, he caused a bridge of ships to be made across the Thracian Bosporus, for the passage of his army; and a similar bridge to be made across the Danube not far from its mouth.

The great levy of nations organized by Xerxes for his expedition against Greece, in order to avenge the affront of Marathon, is amply described by Herodotus, who estimates his army at 2,317,000 men, and his fleet at 4207 ships, when he crossed the Hellespont. With its subsequent accessions, and with the attendants and camp-followers, he supposes the entire numbers of the host of Xerxes, by land, to have amounted to 5,283,222 men.(147)

---

(146) Mr. Grote remarks, Hist. of Gr. vol. v. p. 30, that 'the men who excavated the canal at Mount Athos worked under the lash; and these, be it borne in mind, were not bought slaves, but freemen, except in so far as they were tributaries of the Persian monarch.' 'We shall find (he adds) other examples as we proceed of this indiscriminate use of the whip, and full conviction of its indispensable necessity, on the part of the Persians, even to drive the troops of their subject-contingents on to the charge in battle.' No Persian subject was properly a freeman; they were all slaves of the great king.

In the representation of building work in progress, from a temple at Thebes, among the drawings of the Prussian expedition, in Brugsch, Hist. d'Egypte, vol. i. p. 106, an overseer with a stick is represented as sitting down while the men are at work.

(147) See Mr. Grote's comments upon these numbers, Hist. of Gr. vol. v. p. 44—51. He arrives at the conclusion that 'the numbers of Xerxes were greater than were ever assembled in ancient times, or perhaps at any known epoch of history,' p. 49. The army of Xerxes consumed 7 entire days and nights in marching across the Hellespont, Herod. vii. 56. Æschylus,

Xerxes, in imitation of Darius, caused a bridge of ships to be constructed across the Hellespont, and to be renewed after it had been once broken by the wind, merely to save the transport of his army in vessels impelled by sails or oars: he likewise caused a ship-canal to be dug across the neck of the promontory of Athos, merely in order to save his fleet from the navigation round the cape:([148]) the traces of this canal are still visible.

These examples show that the construction of great works, even for a temporary purpose, was familiar to the mind of an Oriental prince. The consciousness of his power, combined with unbounded pride and entire irresponsibility, likewise begot a wanton use of it, and induced the monarch to bestow a vast expenditure of labour upon inadequate objects. We may easily conceive that the same power which enabled Darius and Xerxes to collect their vast hosts on the Bosporus, would have enabled them to congregate an enormous body of labourers upon any spot, and to supply them with food and the materials of building. We may likewise easily conceive that the same state of mind which led to the fabrication of the bridges over the Bosporus and Hellespont, and to the excavation of the canal of Athos, might have led to the construction of the walls of Babylon for the defence of the city, and of the wall of Media for the defence of the province;([149]) to the erection of temples for the worship and honour of the gods, or to the formation of pyramids, labyrinths, and palaces, for the enjoyment and glorification of the kings themselves.

---

who was a grown-up man at the time of Thermopylæ and Platæa, describes the Persian army, on its return, as principally destroyed by hunger, on account of its excessive numbers, Pers. 482—91, 794.

([148]) Herodotus attributes the excavation of the canal of Athos by Xerxes to a desire of exhibiting his power, and of leaving a memorial of himself; ἐθέλων τε δύναμιν ἀποδείκνυσθαι καὶ μνημόσυνα λιπέσθαι, vii. 24.

([149]) Concerning the wall of Media, see Grote, Hist. of Gr. vol. ix. p. 85—90. The wall of Media is described by Xenophon, who saw it, as built of baked bricks cemented with asphalt, and as 20 feet in thickness, 100 feet in height, and 20 parasangs (=75 miles) in length, Anab. ii. 4, 12. It cut off the country between the Tigris and the Euphrates.

We can trace the same system of great constructive works in operation in other Oriental countries, according to the measure of their means. Herodotus says that the tomb of Alyattes, king of Lydia, a mound of earth, resting on a basement of large stones above three-quarters of a mile in circumference, was the greatest work after those of Egypt and Babylon.[150] The Temple of Jerusalem was likewise a great enterprize for the comparatively limited kingdom of Solomon.[151] This temple, after passing through various casualties, had grown to an enormous size and strength, when the city was besieged by Titus.[152]

The great wall of China, which is twenty feet in height, and twenty-five feet in thickness at the base, and which extends for 1400 miles, was constructed about two hundred years before the Christian era. Its utility in defence is unimportant, and it seems to have been dictated by the caprice of a powerful despot. Two hundred thousand men are said to have perished in the work.[153]

The Taj Mahul near Agra, in Northern India, erected by Shah Jehan, as a mausoleum for himself and his queen, in the seventeenth century, is an immense and splendid edifice. Its cost is reported to have exceeded three millions sterling; and the work to have occupied twenty thousand men for twenty-two years.[154] The Kuth Minar, the highest column in the world,

---

(150) i. 93. The remains of this barrow are still extant; see Rawlinson's Herodotus, vol. i. p. 232. Mr. Hamilton estimates the circumference at nearly half a mile. The story of the diversion of the Halys appears to be fabulous, Herod. i. 75.

(151) The account given in 1 Kings v. 13—16, and 2 Chron. ii. 2, is that Solomon assigned the cutting of the timber to 30,000 men, and that they were divided into bodies of 10,000 men, each of which worked for one month out of 3. He further employed 70,000 men as carriers, and 80,000 men as hewers of stone in the quarries. The overseers were 3300 in number.

(152) See Joseph. Bell. Jud. v. 5.

(153) Concerning the great wall of China, see Anderson's Narrative of the British Embassy to China in 1772, 3, and 4, ed. 2, 8vo, p. 196.

(154) See Sleeman's Recollections of an Indian Official, vol. ii. p. 27—37; Tavernier, Voyages des Indes, liv. i. c. 7.

being 242¼ feet high, and only forty-eight feet two inches in diameter at the base, stands near Delhi. It was built about the beginning of the thirteenth century.(155) There are also colossal tombs near Golconda, whose antiquity does not exceed 300 years.(156)

We are not therefore driven to the necessity of supposing the lapse of a long period of time to account for the great constructive works of Assyria and Egypt. The architectural legends of Herodotus and Diodorus do not, indeed, deserve much attention; but we may observe, that these writers do not assign a remote antiquity to the pyramids and other great buildings of Egypt. According to the Egyptian chronology of Herodotus, so far as it can be determined from his account, the three pyramid kings — Cheops, Chephren, and Mycerinus — reigned from about 913 to 813 B.C.(157) He fixes the construction of the labyrinth, which he considers a greater work than even the pyramids, at the period of the Dodecarchy, 680—670 B.C.(158) The pyramid kings are likewise placed by Diodorus near the end of his series, though their exact chronological place in his system cannot be assigned. Those who attributed the construction of one of the pyramids to Rhodopis, a contemporary of Sappho, supposed a still later date.(159) On the other hand, there were other stories which gave them an earlier origin.(160) But the Egyptian priests, from whom Herodotus derived his information, evidently did not seek to impress him with the remote antiquity of either pyramids or labyrinth, though they counted their divine and semi-divine dynasties by thousands of years, in order to magnify the age of their nation. The canal from the Nile to the Red Sea, which was commenced and left incomplete by Neco, and on which 120,000 men are said to have perished,(161) is likewise of no great antiquity, as Neco

---

(155) See Thornton's East India Gazetteer, art. *Delhi*.
(156) Thornton, art. *Golconda*.   (157) Above, p. 321.
(158) Above, p. 325.   (159) Above, p. 355.
(160) Above, p. 354.   (161) Above, p. 317.

reigned from 616 to 600 B.C., and was contemporary with Alcæus and Sappho.

Egyptian Thebes was known to Homer, as distinguished for its wealth, and for its hundred gates, through each of which two hundred charioteers went forth to battle.(162) The size which the poet intended to ascribe to Thebes in Egypt may be conjectured from the circumstance that Thebes, in Bœotia, was supposed to have only seven gates.(163) According to Herodotus, Babylon had a hundred gates, all of brass.(164) Egypt is alluded to several times in the Odyssey; Menelaus describes at length his visit to its shores.

Homer is acquainted with the Nile, which he calls the divine river Ægyptus;(165) but, with the exception of the allusion to the hundred-gated Thebes, there is nothing to indicate that he had heard of any large constructive works in the country. He speaks of the voyage from Greece to Egypt as long and difficult.(166)

Taking into consideration all the evidence respecting the buildings and great works of Egypt extant in the time of Herodotus, we may come to the conclusion that there is no sufficient ground for placing any of them at a date anterior to the building of the Temple of Solomon, 1012 B.C.(167) A similar conclusion applies to the walls and great buildings of Ba-

---

(162) Iliad ix. 381—4; Od. iv. 127. Aristotle, in his Meteorologics, i. 14, refers to the mention of Egyptian Thebes by Homer, and speaks of him as quite recent in comparison with the physical changes of Egypt produced by the Nile. δηλοῖ δὲ καὶ Ὅμηρος οὕτω πρόσφατος ὢν ὡς εἰπεῖν πρὸς τὰς τοιαύτας μεταβολάς. He comments on the silence of Homer respecting Memphis, and remarks that, being lower down the Nile than Thebes, it was probably of later origin.

(163) Iliad iv. 406; Od. xi. 263.   (164) i. 179.

(165) Od. iv. 477, 481, xiv. 258, xvii. 427.

(166) Od. iv. 483, xvii. 427. In Od. xvii. 448, Egypt is called πικρή, in reference to the previous story told by Ulysses: 'the country in which you narrowly escaped the lot of a slave.'

(167) The building of Solomon's temple occupied 7 years, 1 Kings vi. 38. It was burnt by Nebuchadnezzar in 588 B.C., after having stood about 418 years. The second temple was completed in 516 B.C. This temple, with great enlargements by Herod, was in existence at the siege of Jerusalem by Titus. See the description of Josephus, Bell. Jud. v. 5.

bylon. With regard to the great temple of Belus in the centre of Babylon, we may remark that Herodotus, who visited Babylon after the reign of Xerxes, speaks of it as still extant;[168] whereas Arrian states that it was demolished by Xerxes after his return from Greece, and that Alexander conceived the intention of rebuilding it on its old foundations.[169]

The polity of the Greeks, even in the Homeric times, was not consistent with forced labour by the command of the government on a large scale; but it is creditable to their taste and good sense that they did not attempt, with the means at their command, any works on the Oriental scale. The only Greek works which Aristotle classes with the pyramids were executed under a despotic regimen, and these (such as the statues dedicated by the Cypselidæ) were on a different scale from the colossal structures of Egypt. The Cretan labyrinth is a fabulous building, invented by mythologists, which never had any real existence.[170]

The great tomb of Porsenna at Clusium, of which an exaggerated description has been preserved by Pliny,[171] seems to furnish a confirmation of the Asiatic origin assigned by Herodotus to the Etruscans. It may be added, that their language, devoid of all affinity with any Italian or Greek dialect, their

---

[168] i. 181.

[169] Anab. vii. 17. In the book of Genesis the foundation of Babel is attributed to Nimrod, and the foundation of Nineveh to Asshur, in the second generation after the Noachian deluge, x. 10, 11. The building of the lofty tower, and the confusion of tongues, are the subject of another narrative, and are not connected with Nimrod, xi. 1—10. Dante follows the common view in connecting Nimrod with the Tower of Babel:—

> Vedea Nembrotto appiè del gran lavoro
> Quasi smarrito, e riguardar le genti,
> Che 'n Sennaar con lui insieme foro.
>   Purgatorio, xii. 34.

[170] See Höck's Kreta, vol. i. p. 56—68; Pashley's Travels in Crete, vol. i. p. 208—9.

[171] Plin. xxxvi. 19. See Müller, Etrusker, vol. ii. p. 224; Dennis, Cities and Cemeteries of Etruria, vol. ii. p. 385—91. Fergusson, Handbook of Architecture, p. 283, remarks that the Greeks were not tomb-builders.

intellectual torpor, their addiction to a ritual religion, and their form of government, accord with the supposition of an Oriental extraction.

There is a constant disposition to attribute a high antiquity to buildings when their true origin has been forgotten. Ruins of mediæval castles, both in England and on the Continent, have been called by the name of Cæsar. The Round Towers of Ireland, which have now been proved to be Christian edifices, not earlier than the fifth century, have been ascribed, by Irish antiquaries, to the Persians, and have been supposed to be emblems of their fire worship, or they have been traced to a Phœnician origin.([172])

The ancient walls of rude colossal masonry, called Cyclopean, of which many remains are extant in Asia Minor, Greece, and Italy, have been referred to the same primitive age, and have been attributed to the semi-fabulous Pelasgian race, which is supposed to have once inhabited all this region. Nevertheless, it is certain that the Cyclopean([173]) remains are of different

---

([172]) See Petrie's Ecclesiastical Architecture of Ireland, vol. i. p. 12. The conclusions as to the recent date of the Round Towers established by Mr. Petrie are stated in pp. 353—4.

([173]) The Cyclopes were conceived by the Greeks under a triple aspect. 1. As a tribe of one-eyed savage giants, living separately in caves, according to the description in the Odyssey, repeated in the Cyclops of Euripides, in Theocritus, Id. xi. 31, and in Virg. Æn. iii. 616—654. The Cyclopes in this form were regarded by the philosophers as the type of primitive savage life. 2. As workers in iron, who forged the thunderbolts of Jupiter. Hesiod describes them under this type, and makes them only 3 in number, Theog. 139—147. Similarly Apollon. Rhod. i. 730. Afterwards they were associated with Vulcan, and were supposed to work in his smithy under Ætna or Lipara, Callim. Dian. 46—86; Virg. Geor. iv. 170; Æn. viii. 416—453. 3. Lastly they were conceived as constructors of gigantic works of masonry. See Eurip. El. 1167; Herc. Fur. 948, and other passages in Müller, Arch. der Kunst, § 45; Dennis, Cities and Cemeteries of Etruria, vol. ii. p. 280. Aristotle ap. Plin. viii. 56, attributes to them the invention of towers. They are supposed to have built the walls of Tiryns, Apollod. ii. 2, 1, alluded to by Statius, Theb. iv. 150.

Rarus vacuis habitator in arvis
Monstrat Cyclopum ductas sudoribus arces.

They were likewise said to have built the walls and gate of Mycenæ, Paus. ii. 16, 4. Strabo states that the building Cyclopes were seven in number, and came from Lycia: he adds that they were called γαστερό-χειρες, because they lived by their craft, viii. 6, § 11.

Besides the single eye (to which the name was held to allude), the only

ages; and that some of them belong to comparatively recent periods, and to nations of a purely historical character. It has been proved that some of the Cyclopean walls of Italy are of Roman origin;(174) and some of the Cyclopean walls of Asia Minor are subsequent to the Peloponnesian War;(175) while those of Tiryns are anterior to Homer.(176) The occurrence of the arch in some of the Cyclopean remains of Asia Minor and Greece is also a conclusive proof of a comparatively late date.(177) The date of the Nuraghi of Sardinia is in like manner quite indeterminate.(178)

As an additional proof that the mere style of masonry affords no decisive indication of the date of construction, we may mention the remarkable fact that many of the remains of ecclesiastical edifices in Ireland, subsequent to the fifth century, closely resemble the Cyclopean remains of Greece and Italy.(179) The

---

attribute common to all three ideas is that of gigantic size and strength. The attribute of skill in workmanship is common to the last two: compare the verse of Hesiod, ἰσχύς τ' ἠδὲ βίη καὶ μηχαναὶ ἦσαν ἐπ' ἔργοις, Theog. 146.

(174) See the excellent paper of Mr. Bunbury, in the Classical Museum, vol. ii. p. 147. The Cyclopean remains of Greece and Italy are faithfully portrayed in the splendid posthumous work of Mr. Dodwell.

(175) Mr. Charles Newton, in a communication which he has had the kindness to make to me on this subject, remarks: 'The walls of Halicarnassus can hardly be earlier than the time of Mausolus, for he greatly enlarged the city. These walls present polygonal masonry in the parts built of limestone, and isodomous masonry in the parts built of freestone. The adoption of polygonal masonry in cases where limestone is the building material, is obviously caused by the convenience of shaping the stone by the law of cleavage. At Cnidus the city walls are also partly of polygonal and partly of isodomous masonry. Now Cnidus, according to Thuc. viii. 35, was as yet unwalled towards the close of the Peloponnesian War. The polygonal masonry in these walls must therefore be of a later period. To the east of Cnidus is a necropolis full of tombs chiefly built of polygonal blocks. There can hardly be a doubt that all these tombs are of the Roman period.' Concerning the walls of Cnidus, see Dr. Smith's Dict. of Anc. Geogr. in v.

(176) Iliad ii. 557.

(177) See Dennis, Cities and Cemeteries of Etruria, vol. ii. p. 275.

(178) See Müller's Etrusker, vol. ii. p. 227.

(179) See Petrie's Ecclesiastical Architecture of Ireland (Dublin, 1845), vol. i. p. 166—7, 170—1, 184, 250, 316, 396, 398, 409. With respect to the doorway of the Church of St. Fechin, at Fore, in the county of Westmeath, probably erected in the 7th century, Mr. Petrie states that 'the late eminent antiquarian traveller, Mr. Edward Dodwell, declared to

Cyclopean remains of Peru likewise demonstrate that this style of building is not peculiar to any people or period.(¹⁸⁰)

If we adopt the hypothesis of the ancients, and suppose the early generations of mankind to have been in a state scarcely superior to that of the brute animals,(¹⁸¹) they would have left no historical traces of their existence. They would have bequeathed to posterity no historical record, or any architectural monument which would have attested their labour. Such would now be the fate of the negro nations of Africa, if they were extinguished by any catastrophe. They have no writings, and their buildings are mere perishable huts, which would not long survive their authors. A nation in the negro state of civilization might therefore subsist for centuries without furnishing any historical evidence of its duration. The rude flint weapons, entitled 'celts,' which have been found in superficial beds of gravel and caves in France and England, mixed with the bones of large extinct mammalia,(¹⁸²) prove the existence of man in these countries at a time when the species in question were still extant. This fact, however, does not determine the time; we know that several species of large animals have been extinguished by man during the period of historical memory, either wholly or in particular countries. The bonasus has become

---

him that this doorway was as perfectly Cyclopean in its character, as any specimen he had seen in Greece,' p. 171.

(180) See Fergusson's Handbook of Architecture, p. 155—9, who says: 'Examples occur of every intermediate gradation between the house of Manco Capac, and the Tambos, precisely corresponding with the gradual progress of art in Latium, or any European country where the Cyclopean or Pelasgic style of building has been found. So much is this the case, that a series of examples collected by Mr. Pentland from the Peruvian remains might be engraved for a description of Italy, and Dodwell's illustrations of those of Italy would serve equally to illustrate the buildings of South America.'

(181) See Æsch. Prom. 451—515; Eurip. Suppl. 201—215; Moschion, ap. Stob. Ecl. Phys. i. 8, 38 (a passage consisting of pure iambi, without any resolved feet. In v. 13 read κοὐ τροφὴν φέρουσα γῆ, and in the last verse the sense requires δυσσεβοῦς for δυσσεβές); Lucret. v. 923, to the end of the book; Horat. Sat. i. 3, 99; Manil. i. 66—94; Juvenal. xv. 151.

(182) See Owen's Palæontology, p. 401 (Edinburgh, 1860); Phillips, Life on the Earth, p. 49.

extinct in Pæonia, the wolf has become extinct in the British isles, the lion has become extinct in Northern Greece and Western Asia. The great fossil deer, commonly called the Irish elk, must have also been alive at no very distant period.[183] The alces and urus, moreover, which Cæsar describes as still living in the forests of Germany, have disappeared since his time.[184]

The conjectural arguments, founded upon uncertain astronomical records, by which a high antiquity is assigned to the earth, have been rejected by Cuvier, and are now generally abandoned.[185] Many of them have been examined in the course of the present treatise, and have been shown to be destitute of foundation.[186]

---

[183] See Owen, ib. p. 372.   [184] Bell. Gall. vi. 27, 28.
[185] See his essay on the Theory of the Earth, p. 209 sq. Eng. tr.
[186] Macrobius argues in favour of the comparative recency of the world upon historical grounds : 'Quis facile mundum semper fuisse consentiat, cum et ipsa historiarum fides multarum rerum cultum emendationemque vel ipsam inventionem recentem esse fateatur, cumque rudes primum homines et incuriâ silvestri non multum a ferarum asperitate dissimiles meminerit vel fabuletur antiquitas, tradatque nec hunc eis quo nunc utimur victum fuisse, sed glande prius et baccis altos sero sperasse de sulcis alimoniam, cumque ita exordium rerum et ipsius humanæ nationis opinemur, ut aurea primum secula fuisse credamus, et inde natura per metalla viliora degenerans ferro sæcula postrema fœdaverit? Ac, ne totum videamur de fabulis mutuari, quis non hinc æstimet mundum quandoque cœpisse ; et nec longam retro ejus ætatem, cum abhinc ultra duo retro annorum millia de excellenti rerum gestarum memoriâ ne Græca quidem extet historia? Nam supra Ninum, a quo Semiramis secundum quosdam creditur procreata, nihil præclarum in libros relatum est. Si enim ab initio, immo ante initium, fuit mundus, ut philosophi volunt; cur per innumerabilem seriem seculorum non fuerat cultus quo nunc utimur inventus? non literarum usus, quo solo memoriæ fulcitur æternitas? cur denique multarum rerum experientia ad aliquas gentes recenti ætate pervenit, ut ecce Galli vitem vel cultum oleæ Româ jam adolescente didicerunt, aliæ vero gentes multa nesciunt quæ nobis inventa placuerunt?' In Somn. Scip. ii. 10, § 5—8.

Chapter VIII.

NAVIGATION OF THE PHŒNICIANS.

§ 1 WE have already remarked that the origins of astronomy and arithmetic were sometimes traced to the Phœnicians.(¹) The commercial wants of this manufacturing and seafaring people led them, it was supposed, to invent the one science, and their nocturnal navigation the other.(²) For a similar reason, they were reported to have been the inventors of coined money.(³)

The Phœnicians appear to have been the earliest navigators of the Mediterranean: their commercial ships had penetrated to the Adriatic Sea, and had sailed along the African coast as far as the opening of the great Western Ocean, at a time when the Greek navigation was still confined to the coasts of Asia Minor and Greece.(⁴) Their nautical skill was, in later

---

(1) Above, p. 262.

(2) Strab. xvi. 2, § 24, xvii. 1, § 3. Dionys. Perieg. 907—9, attributes the invention of navigation, commerce, and astronomy to the Phœnicians. Pliny, N. H. v. 12, says: 'Ipsa gens Phœnicum in magnâ gloriâ litterarum inventionis et siderum navaliumque ac bellicarum artium.' In vii. 57, he says that the Phœnicians introduced the observation of stars in navigation, and that they were the first to engage in commerce. 'Mercaturas Pœni . . . siderum observationem in navigando Phœnices.' Mela, i. 12, ascribes the invention of navigation to the Phœnicians. 'Prima ratem ventis credere docta Tyros,' Tibull. i. 7, 20. Compare Movers, Phön. Alt. iii. 1, p. 14, 149.

(3) Alcidamas, Ulyss. § 26, ed. Bekker. Æschylus, Prom. 467, attributes the invention of ships, together with other inventions, to the philanthropic Prometheus. It may be remarked that the absence of navigation and maritime commerce was one of the characteristics of the Golden Age, Arat. 110; Tibull. i. 3, 37—40; Virg. Ecl. iv. 32, 38; Horat. Carm. i. 3; Ovid, Met. i. 94—6.

(4) The skill of the Sidonian women in embroidery is mentioned, Il. vi. 290. The metallic work of the Sidonians and the navigation of the Phœnicians are mentioned, ib. xxiii. 743—4. The decorative skill of the Phœnicians, and their seafaring habits, are alluded to in several passages of the Odyssey; see particularly Od. xv. 415—484. Compare Pindar, Pyth. ii. 125; Soph. fragm. 756, Dindorf.

times, considered as evidence of their astronomical knowledge. The navigator, steering his bark by night along the trackless sea, guided his course by a northern constellation;(⁵) even travellers, passing through an unknown country,(⁶) and cameldrivers journeying across the desert,(⁷) used the same indication.

The Greeks are reported to have steered by the Great, and the Phœnicians by the Little Bear.(⁸) Thales is supposed to have taught his countrymen the use of the Little Bear in navigation; but the statements on this head, referred to in a former chapter, are confused, and probably inaccurate.(⁹)

The connexion between navigation and astronomy was considered as intimate;(¹⁰) Virgil supposes that sailors were the first to give names to the stars.(¹¹) A treatise on Nautical Astronomy was attributed to Thales; and the works of Eudoxus and Aratus on the stars are stated to have been intended for

---

(5) Il. xviii. 485; Od. v. 272.

(6) Soph. Œd. T. 795.

(7) πρότερον μὲν οὖν ἐνυκτοπόρουν πρὸς τὰ ἄστρα βλέποντες οἱ καμηλέμποροι, καὶ, καθάπερ οἱ πλέοντες, ὥδευον κομίζοντες καὶ ὕδωρ, Strab. xvii. 1, § 45. Diod. ii. 54, says that travellers in the Arabian deserts directed their course by the Bears.

(8) Aratus, v. 42—44, says that the Sidonians steer by the Little Bear, and that it is preferable to the Great Bear, as being nearer the North Pole.

Magna minorque feræ, quarum regis altera Graias,
Altera Sidonias, utraque sicca, rates.—Ovid. Trist. iv. 3, 1.

Esse duas Arctos, quarum Cynosura petatur
Sidoniis, Helicen Graia carina notet.—Fast. iii. 107.

Majoremque helice major decircinat arcum,
Septem illam stellæ certantes lumine signant,
Quâ duce per fluctus Graiæ dant vela carinæ.
Angusto cynosura brevis torquetur in orbe,
Quam spatio, tam luce minor; sed judice vincit
Majorem Tyrio.   Manil. i. 304—8.

Compare Hygin. Poet. Astr. ii. 2, who states that the Little Bear bore the name of Φοινίκη.

(9) Above, p. 83, n. 59.

(10) Pliny informs us that the inhabitants of Taprobane did not steer by the stars, but that, in order to sail to India, they let out birds, and followed their course, vi. 24.

(11)   Navita tum stellis numeros et nomina fecit,
Pleïadas, Hyadas, claramque Lycaonis Arcton.
                    Georg. i. 137—8.

the use of mariners.(12) It becomes, therefore, a matter of interest, as bearing indirectly upon the present inquiry, to ascertain what foundation there may be for the statements and conjectures, which have obtained a favourable acceptance among some modern writers of high authority, as to the distant voyages of the Phœnicians at an early period of history. The circumnavigation of Africa by a Phœnician ship, in the reign of Neco, about 610 B.C., is credited by Alex. von Humboldt, Rennell, Heeren, Mr. Grote, and Mr. Rawlinson, and is generally received as an historical fact. The voyages of the Phœnicians to Cornwall for tin, and to the southern coast of the Baltic for amber,(13) pass as almost equally certain; and some modern writers have even gone the length of supposing that these enterprising sailors had anticipated Columbus by more than 2000 years, and had discovered America.(14)

§ 2 The accounts handed down by the Greek and Roman writers agree in representing Gadeira, or Gades—the modern Cadiz,—as an ancient foundation of the Phœnicians of Tyre.(15) Its peculiar position—an island, or peninsula, easy of defence, and convenient for trade, lying at the mouth of the Mediterranean, communicating with a fertile and metalliferous region, and washed by a sea abounding in fish—marked it out as an advantageous spot for a commercial station. Velleius states that it was founded by the Tyrians before Utica;(16) while the

---

(12) Above, p. 83; Leontius, ap. Buhl. Arat. vol. i. p. 285.

(13) Link, Die Urwelt und das Alterthum, vol. ii. p. 305 (Berlin, 1821—2) thinks that the Phœnician ships penetrated to the coast of Prussia, and carried on a trade in amber. Heeren, Ideen, ii. 1, p. 178, advances the same opinion. Compare i. 2, p. 85.

(14) See Pauly, art. *Geographia*, vol. iii. p. 737; art. *Navigatio*, vol. v. p. 445. Niebuhr, Hist. of Rome, vol. i. p. 281, Eng. tr., connects the primitive astronomy of Europe with that of America, and therefore must suppose the latter country to have been discovered.

(15) See Diod. v. 20; Strab. iii. 5, 5; Appian, Hisp. 2; Scymnus, v. 160. According to Movers, the Punic word *gadir* meant a walled enclosure or fort. A considerable portion of the materials for this chapter had been previously used by the author for articles which have appeared in Notes and Queries.

(16) i. 2.

author of the Aristotelic Collection of Marvellous Reports, cites Phœnician histories as declaring that Utica was founded 287 years before Carthage.([17]) If this comparative chronology is correct, the Phœnicians founded the distant colony of Gades before they founded the cities of Utica and Carthage, which lay on the African coast on the way to Gades. The foundation of Gades is placed by Mela at the time of the siege of Troy. Justin describes Gades as having been founded by the Tyrians, but as having been subsequently annexed by the Carthaginians to their Empire.([18]) Its fidelity to Carthage seems to have been ambiguous; for there was a party in it which was in traitorous correspondence with the Romans during the Second Punic War.([19])

Strabo says that the Phœnicians occupied the productive district of Southern Spain, from a period earlier than Homer down to the time when it was taken from them by the Romans.([20]) Their presence can be clearly traced westwards along the coast inhabited by the Bastuli, as far as the Pillars of Hercules, and from the Pillars along the Turdetanian coast as far as the Anas or Guadiana, or perhaps as far as the Sacred Promontory, the south-western extremity of Lusitania (Cape St. Vincent).([21]) Ulysippo, the modern Lisbon, is treated by Greek legends as a foundation of Ulysses. This is a mere etymological mythus; and the conjecture of Movers, derived from the occurrence of the termination *-ippo* in other proper names, that this is a Phœnician form, is probable.([22]) Scylax, whose Periplus was composed about 340 B.C., mentions many factories of the Carthaginians to the west of the Pillars of Hercules, apparently on the European side.([23]) But whatever

---

(17) C. 134. (18) Mela, iii. 6; Justin, xliv. 5.
(19) Livy, xxviii. 23, 30. (20) iii. 2, 14.
(21) See Movers, Phönizier, vol. ii. p. 615—647.
(22) Ib. p. 639.
(23) ἀπὸ Ἡρακλείων στηλῶν τῶν ἐν τῇ Εὐρώπῃ ἐμπόρια πολλὰ Καρχηδονίων καὶ πηλὸς καὶ πλημμυρίδες καὶ πελάγη, c. 1. The last words show that Scylax considered the ocean beyond the Pillars of Hercules as unknown.

factories or colonies the Phœnicians, either of Tyre or Carthage, may have established on the western coast of Iberia, they must have been obscure and unimportant, and have perished without leaving any historical vestiges of their origin.

From Gades the Phœnician navigators are supposed to have steered their enterprising ships along the shores of the Atlantic Ocean, to have procured amber and tin from Northern Europe, and to have afterwards sold their merchandize to the nations at the eastern extremity of the Mediterranean.

§ 3 Both these substances appear to have been known to the Homeric Greeks. The metal κασσίτερος occurs several times in the Iliad, as an ornament of arms and chariots; and it is placed in juxtaposition with gold. It receives the epithets 'white' and 'shining.' The word does not occur in the Odyssey. Beckmann[24] and Heyne[25] think that κασσίτερος was originally a mixture of silver and lead; and that, on account of the resemblance of the colour, the name was afterwards applied to tin. It is, however, most probable that the metal signified by κασσίτερος in the Iliad is tin; and it is so understood by Pliny.[26]

The question arises, whence the tin used by the Greeks of the Homeric age was obtained. The statement of Pliny is, that tin was fabled to be imported from some islands in the Atlantic Sea, but that it was known in his time to be produced in Lusitania and Gallæcia; the ore being a dark-coloured sand, found on the surface of the earth, and recognised by its weight. He adds, that lead is not found in Gallæcia, though it abounds in the neighbouring country of Cantabria.[27] Diodorus likewise states that tin occurred in many parts of Iberia; and that it was not found on the surface, but was mined and melted like silver and gold.[28]

---

(24) Hist. of Invent. vol. iv. p. 20.
(25) Iliad, vol. vi. p. 120.   (26) N. H. xxxiv. 47.
(27) N. H. xxxiv. 47. Concerning the tin trade, see Kenrick's Phœnicia (Lond. 1855), p. 209—225.
(28) v. 38.

Other reports also connect tin with Iberia. Posidonius speaks of tin being worked in the country of the barbarians beyond Lusitania.(29) Dionysius Periegetes says that the western islands, where tin was produced, were inhabited by Iberians.(30) Several writers speak of a river, by which the Bætis is meant, as carrying down tin to Tartessus.(31) In the *Ora Maritima* of Avienus, there is a description of certain places on the western coast of Spain, near the Bay of Tartessus; among these Mount Cassius is enumerated, from which the Greeks derived the name κασσίτερος.(32) This is a rude and childish etymological mythus, not much superior to that which derived Britain from Brutus the Trojan, son of Ascanius. Tin-ores are, however, stated to be still found in Galicia, and it is possible that supplies of this metal may have been brought in antiquity from the western parts of the Iberian peninsula to Gades, and have thus passed into Greek consumption.

But it cannot be doubted that Britain was the country from which the tin sold by the Phœnicians to the Greeks was chiefly procured. Herodotus had heard of the Cassiterides, or Tin Islands, from which tin was brought to Greece, but was unable to ascertain anything as to their existence;(33) and it was not till the time of Cæsar, when the Romans crossed into Britain, that the nations of Southern Europe obtained any authentic information respecting the country which produced this metal. Strabo describes the Tin Islands as ten in number, situate in the open sea to the north of the country of the Artabri (Cape Finisterre). In early times (he says) the Phœnicians carried on the tin trade from Gadeira (Cadiz), and retained the monopoly by concealing their course. On one occasion the Romans, desirous of discovering the port where the tin was shipped, followed a Phœnician vessel; but the captain intentionally steered his ship into shallow water, and both it and the Roman

---

(29) Fragm. 48.   (30) v. 561—4.
(31) Eustath. ad Dion. Perieg. 337; Scymnus, v. 162; Steph. Byz. in Ταρτησσός.
(32) v. 259.   (33) iii. 115.

ship were lost; he himself escaped on a fragment of the wreck, and received from the State the value of his cargo. The Romans, however, after many attempts discovered the secret. Since Publius Crassus visited the islands, and ascertained that tin was found near the surface, and that the inhabitants were peaceable, the voyage has been frequently made, though it is longer than the passage from Gaul to Britain.[34]

The Publius Crassus alluded to in this passage must be the youngest son of the triumvir, who was Cæsar's lieutenant in Gaul from 58 to 55 B.C.[35] By the Phœnicians the Carthaginians appear to be meant. This story is not very intelligible, nor is it easy to fix a date for the occurrence; for the Romans were not a seafaring people, and they were not likely to attempt voyages beyond the Pillars of Hercules before the destruction of Carthage in 146 B.C., whereas after that time the Carthaginians had no ships or factories; Gades had been sixty years in the hands of the Romans; and ever since the end of the Second Punic War the Romans had been able to extort the secrets of the Carthaginians without resorting to stratagem.

Moreover, the account of P. Crassus opening the navigation with the Tin Islands cannot fairly be reconciled with the fact (to which we shall presently advert) that, before and during Cæsar's life, the trade in British tin was carried on through Gaul. The story in question doubtless originated in the known commercial jealousy of the Carthaginians, and in the rigour with which they maintained the exclusion of competitors with their trade. They are stated by Strabo to have sunk any strange ship which sailed even as far as Sardinia or Gades.[36]

Diodorus describes Britain as being, like Sicily, triangular, but with sides of unequal length. The promontory nearest the

---

[34] iii. 5, § 11.

[35] See Cæsar, B. G. ii. 64; Drumann, Geschichte Roms, vol. iv. p. 116.

[36] xvii. 1, 19; Redslob, Thule (Leipzig, 1855), p. 22, discredits the story of the Gaditan navigator. He thinks that one ship could not follow another, as a traveller could track another by land. He also thinks that the Romans never went to Britain by the Straits of Gibraltar.

mainland was called Cantium (Kent); that at the opposite extremity was called Belerium; that turned towards the sea was named Orca (a confusion with the Orcades). The inhabitants of the promontory of Belerium were hospitable, and, on account of their intercourse with strangers, civilized in their habits. It is they who produce tin, which they melt into the shape of astragali, and they carry it to an island in front of Britain, called Ictis. This island is left dry at low tides, and they then transport the tin in carts from the shore. Here the traders buy it from the natives, and carry it to Gaul, over which it travels on horseback, in about thirty days, to the mouths of the Rhone.[37]

Timæus mentioned an island of Mictis, within six days' sail of Britain, which produced tin, and to which the natives of Britain sailed in coracles.[38] The Mictis of Timæus and the Ictis of Diodorus are probably variations of the name Vectis, by which the Roman writers designated the Isle of Wight. The south-western promontory of Britain reappears in Ptolemy under the form of Bolerium.[39]

In another passage Diodorus speaks of the Tin Islands as being in the ocean beyond Iberia; he distinguishes them from Britain, whence he states that tin was imported into Gaul, and carried on horseback by traders to Massilia, and to the Roman colony of Narbo.[40]

Posidonius, the contemporary of Cicero, likewise states that tin was brought from the Britannic islands to Massilia,[41] by which he implies the overland route. The author of the Aristotelic Collection of Marvellous Reports, which treatise was composed about 300 B.C., describes the *Celtic tin* as being more easily melted than lead.[42] This must mean tin brought over

---

(37) v. 21, 22.
(38) Ap. Plin. vi. 16 (fragm. 32, ed. C. Müller). For 'insulam Mictim' in Pliny, Mr. Kenrick reads 'insulam Ictim,' which is not an improbable conjecture.
(39) Geogr. ii. 3, § 3.   (40) v. 38.   (41) Fragm. 48.
(42) ὁ κασσίτερος ὁ Κελτικός, Mirab. Ausc. 50. Bochart interprets this

Gaul, and imported into Greece from a Gallic port; as we speak of *Dutch toys*, which, though exported from Holland, are manufactured in Southern Germany; and of *Leghorn bonnets*, which, though shipped at Leghorn, are made in the interior of Tuscany.

Pliny describes the Cassiterides as a group of several islands, lying opposite Celtiberia, to which the Greeks gave this name from their production of tin. He adds, that six other islands, called the Islands of the Gods, or the Happy Islands, lay off the promontory of the Arrotribæ (or Artabri);[43] the position which Strabo assigns to the Tin Islands. Dionysius Periegetes speaks of the Tin Islands as situated near the Sacred Promontory, the extreme point of Europe.[44] The ten tin islands to the west of Spain are also recorded by Ptolemy.[45] These writers, together with Strabo, conceive the Cassiterides as distinct from Britain. Scymnus, in his geographical poem, composed about 90 B.C., even places two tin islands in the upper part of the Adriatic Sea, opposite the territory of the Istrians.[46]

We learn from Pliny that Hanno, during the prosperous period of Carthage, sailed from Gades to the extremity of Arabia, and left a written account of his voyage. He adds, that Himilco was sent at the same time to examine the external coasts of Europe.[47] The Periplus of Hanno is extant: his voyage was partly for the foundation of colonies, and partly for discovery; he is supposed to have sailed along the coast as far as Sierra Leone; and, according to the best considered conjecture, his expedition took place about 470 B.C.[48] The discoveries of Himilco, as preserved in a written record, are referred to by Avienus, in his geographical poem, the *Ora Maritima*. He de-

---

epithet as referring to Britain. But the Greeks had little or no knowledge of Britain at the time when this treatise was written, and they never considered the Britannic islands as forming part of Celtica.

[43] N. H. iv. 36.
[44] v. 161—4.
[45] Geogr. ii. 6, § 76.
[46] v. 392.
[47] N. H. ii. 67, and see v. 1.
[48] See C. Müller, Geogr. Gr. Min. vol. i.; Prol. p. xxii.; below, § 8.

scribes certain islands, called the Œstrymnian Islands, off the coast of Spain, with which the Tartessians traded, which produced tin and lead, and which were only two days' sail from the islands of the Hibernians and the Albiones. He proceeds to say, that the Carthaginians, both of the mother country and the colonies, passed the Pillars of Hercules, and navigated the Western Sea. Himilco stated from personal experience that the voyage occupied at least four months; and he described the dangers of these unknown waters by saying that there was no wind to impel the ship; that its course was impeded by weed; and that, while in this helpless state, it was surrounded by marine monsters.[49]

If the date of the voyages of Hanno and Himilco is correctly fixed, it follows that at a period subsequent to the expedition of Xerxes, the Carthaginians, though there was a Phœnician establishment at Gades, had not carried their navigation far along the coasts of the Atlantic; and that they sent out two voyages of discovery—one to the south, the other to the north—at the public expense. The report of Himilco, that the voyage from Gades to the Tin Islands (*i.e.* to Cornwall) occupied at least four months; and that navigation in these remote waters was impeded by the motionless air, by the abundance of seaweed, and by the monsters of the deep—fables which the ancient mariners recounted of unexplored seas — would not be very attractive to the traders of the Carthaginian colonies.

On the whole, the accounts preserved by the Greek and Latin writers lead to the inference that the tin supplied in early times to the nations in the east of the Mediterranean came by the overland route across Gaul, and that the Phœnician ships brought it from the mouth of the Rhone, without sailing as far as Britain. Some Iberian tin may perhaps have been obtained from Gades.

A passage of the prophet Ezekiel (about 590 B.C.) might indeed seem to point to a different conclusion. It mentions tin among the articles of merchandize brought to Tyre from Tarshish.

'Tarshish was thy merchant by reason of the multitude of all kind of riches; with silver, iron, *tin*, and lead, they traded in thy fairs.' (xxvii. 12.)

The prevalent opinion among Biblical critics is, that Tarshish is equivalent to the Greek Tartessus, and that the southern coast of Spain to the west of the Straits of Gibraltar is here signified.[50] But the meaning of Tarshish in the Old Testament is not free from doubt; and even if Tarshish, in its strict geographical sense, be equivalent to Tartessus, its meaning in this passage would be satisfied by the supposition that tin was brought by sea from the western parts of the Mediterranean.

There are some traces of a trade in tin with India in antiquity, and it has been conjectured that the Tyrians might have procured their supplies of this metal from the east, and not from the west. Thus Diodorus says, that India contains veins of various metals — namely, much gold and silver, not a little copper and iron, also tin.[51] It is stated by Stephanus of Byzantium, on the authority of the *Bassarica* of Dionysius, that Cassitira was an island in the ocean near India, from which tin was obtained.[52] The *Bassarica* was a poem; and its author, Dionysius, was apparently Dionysius Periegetes, who lived at the end of the third or the beginning of the fourth century of our era. It celebrated the exploits of Bacchus, and, among others recounted his expedition to India, where it enumerated many names of places.[53] Whether this geographical poet knew of tin being imported into Europe from the island of Banca, or whether he considered the Indian island of Cassitira as a tin-island on mere etymological grounds, cannot now be determined; though the latter supposition seems the more probable.

---

(50) See Winer, in *Tharschisch*; Dr. Smith's Dict. of Anc. Geogr. in *Tartessus*.

(51) ii. 36.   (52) In v. Κασσίτιρα.

(53) See Bernhardy, ad Dionys. Perieg. p. 507, 515.

The author of the Periplus of the Erythræan Sea, which bears the name of Arrian, mentions tin as an article of import into the following places—namely, Port Avalites, in Abyssinia, near the entrance of the Red Sea;[54] Canè, on the southern coast of Arabia, whither this metal was brought from Egypt;[55] and two Indian emporia—one, the Port of Barygaza, at the mouth of the Nerbudda, north of Bombay; and the other the Port of Bacarè, on the Malabar coast.[56] The author of this Periplus is proved by internal evidence to have been an Egyptian merchant, who wrote at the end of the first century after Christ.[57]

Movers, in his learned work on the Phœnicians, rejects the hypothesis of an ancient trade in tin between Tyre and India. He allows no weight to the argument which has been drawn in its favour from the resemblance of the Sanscrit *kastira* to the Greek κασσίτερος: and holds that this word, as well as the Aramaic *kastir* and the Arabic *kasdir*, was derived from the Greek; he refers to the passages concerning tin in the Periplus of Arrian, as showing that tin was imported into Arabia and India from Alexandria; and he believes that the Malacca tin had not been worked in antiquity.[58]

§ 4 The other main argument in favour of the distant voyages of the Phœnicians is derived from the early use of amber.

The Greek word *electron* had a double signification; it denoted amber, and also a metallic compound formed by the mixture of gold and silver in certain proportions. Whichever of these significations was the original one, it is certain that the transfer from the one to the other was owing to the tawny colour and the lustre which were common to the two substances.[59]

---

(54) c. 7. (55) c. 28. (56) c. 49, 56, ed. C. Müller.
(57) See C. Müller's Prolegomena to his recent edition of the Geogr. Gr. Min. vol. i. p. 97.
(58) Ib. iii. 1, p. 62—5. The Assyrian bronze is found to be composed of one part tin to ten parts copper. Rawlinson's Herodotus, vol. i. p. 498.
(59) Virgil says of a river, 'clearer than amber,' where a modern poet

The use of the word *electron* in Homer and Hesiod, where it is described as applied to different ornamental purposes, does not determine its meaning. Buttmann, however, in his dissertation on the subject,(⁶⁰) has made it probable that it signifies amber in the early epic poetry; and he derives the word from ἕλκω, in allusion to the attractive properties of amber. The use of the word in the plural number for the ornaments of a necklace in two passages of the Odyssey,(⁶¹) though not decisive, agrees best with the supposition that knobs or studs of amber are meant; as in the passage of Aristophanes, where it denotes the ornaments fastened to a couch.(⁶²) Thales is reported to have held that inanimate substances were endued with souls, arguing from the magnet and amber, (⁶³) where the word used is *electron*. Upon this hypothesis, the acceptation of the word in the sense of pale gold would be derivative and secondary.(⁶⁴)

The notions of the ancients, both as to the nature of amber and the places where it occurred, were singularly conflicting and indistinct, as we learn from the full compilation in Pliny.(⁶⁵) It may, however, be considered as certain, that the amber imported into ancient Greece and Italy was brought from the southern shores of the Baltic, where it is now almost exclusively obtained.(⁶⁶) According to Herodotus, amber was in his time

---

would say 'clearer than crystal.' 'Purior electro campum petit amnis,' Georg. iii. 522. The metaphor is imitated by Milton:—

'There Susa by Choaspes, *amber stream*,
The drink of none but kings.'
Par. Reg. iii. 288.

'And where the river of bliss through midst of heaven
Rolls o'er Elysian flowers *her amber stream*.'
Par. Lost, iii. 358.

(60) Ueber das Elektron, Mythologus, vol. ii. p. 337.
(61) xv. 460, xviii. 295.   (62) Eq. 532.
(63) Diog. Laert. i. 24. The authorities cited are Aristotle and Hippias. Aristotle refers only to the magnet, De An. i. 2. By Hippias, the sophist seems to be meant. Plato, Tim. § 60, p. 80, mentions the attractive properties of ἤλεκτρον.
(64) Compare Boeckh, Metrol. Untersuchungen, p. 129.
(65) H. N. xxxvii. 11.
(66) Tavernier, Voyages des Indes, ii. 23, states that in his time amber

reported to come from a river, called Eridanus by the barbarians, which flowed into the Northern Sea. Herodotus rejects this story; he considers the name Eridanus as being manifestly of Greek origin, and as invented by some poet: he cannot ascertain that such a river exists, or that Europe is bounded by sea on the opposite side. He believes, however, with respect both to amber and tin, that they come from countries at the extremity of the earth.(67) The account of Pytheas, the navigator (about 350 B.C.), as recited to us by Pliny, is, that a shore of the ocean called Metonomon, reaching 6000 stadia (750 miles) in length, was inhabited by the Guttones, a nation of Germany; that beyond this coast, at the distance of a day's sail, the island of Abalus was situated; that amber was thrown upon this island in spring by the waves; and that the natives used it as a fuel, and likewise sold it to their neighbours the Teutoni. The account of Pytheas was, according to Pliny, followed by Timæus; with this exception, that he called the island, not Abalus, but Basilia.(68) The testimony of Timæus is, however, differently reported by Pliny in another place;(69) he there states that, according to Timæus, there was an island one day's sail from the northern coast of Scythia, called Raunonia, into which amber was cast up by the waves in spring. In the same chapter he likewise says, that a large island off the northern coast of Scythia, which others called Baltia, was by Timæus called Basilia. The report of Diodorus is not very different, and is apparently derived from a similar source. He states that Basileia is an island in the ocean oppo-

---

was found exclusively upon the coast of Ducal Prussia in the Baltic. He says that it is cast up by the sea, and that the collection of it is farmed out by the Elector of Brandenburgh for an annual rent of 18,000 or 20,000, and sometimes as much as 22,000 dollars. Sir John Hill, however, (Translation of Theophrastus on Stones, ed. 2, p. 132), affirms that amber is not peculiar to the coast of the Baltic, and that it is sometimes found in England.

(67) iii. 115.
(68) xxxvii. 2. Compare Zeuss, Deutschen, p. 269.
(69) iv. 13.

site the coast of Scythia, beyond Galatia;(70) that amber is cast up by the sea on this island, and that it occurs nowhere else; and that it is here collected and carried by the natives to the opposite continent, whence it is exported to Greece and Italy.(71)

Tacitus informs us, in his *Germania*,(72) that the Æstui, who dwell on the right or eastern shore of the Suevic Sea, find in the shoal water and on the shore, amber, which they call *glesum*. Like other barbarians (he continues), they were incurious about its nature, and it lay for a long time among the substances cast up by the sea; they made no use of it, until Roman luxury gave it value; they now collect it and send it onwards, in an unmanufactured state, and wonder at the price which they receive for it. Tacitus himself believes it to be a gum, which distils from trees in the islands of the west, under the immediate influence of the sun, falls into the sea, and is carried by the winds to the opposite coast. One of the islands in the Northern Ocean is stated by Pliny to have been named by the Roman soldiers Glessaria, from its producing *glessum*, or amber (*glass*); it had been reduced by Drusus, and was called Austravia, or Actania, by the natives.(73) Pliny places it near the island of Burcana, which was between the mouths of the Rhine and the Sala, and was likewise taken by Drusus.(74)

These accounts agree in pointing to the northern coast of Europe as the place where amber was found in antiquity. Pliny, however, adds a statement of a more precise and satisfactory character. Amber was, he says, brought from the shores of Northern Germany to Pannonia; the inhabitants of this province passed it on to the Veneti, at the head of the Adriatic, who conveyed it further south, and made it known in Italy. The coast where it is found had (he says) been lately seen by a Roman knight, who was sent thither by Julianus, the curator

---

(70) By Galatia, Diodorus means Celtica, that is, central and western Europe.

(71) v. 23.   (72) c. 45.   (73) iv. 13, xxxvii. 3.

(74) Strab. vii. 1, 3. Compare Tac. Germ. 34. The date of this expedition is 12 B.C. See Merivale's Roman Empire, vol. iv. p. 228.

of the gladiatorian shows for the Emperor Nero, in order to purchase it in large quantities. This agent visited the coast in question, having reached it by way of Carnuntum, the distance from Carnuntum to the amber district being nearly 600 miles; and he brought back so large a supply, that the nets in the amphitheatre for keeping off the wild beasts were ornamented with amber at the knots; and the arms, the bier, and all the apparatus for one day were decorated with the same material. He brought with him one lump thirteen pounds in weight.[75]

Carnuntum was a town of Upper Pannonia, on the southern bank of the Danube, between the modern Vienna and Presburg; and after the reduction of Pannonia by the Romans, it would without difficulty have been reached from the head of the Adriatic. The distance from Carnuntum to the coast of the Baltic is not more than 400 miles.[76] Hüllmann has pointed out that in the Middle Ages there was a commercial route from the Upper Vistula to Southern Germany, which, passing through Thorn and Breslau, reached the river Waas, and thus descended to the Danube.[77] A Roman knight, with a sufficient escort of slaves, would doubtless have effected this journey without serious difficulty. The large piece of amber which Pliny reports him to have brought to Rome is exceeded in size by a mass of eighteen pounds, which is stated in M'Culloch's Commercial Dictionary to have been found in Lithuania, and to be now preserved in the Royal Cabinet at Berlin. It appears from Tacitus that Claudius Julianus had still the care of the gladiators under Vitellius, in 69 A.D.[78] He was murdered in the struggle which accompanied the downfall of that Emperor.

Hüllmann [79] justly points out the improbability that the

---

(75) xxxvii. 3.

(76) See Cluvier, Germ. Ant. p. 692. A traditionary account has been preserved in Prussia that in the Roman times amber was carried on horseback through Pannonia to Italy, Hüllmann, Städtewesen des Mittelalters, vol. i. p. 344.

(77) Handelsgeschichte der Griechen, p. 77.

(78) Hist. iii. 57, 76. (79) Ib. p. 76.

Phœnician navigators, however enterprising they may have been, should have sailed through the Sound, and have carried on a trade with the southern coasts of the Baltic. He makes the remark that, in very early times, trade with remote regions was always conducted, not by sea, but by land. This opinion is doubtless well founded: one reason was, the helplessness, timidity, and unskilfulness of the ancient navigation; but another and more powerful one was, that land-traffic could be carried on by native travelling merchants, such as those mentioned by Livy as visiting different parts of Italy;[80] whereas navigators were foreigners, who came in a foreign ship, and were as such liable to all the dangers and disadvantages to which this class of persons were exposed in antiquity.

Dr. Vincent, whose learned and judicious researches into the voyages of the ancients give great weight to his opinion, conceives it 'to be agreeable to analogy and to history, that merchants travelled before they sailed;' and he refers to the transport of silk by land for a distance of more than 2800 miles.[81] Gibbon likewise remarks, with reference to the ancient caravan trade in the same commodity, that 'a valuable merchandize of small bulk is capable of defraying the expense of land carriage.'[82] This observation applies with peculiar force to amber, which combines a great value with a small bulk and a small weight. It likewise applies, though with less force, to tin.

Brückner, in his Historia Reipublicæ Massiliensium,[83] adopts the view that amber was brought by an overland journey to the Mediterranean; but he conceives Massilia to have been the point with which the connexion was established. It seems, however, much more probable that the more direct route to the head of the Adriatic was preferred; and that even in the time of Homer amber had reached the Mediterranean, and had been diffused over the Grecian world, by this channel. The Phœni-

---

(80) iv. 24, vi. 2.
(81) Commerce and Navigation of the Ancients in the Indian Ocean, 1807, vol. ii. p. 365, 589.
(82) c. 40.   (83) p. 60.

cians were probably the intermediate agents by which this diffusion was effected. An embassy from the Æstii, on the southern shores of the Baltic, who visited Theodoric in the sixth century, and who brought him a present of amber, appears to have travelled to Italy by this route.[84] Otfried Müller concurs in rejecting the hypothesis that amber was carried to Greece from the Baltic in Phœnician ships, and in supposing that it found its way over land to the northern shores of the Adriatic. He conjectures, however, that it was conveyed to Greece by the Etruscans, and not by the Phœnicians; a hypothesis less reconcilable with such historical traces as remain, than that to which he prefers it.[85]

The fable of the daughters of the sun being changed into poplars on the banks of the river Eridanus, and their tears for the death of their brother Phaethon being converted into amber, though posterior to Homer, occurred in the poetry of Hesiod, and was adopted by Æschylus and other Attic tragedians.[86] The Eridanus was originally, as Herodotus perceived, a purely poetical stream, without any geographical position or character. By degrees, however, it obtained a footing in positive geography, though its identification was at first unfixed. The Theogony of Hesiod places it in the category of real rivers.[87] Herodotus conceived it as falling into the external ocean, to the north of Europe. Æschylus regarded it as an Iberian stream, and is said to have confounded it with the Rhone.[88] At an early period the Eridanus was identified with the Po. This identification is traced as high as Pherecydes;[89] and it is clearly ex-

---

[84] See the king's curious rescript of thanks, Cassiod. Var. v. 2. Compare Zeuss, Deutschen, p. 667.

[85] Etrusker, vol. i. p. 280—285.

[86] Markscheffel, Frag. Hesiod. p. 355. The Heliades of Æschylus and the Phaethon of Euripides were founded on this fable.

[87] v. 338.

[88] Plin. xxxviii. 2. Æschylus in Iberia, hoc est in Hispania, Eridanum esse dixit, eundemque appellari Rhodanum. In Heliad. fragm. 63, Dindorf. however, he appears to place the fall of Phaethon on the shores of the Adriatic.

[89] Fragm. 33 c. ed. C. Müller.

pressed by Euripides.(⁹⁰) Apollonius introduces the fable into his epic poem. He represents the Argonauts as passing along the Eridanus, in their voyage from the Ister to the Rhone, and as there hearing the lament of the Heliades, and seeing their amber tears. He likewise mentions another version, which he attributes to the Celts, that the tears were those of Apollo, shed for his son Æsculapius, when he was on his way to the sacred Hyperboreans.(⁹¹) Ovid relates the fable in its original form of a metamorphosis, and shows how the tears of the Heliades, hardened by the sun, and falling into the Eridanus, produced ornaments for the Roman ladies :—

> Cortex in verba novissima venit.
> Inde fluunt lacrimæ, stillataque sole rigescunt
> De ramis electra novis, quæ lucidus amnis
> Excipit, et nuribus mittit gestanda Latinis.(⁹²)

The story of the amber tears of the Heliades is treated as fabulous by Polybius and Strabo;(⁹³) and Lucian ridicules it in a short piece,(⁹⁴) in which he describes himself as having been rowed up the Po, and having in vain inquired of the wondering boatmen if they could show him the poplars which distilled amber. Others had previously rationalized the fable. The Collection of Marvellous Stories ascribed to Aristotle, written about 300 B.C., describes amber as a gum which liquefied from poplars near the Eridanus, in the extremity of the Adriatic, and which, being hardened into the consistency of stone, was collected by the natives, and exported into Greece.(⁹⁵)

But the identification of the imaginary Eridanus with the Po was not accidental. The Greek mythology connected the amber tears of the Heliades with the Eridanus; and as amber was imported into Greece from the upper extremity of the Adriatic, it was naturally identified with the great river of Northern Italy, which falls into this part of the Adriatic.

---

(90) Hipp. 735—741.
(91) iv. 595—626. Compare Scymnus, v. 395—401.
(92) Met. ii. 363—6. (93) Polyb. ii. 16, 17; Strab. v. 1, 9.
(94) De Electro. (95) Mir. Ausc. c. 81.

Amber was supposed to be so abundant in this region, that according to Pliny the peasant women of the Transpadane district wore amber necklaces in his time. The statement of Theophrastus, that amber was found in Liguria,[96] must be explained on the supposition that he had heard of its derivation from Northern Italy.

An unnecessary attempt has been made by some writers to identify the Eridanus with some real river in the north of Europe, having a name of similar sound;[97] but Heeren has remarked with justice, that the Eridanus is a fabulous stream, which existed only in popular legend and in the imagination of poets, and that nothing is gained by explaining it to mean the Rhine or the Raduna.[98]

The story of amber being found near a river, as in the mythological fable, or in an island, as in the accounts of Pytheas and Timæus, does not rest on any foundation of fact. Even the *insula Glessaria*, which must be one of the islands to the east of the Helder, off the coast of Holland and Friesland, appears to have received its name from some accidental connexion with amber; as the islands on this coast are not known to have yielded that substance. The notion of amber being found in islands gave rise to the belief in the existence of the Electrides at the mouth of the Po, at the extremity of the Adriatic.[99] Both Strabo and Pliny remark that the Electrid islands are a fiction, and that none such exist on the spot indicated. It may be remarked that the obscurity of vision, caused by distance, multiplied Britain into a group of tin islands.

The Greeks were for centuries acquainted both with tin and amber, without obtaining any certain knowledge of the places from which they came. Their incurious ignorance, however,

---

(96) De Lapid. § 16, ed. Schneider.

(97) See Bayer, de Venedis et Eridano Fluvio, in Comm. Acad. Petrop. 1740, vol. vii. p. 351.

(98) Ideen, ii. 1, p. 179.

(99) Aristot. Mir. Ausc. 81; Steph. Byz. in v.; Mela, ii. 4. Apollonius Rhodius places the sacred Electrid island near the river Eridanus, iv. 505. Scymnus, v. 374, places the Electrid islands in the Adriatic Sea.

was not confined to the two articles in question; it extended likewise to ivory. That ornamental and useful substance was known to the Jews in the time of Solomon, about 1000 B.C.,[100] and to the Greeks in the time of Homer, probably about 200 years later. It reached the shores of the Mediterranean through various hands, from India, and the remote parts of Africa.[101] The early Greeks knew nothing of the animal to which it belonged. The word *elephas*, with them, meant simply ivory. Herodotus, indeed, mentions the elephant as an animal, and describes it as occurring in the western extremity of Africa:[102] but Ctesias appears to have been the first Greek who spoke of the elephant from personal knowledge; he had seen the animal at Babylon.[103] It was not, however, till the expedition of Alexander that the elephant became generally known to the Greeks; in consequence of their acquaintance with his military capacities, the successors of Alexander first used the Asiatic elephant in war, and the Egyptian kings and the Carthaginians afterwards used the African elephant for the same purpose.[104] The natural history of the silkworm was known to Aristotle;[105] but the Romans seem to have thought that silk, like linen and cotton, was a vegetable product; Virgil describes it as the delicate fleece which the Seres, or Chinese, combed from the leaves of trees,[106] and the naturalist, Pliny, gives a similar account.[107] The modern nations of Europe likewise long received spiceries and other commodities from the islands of the Indian Ocean, before they were acquainted with the countries in which these articles of commerce were produced.[108]

§ 5 The hypothesis of a Phœnician trade with the northern shores of Europe in tin and amber would receive material con-

---

(100) 1 Kings x. 22. See Dr. Smith's Bib. Dict. art. *Ivory*.
(101) Paus. i. 12, 4, v. 12, 3. (102) iv. 191.
(103) Ælian, Hist. An. xvii. 29; Bähr ad Ctes. p. 268, 352.
(104) Armandi, Histoire Militaire des Eléphants, Paris, 1843, pp. 39—43, 64, 85, 134.
(105) Hist. An. v. 19. (106) Georg. ii. 121. (107) N. H. vi. 17.
(108) See Robertson's Hist. of America, b. i. vol. vii. p. 39.

firmation from the alleged voyage of Pytheas to the remote island of Thule, if it is to be received as a historical fact. The reality of this voyage therefore requires investigation.

The existence of the island of Thule was first announced to the Greeks by the navigator Pytheas, of Massilia, who lived about the time of Alexander the Great, and published an account of a voyage of discovery made by himself in the northwestern seas of Europe. Pytheas had doubtless passed the Pillars of Hercules, and had sailed along some of the external coast of Europe; but in relating what he professed to have seen and discovered, he, in common with other early navigators, thought himself privileged to magnify his own exploits by recounting as facts marvellous stories invented by himself, or collected from common rumour in remote places which he had visited. Both Polybius and Strabo treat him as a mere impostor, whose reports are wholly undeserving of belief.

Polybius not only argued in detail against the reality of his supposed discoveries, as we learn from the citation of Strabo;[109] but in an extant passage of his history states broadly that the whole of Northern Europe, from Narbo in Gaul to the Tanais in Scythia, was unknown in his time; and that those who pretended to speak or write on the subject were mere inventors of fables.[110] Elsewhere, too, he remarks that the Strait at the Pillars of Hercules was rarely passed by the dwellers upon the Mediterranean, owing to their want of intercourse with the nations at the extremities of Europe and Africa, and to their ignorance of the external sea.[111] Strabo declares that the account which Pytheas had given of Thule, and other places to the north of the British Isles, was manifestly a mere fabrication; 'his descriptions (Strabo adds) of countries within our knowledge are for the most part fictitious, and we need not doubt that his descriptions of remote countries are even less trustworthy.'[112] One of these fabulous stories respecting

---

[109] ii. 4, 1.   [110] iii. 38.
[111] xvi. 29.   [112] iv. 5, 5.

countries lying within the horizon of Greek knowledge has been accidentally preserved. Pytheas, it seems, stated that if any person placed iron in a rude state at the mouth of the volcano in the island of Lipari, together with some money, he found on the morrow a sword, or any other article which he wanted, in its place. The fable was founded on the Greek idea that Ætna and the neighbouring volcanoes were the workshop of Vulcan. He likewise stated that the surrounding sea was in a boiling state.([113]) A navigator who could venture to recount as true such marvels respecting an island close to Italy and Sicily, was not likely to be very veracious in his relations of his own discoveries in the far north. In another place Strabo states, that Pytheas the navigator has been convicted of extreme mendacity; and that those who have seen Britain and Ireland say nothing of Thule, reporting only the existence of small islands near Britain.([114]) Strabo is not quite consistent in his views respecting Thule; in the latter words he appears to treat its existence as a mere fiction; but in the chapter before quoted he regards it as a real place, indistinctly known on account of its remoteness: he proposes to apply to it, by conjecture, the characteristics of cold northern climates known to the Greeks by authentic observation.

The tendency of the ancient geographers to invent fables respecting remote countries is enlarged upon by Polybius;([115]) and it is satirized by Lucian in the introduction to his *Vera Historia;* where he says of Ctesias, that the things which this historian relates of India are such as he had not seen himself, nor heard from the testimony of others.([116])

([113]) Schol. Apollon. Rhod. iv. 761; Schol. Callim. Hymn. Dian. 47. The contrivance is similar to that described in Milton's Allegro:—

> Tells how the drudging goblin sweat,
> To earn his cream-bowl duly set,
> When in one night, ere glimpse of morn,
> His shadowy flail hath threshed the corn
> That ten day-labourers could not end.

([114]) i. 4, 2. ([115]) iii. 58.
([116]) Compare the passage of Aristotle cited in Athen. i. p. 6 D. Strabo

The account of Thule given by Pytheas was, that it was an island six days' sail to the north of Britain, near the Frozen Sea; in which there was neither earth, air, nor water in a separate state, but a substance compounded of the three, similar to the *pulmo marinus;* that it served as a bond of all things, and could be crossed neither on foot nor in ships; having, as it appears, neither sufficient solidity to support the weight of a man, nor sufficient fluidity to admit of navigation. He had seen the substance like the *pulmo marinus,* but related the rest on hearsay report.([117]) He also affirmed that six months of the year were light, and that six months were dark, without distinction of day and night.([118]) From this account it would appear that Pytheas did not represent himself as having visited the island of Thule. The πλεύμων θαλάττιος, or *pulmo marinus*—still called *pulmone marino* in Italian — is a mollusca which appears to abound in the Mediterranean.([119])

The account of Tacitus is, that the Roman fleet first circumnavigated Scotland in the time of Agricola; and that it discovered and subdued the Orcades, islands hitherto unknown. Thule was only just distinguished; for the fleet was ordered not to go further, and winter was approaching; but the sea was sluggish, and offered resistance to the oar; it was said not to be even movable by wind.([120]) This, the only account of Thule

---

remarks that the historians of Alexander indulge in fiction on account of the remoteness of the countries which he invaded, τὸ δὲ πόρρω δυσελεγκτον, ix. 6, 4.

([117]) Strab. i. 4, 2; ii. 4, 1; Plin. ii. 77.   ([118]) Plin. ib.

([119]) Compare Pliny, xviii. 65. An engraving of the *pulmo marinus* is given by Fuhr, De Pythea Massiliensi dissertatio (Darmstadt, 1835). This dissertation contains a complete collection of the passages of the ancients relating to Pytheas, and references to the numerous modern authors who have written concerning him. Gossellin, in his Recherches sur la Géographie des Anciens (4 vols. Paris, 1813), after a careful analysis of the supposed facts reported by Pytheas, comes to the conclusion that Pytheas had never been near the British Isles; that he collected, either at Gades or at some other port frequented by the Carthaginians, some vague notions on the northern seas and regions of Europe, and that he passed them off upon his countrymen for his own discoveries, using his astronomical knowledge for the purpose of giving them currency (vol. iv. p. 178).

([120]) Agric. 10.

which professes to rest on actual inspection, is tinged with fable, and cannot be admitted as sufficient evidence. The distant land, supposed to be Thule, was probably not more real than Croker's Mountains in the Arctic Seas, which were afterwards sailed over by Sir Edward Parry.

The notion of remote seas being impassable by ships, either from their shoals, or from the obstacles to navigation produced by the semi-fluid and muddy properties of the water, frequently recurs among the ancients, and was probably invented by sailors, as a reason why their further progress had been arrested. Thus the voyage of the Egyptian Sesostris into the Eastern seas was, according to Herodotus, arrested by shoals.[122] Plato describes the Atlantic Ocean as impermeable by vessels, on account of the depth of mud, which he attributes to the subsidence of the island of Atlantis.[123] Himilco the Carthaginian affirmed that the Northern Sea beyond the Pillars of Hercules could not be navigated: the obstacles were, the absence of wind, the thickness of the sea-weed, the shallowness of the water, and the monsters with which it was infested.[124] The muddy nature of the sea beyond the Pillars of Hercules is also mentioned by Scylax in his Periplus.[125] Tacitus himself describes the Northern Sea near the Suiones in Germany as 'sluggish, and nearly motionless.'[126] Even the scientific Aristotle believed the current fable: 'The waters beyond the Pillars of Hercules are (he says) shallow from mud, and unmoved by winds, as being in the hollow of the sea.'[127] Cleomedes declares that we cannot reach our perioeci,—those on the same parallel under the earth—because the ocean which separates us from them is innavigable, and full of monsters.[128] Alexander the Great is likewise reported to have been prevented

---

[122] Herod. ii. 102; below, p. 499.   [123] Tim. § 6.
[124] Avienus, Ora Maritima, v. 117—129; and compare v. 192, 21 362. See above, p. 455.
[125] § 1.   [126] Pigrum ac prope immotum, Germ. 45.
[127] Meteor. ii. i. § 14.   [128] i. 2.

by the fear of similar obstacles from embarking on the great Eastern Ocean.([129])

According to Pliny, Thule was an island situate beyond Britain, at the distance of one day's sail from the Frozen Sea; at the summer solstice it had no night, and at the winter solstice no day.([130]) The account of Solinus is that Thule is five days' and nights' sail from the Orcades; that at the summer solstice it has scarcely any night, at the winter solstice scarcely any day; that it abounds with fruits; that its inhabitants live in spring upon grass, like cattle; afterwards on milk, and in winter on dried fruit; they have no marriages, and their women are in common. Beyond this island the sea is motionless and frozen.([131])

The current notion respecting Thule as a remote island in the northern sea, is repeated by the later geographers, but without adding anything to the evidence of its existence. Thus Mela, who wrote under the first Cæsars, speaks of Thule as opposite the coast of the Belgians, and as celebrated by Greek and Latin poets. He states that the nights are short and semi-obscure in summer; and that at the summer solstice, the sun never sets, and there is no night.([132]) According to Dionysius Periegetes, Thule is an island beyond Britain where the sun shines both day and night.([133]) Stephanus of Byzantium says that Thule is a large island in the Hyperborean regions, where in summer the day is of 20 hours, and the night of 4, and in winter the reverse.([134]) Cleomedes, in his As-

---

(129) Stat immotum mare, et quasi deficientis in suo fine naturæ pigra moles, novæ ac terribiles figuræ, magna etiam oceano portenta, quæ profunda ista vastitas nutrit, confusa lux altâ caligine, et interceptus tenebris dies, ipsum vero grave et devium mare, et aut nulla aut ignota sidera, Seneca, Suas. i. p. 3, ed. Bipont. Lower down, p. 4, 'fœda belluarum magnitudo, et immobile profundum,' are mentioned. Curtius, ix. 4, speaks of 'caliginem ac tenebras ac perpetuam noctem profundo incubantem mari, repletum immanium belluarum gregibus fretum, immobiles undas, *in quibus emoriens natura defecerit.*' This last idea seems to be borrowed from Seneca.

(130) N. H. iv. 30.   (131) c. 22.
(132) iii. 6.   (133) v. 580—6.
(134) In v. Θούλη.

tronomical Treatise, speaks in a doubting manner of the visit of Pytheas to Thule: he says that in this island the tropic of Cancer coincides with the Arctic circle; and that when the sun is in Cancer it never sets, and there is continuous light for a month.(135) Geminus, without mentioning Thule, cites the testimony of Pytheas that he had visited places in the north where the night was only of two or three hours.(136) Achilles Tatius speaks of light lasting for 80 days in the region north of Thule.(137) Servius likewise refers to the long days of Thule, when the sun is in Cancer.(138) According to Martianus Capella, the testimony of Pytheas was that in Thule half the year was day, and half was night.(139)

Although Mela describes Thule as having been celebrated by both Greek and Latin poets, its name occurs in no extant Greek verse with the exception of the geographical poem of Dionysius. By the Latin poets it is often mentioned; but for the most part in the vague sense of a remote and unknown island, without any positive attributes savouring of geographical reality. Thus Virgil, in the elaborate flattery of Augustus which he places near the beginning of the Georgics, represents him as god of the sea; and in this character as ruling over Thule at the extremity of the ocean, and espousing a daughter of Tethys.(140) The celebrated verses of Seneca, which have been supposed to contain a prediction of the discovery of America, likewise refer to the remote position of Thule.(141) Juvenal ironically describes the progress of Greek and Roman literature towards the barbarous north, by saying that the Bri-

---

(135) De Met. i. 7, p. 47, 48, ed. Bake.
(136) Elem. Astr. c. 5, p. 13, Petav.
(137) c. 35, p. 92, Petav.
(138) Ad Georg. i. 30. Speaking of the islands near Britain, Cæsar says: "De quibus insulis nonnulli scripserunt, dies continuos triginta sub brumâ esse noctem. Nos nihil de eo percunctationibus reperiebamus, nisi certis ex aquâ mensuris breviores esse quam in continente noctes reperiebamus, B. G. v. 13.
(139) vi. § 595, 666. ed. Kopp.   (140) Georg. i. 29.
(141) Med. 374.

tons had learnt eloquence from the Gauls; and that even Thule thinks of hiring a rhetorician.(142) Similar passages occur in Statius, who speaks of Thule as a distant island, enveloped in darkness, and lying beyond the course of the sun.(143)

The imaginary character of Thule was preserved by Antonius Diogenes, the author of τὰ ὑπὲρ Θούλην ἄπιστα, 'The Marvels of the Parts north of Thule,' in 24 books, an abridgment of which work is extant in the Bibliotheca of Photius.(144) This romance, which Photius declares to have been highly amusing, and full of wonderful stories related in a plausible manner, belonged to the class of *Voyages Imaginaires*. Dinias and his son Demochares were described in it as travelling by the Black Sea and the Caspian to the Rhipæan mountains and the river Tanais, until, on account of the severe cold, they made for the Scythian Ocean. Here they wandered a long time, and first navigated the Eastern Sea, and reached the rising of the sun; afterwards they visited the island of Thule, which they used as a station during their peregrinations in the north. At Thule, Dinias meets a noble Tyrian woman named Dercyllis, with whom he falls in love; her adventures, and those of other persons, were related at length, so that the first 23 books, says Photius, contained little or nothing about Thule. In the 24th book was an account of the visit of Dinias, with two companions, named Carmanes and Meniscus, to the regions north of Thule, where they find plenty of marvels, and at last succeed in reaching the moon. This part of the novel seems to have resembled the *Vera Historia* of Lucian, and some of the modern fictions imitated from it. Diogenes supposes the story to have been written on tablets of cypress wood, enclosed in a box, which was discovered in a subterranean deposit near the city of Tyre by Alexander the Great.(145)

---

(142) xv. 110.
(143) Sylv. iii. 5, 19, iv. 4, 62, v. i. 90, v. 2, 54.
(144) Cod. 166. See above, p. 270, n. 87.
(145) Concerning this romance, see Dunlop's History of Fiction,

Photius declares himself ignorant of the age of this Diogenes, but conjectures him to have been not much later than the age of Alexander. It is more probable, as modern critics have supposed, that he was of far more recent date, and that he wrote in the second or third century after Christ.(146) The name of Thule originated with Pytheas, and it does not seem to have become current in Greek and Latin literature till the Augustan age.

Some of the Roman poets, seeking to find a local habitation for the name of Thule, identify it with Britain. Thus Silius describes the campaign of Vespasian, and Claudian the exploits of Theodosius the elder, in Britain, as having occurred in Thule.(147) Probus, in his Commentary on Virgil, designates Thule as the furthest of the Orcades.(148) According to Servius, Thule is an island in the ocean, to the north-west, beyond Britain, near the Orcades and Hibernia.(149) Orosius, however, who wrote in the 5th century, still uses it in the ancient indeterminate acceptation. He describes Thule as separated by an infinite distance from the Orcades, lying towards the north-

---

vol. i. p. 8. A similar story respecting the discovery of his work is told by the author of Dictys.

(146) Meiners, Gesch. der Wiss. vol. i. p. 253, places him in the first half of the third century after Christ. Servius, ad Georg. i. 30, says that various marvels are related of Thule, both by Greek and Latin writers, by Ctesias and Diogenes among the former, and by Sammonicus among the latter. The reference to Ctesias cannot be explained; Sammonicus is Sammonicus Serenus, who was murdered by the command of Caracalla in 212 A.D.; see Dr. Smith's Dict. of Anc. Biogr. in *Serenus*. Suidas, in Θούλις states that a great Egyptian king, Thulis, gave his name to the island of Thule. The origin of this idea does not appear.

(147) Silius, iii. 597. In the following passage Silius likewise uses Thule as a synonym for Britain. The native custom of colouring the body with blue, and the war chariots of the Britons, are alluded to:—

Cœrulus haud aliter, quum dimicat, incola Thules
Agmina falcifero circumvenit arcta covino.—xvii. 416.

Claudian identifies Thule with Britain in De Tert. Cons. Hon. 51—3; De Quart. Cons. Hon. 32. In the following passages of the same poet it bears the more general meaning of a remote island in the Northern Ocean; in Rufin. ii. 239—41; in Secund. Cons. Stilich. 156; De Bell. Get. 201—4.

(148) Ad Georg. i. 30, vol. ii. p. 358, ed. Lion.

(149) Ad Georg. i. 30.

west, in the middle of the ocean; and as hardly known even to a few persons.(¹⁵⁰)

The campaigns of Cæsar opened Gaul and Britain to the Romans, and after a time their knowledge extended to northern Germany and to the Scandinavian peninsula, which, however, they supposed to be a group of islands. The German Ocean was first navigated by Drusus, who in 12 B.C. reached the sea by the Rhine, and landed on the coast of Friesland.(¹⁵¹) Sixteen years afterwards (4 A.D.) Tiberius sent a flotilla down the Rhine, with orders to follow the coast eastward, and to sail up the Elbe, until he himself effected a junction by his land forces with his naval armament. This junction was successfully accomplished, and is celebrated with merited praises by Velleius, who speaks of this fleet sailing to the Elbe through a sea previously unknown and unheard of.(¹⁵²)

Strabo declares that all the region beyond the Elbe, adjoining the ocean, was unknown in his time. 'No one (he adds) is recorded to have navigated along this coast eastward as far as the mouths of the Caspian Sea; the Romans have not penetrated beyond the Elbe; and no one has made the journey by land.'(¹⁵³) It will be observed that Strabo seems quite ignorant of any voyages of Phœnician ships to the west of the Elbe.

The original belief was, that the ocean flowed from Scythia, round the north of Germany and Gaul, to Iberia and the Pillars of Hercules; and that in this northern ocean there were many large islands. Pliny mentions that islands of vast size, lying off the coast of Germany, had been recently discovered in his

---

(150) i. 2. On the supposed situation of Thule, see Humboldt, Examen de la Géogr. du Nouv. Cont. tom. ii. p. 214.

(151) Tac. Germ. 34; Merivale's Romans under the Empire, vol. iv. p. 229. [Classis Romana] ab ostio Rheni ad solis orientis regionem usque ad [orbis extrem]a navigavit, quo neque terra neque mari quisquam Romanorum ante id tempus adît, Tab. Ancyr.

(152) ii. 106; Merivale, ib. p. 309.

(153) vii. 2, 4.

time.(154) Xenophon of Lampsacus, a geographer—whose date is unknown, but who probably lived about the Augustan age—stated that at a distance of three days' sail from the shore of Scythia was an island of enormous size, called Baltia.(155) Mela speaks of the Codanus Sinus—the Cattegat, or southern part of the Baltic—as a large bay beyond the Albis (Elbe), full of great and small islands.(156) The largest island in the bay, inhabited by the Teutoni, he calls Codanonia.(157) The peninsula of Jutland was likewise known to the Romans at the same period, and was named the Cimbric Chersonese.(158)

One of the great islands in this part of the Northern Ocean was called Scandia, or Scandinavia. According to Pliny, Scandinavia was the most celebrated island in the Codanus Sinus; its size was unknown. The portion of it which was known was inhabited by the Hilleviones, a nation containing 500 pagi, who regarded it as another quarter of the world.(159) Another account preserved by Pliny describes Scandia as an island beyond Britain.(160) Agathemerus mentions Scandia as a large island near the Cimbric Chersonese, extending to the north of Germany; and he couples it with the island of Thule.(161) According to Ptolemy, there were to the east of the Cimbric Chersonese four islands called Scandia, viz. three small ones, and a large one, furthest to the east near the mouths of the river Vistula.(162) Between the times of Strabo and Ptolemy, therefore, discovery had advanced from the Elbe to the Vistula. It may be added that the island of Scanzia is mentioned by Jornandes, who lived in the sixth century.(163)

Another writer, who also lived in the sixth century, having occasion to mention the island of Scandinavia, gives it the ap-

---

(154) Nam et a Germaniâ immensas insulas non pridem compertas cognitum habeo, N. H. ii. 112.
(155) Plin. iv. 27. (156) iii. 3. (157) iii. 6.
(158) Strab. vii. 2, § 1; Plin. iv. 27; Compare Zeuss, die Deutschen, p. 144.
(159) Ib. (160) iv. 30. (161) De Georg. ii. 4.
(162) ii. 11, § 33, 34. Compare viii. 6, § 4.
(163) De Reb. Get. c. 3.

pellation of Thule. Procopius, in his 'History of the Gothic War,' describes the course of the Heruli across central Europe. He states that, defeated by the Lombards, they first crossed the country of the Sclaveni (near the Danube), and afterwards that of the Varni (Saxony); that they next overran the Danes, from whose country they reached the ocean; and having embarked in ships, they sailed to the island of Thule, where they remained.[164]

In this passage, Procopius, wishing to designate the great island which (as he believed) lay to the north of Germany, applied to it the vague appellation of Thule, familiar indeed to the Greeks, but never hitherto used as the name of any real country. He then proceeds to describe this island:—

'Thule,' he says, 'is an island of great size, more than ten times as large as Britain, and lies at a distance from it, to the north. Most of the land is barren, but there are thirteen large nations in the cultivated regions, all governed by kings. For forty days about the summer solstice the sun does not set, and for the same time at the winter solstice it does not rise. The latter period is passed by the inhabitants in dejection of spirits, as they are unable to communicate with each other. Although (adds Procopius) I much wished to visit this island, and to see these phenomena with my own eyes, I have never been able to accomplish my desire. Nevertheless, I have heard a credible account of them from natives of the country who have travelled to these parts. During the period when the sun never sets, they reckon the days by the motion of the sun round the horizon. During the period when the sun never rises, they reckon the days by the moon. The last five days of the dark period are celebrated by the Thulitæ as a great festival. These islanders are perpetually haunted with a fear that the sun should on some occasion fail to return, although the same phenomenon recurs every year.'

---

(164) On the course of this migration, see Buat, Hist. Anc. des Peuples de l'Europe, tom. ix. p. 388; Zeuss, ib. p. 481.

'The Scrithifini, one of the nations of Thule, are in a savage state, wearing no clothes or shoes, not drinking wine, or eating any vegetable product. They never cultivate the ground, but both men and women follow the chase. They live on the animals thus killed, and use the skins of beasts as clothes. Their infants are nourished not with milk, but with the marrow of wild animals.'

'The remaining Thulitæ scarcely differ from other men. They worship a variety of gods in heaven, earth, and sea, and particularly in springs and rivers, and they sacrifice human victims, killing them with frightful tortures. The largest nation is the Gauti, to whom the Heruli came.'(165)

The Scrithifini mentioned in this passage are more correctly called Skridefinni by other writers. They were sometimes called simply Fins; they inhabited part of Sweden and Norway.(166) The Gauti are a nation of Goths, dwelling in this region, whose name is preserved in the island of Gothland. According to Ptolemy,(167) the Goutæ (Γοῦται) occupied the southern part of Scandia: this nation is doubtless identical with the Gauti of Procopius, and this coincidence affords an additional proof that Thule is used by him as synonymous with Scandia.(168) The mention of the Scrithifini, who are expressly placed by other writers in the Scandinavian peninsula, likewise indicates the sense which he assigns to the old fabulous name of Thule.

Dicuil, an Irish monk, who composed his geographical treatise *De Mensurâ Orbis* in the year 825, describes the island of Thule upon the authority of certain *clerks*, who had visited it 30 years previously, and had remained in the island from February to August. According to their account, the nights were of extreme shortness about the time of the summer solstice.(169) Letronne is of opinion that Iceland must be in-

---

(165) Bell. Goth. ii. 15.   (166) Zeuss, ib. p. 684.
(167) Ubi sup.   (168) Zeuss, ib. p. 158, 511.
(169) 'Trigesimus nunc annus est a quo nuntiaverunt mihi clerici, qui, a kalendis Februarii usque kalendas Augusti, in illâ insulâ manserunt, quod,

tended, and he considers this passage to prove that Iceland was known to the Irish at the end of the eighth century.(170)

The preceding investigation respecting Thule, shows that the name, as first promulgated by the impostor Pytheas, denoted a fictitious, but not properly a fabulous or mythical island; that is to say, although the story of this island was a fabrication, yet he desired to make it pass for truth. The poets employed it, for the most part, in a general and almost abstract sense, for a remote unknown island in the Northern Sea; while the novelist Diogenes gave it a purely fabulous character, and made it as unreal as Lilliput and Brobdignag in Gulliver's Travels. Silius indeed and Claudian use Thule as synonymous with one of the Britannic islands; Probus makes it one of the Orcades; Procopius identifies it with Scandinavia, and Dicuil uses it as the designation of Iceland. The Roman fleet, in the time of Agricola, believed that they saw Thule, in the dim distance, beyond the northern extremity of Scotland, and the geographical writers attempt to assign it some fixed locality in the northern seas. They particularly connect with it the long night of the polar winter, and the long day of the polar summer,—phenomena of which they had obtained authentic information, and which were explained by the theories of the astronomers. But the name Thule never acquired any fixed geographical signification; it never came to be used, either by the natives or by the geographers, as the appellation of any real island.

Pytheas affirmed that in returning from his great northern voyage, in which he first obtained accounts of the remote island of Thule, he had sailed along the entire coast of the ocean between Gadeira and the Tanais;(171) that is from Cadiz

---

non solum in æstivo solstitio, sed in diebus circa illud in vespertinâ horâ, occidens sol abscondit se quasi trans parvulum tumulum; ita ut nihil tenebrarum in minimo spatio ipso fiat; sed quicquid homo operari voluerit, vel pediculos de camisiâ abstrahere, tanquam in præsentiâ solis potest; et si in altitudine montium ejus fuissent, forsitan nunquam sol absconderetur ab illis,' De Mens. Orb. Terr. c. 7, § 2.

(170) Recherches sur Dicuil (Paris, 1814), p. 133—147.
(171) Strab. ii. 4. 1.

round Spain, Gaul, Germany, and Scythia, to the river Don, which was considered by the ancients as the boundary of Europe and Asia. This statement furnishes an additional proof of the mendacity of Pytheas, because it is founded on the belief, received in his time, that Europe did not project far to the north, and that the ocean swept along its shores to the north of Scythia and India. Of this great Northern Ocean, the Caspian was supposed to be a gulf, connected with it by a narrow strait. The erroneous idea that the Caspian Sea was a gulf of the ocean was not dispelled by the expedition of Alexander.[172] Arrian represents Alexander as assuring his soldiers that, if they will continue their march eastward, they will discover the great Eastern Sea to be continuous with the Caspian.[173] Strabo repeats this belief.[174] Both Mela and Pliny state that the Caspian is connected with the Northern Ocean by a long and narrow channel.[175] Dionysius Periegetes, following the ancient belief, makes it a gulf of the Northern Ocean, opposite the Persian Gulf, and one of the four great gulfs of the external sea.[176]

Pliny and Mela mention in proof of an external sea connecting the northern shores of Germany with India, that Q. Metellus Celer, when proconsul of Cisalpine Gaul, in 62 B.C. received as a present from the king of the Suevi some Indians, who were said to have sailed from India for purposes of trade, and to have been carried by adverse winds to Germany.[177] The Suevi were a German tribe who inhabited the country on the eastern bank of the Rhine. The Indians, of whom this fable was narrated, by whatever road they reached Germany, must have been sent to Metellus across the Alps. Forbiger

---

[172] Plut. Alex. 44; Strab. ix. 6, 1.
[173] v. 26. Compare vii. 16.   [174] xi. 6, § 1.
[175] Mela, iii. 5; Plin. N. H. vi. 15.
[176] v. 47. These four gulfs, or internal seas, were the Caspian, the Persian Gulf, the Red Sea, and the Mediterranean; see Plut. Alex. 44; Strab. ii. 5, § 18. Macrobius, in Somn. Scip. ii. 9, § 7, enumerates these four gulfs, but says that the connexion of the Caspian with the ocean is denied by some.
[177] Plin. ii. 67; Mela, iii. 5; Cic. ad Div. v. 1, 2.

conjectures that these supposed Indians were inhabitants of Labrador or Greenland, who were mistaken for Indians on account of their dark-coloured skin.([178]) It should be observed that the war-elephants of the Greeks came from India,([179]) and were driven by Indians. Hence Ἰνδὸς was the general name of an elephant-driver.([180])

In the description of Dionysius Periegetes,([181]) a ship which has left the Britannic Islands and Thule, traverses the Scythian Ocean, and thus gains the Eastern Sea, where the Golden Island adjoins the rising of the sun; it there makes a turn, and reaches the island of Taprobane, or Ceylon. By the 'Golden Island,' or Chersonese, the peninsula of Malacca is meant.

§ 6 If the Phœnicians, the Carthaginians, and after them the Romans, had in succession, through several centuries, maintained a direct trade in tin with Britain by the Straits of Gibraltar, the nations of the Mediterranean would doubtless from an early date have become acquainted with that island. On the other hand, if the tin was carried across Gaul to Massilia, and imported thence into Greece and Italy, it would be natural that the Greeks and Italians should have remained in ignorance of the remote island from which it was brought for so many miles by barbarian merchants.

Now the ancient authors are unanimous in declaring that the Britannic Islands remained practically unknown until the invasion of Cæsar. It was stated by Polybius, in a lost portion of his history, that the inhabitants of Massilia, Narbo, and Corbilo on the Loire, being interrogated by Scipio Africanus (who died in 129 B.C.) were unable to give him any information respecting Britain.([182]) With regard to Massilia and Narbo, towns on the Mediterranean, this ignorance is intelligible; though the Britannic tin trade is said to have been carried on

---

(178) Handbuch der Alten Geographie, vol. ii. p. 4.
(179) Aristot. H. A. ix. 2.
(180) See Polyb. i. 40, iii. 46, xi. 1.
(181) v. 587—93.
(182) Ap. Strab. iv. 2, § 1.

through Massilia; but Corbilo, being situated close to the mouth of the Loire, would seem to have lain within a sphere to which a knowledge of Britain would have penetrated, if the ordinary intercourse between that island and Gaul had extended beyond the Channel. Consistently with this statement, Cæsar informs us that the Gauls in his time knew scarcely anything in detail of Britain. It was only, he says, visited by traders, who did not go beyond the coast which lay opposite to their own country.[183] It may be inferred from this passage that the Gauls, who carried on the cross-channel trade between Gaul and Britain, were the inhabitants of the northern coast.

It is remarked by Dio Cassius, that the very existence of Britain was unknown to the Greeks and Romans in early times, and that afterwards they were ignorant whether it was an island or not; various opinions on this point, founded on mere probability, and not on actual examination of the locality, had been promulgated.[184] The insular character of Britain was first demonstratively established by the fleet of Agricola.[185] Livy (as we learn from Jornandes)[186] declared that in his time Britain had not been circumnavigated.

Again, Dio speaks of the pride felt by Cæsar himself, and by his countrymen, at his having actually landed in a country previously unknown, even by report.[187]

Plutarch describes Cæsar's audacity in venturing to cross the Western Ocean with an army, and to sail against Britain over the Atlantic Sea. He attacked an island whose very existence was in question, and advanced the Roman dominion beyond the limits of the inhabited world.[188]

According to Eumenius, in his Panegyric of Constantius, the expression of Britain being another world was used by Cæsar in reporting to Rome his first invasion of Britain.[189] It is

---

(183) B. G. iv. 20.   (184) xxxix. 50.
(185) Tac. Agric. 10.   (186) De Reb. Get. c. 2.
(187) Ib. c. 53. Compare likewise the allusions to the remoteness of Britain, in Antony's speech over the body of Cæsar, ib. xliv. 42, 43, 49.
(188) Cæsar, c. 23.   (189) c. 11.

likewise applied by Velleius and Florus to this achievement of Cæsar.(¹⁹⁰)

The mere name of Britain seems indeed to have been known to the Greeks since the time of Pytheas; who affirmed that he had landed on the island;(¹⁹¹) and it is mentioned by Polybius and Posidonius in connexion with the tin trade. Lucretius, whose poem was published before Cæsar's Commentaries, uses Britain as an example of a country in the extreme north.(¹⁹²)

The boundaries of the known world, according to the præ-Alexandrine Greeks, were the Phasis (or the Pontus) to the East, and the Columns of Hercules to the West.(¹⁹³) Pindar says that everything to the west of Gadeira is inaccessible.(¹⁹⁴) After the age of Alexander, the eastern boundary receded; but the western boundary remained the same.(¹⁹⁵)

But though the name of Britain was known to the Greeks of the post-Alexandrine period, the earliest mention of Ireland

---

(190) Vell. ii. 46; Flor. iii. 18, § 16. Compare Drumann, Geschichte Roms, vol. iii. p. 293. Servius, in his Commentary on the verse in Virgil's First Eclogue, 'Et penitus toto divisos orbe Britannos' (v. 67), states that Britain is an island lying at a distance in the Northern Ocean, and that it is called by the poets, another world.

(191) According to Polybius, he stated that he had travelled over the whole island, ὅλην τὴν Βρεττανικὴν ἐμβαδὸν ἐπελθεῖν, Strab. ii. 4, 1.

(192) vi. 1104. Dr. Lappenberg, in his History of England under the Anglo-Saxon Kings (Thorpe's transl. vol. i. p. 3) commits a serious mistake when he states that 'British timber was employed by Archimedes for the mast of the largest ship of war which he had caused to be built at Syracuse.' He has been misled by the name Βρεττανίας, in Athen. v. p. 208 E, where Camden in his Britannia, has restored Βρεττιανῆς, and Casaubon Βρεττίας. It is not very likely that Hiero should have sent to the remote and almost unknown island of Britain in search of timber, when it could be procured in abundance and perfection on the neighbouring coast of Italy; as Casaubon, in his animadversions to Athenæus, has pointed out. A similar confusion of Bruttia and Britain likewise occurs in Diod. xxi. 21, ed. Bekker, where the movements of Hannibal are described. Compare Diefenbach's Celtica, vol. iii. p. 68.

(193) Herod. viii. 132; Eurip. Hipp. 3, 1350; Plat. Phædon. § 133, p. 109.

(194) Pindar, Ol. iii. 79; Nem. iii. 36, iv. 112; Isthm. iv. 95. Compare Ukert, i. 2, p. 206.

(195) Hence Juvenal says:—

Omnibus in terris, quæ sunt a Gadibus usque
Auroram, et Gangem.—x. 1.

by a writer whose age is ascertained occurs in Cæsar. In his History of the Gallic War, he says that Hibernia lies to the west of Britain, being estimated at less than half its size: the distance between the two islands is the same as that between Britain and Gaul. He likewise states that the island of Mona is situated midway between Hibernia and Britain, by which he means Anglesey or Man.([196])

Strabo states that the remotest island to the north of Celtica (or Gaul) is Ierne, beyond Britain, occupied by savages, and barely habitable from cold.([197]) In another place he says that Ierne lies to the north of Britain, being long in proportion to its width. Its inhabitants are more savage than those of Britain; they are cannibals, and they likewise feed on grass; they eat the bodies of their fathers after death; and in their relations with women they set at nought the rules observed by civilized nations. These latter accounts, however, he adds, do not rest on authentic testimony.([198])

Diodorus remarks that the tribes dwelling in the north, in the vicinity of Scythia, are wholly savage and uncivilized, and that some of them are said to be cannibals, such as the Britons who inhabit the country called Iris.([199])

The account of Mela is, that Iverna lies beyond Britain; that its climate is unfitted for ripening grain, but so abundant in grass, not only of rapid growth, but also of sweet taste, that cattle eat to satiety in a small part of the day, and if they are not driven from the pasture, burst from the excess of food. Its inhabitants are uncivilized; they are ignorant of every virtue, and remarkably free from humanity.([200])

Pliny states that the island of Hibernia was situated beyond Britain, at a distance of thirty miles from the coast of the Silures.([201])

---

(196) v. 13.
(197) ii. 1, § 13, ii. 5, § 8.
(198) iv. 5, § 4.
(199) v. 32.
(200) The words 'pene par spatio, sed utriusque æquali tractu litorum oblonga' are corrupt. The meaning seems to be similar to that expressed in Strabo, that the length is equal to that of Britain, but not the width.
(201) iv. 30.

Tacitus, in his Life of Agricola,(202) describes Hibernia as lying between Britain and Iberia, and as exceeding the islands of the Mediterranean in size. He says that, in its soil and climate, and in the character and civilization of the natives, it differs little from Britain. Its harbours and approaches were known by the reports of traders. Tacitus adds, that he had often heard Agricola say that Hibernia might be subdued with one legion and a few auxiliaries. In the Annals, the same historian mentions the proprætor P. Ostorius, in 50 B.C., approaching the sea which divided Britain and Hibernia.(203)

The poet Juvenal, who wrote about 100 A.D., speaks of Ireland in connexion with Britain and the Orkneys:—

> ' * * * * Arma quidem ultra
> Littora Juvernæ promovimus, et modo captas
> Orcadas, ac minimâ contentos nocte Britannos.'(204)

The recent capture of the Orkneys alludes to the expedition of the Roman fleet round Britain in the time of Agricola.(205)

In the Geography of Ptolemy, who lived about the middle of the second century, the two Britannic Islands, Albion and Ivernia, are described at length. A large number of towns, rivers, and promontories belonging to the latter island are specified by name.(206)

The account of Ireland given by the geographer Solinus, who is supposed to have lived about the middle of the third century after Christ, is copious and detailed.

'Of the islands surrounding Britain (he says), Hibernia is the nearest to it in size. The manners of its inhabitants are rude and savage. Its pastures are so excellent, that unless the cattle are sometimes driven from them, they are in danger of dying from repletion. The island has no snakes and few birds. The people are inhospitable and warlike. When they are victo-

---

(202) c. 24.   (203) xii. 32.
(204) ii. 159—161.   (205) Tac. Agric. 10.
(206) ii. 2. The same form of this name, Ἰουερνία, recurs in the Periplus of Marcianus, c. 42, and in Stephanus of Byzantium. The form used by Juvenal, *Juverna*, is similar.

rious, they both drink and smear their faces with the blood of the enemy. They know no distinction between right and wrong. When a woman has produced a male child, she places its first food on the point of her husband's sword, and thus introduces it gently into the mouth of the infant. Prayers are offered up, on behalf of the family, that he may meet his death in war. Those who study ornament, decorate the hilts of their swords with the teeth of marine animals; the chief glory of the men is in the brilliancy of their arms. They do not possess bees; and if a pebble or some earth, brought from Hibernia, is thrown into a hive, the bees will desert it. The sea between Hibernia and Britain is disturbed and stormy during the whole year, and can only be navigated for a few days. The boats are of wicker covered with the hides of oxen; whatever time the passage may occupy, the mariners abstain from food while they are at sea. The width of the Strait is estimated at 120 miles.'[207]

For 120 Salmasius corrects twenty miles, comparing Pliny, who states the distance at thirty miles.

In the Orphic Argonautics, the speaking ship warns the heroes to avoid the Iernian islands, and to steer for the Sacred Promontory (on the Lusitanian coast), lest she should be carried out into the Atlantic Sea; and Ancæus, the pilot, obeys the injunction.[208] This poem may be placed between the second and fourth centuries after Christ.[209] There is no mention of Ierne or of the Britannic Islands in the Argonautics of Apollonius Rhodius.

The Ora Maritima of Avienus (who appears to have lived at the end of the fourth century) describes the Sacred Island inhabited by the Hibernians, near the island of the Albiones, as separated by two days' sail from the Œstrymnian islands, off the coast of Spain, where tin and lead are found.[210]

The belief of the severe cold of Ireland entertained by some

---

[207] c. 22.   [208] v. 1170—1190.
[209] See Bernhardy, Grundriss der Griechischen Litteratur, vol. ii. p. 267—272; Hermann, Orph. p. 798.
[210] v. 94—112.

of the ancients, was founded on the vague idea of its position in the extreme north. The accounts of the savage manners of its inhabitants are, doubtless, strongly tinctured with fable; but it is probable that, having less intercourse with the continent than the inhabitants of Southern Britain, they were less civilized in their customs. In some of these passages a knowledge of the rich pasturage of the Emerald Isle, which must have been derived from the reports of eye-witnesses, is perceptible.

Such are the most ancient testimonies respecting Ireland which occur in the works of writers whose age is ascertained. One testimony, however, which, by Mannert, Dr. Latham, and other modern writers, has been considered as containing the earliest mention of this island, remains to be noticed.

The author of the Aristotelic Treatise concerning the Universe (περὶ Κόσμου), adopting the received notion of the Greeks, which descended from the Homeric age, describes the inhabited world as surrounded by the ocean. He first traces the ocean from the Pillars of Hercules along the Mediterranean to the Pontus and the Palus Mæotis; and he then follows its eastern course along the shores of Asia.

'In one direction (he says) it forms the Indian and Persian gulfs, with which the Red Sea is continuous; in the other, it passes through a long and narrow channel, and widens into the Caspian Sea. Further on it encircles the space beyond the Palus Mæotis. Then stretching its course above Scythia and Celtica, it encompasses the inhabited world, in the direction of the Gallic Gulf and the Pillars of Hercules. In this part of the ocean there are two great islands, larger than any in the Mediterranean, called the Britannic Islands, Albion and Ierne, situated beyond the country of the Celts. Equal in size to these are Taprobane, on the further side of India, turned obliquely to the mainland, and the island named Phebol, lying near the Arabian Gulf. Many small islands likewise are placed around this continent, near the Britannic Islands and Iberia.'[211]

---

(211) c. 3, p. 393, ed. Bekker.

Taprobane became known to the Greeks through the expedition of Alexander; but what this writer can mean by *Phebol*, an island near the Arabian Gulf, as large as Britain, is an enigma. The passage of the Aristotelic Treatise is repeated in Latin by Apuleius de Mundo,(²¹²) where the name *Phebol* reappears. It is likewise cited in Stobæus,(²¹³) where, however, Φοβέα καὶ Εὔβοια is read for Φέβολ, being an attempt at emendation. It has been conjectured that Socotra, or Madagascar, is signified by this unknown name; but Salmasius(²¹⁴) is doubtless right in treating it as corrupt, and in substituting for it Ψεβώ, the name of a lake and an island beyond Meroe in Upper Egypt.(²¹⁵) This lake is identified with Lake Tsana in Abyssinia, which is stated to contain eleven islands. The Psebæan mountains and promontory on the Æthiopian side of the Red Sea, are mentioned by Diodorus.(²¹⁶) Theophrastus(²¹⁷) mentions the district of Ψεφώ, in conjunction with Syene; here Ψεβώ has been rightly corrected. No corruptions are so common in manuscripts as those of proper names, particularly if the name is not well known. The form of the name in Strabo is Ψεβώα, which was probably that used by the author of the treatise *de Mundo*, and ΨΕΒΩΑ might have been easily corrupted into ΦΕΒΟΛ. What inaccurate reports could have induced this writer to believe that the island in Lake Pseboa was as large as Britain, cannot now be ascertained. The ten largest islands and peninsulas, according to the received belief of his time, are enumerated in their order by Ptolemy; he places Taprobane first, and Albion second; but he says nothing of Pseboa.(²¹⁸)

---

(212) P. 716, ed. Oudendorp.
(213) Ecl. Phys. i. 34, 2, vol. i. p. 258, Gaisford.
(214) Exerc. Plin. ad Solin. c. 53, p. 782.
(215) See Strab. xvii. 2, § 3: Steph. Byz. in v.
(216) iii. 41. The Psebæan mountains in this region are also mentioned by Agatharchides, de Mari Rubro, c. 84; Ap. Geogr. Gr. Min. vol. i. p. 174, ed. C. Müller.
(217) De Lapid. § 33, vol. i. p. 695, iv. p. 565, ed. Schneider.
(218) Geogr. vii. 5, § 11. Measurements of the largest islands are given in Agathem. c. 5, and in a fragment of a geographer in Müller's Geogr. Gr. Min. vol. ii. p. 509.

It has been already mentioned that the Aristotelic treatise *de Mundo* belongs to the last century B.C.(²¹⁹) The mention of Ierne shows that it is posterior to the expedition of Cæsar; and the form Ἄλβιον seems to betray a Latin derivation.

§ 7 The idea of a sacred island, rising amidst the waves, removed from all contentions and wars, the abode of quiet and purity, the secure refuge of men buffeted by the storms of the world, seems naturally to suggest itself to the human mind. By an easy transition this residence of a pious and holy race becomes an Elysian field; it is endowed with perpetual spring; the ground produces its fruits without labour; there are no serpents or wild beasts within its hallowed precinct; its inhabitants are no longer a sacred colony of living men, but the souls of the departed, translated to a region of bliss.

The notion of holy islands first occurs in Hesiod. He describes the race of heroes, who form the fourth age of mankind, as residing after death apart from the world, in the Islands of the Blest, near the ocean, free from care, and enjoying three harvests in the year.(²²⁰) Pindar, in like manner, conceives the Islands of the Blest as the abodes of the just and virtuous after death.(²²¹) On the other hand, Horace supposes his countrymen to seek an escape from the horrors of the civil war, in the Happy Islands, where peace and plenty will be their permanent lot.(²²²)

The Canary Islands became known to the Romans after the war of Sertorius, and were identified with the happy region at the extremity of the world, described in the Odyssey.(²²³) Mela accordingly describes the Fortunate Islands as really existing in the Atlantic;(²²⁴) and Strabo identifies them with some islands not far from the promontory of Maurusia, opposite Cadiz;(²²⁵) while Philostratus places them at the extremity of Libya, near the uninhabited promontory.(²²⁶)

---

(219) Above, p. 158, 218.   (220) Op. 166—171.
(221) Olymp. ii. 68.   (222) Epod. xvi.
(223) Plut. Sert. 8; Plin. vi. 37; Dr. Smith's Dict. of Geogr. art. *Fortunatæ Insulæ*.
(224) iii. 10.   (225) iii. 2, § 13.   (226) Vit. Apoll. v. 3.

The marvellous islands in the Odyssey, the island of Ogygia, inhabited by Calypso, Ææa, the island of Circe, and the Æolian island, furnish other examples of the tendency to invest islands with supernatural attributes.

There was a constant disposition in the Greek mind to realize the ideals of their ancient mythology and poetry, and therefore to identify imaginary with actually existing places. But as the horizon of their geographical knowledge extended, and as positive science expelled fiction, the province of fable receded, and the marvels of fancy were banished into distant regions of the earth, unknown by name to the generations with which the stories originated.[227]

In remote antiquity, the countries and waters of Northern Europe were wholly unknown to the dwellers upon the shores of the Mediterranean; and even in later times, after Cæsar had invaded Britain, their acquaintance with these regions was limited. When, therefore, the western parts of the Mediterranean had been explored, and became familiar to the Greeks, the north of Europe afforded a convenient field for supernatural and marvellous stories.

The first trace of this tendency occurs in the account of Hecatæus of Abdera, a contemporary of Alexander the Great, who wrote a work concerning the Hyperboreans. This fabulous nation were originally conceived to be under the immediate care of Apollo, and to pass their time in uninterrupted enjoyment, inhabiting a region beyond the origin of the north wind, and therefore exempt from the cold of winter.[228] By Hecatæus they were represented as dwelling in an island, as large as Sicily, in the ocean opposite Celtica, which was endowed with a mild climate, and yielded two harvests in the year.[229]

It was, however, in the writers of the first five centuries after Christ that the transposition of imaginary islands and countries

---

(227) See Ukert, i. 2, p. 345.
(228) See Müller, Dor. b. ii. c. 4.
(229) Fragm. Hist. Gr. vol. ii. p. 286.

to this part of the world chiefly occurs. Plutarch, in his treatise
'De Facie in Orbe Lunæ,'(230) describes the Homeric Island of
Ogygia as situated five days' sail to the west of Britain, together
with three other islands in the same direction, at equal distances
from each other. In one of these islands, he proceeds to say,
Saturn is related to be imprisoned by Jupiter; whence the
neighbouring sea is called the Cronian, or Saturnian. The
great external continent, lying beyond the circumfluous ocean,
is at a distance of 5000 stadia (or 625 miles) from Ogygia, which
is the farthest from it of the four islands. The intermediate sea
is difficult to navigate, on account of its muddy properties;
whence it has been believed to be frozen. On the shore of the
external continent there are Greeks, dwelling round a gulf equal
in size to the Palus Mæotis, the mouth of which lies directly
opposite to the mouth of the Caspian Sea. The inhabitants of
this continent consider our earth as an island, because it is surrounded by the ocean. They believe that this Hellenic population is composed of the original subjects of Saturn, subsequently
reinforced by some of the companions of Hercules. Hence they
pay the principal honours to Saturn, and after him to Hercules.
When the planet Saturn is in the sign of Taurus—a coincidence
which occurs every thirtieth year—they send out a body of men,
selected by lot, to seek their fortunes across the sea. A band
of this description, having escaped from the dangers of the sea,
landed on one of the above-mentioned islands, which are inhabited by Greeks, the descendants of former colonists from
the same continent: after a residence of ninety days, during
which they were entertained with honour and hospitality, and
regarded and called sacred, they sailed onwards in their course.
It is permitted to dwellers in these islands to return to their
original country after a series of years; but the majority prefer
to remain, either from habit or because the climate is mild and
the soil produces everything in abundance without toil. They

---

(230) c. 26. In this passage, Boeckh alters ὧν ἐν μίᾳ into ἐν δὲ τῇ
'Ωγυγίᾳ, or ὧν ἐν τῇ πρώτῃ, an alteration which does not seem to be necessary.

pass their time in sacrifices and choral solemnities, or in literature and philosophy. To some the deity has appeared in a visible form, not in dreams or signs; and, addressing them as friends, has prevented their departure from the island. Saturn himself is confined in a deep cavern, sleeping on a gold-coloured rock, sleep being contrived by Jupiter as his chain. From the top of the rock birds fly down, and bring him ambrosia, which fills the whole island with fragrance. There are likewise genii, the companions of Saturn when he reigned over men and gods, who minister to him; these divine beings are endued with prophetic powers, and their most important predictions are communicated to Jupiter as the dreams of Saturn, and they become the foreknowledge of Jupiter.

In the treatise, 'De Defectu Oraculorum,'(231) by the same writer, one of the interlocutors says that among the many desert isles near Britain some are believed to be the seats of genii and heroes. That he had himself, being on a mission from the Emperor, sailed to one which was next to them, from motives of curiosity; it had few inhabitants, but they were deemed holy by the Britons, and their persons and property were respected as inviolable. When he lately visited the island, the air was shaken, and there were many portents, with hurricanes and lightning. When quiet was restored, the islanders said that one of the supernatural beings had passed away—an event which caused a disturbance in nature. Plutarch adds, similarly to the other passage, that in one of the islands Saturn is confined; that he sleeps under the custody of Briareus; that sleep is his chain, and that he is attended by ministering genii.(232)

These passages involve the idea of a great open sea encircling the north of Europe, and connected with the Caspian. Other writers had relegated Saturn into these distant regions: the Frozen Sea, in the extreme north, is often called Cronian.(233)

---

(231) c. 18.
(232) Compare Hesiod, Theog. 734—5.
(233) Plin. iv. 30; Dionys. Perieg. 32, with C. Müller's note; Orph. Argon. 1084. Claudian, Laud. Stilich. 198.

An island in the Western Sea, sacred to Saturn, is likewise mentioned by Avienus, in his 'Ora Maritima.'(234) Saturn is the king of the Golden Age, and Pindar connects him with the Islands of the Blest.(235)

Other sacred islands were likewise found in the Western and Northern Seas. According to Pliny, six islands, called the islands of the gods, or the Happy Islands, lay off the promontory of the Arrotribæ, or Cape Finisterre.(236) Dionysius places the Western Islands, which produced tin, near the Sacred Promontory, which was the extremity of Europe.(237) This Sacred Promontory, at the western point of Europe, is also mentioned by Strabo.(238) The sacred promontory of Bacchus, on the coast of the Atlantic, occurs in the Orphic Argonautics, as well as the wooded island of Ceres in the Western Ocean.(239)

According to Artemidorus, who lived about 100 B.C., there was an island near Britain, in which rites were celebrated in honour of Ceres and Proserpine, similar to those celebrated in the island of Samothrace.(240) An island off the mouth of the Loire, where orgies and initiations were performed to Bacchus by the *Samnite* women, is described by Strabo.(241) In Dionysius, certain islands near Britain are the seat of these Bacchic rites, and the nation is designated as that of the *Amnitæ*.(242)

Mela describes an island, named Sena, in the Britannic Sea, opposite the country of the Osismii (Bretagne), as renowned for the oracle of a Gallic deity; its priestesses were nine virgins, who were endued with the power of raising storms by their incantations; of changing themselves into the shapes of animals; of curing diseases incurable by human art; and of predicting the future.(243)

According to Avienus, the island of the Hibernians was

---

(234) v. 165.
(235) παρὰ Κρόνου τύρσιν, Olymp. ii. 70.
(236) iv. 36. (237) v. 561.
(238) ii. 5, 14, iii. 1, 2. (239) v. 1172, 1192, 1250.
(240) Ap. Strab. iv. 4, 6. (241) Ib.
(242) v. 570. (243) iii. 6.

called the Sacred Island.(244) Ptolemy states that the southern promontory of Ireland was called Sacred.(245)

Procopius, in his 'History of the Gothic War,'(246) describes Brittia as an island opposite the mouths of the Rhine, at a distance of 200 stadia (25 miles); situated between the islands of Thule and Brettania, and inhabited by the three nations of Angili, Frissones, and Britons. For the position of Thule he refers to a former passage,(247) where he identifies it with Scandinavia. With respect to Brettania, he represents it as lying to the west, opposite the extremity of Spain, and divided from the Continent by an interval of 4000 stadia (500 miles); whereas Brittia lies opposite the coast of Gaul, to the north of Spain and Brettania. Grimm, in his 'Deutsche Mythologie,'(248) thinks that the Brettania of Procopius is the extremity of Gaul, the modern Brittany; but Procopius conceives it as an island; and there seems no doubt that by Brittia he means Britain, and by Brettania Ireland, which are the two Britannic islands.(249)

After recounting some marvels respecting the natural history of Brittia, Procopius proceeds to say that he is unwilling to pass over in silence a story related of this island; for although it has a fabulous appearance, it is repeated by numerous persons, who affirm that they have both seen and heard the circumstances described. He declares himself to have frequently heard it from natives of the place, who believed its reality; though he himself conceives it to be a phenomenon of dreams. The story is as follows:—

Along the shore of the ocean, opposite the island of Brittia, there are numerous villages inhabited by fishermen, cultivators, and seafaring men, who carry on the trade with this island. They are subject to the Franks, but are exempt from tribute, in

---

(244) Ora Marit. 108.
(245) ii. 2, 6.
(246) iv. 20.
(247) ii. 15.
(248) P. 482, ed. 1.
(249) Ptolemy, Synt. ii. 6, p. 85—6, speaks of Great Britain and Little Britain, μεγάλη and μικρὰ Βρεττανία. By the latter he appears to mean Ireland.

consideration of a service which they render. This service is the duty of ferrying over the souls of the dead. Those whose turn it is to be on duty for the ensuing night come back to their homes at the hour of darkness, and betake themselves to sleep, awaiting the visit of the superintendent. In the dead of the night they hear a knocking at their door, and a voice calling them to their work. Without a moment's delay, they rise from their beds, and walk to the sea shore, impelled by an irresistible necessity. Here they find empty boats, different from their own, ready for their reception, which they enter, and proceed to row. These boats are so weighed down by the number of passengers, that the water rises to within an inch of the edge; no one, however, is perceptible to the sight. In an hour they effect the passage to Brittia; and yet, when they make the passage in their own barks, it takes a day and a night. When they have reached the island, and discharged their cargo, they return with boats so lightened that the keel alone sinks in the water. They see no one either remaining in the boat or leaving it, but they hear a voice calling over the names of the passengers, and repeating the dignities and patronymic of each. In the case of women, the names of their husbands are mentioned.([250])

Claudian, who preceded Procopius by a century and a half, describes necromantic rites performed by Ulysses on the coast of Gaul:—

> Est locus, extremum pandit qua Gallia litus,
> Oceani prætentus aquis, ubi fertur Ulysses
> Sanguine libato populum movisse silentem.
> Illic umbrarum tenui stridore volantum

([250]) This story is repeated by Tzetzes on Lyc. 1204, and on Hesiod, Op. 169. Compare Plutarch, Prov. tom. v. p. 764, ed. Wyttenbach. In the text of Procopius, vol. ii. p. 567, ed. Bonn, the words near the beginning of the passage, descriptive of the shore in question are: παρὰ τὴν ἀκτὴν τῆς κατὰ τὴν Βριττίαν τοῦ ὠκεανοῦ νήσου. In Tzetzes, on Lycophron, the corresponding words are: περὶ τὴν ἀκτὴν τοῦ περὶ τὴν Βρεταννίαν νῆσον ὠκεανοῦ. There is a various reading, νῆσον, in Procopius. The text of Procopius seems to be corrupt, and the sense requires the reading preserved by Tzetzes. According to Procopius, the shore would be that of an island opposite to Britain.

> Flebilis auditur questus. Simulacra coloni
> Pallida, defunctasque vident migrare figuras.
> Hinc dea prosiluit, Phœbique egressa serenos
> Infecit radios, ululatuque æthera rupit
> Terrifico. Sensit ferale Britannia murmur,
> Et Senonum quatit arva fragor, revolutaque Tethys
> Substitit, et Rhenus projectâ torpuit urnâ.(251)

The passage occurs in the poem against Rufinus. Megæra is described as ascending from the infernal regions to the light of day at the seat of these necromantic rites, in order to visit Rufinus, whose native place was Elusa, in Aquitania. Necromancy was conceived by the ancients as connected with Hades;(252) and the place where Ulysses evoked the souls of the dead was a natural outlet for a Stygian deity, as the mephitic cavern of Amsanctus in Italy was, for a different reason, a proper channel for Alecto to return to hell in the Æneid.(253) Cumæ, where there was a mephitic cavern by which Æneas descended to hell, was one of the localities at which the necromancy of Ulysses was fixed.(254)

Homer describes the land of the Cimmerians, where Ulysses evoked the souls of the dead, as being on the furthest extremity of the ocean;(255) and when the localities of fiction receded, and Ogygia was placed five days' sail to the west of Britain, it was natural that Claudian, seeking for a subterranean communication with Hades, by which Megæra might emerge in order to visit Rufinus at Elusa, should suppose Ulysses to have performed his necromantic ceremonies at the extremity of Gaul in the far west.(256)

Ulysses is related by Strabo to have penetrated beyond the

---

(251) In Ruf. i. 123.

(252) Nitzsch on the Odyssey, vol. iii. p. 152, 355. Compare the description of the cave of the witch Erichtho in Lucan, vi. 642, where she performed her necromantic rites.

(253) vii. 568.

(254) Strab. v. 4, § 5; Serv. Æn. vi. 106.

(255) Od. xi. 13.

(256) Nitzsch, ib. p. 187.

Pillars of Hercules.(257) Ulysippo, the modern Lisbon, was considered his foundation. Tacitus carries him to Germany, where there were monuments and inscriptions testifying his presence;(258) and Solinus as far as Caledonia.(259)

Grimm connects the passage of Claudian with the singular story in Procopius; but the latter appears to be derived from some local legend; whereas the former is nothing but an application of the classical ideas respecting the wanderings of Ulysses, and the connexion of necromantic evocation with the subterranean passages to Hades.

§ 8 The views of those who maintain the probability of voyages by the Phœnicians to distant lands—who suppose them to have sailed to the amber-coast of the Baltic, and even hint at their having reached America—receive some confirmation from the accounts, preserved by the ancients, of the circumnavigation of Africa. These accounts lie within a small compass, and deserve a separate examination.

The accurate knowledge of the Greeks respecting Egypt began with the reign of Psammitichus,(260) and the reign of his successor Neco is fixed, on apparently authentic evidence, at 616—600 B.C. According to the account of Herodotus, Neco began to dig a canal connecting the Nile with the Red Sea; and 120,000 men had perished in its formation, when he desisted from the work, in consequence of the admonition of an oracle. He afterwards turned his attention to military affairs; he built vessels of war both in the Red Sea and in the Mediterranean; and he invaded Syria.(261) But soon after the abandonment of

---

(257) iii. 4, 4. (258) Germ. 3.
(259) c. 22. (260) See above, p. 318.
(261) ii. 158—9. The attempt of Neco is likewise mentioned by Diod. i. 33. Aristotle, Meteor. i. 14, § 27, states that Sesostris and afterwards Darius attempted to connect the Nile with the Red Sea; but discovered that the level of the sea was higher than that of the river, and therefore desisted from the enterprise. This account is repeated by Strabo, xvii. 1, § 25, with the insertion of Neco between Sesostris and Darius. He adds that the Ptolemies executed the work, and proved the fear of the irruption of the sea to be groundless. Pliny, vi. 29, gives the same account, but with the omission of Neco.

the canal, and with a view, as it appears, of accomplishing the same object by different means, he sent some vessels, navigated by Phœnicians, to circumnavigate Africa, ordering them to commence their voyage from the Red Sea, and to reach Egypt by the Pillars of Hercules and the Mediterranean. If this voyage could be effected, a ship would sail between the Red Sea and the Mediterranean;([262]) to connect which was the object of the canal. Herodotus proceeds to state that the Phœnicians, starting as they were ordered, sailed along the Southern Sea; and, whatever part of Africa they had reached, when autumn arrived, they landed, sowed the ground, and awaited the harvest; and having gathered the corn, they then continued their voyage: that having thus consumed two years, in the third year they passed the Pillars of Hercules, and returned to Egypt. 'The account which they gave,' says Herodotus, 'which others may, if they think fit, believe, but which to me is incredible, is that when they were sailing round Africa, they had the sun on their right hand.' Herodotus adds that the Carthaginians at a later period maintained that Africa could be circumnavigated; and he subjoins a story of Sataspes, a Persian nobleman, who, in the reign of Xerxes (485—465 B.C.) was relieved from a sentence of crucifixion, upon the singular condition that he should circumnavigate Africa. Herodotus tells us that Sataspes obtained a ship and sailors in Egypt; passed the Pillars of Hercules, and having rounded the western promontory of Africa, called Soloeis, pursued his voyage to the south; but after sailing many months, and finding that he was still far from the Red Sea, he turned back, and came again to Egypt. The account which he gave to Xerxes on his return was that, at the extremity of his voyage, he sailed by little men, clad in garments of palm-leaves, who, when he landed, left the towns and fled to the mountains; that his crew used to take nothing, except some sheep; and that

---

(262) It may be observed that Herodotus here calls the Mediterranean the βορηίη θάλασσα, as opposed to the νοτίη θάλασσα, the sea to the south of Libya, ii. 158, iv. 42.

the reason why he did not proceed further was, that the ship stuck fast, and would not move. Xerxes did not believe this story, and, as Sataspes had not fulfilled the required condition, ordered him to be crucified. Herodotus adds that an eunuch of Sataspes, when he heard of his master's death, fled to Samos with a large sum of money; and that this money was dishonestly retained by a Samian, with whom it had doubtless been deposited. 'I know the name of this Samian' (says Herodotus), 'but suppress it out of regard for his memory.'[263] It will be observed that Herodotus resided at Samos during the early part of his life, and thus might have an opportunity of becoming acquainted with a circumstance which must have occurred within his lifetime.

The next reference to this subject occurs in Strabo. This geographer quotes Posidonius as treating of the circumnavigation of Africa, and as referring to the expedition mentioned by Herodotus (which is by an error of memory attributed to Darius instead of Neco), as well as to a certain Magus who was represented by Heraclides Ponticus to have assured Gelo (485—478 B.C.) that he had performed this voyage. Posidonius declared that these voyages were unauthenticated by credible testimony; but he related the following story of a certain Eudoxus, who lived in the second century before Christ, as deserving of belief. Eudoxus of Cyzicus (he said), being in Egypt in the reign of Ptolemy Euergetes the Second (170—117 B.C.), accompanied this king in voyages up the Nile; on one of these occasions, an Indian was brought to Ptolemy by the guards of the Red Sea, who said that they had found him alone and half dead in a ship. By the king's command, the Indian was taught Greek; whereupon he offered to steer a ship to India: the voyage was made under the guidance of this Indian, and Eudoxus went out and returned with the ship; but the king took away all the precious stones which he brought back. In the following reign of Queen Cleopatra (117—89 B.C.) Eudoxus was sent on a second voyage

---

[263] iv. 42, 43.

to India with a larger expedition; but on his return he was carried by adverse winds beyond Æthiopia, along the eastern coast of Africa. Having landed at different places, he communicated with the inhabitants, and wrote down some of their words. He here met with a prow of a ship, saved from a wreck, with a figure of a horse cut in it; and having heard that it was a part of a vessel which had come from the west, he brought it away. On his return to Egypt, he found that Cleopatra had been succeeded by her son (Ptolemy Soter II. Lathyrus, 89—81 B.C.), who again deprived him of all his profits in consequence of an accusation of embezzlement. Eudoxus showed the prow which he had brought with him to the merchants in the harbour; they immediately recognised it as belonging to a ship of Gadeira; and one ship-captain identified it as having formed part of a vessel which had sailed along the western coast of Africa beyond the river Lixus, and had never returned. Eudoxus hence perceived that the circumnavigation of Africa was possible; he then took with him all his money, and sailed along the coast of Italy and Gaul, touching at Dicæarchia (or Puteoli), Massilia, and other ports, on his way to Gadeira; at all which places he proclaimed his discovery, and collected subscriptions: by these means he procured a large ship and two boats, and having taken on board some singing boys, physicians, and other professional persons, he steered his course through the Straits for India. After some accidents in the voyage, they reached a part of the African coast, where they found men who used the same words as those which he had written down in his former course from the Red Sea; whence he perceived that the tribes which he had reached from the west were of the same race as those which he had reached from the east, and that they were conterminous with the kingdom of Bogus (Mauretania). Eudoxus, having ascertained this fact, turned back his ship; when he had arrived at Mauretania, he attempted to persuade King Bogus to send out another expedition. The final results of this attempt were not, however, known to Posidonius.[264]. The King Bogus here

---

[264] Strab. ii. 3, 4.

mentioned is either the king of Western Mauretania, who, with Bocchus, was confirmed by Julius Cæsar in 49 B.C., or he is an earlier king of the same name.[265]

The voyage of Eudoxus was likewise reported by Cornelius Nepos, who stated that, in his own time, Eudoxus, in order to escape from Ptolemy Lathurus, had sailed from the Red Sea, and had reached Gades.[266] The historian Cælius Antipater, who lived about 120 B.C., also declared that he had seen a man who had made the voyage from Spain to Æthiopia for commercial purposes.[267]

Before examining these accounts in detail, it is necessary to ascertain the notion formed by the ancients respecting the geography of Africa.

Strabo says, that although the world is divided into the three continents of Europe, Asia, and Africa, the division is unequal: for that Europe and Africa put together are not equal in size to Asia; and that Africa appears to be smaller even than Europe. He describes Africa as forming a right-angled triangle; the base being the distance from Egypt to the Pillars of Hercules; the other side of the right-angle being the line of the Nile to the extremity of Æthiopia, and the hypotenuse being the line connecting the latter point with the Pillars of Hercules.[268]

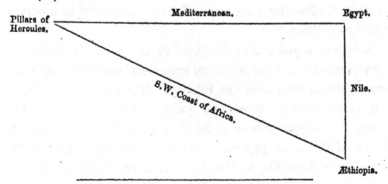

[265] The Latin writers call him Bogud: Dio Cassius writes his name Βογούας. Pliny says that the two divisions of Mauretania, eastern and western, were respectively named after their kings Bocchus and Bogud. Compare Strab. xvii. 3, 7.

[266] Mela, iii. 9; Plin. ii. 67; Martianus Capella, vi. § 621, ed. Kopp.
[267] Plin. ib.   [268] xviii. 3, 1.

Elsewhere he likens Africa to a trapezium, which figure is formed by supposing that the eastern extremity of the south-western coast is parallel to the northern coast.([269])

Mela has a similar notion of the form of Africa. He says that its length from east to west is greater than its width from north to south; and that its greatest width is at the part where it adjoins the Nile.([270])

As the ancients believed that the Northern Ocean swept across the back of Europe, from the vicinity of the Caspian and the Palus Mæotis, along the shores of Scythia, Germany, and Gaul, to the Pillars of Hercules—thus suppressing the Scandinavian peninsula and the chief part of Russia—so they believed that the Southern Ocean extended in a direct line from the Pillars of Hercules to the extremity of Æthiopia beyond Egypt; and hence they called the Negro tribes on the western coast of Africa Æthiopians, and brought them into connexion with the Æthiopians of the Upper Nile. According to the statement of Scylax, some persons thought that the Æthiopians of the northern shores of Africa were continuous with those who inhabited Egypt; that Africa was a peninsula stretching to the west, and that the sea was uninterrupted from its western extremity to the Egyptian side. According to Juba, the Atlantic Sea began with the Mossylian promontory, near the south-eastern extremity of the Red Sea; and the navigation thence to Gades, along the coast of Mauretania, was in a north-westerly direction.([271])

Aristotle, arguing that the form of the earth is spherical, explains upon this hypothesis the opinion of those who not only connect the country near the Pillars of Hercules with India, as well as the seas in those two quarters, but account for the presence of elephants both in Africa and India by the resemblance of the most remote extremes. The true explanation, according to Aristotle, is, that India is near the north-western

---

([269]) ii. 5, 33.   ([270]) i. 4.
([271]) Scylax, § 112. Compare Geminus, c. 13. Plin. vi. 34.

coast of Africa, because the earth is a sphere.(272) So Eratosthenes expressed an opinion that, if it were not for the great size of the Atlantic (or external) Sea, a ship might sail along the same parallel from Iberia to India.(273) On the other hand, Seneca thought that this distance was not great, and that the voyage could with favourable winds be made in a short time.(274)

The belief as to the affinity between the extreme east and the extreme west explains some of the mythological stories respecting the population of Africa: thus the Maurusii are said to have been Indians who accompanied Hercules to the west of that continent.(275)

These opinions as to the shape of Africa, though predominant, were not universal: for Polybius considers it to be unascertained whether the sea passes round it to the south.(276) According to Mela, the question long remained doubtful, but it was settled by the voyages of Hanno and Eudoxus.(277)

Such being the notions of the ancients respecting the shape of Africa, the next point to be ascertained is, how far their geographical exploration of the coast can be proved, by sure evidence, to have extended.

The entire northern coast of Africa had, from a remote period, been visited by the Phœnician navigators: who, together with their colonists the Carthaginians, likewise established themselves in force on the southern coast of Spain, and used their establishments at Gades and its neighbourhood as starting-places for ulterior discovery. Their efforts seem to have been directed principally towards the opposite coast of Africa, and not to the Lusitanian coast—a policy connected with the na-

---

(272) De Cœlo, ii. 14. Compare Ideler, ad Aristot. Meteorol. vol. i. p. 569. Above, p. 167.

(273) Ap. Strab. i. 4, 6.

(274) Quantum enim est, quod ab ultimis litoribus Hispaniæ usque ad Indos jacet? Paucissimorum dierum spatium, si navem suus ventus implevit, Nat. Quæst. 1, Præf. § 11.

(275) Strab. xvii. 3, 7.   (276) ii. 38.   (277) iii. 9.

tural views for the extension of the Carthaginian Empire. Tingis, the modern Tangier, and Lixus and Thymiateria lying to the south on the same coast, are expressly mentioned as Carthaginian foundations: we also hear of a large number of Tyrian or Carthaginian towns on the western coast of Mauretania, which, having once amounted to 300, were destroyed by the neighbouring barbarians. These extensive settlements are indeed discredited by Strabo[278] and Pliny;[279] but it cannot be doubted that the Phœnicians, both of Tyre and Carthage, used their important port and factory of Gades as a means of extending their dominion on the opposite coast of Africa.[280]

An authentic record of the most important of these attempts still remains in the Periplus of Hanno, whose voyage is conjecturally fixed at 470 B.C. The extant narrative is probably an exact transcript of the original, which (like the bilingual inscription of Hannibal),[281] may have been engraved on brass, both in Punic and Greek. The expedition was partly for colonization, partly for discovery. The most distant settlement was not far from the Straits; the extent of the exploring voyage cannot be fixed with certainty. Gossellin takes it only as far as Cape Nun; the more prevailing opinion extends it to a point near Sierra Leone. The numbers of the expedition appear to be exaggerated; but its strength was such as to enable it to master all opposition of the natives. Some of the circumstances related in the exploring part of the voyage are manifestly fabulous; but there is no reason for doubting the general truth of the account.[282]

We are informed by Pliny, that when Scipio was in command in Africa (about 146 B.C.), he employed Polybius the historian to explore the western coast of that continent, and furnished him with a fleet for the purpose. Pliny gives a summary of the extent of coast examined by Polybius; the furthest

---

(278) xvii. 3, 3.   (279) v. i.
(280) Movers, vol. ii. p. 521—554.
(281) Livy, xxviii. 46.   (282) See above, p. 454.

point which he visited was the river Bambotus, in which were crocodiles and hippopotami.([283]) This voyage is referred to by Polybius in an extant passage of his history.([284]) Pliny's account of the places which he visited is analysed by Gossellin, who identifies the Bambotus with the Nun.([285]) Gossellin thinks that the ancients never passed Cape Boyador.

Another proof of the voyages of the Gaditane navigators to the south, along the African coast, is the fact that they had discovered the Canary Islands, certainly before the time of Sertorius, about 82 B.C., and probably at a much earlier period.([286])

On the eastern coast of Africa, the ancients had, from an early period, navigated the Red Sea, and had made considerable progress along the southern coast of Asia. Herodotus indeed informs us that Darius (521—485 B.C.) hearing that the Indus, as well as the Nile, contained crocodiles,([287]) wished to ascertain whether that river joined the sea. He accordingly sent Scylax of Caryanda, and other persons whom he could trust, to ascertain the truth. They started from the city of Caspatyrus and the land of Pactya, and sailed down the Indus to the east, until they reached the sea. They then sailed by sea to the west, and in the thirtieth month reached the point from which Neco had sent the Phœnicians to circumnavigate Africa. After this voyage, adds Herodotus, Darius subdued the Indians, and navigated the intermediate sea.([288])

The Scylax of Caryanda, here mentioned by Herodotus, is cited by Aristotle and other writers as having left a work containing geographical and ethnographical notices of India; but

---

([283]) Plin. v. 1.   ([284]) iii. 59.
([285]) Recherches sur la Géographie des Anciens, tom. i. p. 106.
([286]) Plut. Sert. 8; Diod. v. 19, 20; Aristot. Mir. Ausc. 84; Dr. Smith's Dict. of Geogr. art. *Fortunatæ Insulæ*.
([287]) Alexander the Great, finding that there were crocodiles in the Indus, and that a bean grew on the banks of the Acesines, which fell into the Indus, similar to the Egyptian bean, concluded that the Indus and the Nile were the same river; and wrote word to his mother Olympias, that he had discovered the sources of the Nile; Arrian, Anab. vi. 1.
([288]) iv. 44. Compare iii. 101.

the account of his voyage down the Indus, and from the mouth of the Indus to the Persian Gulf, is discredited by Dr. Vincent, on grounds which deserve attentive consideration, and which are regarded as conclusive by C. Müller, in his recent edition of the Minor Greek Geographers.[289]

Whatever may be the authenticity of the Persian expedition under the command of Scylax, it is certain that the ancients had, at an early period, navigated the Red Sea. They were acquainted with the island of Socotra, which they called Dioscoridis Insula; and the Periplus of the Erythræan Sea, attributed to Arrian, which was composed in the first century of our era, describes the southern coast of that gulf as far as the northeastern promontory of Africa (Cape Guardafuy). From this point the description of the eastern coast of Africa is carried, according to Gossellin, as far as the island of Magadasko, in lat. 2° N.; but according to Dr. Vincent,[290] who is followed by C. Müller, in his recent edition, as far as the island of Zanguebar, in lat. 6° S. 'Beyond this point (says the Periplus) the ocean is unexplored; but it is known to turn to the west, and, stretching away along the south towards the regions of Æthiopia, Libya, and Africa on the opposite side, to unite with the western sea.'[291]

Such being the geographical limits which the knowledge of Africa possessed by the ancients can be ascertained to have reached, the question remains whether the accounts of the entire circumnavigation of this continent in the single cases above adverted to are worthy of belief.

In the first place, the story of the Magus reported by Heraclides Ponticus may, with Posidonius, be safely rejected; neither is any credit due to the merchant who assured Cælius Antipater that he had sailed round Africa. These stories, doubtless, did

---

(289) Commerce and Navigation of the Ancients in the Indian Ocean, vol. i. pp. 303—311, vol. ii. p. 13—15, ed. 1807; Geogr. Gr. Min. vol. i. Prol. p. xxxv.

(290) vol. ii. p. 178—180.

(291) § 18, ed. C. Müller; Vincent, ib. p. 186.

not rest on any firmer basis of reality than the exploit of Menelaus, whose voyage of eight years, mentioned in the Odyssey—in which he visited the Æthiopians, the Sidonians, the Erembi, and Libya—was interpreted by one of the ancients as referring to a circumnavigation of Africa from the Pillars of Hercules to the Indian Ocean.([292])

The account of Eudoxus of Cyzicus was accepted by Posidonius; but it is discredited on sufficient grounds by Strabo, who subjects it to a detailed examination.([293]) The story of the Gaditane prow, found on the eastern coast of Africa, and identified by a ship-captain as belonging to a particular vessel, is an evident fabrication, resting on the erroneous belief that the distance between the coasts of Abyssinia and Morocco is inconsiderable. This seems to have been a favourite mode of proving the circumnavigation of Africa; for Pliny states that when Caius Cæsar (Agrippa), the son of Augustus, was in the Red Sea (during his command in Asia Minor), a part of a wreck was found there, which was recognised as belonging to a Spanish ship.([294]) It should be added that, according to Cornelius Nepos, Eudoxus effected the entire circumnavigation from the Red Sea to Gades; which is not affirmed in the detailed narrative of Posidonius. In like manner Pliny states that Hanno sailed round Africa as far as Arabia:([295]) whereas his extant account shows that he made no great progress along the western coast.

A certain Euthymenes of Massilia described himself as having passed the Pillars of Hercules, and as having sailed along the African coast of the Atlantic Ocean, until he reached the source of the Nile. He represented the Nile as being fed from the waters of the ocean, which were sweet in these quarters, and contained monsters similar to the Nilotic crocodiles and hippo-

---

([292]) Strab. i. 2, 31. Compare Od. iv. 84.
([293]) ii. 3, 5; See Fragm. Hist. Gr. vol. iv. p. 407.
([294]) ii. 67.   ([295]) Ib.

potami.(296) It is evident that this is a fabulous account, founded upon the belief that the shape of Africa was similar to that conceived by Strabo.

There remains only the account of the expedition in the time of Neco, given by Herodotus. This account has attracted much attention, and has been considered credible by many modern writers,(297) particularly by Major Rennell,(298) Prof. Heeren,(299) and, lastly, by Mr. Grote.(300) Before we yield to the arguments advanced by critics of such high authority, we must give due weight to the circumstances which detract from the credibility of the narrative of Herodotus. Many of these are stated by Gossellin, who, in the first volume of his work on ancient geography, has subjected this question to a systematic investigation.(301) The objections to it are, however, set forth with the greatest force and completeness by Dr. Vincent in his valuable work already cited.(302)

In the first place, it must be remarked that the interval between the last year of the reign of Neco and the birth of Herodotus was 117 years; and therefore that at least a century and a half must have elapsed between the time of the supposed

---

(296) See Fragm. Hist. Gr. vol. iv. p. 408.
(297) See Gossellin, ib. vol. i. p. 199.
(298) Geogr. Syst. of Herod. vol. ii. p. 348, ed. 8vo.
(299) Ideen, i. 2, p. 79—85.
(300) Hist. of Gr. vol. iii. p. 377—385. Hüllmann, Städtewesen des Mittelalters, vol. iv. p. 377, takes the same view.
(301) The arguments of Gossellin against the circumnavigation of Neco are controverted by Larcher on Herod. iv. 42, tom. iii. p. 458. He thinks that the Phœnicians gave no written report of their voyage, ib. 462.
(302) vol. ii. p. 186—205. See also Ukert, i. 1, p. 46, ii. 2, p. 35. Forbiger, vol. i. p. 64; Kenrick, Egypt under the Pharaohs, vol. ii. p. 405; Niebuhr, Lect. Anc. Hist. vol. i. p. 118, Engl. transl.; and the art. *Libya*, in Dr. Smith's Dict. of Anc. Geogr. vol. ii. p. 177. The story of the circumnavigation of Africa in the time of Neco is likewise discredited by Col. Leake, On Some Disputed Questions of Ancient Geography (Lond. 1857), p. 1—8, chiefly on the ground that the time was not sufficient for so long a voyage. He says that 'Ptolemy, whose work comprehends everything that was known of Africa in the Greek and imperial times of Egypt, denied the junction of the Atlantic and Indian seas, and must therefore have believed that Africa was not a peninsula,' p. 8.

voyage and the time when Herodotus collected materials for his history. The reign of Neco is contemporary with Pittacus and Periander, and is anterior to the legislation of Solon; it is a period as to which our knowledge even of Greek history is faint and imperfect; and we are not entitled to suppose that the tradition of such an event in Egyptian history, resting doubtless on oral repetition, could have reached Herodotus in an accurate shape. No particulars are given as to the persons who commanded the expedition, or as to the number or character of the ships concerned; and we are not informed how the difficulties which must have surrounded such an enterprise were overcome.

The general system of navigation in antiquity, whether the vessel was impelled by sails or by oars, was to keep close to the shore, and never to venture into the open sea, except in order to reach an island, or to cross a channel of moderate width. Navigation was moreover suspended during the winter months.[303] A modern vessel takes water and provisions for the whole or a large part of its voyage, and stands out to sea, steering its course by the compass, and by astronomical observations: it is likewise assisted by charts. An ancient vessel crept along the shore; advanced merely from one port or landing-place to another; stopped at night, when the difficulty of steering was greater; and took in water and food at the successive stations. The mean rate of a day's sail (exclusive of the night) is estimated by Rennell at about thirty-five miles,[304] and at every interval of this length it put into land. It was therefore dependent on its communications with the coast, and its successful progress could only be insured under one of two conditions: either that the coast was friendly, or that, if the coast was unfriendly, it had sufficient force to overawe the natives. The first of these cases was the ordinary state of navigation in the Mediterranean; either when a Phœnician ship sailed along the northern coast

---

[303] Plin. ii. 47; Veget. de Re Mil. v. 9. Compare Ideler, Chron. vol. i. p. 253

[304] Ib. p. 360.

of Africa, or when a Greek ship made its way along the coasts of Greece and Italy. The second case is exemplified by the early voyages of the Phocæans, which they are said to have made in long narrow ships of war, and not in merchant vessels built for carrying a cargo.[305] Other examples are found in the expedition of Nearchus from the mouth of the Indus to the head of the Persian Gulf, whose relations with the natives are described throughout as hostile and suspicious, and who chiefly obtained food by the method of plunder;[306] in the expedition of Hanno, who sailed along the western coast of Africa with a fleet which (according to his own account) consisted of sixty war penteconters, and 60,000 men and women; and in the voyage of Polybius along the same coast, who is expressly stated to have been furnished by Scipio with a *fleet* for the purpose.[307]

Major Rennell, proceeding from the remark that 'the difficulties of coasting-voyages do not, in respect of their length, increase beyond arithmetical proportion,' inquires, 'What should have prevented Scylax, Hanno, or the Phœnicians, from extending their voyages, had their employers been so inclined, and preparations had been made accordingly?'[308]

It is true that a coasting-voyage might have been indefinitely lengthened under the conditions favourable to its performance: for example, it is quite conceivable that an ancient ship, starting from a port of Syria, might have followed the coasts of Asia Minor, Greece, and Italy, as far as Massilia, and have repeated this course continuously, backwards and forwards, until it had completed as great a distance as would be necessary for the circumnavigation of Africa. But these were not the conditions under which the voyage of the Phœnicians, ordered by Neco, was undertaken. We are not informed that they were provided with a sufficient force to compel submission at the places where they landed: on the contrary, the account of

---

(305) Herod. i. 163.   (306) Arrian, Ind. c. 20, sqq.
(307) Ab eo acceptâ classe, Plin. v. 1.
(308) Ib. p. 354.

their landing in the autumn, in order to sow their corn, and of their waiting until the harvest, implies that they relied for food upon their own resources. It seems incredible that a few vessels, thus situated, could have made their way from the Red Sea to the Straits of Gibraltar. The probability is, that the crews would have fallen victims to the jealousy and hostility of the barbarous natives. Navigation in early times was generally connected with piracy; and an unknown ship arriving on a coast would not fail to be regarded as an enemy. The mere difficulty of language would in such length of coast as that in question, and with so vast a succession of different savage tribes, have rendered friendly communication impossible. The Periplus of Hanno mentions that he took with him interpreters; but even his limited expedition reached a point at which his interpreters could not understand the language of the natives.[309] He assigns the failure of food as the reason for turning back.

The length of time mentioned by Herodotus seems likewise insufficient, if we subtract the intervals between seed-time and harvest, and allow for the other casualties of such a navigation. Herodotus states that the expedition of Scylax occupied thirty months in its voyage down the Indus, and thence to the Red Sea; whereas the time allowed for the circumnavigation of Africa is under three years, with a further deduction for the periods requisite for bringing the crops to maturity. It may be added that the Phœnicians could not have provided themselves with seeds proper for the different climates and soils to be passed over; and as they could as easily have obtained provisions from the natives, as information respecting the proper seed and the seed itself, it is difficult to understand how the mode of procuring food to which they are described to have had resort could have been successful. Moreover, the proper time for sowing would not have fallen in autumn in the southern hemisphere, as Gossellin has remarked. It may be considered as

---

[309] § 11, 14.

certain that neither Neco nor Herodotus had any idea of the great length of the voyage from the Red Sea to the Straits of Gibraltar, and that they both believed Africa to be a peninsula of which the Nile was the base.(310)

The only circumstance in the account which invests it with credibility, is the report of the navigators, disbelieved by Herodotus himself, that they had the sun on their right hand: the most obvious interpretation of which supposes them to have reached the southern hemisphere. Upon this statement, however, which is the main title of the story to acceptance, two remarks may be made. In the first place, Herodotus himself ascended the Nile as far as Elephantine;(311) and Elephantine is opposite Syene, which is nearly within the tropic, and which contained afterwards the celebrated well. Now if Herodotus himself had visited a place where the shadows were vertical at the solstice, it is not unlikely that he may have obtained the story of Neco's expedition from persons who might conceive that a sufficient progress southward would bring the navigator to a region where the shadows at noon inclined from north to south. In the next place, Nearchus, the admiral of Alexander the Great, in the description of his coasting-voyage from the mouth of the Indus to the Persian Gulf, stated that in a part of his course the shadows were either vertical or fell to the south.(312) Now, when we consider that Nearchus could not have been south of 25° north latitude, which is north of the tropic, and of the latitude of Elephantine (24° N.), we can easily conceive that the informants of Herodotus may have imagined for the Phœnician navigators of Neco a physical phenomenon to which the Nile above Elephantine afforded an approximation, and which Nearchus declared himself to have actually witnessed at a higher latitude.(313) Onesicritus, who accompanied Alexander in his expedition, likewise stated that there were certain parts

---

(310) Compare Vincent, vol. ii. p. 565.
(311) ii. 29.   (312) Arrian, Ind. c. 25.
(313) See Vincent, ib. vol. i. p. 222, 304.

of India—he specified one to the north of the Hyphasis or Sutledge—where the sun was vertical at the solstice, and there were no shadows. (These places were called by him ἄσκιοι.) He declared, moreover, that in these districts the constellation of the Great Bear was never visible.(314) Pliny also reports that at Mount Maleus, in the territory of the Oretes in India, the shadows fall to the south in summer, and to the north in winter; that at the port of Pattala (Tatta on the Indus) the sun rises to the right, and the shadows fall to the south.(315) Eratosthenes affirmed that in the country of the Troglodytes, on the south-eastern coast of the Red Sea, the shadows fell to the south for forty-five days before and for the same period after the solstice.(316)

Some ambassadors from the island of Taprobane, or Ceylon, who came to Rome in the time of the Emperor Claudius, are represented by Pliny as having expressed their wonder that the shadows fell to the north and not to the south; and that the sun rose to the left, and not to the right;(317) although, as Dr. Vincent remarks, they must have annually witnessed that phenomenon, when the sun was south of the equator.(318)

These examples prove that the imagination of the ancients was active in conceiving the solar phenomena of the northern hemisphere to be reversed, even in districts which lay to the north of the tropics. It may be observed that the ancients had likewise heard accounts of the long polar nights, which they transferred to latitudes in which this phenomenon did not exist. Thus Cæsar states that the smaller islands near Britain had been reported by some writers to be continually dark for thirty days at the winter solstice. He adds, that on inquiry he was unable to confirm this statement; but he ascertained by means of water-clocks that the nights in Britain were shorter than on

---

(314) Plin. ii. 75, vii. 2. Places under the equator were called ἀμφίσκιοι, because they threw their shadows both south and north, Cleomed. i. 7; Achill. Tat. c. 31. The latter also mentions the ἄσκιοι.

(315) ii. 75.   (316) ii. 75—6, vi. 34.
(317) Plin. vi. 24.   (318) vol. ii. p. 492.

the Continent.(319) One of the stories of Pytheas, respecting his fictitious island of Thule, was that it had six months of continual light, and six months of continual darkness.(320)

It may be remarked that the Romans under the Empire are said to have penetrated very far into Africa by land: thus, P. Petronius, prefect of Egypt in the time of Augustus, is stated to have marched 970 miles south of Syene;(321) Ptolemy likewise describes two other Roman officers as having, by marches of three and four months respectively, reached a district south of the equator.(322) It is not impossible that the Egyptians may at an early time have ascended far into the interior of Africa; and in navigating the Red Sea, they would soon have passed the tropic.

Ptolemy states distinctly that the country under the equator has never been visited by any inhabitant of the northern hemisphere; so that he cannot have believed the accounts respecting the circumnavigation of Africa.(323) Cleomedes declares that we cannot visit the temperate zone of the southern hemisphere, as it is impossible to cross the torrid zone.(324) Macrobius states that Meroe lies 3800 stadia (450 miles) south of Syene, which is under the tropic; that the cinnamon region reaches 800 stadia (100 miles) to the south of Meroe; but that beyond this region the country is inaccessible in consequence of the heat.(325) The account of Geminus is different. He says that the southern temperate zone is unknown, but that the torrid zone had been recently explored by the Greek kings of Egypt, and it had been ascertained to consist of land, and not to be all

---

(319) Above, p. 472, n. 138.   (320) Plin. ii. 77; above, p. 469.

(321) Plin. vi. 35. An expedition to explore the source of the Nile sent by Neco, is mentioned in this passage, and also in Sen. Nat. Quæst. vi. 8.

(322) Geogr. i. 8, 5; Vincent, vol. ii. p. 243.

(323) τίνες δέ εἰσιν αἱ οἰκήσεις οὐκ ἂν ἔχοιμεν πεπεισμένως εἰπεῖν. Ἄτριπτοι γάρ εἰσι μέχρι τοῦ δεῦρο τοῖς ἀπὸ τῆς καθ' ἡμᾶς οἰκουμένης, καὶ εἰκασίαν μᾶλλον ἄν τις ἢ ἱστορίαν ἡγήσαιτο τὰ λεγόμενα περὶ αὐτῶν, Synt. ii. 6, p. 78, Halma. Compare Delambre, Hist. Astr. Anc. tom. ii. p. 83.

(324) Cleomed. i. 2, p. 20.

(325) In Somn. Scip. ii. 8, 3.

ocean, as some had supposed, and to be inhabited.(³²⁶) Several of the philosophers, including Cleanthes, held that the chief part of the great, or external, sea lay under the torrid zone.(³²⁷) That the torrid zone was inhabited was affirmed by Panætius the Stoic, and Eudorus the Academic.(³²⁸)

We have the authentic history of the circumnavigation of the Cape of Good Hope by the Portuguese; and when we observe the time which the successive attempts occupied, and the number of failures which preceded the ultimate success, even at a period subsequent to the discovery of the compass,(³²⁹) the improbability of the alleged Phœnician voyage becomes more apparent.

On the whole, we may safely assent to the position of Dr. Vincent, that 'a bare assertion of the performance of any voyage, without consequences attendant or connected, without collateral or contemporary testimony, is too slight a foundation to support any superstructure of importance;'(³³⁰) and we may conclude that the circumnavigation of Africa in the time of Neco is too imperfectly attested, and too improbable in itself, to be regarded as a historical fact.

---

(326) c. 13, p. 31.

(327) Cleomed. i. 6, p. 42; Gemin. c. 13, p. 31.

(328) Fragm. 6, p. 96, Petav. Panætius lived in the second century before Christ. Eudorus wrote commentaries upon Aristotle, and a treatise on the Nile; he was contemporary with Strabo, see xvii. 1, 5.

(329) See Robertson's History of America, b. i.; Vincent, ib. vol. ii. p. 217—9, 229. The polarity of the magnet, and its use in navigation, were known at least as early as the 13th century. See Hüllmann, Städtewesen des Mittelalters, vol. i. p. 125—137.

(330) vol. i. p. 307.

CORRECTIONS AND ADDITIONS.

Page 3, n. 5. In Herod. iv. 36, Bekker for ἔχοντας receives ἐχόντως, the conjecture of Dobree, Adv. vol. i. p. 30, who compares Plat. Phileb. p. 64, A.

Page 6, n. 15, *add*, Flumen Carmaniæ Hyctanis, portuosum et auro fertile. Ab eo primum septemtriones apparuisse adnotavere, arcturum nec omnibus cerni noctibus, nec totis unquam. Plin. vi. 26.

Page 9, n. 29, *add*, Gregory the Great, Dial. iv. 30, reports the following story which he had heard from Julianus, the second Defensor of the Church of Rome, who had died seven years previously. Julianus stated that, in the reign of king Theodoric, his wife's grandfather was returning by sea from Sicily to Italy. The ship stopped at one of the Lipari islands, where a hermit told him that Theodoric was dead. The hermit knew the fact from having seen the king, on the previous day, dragged between John the Pope and Symmachus the patrician, ungirt, unshod, and in chains, and thrown into the crater of the volcano. The kinsman of Julianus made a note of the day, and found, on his arrival in Italy, that Theodoric had died at the time of the appearance described by the hermit. John and Symmachus had been put to death by Theodoric. The death of Theodoric occurred on Aug. 30, 526 A.D. See Clinton, Fast. Rom. ad ann. Gregory was born in 540 A.D.

Page 9, n. 30, *add*, see Æsch. Prom. 366.

Page 54, l. 22, *for* 304 years, *read* 304 days.

Page 54, l. 23, *for* 1825 and 1826 days, *read* 1824 and 1825 days.

Page 69, n. 269, *add*, Philoponus ad Aristot. Meteor. vol. i. p. 204, ed. Ideler, explains the fable of Atreus as signifying that the motion of the seven planets is contrary to that of the stars.

Page 74, n. 11, *for* Ælian. Hist. An. ii. 18, *read* Ælian. Nat. An. ii. 18.

Page 85, n. 71, *add*, Herod. i. 103, says of Cyaxares, οὗτος ὁ τοῖσι Λυδοῖσί ἐστι μαχεσάμενος, ὅτε νὺξ ἡ ἡμέρη ἐγένετό σφι μαχομένοισι.

Page 87, l. 15, *for* partly visible, *read* partly obscured.

Page 87, n. 82, Theophanes, Chronograph. vol. i. p. 732, ed. Bonn., states that, in consequence of the outrage committed by Irene in depriving her son Constantine of his eyes, the sun was obscured for seventeen days, and the darkness was such that ships wandered from their course (789 A.D.). This instance of fabulous exaggeration occurs in a contemporary writer. See Gibbon, c. 48.

Page 101, n. 157. The later Greek writers use κρύσταλλος ὕδατος for ice, in order to distinguish it from crystal. See Salmas. Exerc. Plin. p. 92, C.

Page 104, n. 173, *read* Theodoret. Affect. Græc. lib. ii. p. 79, ed. Gaisford.

Page 104, n. 175, *read* Simplicius ad Aristot. de Cœl. p. 491, a.; 505, a.; ed. Brandis.

Page 108, n. 105. The prosecution of Anaxagoras is placed by Diodorus in 431 B.C.

Page 113, n. 110. All the manuscripts read νυκτοθηρῶν, and this reading has been restored by L. Dindorf, in the recent edition of the Memorabilia printed at the Oxford University Press (1862). The capture of fish by night is supposed to be alluded to. See Plat. Sophist. 12, p. 220; Oppian. Halieut. iv. 640. The netting of birds by night was probably also practised by the Greeks.

Page 132, n. 277. Compare the passage of Victorinus, de arte gramm. ap. Gramm. Lat. Auct. vol. i. p. 2502, ed. Putsch. 'Hymnis vel dithyrambis supercini moris est, quæ epodicis carminibus, si quando præponuntur, προϋμνια, si autem post antistrophon collocentur, μεθύμνια nuncupabuntur. Hoc genus in sacris cantilenæ ferunt quidam instituisse Theseum, qui occiso Minotauro cum apud Delum solveret vota, imitatus intortum et flexuosum iter labyrinthi, cum pueris virginibusque cum queis evaserat cantus edebat, primo in circuitu, dehinc in recursu, id est, στροφῇ et antistropho. Alii tradunt hunc sacrorum concentum mundi cantum cursumque ab hominibus imitari. Namque in hoc quinque stellæ quas erraticas vocant, sed et sol et luna, ut doctiores tradunt philosophorum, jucundissimos edunt sonos, per orbes suos nitentes. Igitur concentum mundi cursumque imitans chorus canebat: dextrorsumque primo tripudiando ibat (quia cœlum dextrorsum ab ortu ad occasum volvitur) dehinc sinistrorsum redibant, quandoquidem sol lunaque, et cetera erratica sidera, quæ Græci πλανήτας vocant, sinistrorsum ab occasu ad ortum feruntur. Tertio consistebant canendo, quia terra (circa quam cœlum volvitur) immobilis medio stat mundo. De quâ re Varius sic tradit: 'Primum huic nervis septem est intenta fides, variique additi vocum modi, ad quos mundi resonat tenor, sua se volventis in vestigia.' Idem et Varro:—

> Vidit et ætherio mundum torquerier axe,
> Et septem æternis sonitum dare vocibus orbes,
> Nitentes aliis alios, quæ maxima divis
> Lætitia est: at tum longe gratissima Phœbi
> Dextera consimiles meditatur reddere voces.'

The latter fragment is from the Chorographia or Cosmographia of P. Terentius Varro Atacinus, who was contemporary with the more celebrated Varro the antiquarian. The passage is exhibited according to the text of Burmann: see Wernsdorf's Poet. Lat. Min. vol. v. part iii. p. 1402. It is alluded to by Licentius, Carm. ad Augustin. 7, vol. iv. p. 518, ed. Wernsdorf. The author of the former passage appears to be L. Varius Rufus, who is otherwise only known as a poet.

Page 157, n. 56. The occultation of Mars by the moon, observed by Aristotle, is referred by Kepler to April 4, 357 B.C. (cited in Chambers' Handbook of Astronomy, p. 147). Aristotle was at that time twenty-seven years old.

Page 169, n. 98, *after* 'Plutarch,' *insert*, the same statement is made by Olympiodorus ad Aristot. Meteorol. vol. i. p. 198, ed. Ideler.

Page 192, n. 169. Roberval, a French mathematician of the seventeenth century, published a work under the name of Aristarchus, in which the heliocentric system was adopted. The following is its title: Aristarchi Samii de Mundi Systemate, partibus et motibus ejusdem, liber singularis. Paris. 1644. 12mo. See Biogr. Univ. art. Roberval; Delambre, Hist. Astr. Mod. vol. ii. p. 517.

Page 196, n. 194. This use of the word *servare*, is doubtless borrowed from that of the word τηρεῖν.

Page 222, n. 42. An alarming eclipse of the sun, visible at Thebes, is alluded to in an extant fragment of Pindar, Hyporchem. fr. 74, ed. Bergk. It is conjectured by Ideler to be the eclipse of April 30, 465 B.C. The calamities which it portends (viz., war, sedition, failure of crops, excessive snow, inundation, and drought) are enumerated.

Page 241, n. 110. Concerning the Syromacedonian notation of months used by Josephus, see Noris, Ann. et Epoch. Syromaced. p. 44, ed. Lips. 1696.

Page 252. Terra mater est in medio, quasi ovum corrotundata. Petron. Sat. c. 39. The spherical earth, fixed immovably at the centre of the spherical heaven, is described as the received opinion by Lactant. Div. Inst. iii. 24, at the beginning of the fourth century.

Page 265, n. 50, Isocrat. Busir. § 24, p. 226, states that the younger portion of the Egyptian priests apply their minds to astronomy and geometry.

Page 275, n. 114. Isocrat. Busir. § 18, p. 225, says that the Lacedæmonians borrowed several important institutions from Egypt.

Page 286, n. 163. Julian, Orat. 4, says that the Chaldæans and Egyptians invented astronomical tables, and that Hipparchus and Ptolemy perfected them.

Page 305, n. 264. The Hermetic verses on the seven planets, ap. Stob. Phys. i. 5, 14. Anth. Pal. App. n. 40, represent Saturn as presiding over tears, Jupiter over birth, Mars over anger, Venus over appetite, Mercury over reason, the sun over laughter, and the moon over sleep. An affinity between the seven planets and certain metals was likewise discovered in antiquity. Lead was assigned to Saturn, electrum (a mixture of gold and silver, see above, p. 457) to Jupiter, iron to Mars, copper to Venus, tin to Mercury, gold to the sun, and silver to the moon. Olympiodorus ad Arist. Meteor. iii. vol. ii. p. 163, ed. Ideler.

Page 313, n. 301. The verses of Philemon, ap. Stob. Phys. i. 5, 11 (Meineke Fragm. Com. Gr. vol. iv. p. 6), declare that there is no general destiny, but that a destiny peculiar to each individual man is born with him.

Page 318, n. 16. Desar, or Esar, a town in Æthiopia, founded by Egyptians, who fled from Psammitichus, is mentioned by Plin. vi. 35, compare Ptol. Geogr. iv. 7, 21.

Page 340, n. 90. Herod. ii. 92, says that monogamy was an institution common to all the Egyptians. Diodorus, on the other hand, says that the

Egyptians were polygamists, and that the rule of monogamy applied only to the priests, i. 80.

Page 350, n. 115. Josephus, Apion. ii. 11, calls Sesostris a fabulous king.

Page 351, n. 118. Sesostris led his army as far as the Mossylian promontory, on the northern coast of Æthiopia, east of the mouth of the Red Sea, Plin. vi. 34; compare Ptol. Geogr. iv. 7, § 10. Sesostris employed a Greek artist, named Bryaxis, to make a statue of Sarapis, according to Athenodorus ap. Clem. Alex. Protrept. iv. § 48, Fragm. Hist. Gr. vol. iii. p. 487. Athenodorus was a preceptor of Augustus Cæsar.

Page 352, n. 124. Dicantur obiter et pyramides in eâdem Ægypto, regum pecuniæ otiosa et stulta ostentatio, quippe cum faciendi eas causa a plerisque tradatur, ne pecuniam successoribus aut æmulis præberent, aut ne plebs esset otiosa. Plin. xxxvi. 12. The last reason is borrowed from Arist. Pol. v. 11.

Page 365, n. 191. The numbers in Augustine are taken from the Septuagint version: see Rosenmüller, Schol. ad Vet. Test. vol. i. p. 146. The numbers in the received text do not admit of the explanation reported by Augustine; they would make Adam only thirteen years old at the birth of his son Seth, and Seth only ten and a half years old at the birth of his son Enos.

Page 372, n. 209, *after* p. 260, *insert* p. 337, n. 82.

Page 380, n. 231. Clinton, Fast. Rom. vol. ii. p. 265, points out that the Emperor Tiberius, not Nero, is referred to by Suidas. Tiberius was born in 42 B.C., which agrees with the date of Chæremon's journey in Egypt.

Page 380, n. 232. See also Voss. de Hist. Gr. ed. Westermann, p. 209. Clinton, Fast. Rom. vol. ii. p. 265.

Page 409, n. 43. According to a story preserved by a late Latin poet, Semiramis, not contented with her own large dominions, expelled her stepson Trebes from the hereditary kingdom of Ninus; and Trebes, flying from Asia to Europe, became the founder of Treves.

> Nini Semiramis, quæ tanto conjuge felix
> Plurima possedit, sed plura prioribus addit,
> Non contenta suis, nec totis finibus orbis,
> Expulit a patrio privignum Trebeta regno,
> Insignem profugus nostram qui condidit urbem.

Incertus de Treveris, ap. Wernsdorf. Poet. Lat. Min. vol. v. p. 1382.

Page 413, n. 59. The practice of making female eunuchs was invented by the Lydians, Xanthus, fr. 19, ap. Fragm. Hist. Gr. vol. i. p. 39.

The author may be permitted to express a hope that the useful and well-edited series of the Greek Classics, now in course of publication at the press of Didot in Paris, will comprehend a volume of *Scriptores Astronomici*. The only modern edition of the great work of Ptolemy is the expensive publication of the Abbé Halma: it may be added that the Uranologium of Petavius is printed in a form inconvenient for study and reference, and is now a scarce work.

INDEX.

A

Actis, of Rhodes, taught astronomy to the Egyptians, 72
Æthiopians, the inventors of astronomy, 261; were identified with the Negroes, 416, 502
Africa, its circumnavigation in the reign of Neco, 498, 508; its form, as conceived by the ancients, 501
Agathemerus, on the original conception of the earth, 3
Alcmæon, 132
Amasis, 316
Amber, whence brought to Greece, 458
Anaxagoras, his lifetime, 102; his astronomical doctrines, 104; his prosecution for impiety, 108
Anaximander, his date, 91; his astronomical opinions, 92
Anaximenes, his date, 94; his astronomical opinions, 95
Ancient chronicle, on Egyptian chronology, 338
Annus, meaning of the word, 10
Antichthon, 124, 126
Anysis, 324
Apollo, not originally god of the sun, 7, 62
Apollonius of Perga, 200
Aratus, on the mythology of the Bears, 64; he versified the work of Eudoxus, 148
Arbaces, 400
Archimedes, his contributions to astronomy, 194
Arcturus, 67
Aristarchus of Samos propounded the heliocentric theory, 189, 204
Aristophanes, his representation of Socrates, 111
Aristotle on the circular figure of the earth, 5; on the course of the sun from west to east, 8; mentions the zodiac, 68; his astronomical writings, 158; his astronomical doctrines, 159; spurious treatise *De Mundo*, 158, 218, 487; his hypothesis of the planetary motions, 163; his geocentric doctrine, 166; his doctrine respecting comets, 168; and the milky way, 169
Aristyllus, an Alexandrine astronomer, 195
Assyrians, the originators of astronomy, 256; their astronomical observations, 263, 286; scientific pursuits of their priests, 265; their year, 267; their chronology, 397; their kings, 401; Assyrian kings mentioned in the Old Testament, 427
Astrology, not originally known to the Greeks, 71; its Chaldæan origin,

292; its introduction into Greece, 296; and Rome, 298; condemned by the Roman emperors, 300; and by the church, ib.; its introduction into Egypt, 301; founded upon births, 308; causes of its diffusion in Greece, 309

Astro-meteorology of the Greeks, 309
Astronomical canon, its list of kings, 404, 428
Asychis, 323
Atlas, 72
Atossa, 413
Atreus, story of, rationalized, 69
Augustus, the month, 23
Autolycus, his astronomical writings, 184
Axis of the universe, 173, 202

B

BABYLON, its founder, 409, 413; its empire, 423; its buildings, 435
Babylonians, see Assyrians
Bear, Great, is mentioned by Homer and later poets, as never setting, 6, 58; origin of the name, 64; the Greeks steered by the Great, the Phœnicians by the Little Bear, 83, 447
Belus, 398, 413
Berosus, 296, 360; his account of Assyrian history, 400
Bissextile year, 238
Bocchoris, 330, 336; his lamb, 348; conflicting evidence respecting him, 353
Boötes, 59
Britain, 452, 474, 481, 494
Buildings of Egypt and Babylon, their probable antiquity, 439
Bunsen, Baron, his treatment of Egyptian chronology, 367; his Egyptian alphabet and vocabulary, 385
Busiris, 356

C

CÆSAR, Julius, his reform of the Roman calendar, 236
Calendars of the Greek states, founded on their respective religious celebrations, 22, 234; Roman Calendar, its confusion, 236; reformed by Julius Cæsar, 237
Callippus, his cycle, 122; his hypothesis of the planetary motions, 163
Canicular period, 281
Cassiterides, 451
Chæremon, 280; on hieroglyphics, 380
Chaldæans, the authors of astrology, 292; they gave their name to the craft, 299. See *Assyrians*
Champollion, his system of interpreting the hieroglyphic writing, 383
Cheops, 323, 336
Chephren, 323, 336
Child-bearing, period of, fixed by the ancients at 10 months, 21
Chiron, 73, 74
Chonuphis, 146
Cleomedes, 215

Cleostratus, 118
Clepsydra, 182, 243
Clocks, 242
Coma Berenices, 197
Conon of Samos, 196
Consuls, the Romans dated the year by them, 27
Coptic language and literature, 387
Costard, Reverend George, his History of Astronomy, 2
Ctesias, his Assyrian history, 398, 406
Cyclopean buildings, 442

D

Dances of the stars, 61
Day, its divisions in Greece, 179; at Rome, 181
Days of the week, their planetary names, 304
Delambre, histories of astronomy, 1; on Hipparchus, 208, 214
Democritus, his date, 137; his journeys, ib. 272; his astronomical doctrines, 138
Diana, not originally the goddess of the moon, 7, 62
Dicuil, on Thule, 478
Dies Ægyptiaci, 306
Diodorus, on Egyptian Chronology, 331; his sources of information, 342
Diogenes, Antonius, his romance on the parts beyond Thule, 473
Diogenes of Apollonia, his astronomical opinions, 109
Dog-star, 59, 61, 67, 311
Dositheus, 200

E

Earth, the, originally conceived as a circular plane, 3; its immobility held by the ancients, 170, 252
Eclipse of Thales, 85; early speculations on the nature of eclipses, 222; eclipse of Nicias, 223; of Pelopidas, 224; other eclipses, 225—230; prediction of eclipses, 231
Ecphantus, his doctrine of the earth's motion, 129, 171
Egypt, visits of the Greek philosophers to, 268; ancient chronology of, 315; period after Psammitichus, 316; its lists of kings, 358; its hieroglyphical writing, 377; its buildings, 434
Egyptian priests, their observations of the planets, 156, 161
Egyptians, the authors of astronomy, 256; their astronomical observations, 264, 287; scientific pursuits of their priests, 265; their year, 32, 266, 279, 363; their ethnological character, 340; their historical records, 341; were not a warlike people, 351
Egyptology, its historical method, 368
Electron, 457
Empedocles, his date, 100; his astronomical opinions, 101; his visit to Egypt, 272
Epicurus, his astronomical doctrine, 218
Era, absence of a chronological, in antiquity, 25; of Nabonassar, 26, 404, 248; of the Olympiads, 26; of the Trojan War, 27; of the foundation of Rome, ib.

INDEX.

Eratosthenes, his astronomical works, 198; his treatment of Egyptian chronology, 331; its source, 342

Eridanus, 459, 463

Etesian winds, 12

Euclid, his treatment of astronomy, 186

Eudemus, his history of astronomy, 174

Eudoxus of Cnidos, on the short year of the Egyptians, 32; his lifetime, 145; his visit to Egypt, 145, 273; his astronomical researches, 147; his Enoptron and Phænomena, 148; his planetary hypothesis, 152; his statement of the periodic times of the planets, 154; and of their synodic periods, 155

Eudoxus of Cyzicus, a navigator, 499, 507

Euthymenes, 507

F

FEBRUARY, the last month of the Roman year, 45

Firmanus, L. Tarutius, 50, 299, 308

G

GADEIRA, or Gades, 448

Galen, on the notation of time of year by the stars, 24

Geminus, on the ancient conception of the earth, 4; his account of the ancient Greek calendar, 18; his astronomical treatise, 216; his account of the torrid zone, 514

Genethliaci, 307

Greswell, Dr., on the nundinal years of Rome, 55

H

HADES conceived as a subterranean vault, 8

Hanno, his Periplus, 458, 504

Heaven, originally conceived by the Greeks as a solid vault, 3; and as the seat of the gods, 159

Heraclides, his doctrine of the earth's motion, 129, 170

Heraclitus, his astronomical opinions, 96

Heraiscus, 361

Hermapion, 381

Hermes Trismegistus, his books respecting Egypt, 375; his astrological writings, 376

Herodotus, ridicules the circularity of the earth, 3; his notation of past time, 25; on the eclipse of Thales, 86; his Egyptian chronology, 320; its source, 342; his Assyrian history and chronology, 397

Hesiod, denotes the time of year by the rising and setting of stars, 60

Hesperus, 62, 144

Hicetas, 127; he holds the rotation of the earth on its axis, 170

Hieroglyphical writing of Egypt, 377

Himilco, his voyage along the western shores of Europe, 455

Hipparchus, his astronomical works and discoveries, 207

Hippocrates of Chios, 168

Homer, his year is the tropical year, 12; he mentions the solstices, 15

Horapollo on hieroglyphics, 381, 382

Hour, use of the word, 178; variable hours, 179, 241, 242
Hyades, 66
Hyperion, the earliest observer of the heavenly bodies, 73

I

INTERCALATION of Numa, 38; trieteric, 114; octaeteric, 116; of the Julian calendar, 239
Ireland, earliest mention of, 483
Islands, sacred, 489

J

JULIUS, the month, 23
Junius, the month, 37

L

LABYRINTH of Egypt, 325, 335, 355
Leo, 361
Leucippus, his astronomical doctrines, 136
Livy, on Numa's reform of the Roman calendar, 41

M

MAIUS, the month, 37
Manetho, his work on Egyptian chronology, 326; his historical character, 360; source of his Egyptian chronology, 342; his treatise περὶ Σώθεως, 284
Martius, the month, 86
Memnon, 416
Menes, 320, 328—332, 354
Menophres, era of, 283
Merkel, on the Roman decimestrial year, 51
Meton, his reform of the calendar, 113; he sets up a sundial at Athens, 178
Mœris, 322, 333
Mommsen, on the Roman decimestrial year, 51
Month, synodical, was fixed at 30 days, 16; periodical, was fixed at 28 days, 20; names of months were derived from names of gods or of festivals, 222
Moon, the, was conceived as driving a chariot, 63; her supposed influence on the weather, 312
Musæus, 74, 77
Mycerinus, 323, 336

N

NABONASSAR, era of, 26, 404, 428
Narrien, John, his history of astronomy, 2
Navigation, its connexion with astronomy, 447
Nausicaa, her supposed mention of the sphere, 74, 78
Necepso, 302
Neco, 317, 497
Newton, Sir Isaac, on the primitive year and month, 19; on the origin of astronomy in the heroic ages of Greece, 73

Niebuhr, on the nundinal year of Rome, 54
Nigidius Figulus, 299
Nile, its inundation, 12, 84, 435; the name probably of Greek origin, 357
Nineveh, its founder, 407; its empire, 423
Ninus, 398, 407
Nitocris, 320, 329, 371, 397
Numa, his reform of the Roman year, 37
Numerical exaggeration of the Orientals, 361
Nundinæ, 56
Nundinal year, 54

O

OCEAN, the, was supposed to flow round the earth, 5, 475, 502
Octaeteric cycle, 38
Oenopides, 132, 135, 272
ὀμφαλὸς of Greece at Delphi, 4
Orion, 67
Orpheus, the author of astronomy, 73

P

PALAMEDES, 73
Palladius, his division of hours, 180
Parmenides, his date, 99; his astronomical opinions, ib.
Phaeïnus, 114
Phaethon, fable of, 7, 463
Philolaus, his system of the universe, 124; his cycle, 135
Phœnicians, their discovery of astronomy, 262, 446; their distant voyages, 448; their voyages to Britain, 450; to the Baltic, 457; their circumnavigation of Africa, 497
Phœnix period, 282
Planets, not observed in early times, 62; their Greek names, 144, 290; Pythagorean doctrine respecting them, 131; theory of revolving spheres, 152, 163; theories respecting their order, 245
Plato, on the regulation of the year by festivals, 19; his astronomical doctrines, 141; his visit to Egypt, 272; on the antiquity of Egypt, 326; his allusion to astrology, 295
Pleiads, 65
Plutarch, on the intercalation of Numa, 39
Posidonius, 214
Precession of the equinoxes, 212
Prometheus, as an astronomical observer, 73, 259
Proteus, 322, 356
Psammitichus, 318
Psebo, Lake, 488
Ptolemæus, Claudius, his system of astronomy, 249
Pyramids, their founders and dates, 354
Pythagoras, his astronomical doctrines, 122; his scientific journeys, 123, 269; his planetary doctrine, 130; his doctrine of the music of the spheres, 131; his doctrine respecting comets and the milky way, 133
Pytheas, 467

Q

QUINTILIS, named Julius, 23

R

RHAMSES, 352
Rhampsinitus, 323
Rhodopis, 371
Rome, its ancient decimestrial year, 34; its calendar, 236; its reception of astrology, 298

S

SABACOS, 324
Sardanapalus, 400, 418
Scandia, 476
Scylax of Caryanda, his eastern voyage, 505
Seasons, their succession, 10; their number, 11
Seleucus, the Babylonian, 192, 289
Semiramis, 397, 399, 407
Sennacherib, 324
Sesostris, 322, 329, 333, 349, 369
Sethon, 324
Sextilis, named Augustus, 23
Sirius. See Dog-star
Socrates, his doctrines respecting astronomy, 109
Solarium, 183
Solinus, on Numa's reform of the Roman calendar, 42
Solon, his regulation of the attic calendar, 20, 90
Sothiac period, 281
Stars were supposed to rise from, and set in, the ocean, 6; were observed at an early period, 58; their affinity with the souls of men, 312
Sun, the, was supposed to rise from, and set in, the ocean, 6; was conceived as driving a chariot, 7, 63; as a universal witness, 7; was fabled to return from west to east in a golden goblet, 8; its place among the planets, 240
Sundial, introduced from Babylon into Greece, 177; attributed to Anaximander and Anaximenes, ib.; its construction, 179; improvements in its construction, 242

T

TARSHISH, 456
Telmessians, they practised divination, 294
Teutamus, 416
Thales, the founder of physical philosophy, 78; predicts an eclipse of the sun, 79, 85; his visit to Egypt, 80, 268; his astronomical doctrines, 81
Theophrastus, on the cosmical system of Plato, 142; his history of astronomy, 174
Thucydides, his notation of current years, 15; his notation of past time, 25
Thule, 467

Timæus, passage in the, respecting the rotation of the earth round the cosmical axis, 142

Timocharis, an Alexandrine astronomer, 195

Tin, whence brought to Greece, 450; whether found in India, 456

Τόρνος, meaning of the word, 3

U

ULYSSES in Gaul, 495

Ulysippo, 449

Uranus, the earliest astronomer, 73

V

Varnish, origin of the word, 197

Venus, the planet, 62

W

WEEK. See days of the week

Whewell, Dr., his History of the Inductive Sciences, 1; on Hipparchus, 210

X

XENOPHANES, his astronomical doctrines, 98

Xenophon, his notation of past time, 26; his account of the opinions of Socrates respecting astronomy, 113

Y

YEAR, the solar, deduced from the seasons, 9; its beginning, 29; antiquity of its use, 30; years of 3 and 4 months, 30; short Egyptian year, 32; Roman year of 10 months, 34; lunar, 116; year of the Egyptians, 266, 279

Z

ZODIAC, its origin, 68; of Tentyra, 289

Zone, torrid, how far known to the ancients, 514

THE END.

CPSIA information can be obtained
at www.ICGtesting.com
Printed in the USA
BVHW080847101219
566202BV00007B/118/P